Abiotic Stress of Crops: Molecular Genetics and Genomics

Abiotic Stress of Crops: Molecular Genetics and Genomics

Guest Editor

Zhaoshi Xu

MDPI

Basel • Beijing • Wuhan • Barcelona • Belgrade • Novi Sad • Cluj • Manchester

Guest Editor
Zhaoshi Xu
Institute of Crop Sciences
Chinese Academy of
Agricultural Sciences (CAAS)
Beijing
China

Editorial Office
MDPI AG
Grosspeteranlage 5
4052 Basel, Switzerland

This is a reprint of the Special Issue, published open access by the journal *Plants* (ISSN 2223-7747), freely accessible at: https://www.mdpi.com/journal/plants/special_issues/AS_Crops.

For citation purposes, cite each article independently as indicated on the article page online and as indicated below:

Lastname, A.A.; Lastname, B.B. Article Title. *Journal Name* **Year**, *Volume Number*, Page Range.

ISBN 978-3-7258-6812-4 (Hbk)
ISBN 978-3-7258-6813-1 (PDF)
https://doi.org/10.3390/books978-3-7258-6813-1

Contents

About the Editor

Zhaoshi Xu

Zhaoshi Xu, Professor, Head of the Research Group of Wheat Molecular Breeding, Institute of Crop Sciences, Chinese Academy of Agricultural Sciences. He was awarded the titles of Changbai Mountain's leading talents and Tianshan Talent. He mainly engages in analysis of molecular regulatory networks of drought-tolerance genes and the synergistic regulation mechanism between Fusarium crown rot resistance and drought tolerance in wheat. He presided more than 10 national projects, obtained more than 50 national invention patents, and published more than 100 papers. He was selected for the "Top 2% of Global Scientists" list in 2025.

plants

MDPI

Article

Genome-Wide Identification of the WRKY Gene Family in Four Cotton Varieties and the Positive Role of *GhWRKY31* in Response to Salt and Drought Stress

Tianyu Dong [1,2], Jiuchang Su [1,2], Haoyuan Li [1], Yajie Du [1], Ying Wang [1], Peilei Chen [1,2] and Hongying Duan [1,2,*]

[1] College of Life Sciences, Henan Normal University, Xinxiang 453007, China; dongty536@163.com (T.D.); 2021034@htu.edu.cn (J.S.); 2104283124@stu.htu.edu.cn (H.L.); gehtbbrt@163.com (Y.D.); wangy32021@163.com (Y.W.); chenpeilei@htu.edu.cn (P.C.)

[2] Henan International Joint Laboratory of Aquatic Toxicology and Health Protection, College of Life Science, Henan Normal University, Xinxiang 453007, China

* Correspondence: duanhy536@163.com

Abstract: The WRKY gene family is ubiquitously distributed in plants, serving crucial functions in stress responses. Nevertheless, the structural organization and evolutionary dynamics of WRKY genes in cotton have not been fully elucidated. In this study, a total of 112, 119, 217, and 222 WRKY genes were identified in *Gossypium arboreum*, *Gossypium raimondii*, *Gossypium hirsutum*, and *Gossypium barbadense*, respectively. These 670 WRKY genes were categorized into seven distinct subgroups and unequally distributed across chromosomes. Examination of conserved motifs, domains, *cis*-acting elements, and gene architecture collectively highlighted the evolutionary conservation and divergence within the WRKY gene family in cotton. Analysis of synteny and collinearity further confirmed instances of expansion, duplication, and loss events among WRKY genes during cotton evolution. Furthermore, *GhWRKY31* transgenic *Arabidopsis* exhibited heightened germination rates and longer root lengths under drought and salt stress. Silencing *GhWRKY31* in cotton led to reduced levels of ABA, proline, POD, and SOD, along with downregulated expression of stress-responsive genes. Yeast one-hybrid and molecular docking assays confirmed the binding capacity of GhWRKY31 to the W box of *GhABF1*, *GhDREB2*, and *GhRD29*. The findings collectively offer a systematic and comprehensive insight into the evolutionary patterns of cotton *WRKYs*, proposing a suitable regulatory framework for developing cotton cultivars with enhanced resilience to drought and salinity stress.

Keywords: WRKY; transgene; gene silencing; molecular docking; yeast one-hybrid

1. Introduction

The increase in worldwide temperatures presents hurdles for the growth and maturation of higher plants, as they encounter diverse abiotic stresses such as extreme temperatures, drought, and high salinity. These stresses not only hinder plant growth but also contribute to a gradual decline in global crop production [1–3]. To address these challenges and acclimate to adverse growth conditions, plants have evolved a series of intricate regulatory mechanisms [4,5]. In this complex system, stress receptor genes, stress-related transcription factors (TFs), and downstream response genes collaborate to create a sophisticated interconnected network [6–9].

The transcription factors (TFs), which play vital roles in regulating gene expression by binding to *cis*-acting elements upstream of the transcription start site (TSS), are critical components in the molecular network of stress response [10–13]. WRKY TFs are one of the most extensive in plants, playing crucial roles in diverse responses to abiotic stressors [14,15]. The WRKY family members have a highly conserved WRKY domain, consisting of the WRKYGQK motif and a $CX_{4-5}CX_{22-23}HXH$ zinc-finger motif [16]. The WRKYGQK motif binds to the W box (TTGACC/T) on the promoter of downstream genes and regulates their expression under various abiotic stresses [17,18]. WRKY TFs are typically classified into seven groups based on the number of WRKY domains (two domains in Group I proteins and one in the others) and the primary amino acid sequence (C_2-H_2 structure in Group IIa-e proteins and C_2HC structure in Group III proteins) [16,18].

Throughout the growth and development of higher plants, encountering stress is unavoidable, necessitating a robust defense mechanism to mitigate its impact. WRKY TFs associated with plant responses to abiotic stress have been extensively documented. For instance, *AtWRKY46* displays rapid induction under water stress conditions and regulates genes involved in reactive oxygen species (ROS) scavenging and cellular osmoprotection. Overexpression of *AtWRKY46* leads to heightened sensitivity to osmotic stress in soil-grown *Arabidopsis* [19]. Conversely, silencing *AtWRKY63* suppresses the expression of stress-responsive genes *RD29A* and *COR47*, thereby compromising drought tolerance in *Arabidopsis* seedlings [20]. The double mutants of *AtWRKY25* and *AtWRKY33* in *Arabidopsis* exhibit heightened susceptibility to salt stress, whereas overexpression of either *AtWRKY25* or *AtWRKY33* enhances salt stress tolerance [21]. The heterologous expression of wheat *TaWRKY146* in *Arabidopsis* enhances drought resistance by facilitating stomatal closure, elevating proline levels, and reducing malondialdehyde (MDA) accumulation [22]. In *Camellia sinensis*, upregulation of *CsWRKY2* under exogenous abscisic acid (ABA) and drought stress enhances drought tolerance by modulating downstream genes in the ABA signaling pathway [23]. Moreover, overexpression of *DgWRKY2/3/4* significantly promotes germination rate and root length in soybean and *Arabidopsis* seedlings under high salinity conditions [24–26]. Collectively, these findings underscore the pivotal role of WRKY family members as key regulators in plants under drought and salt stress scenarios [27].

Cotton, the most vital oilseed and fiber crop, accounts for 35% of the total fiber used worldwide. Previous studies have suggested that all diploid and tetraploid cotton species have evolved from a common ancestor, which subsequently diversified to produce nine groups, including the A-G, K, and AD genomes [28,29]. Cultivated cotton mainly consists of four cotton subspecies, including *G. arboretum* (A2-genome species), *G. raimondii* (D5-genome species), *G. hirsutum* (AD1-genome species), and *G. barbadense* (AD2-genome species). Encouragingly, due to the release of high-quality whole-genome sequences of the four cultivated species [30–33], the genome-wide analysis of the WRKY gene family is feasible and will help to elucidate its regulatory functions in stimuli responses such as osmotic, drought, and salt. Nevertheless, the precise and systematic investigation of WRKY genes in cultivated cotton is largely understudied, and the comprehensive functional validation of *WRKYs* in response to osmotic, drought, and salt stress remains incomplete [34].

In this work, the phylogenetic tree, chromosomal distribution, cis-acting elements, conserved motifs and domains, and collinearity relationship of *WRKYs* were analyzed in cotton. We next confirmed that *GhWRKY31* was up-regulated under salt and drought stress. Functional assays involving heterologous expression and VIGS revealed that *GhWRKY31* contributed to salt and drought tolerance in both *Arabidopsis* and cotton. The drought- and salt-induced expression of genes, such as *GhABF1*, *GhABF2*, *GhDREB2*, *GhRD29*, *GhNAC4*, *GhP5CS*, and *GhSOS1*, was inhibited in *GhWRKY31*-silencing seedlings. Furthermore, the YIH assay confirmed the binding of GhWRKY31 to the W box of *GhABF1*, *GhDREB2*, and *GhRD29*. Our results not only present a comprehensive analysis of the cotton WRKY gene family but also provide new insights for breeding cotton against abiotic stresses.

2. Results

2.1. Identification and Phylogenetic Analysis of the WRKY Gene Family

The shortest WRKY protein, Gbar_D06G009260, comprises 144 amino acids, whereas the longest proteins, Gbar_D12G019910 and Gorai.008G200800, consist of 1340 amino acids. The isoelectric point (pI) values vary from 4.72 (Gbar_D11G016820) to 9.98 (Gorai.004G219300 and Gh_D08G210300). The molecular weights (MW) range from 16,630.56 (Gorai.011G114200) to 151,574.34 (Gorai.008G200800). Subcellular localization analysis using the Plant-PLoc database indicated that WRKY proteins predominantly localize in the nucleus (Supplementary Table S1). A phylogenetic tree was constructed by employing the maximum likelihood method to elucidate the evolutionary relationships among 670 WRKY proteins in cotton (Figure 1). The WRKY proteins were categorized into seven clades, encompassing Group I (107), Group IIa-IIe (134, 74, 122, 86, and 75), and Group III (63), which were further distributed unevenly across seven subgroups.

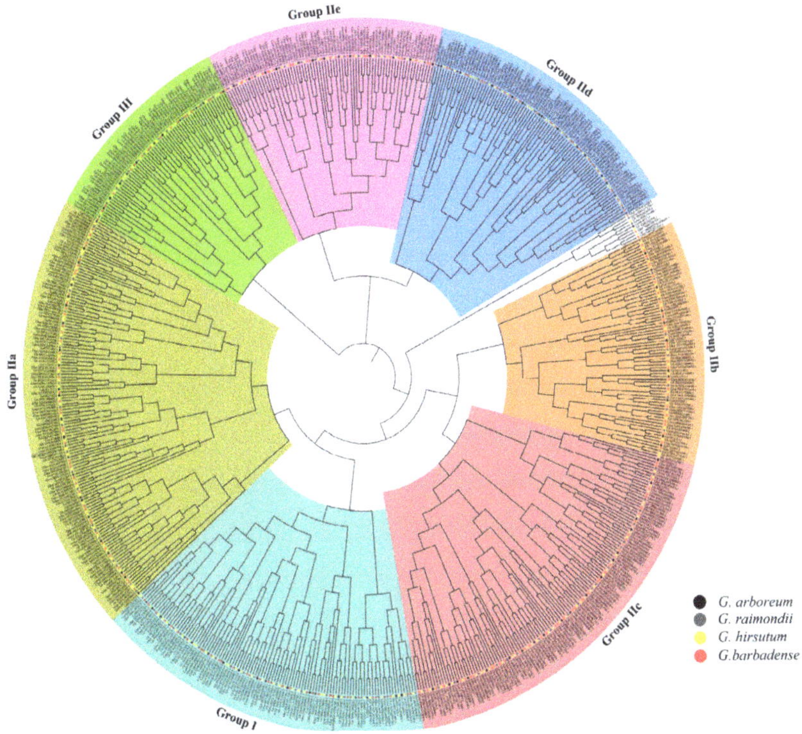

Figure 1. A maximum likelihood (1000 bootstraps) phylogenetic tree of WRKY proteins in *G. arboreum*, *G. raimondii*, *G. hirsutum*, and *G. barbadense*. The 7 color modules represent the 7 subfamilies of WRKY proteins, and no background module indicates unclassified WRKY proteins.

2.2. Chromosome Location of WRKY Genes

To further analyze the distribution of WRKY genes, the chromosome location was mapped (Figure 2). In *Gossypium arboretum* and *Gossypium raimondii*, 109 and 119 WRKY genes were, respectively, localized in the At or Dt sub-genomes. Ga14G1656, Ga14G1714, and Ga14G1560 were identified within contigs (Figure 2A,B). The highest number of WRKY genes in *G. arboreum* was found on the A07 chromosome (13), whereas the lowest was on the A03 chromosome (4) (Figure 2A). For *G. raimondii*, the D01 and D09 chromosomes harbored the highest number of *GrWRKYs* (13), whereas the D02 and D05 chromosomes contained the fewest (4) (Figure 2B). In *Gossypium hirsutum*, chrA05 exhibited the highest WRKY gene count with 16 members, whereas chrA03 had the lowest with three *GhWRKYs*. Notably, Gh_Contig00579G000600, Gh_Contig00383G000300, and Gh_Contig01109G001300 were located in contigs without chromosomal assignments (Figure 2C). In *Gossypium barbadense*, WRKY genes were mapped across chrA01 to chrA13 (four to seventeen genes per chromosome) and chrD01 to chrD13 (three to thirteen genes per chromosome) (Figure 2D). Consequently, WRKY family members were distributed disparately among cotton chromosomes.

Figure 2. Chromosomal distribution of WRKY genes in (**A**) *G. arboretum*, (**B**) *G. raimondii*, (**C**) *G. hirsutum*, and (**D**) *G. barbadense*. The chromosome number is shown at the top of each chromosome, and the scale for the length of chromosomes is megabases (Mb).

2.3. Conserved Motifs and Domains, Cis-Acting Elements, and Gene Structure of WRKYs

To elucidate the detailed characteristics of *WRKYs*, an analysis encompassing gene structure, conserved motifs, domains, and *cis*-acting elements was conducted. Ten conserved motifs were identified in WRKY proteins across four cotton strains. Notably, the majority of WRKY proteins exhibited more than two motifs, with exceptions noted in five GaWRKY, 15 GrWRKY, and 30 GbWRKY proteins. Consistently, motifs 1 and 2 were universally present in all WRKY members (Supplementary Table S2). Furthermore, the identification of at least one WRKYGQK domain in WRKY proteins was observed, with 107 Group I WRKY proteins containing two WRKYGQK domains (Figure 3). Additionally, the presence of basic region-leucine zipper (bZIP) domains (PF00170) was detected in 20 WRKY proteins, and plant_zn_clust (PF10533) structures were predominantly situated at the N-terminus of 80 WRKY members (Figure 3). These findings underscore the evolutionary stability and diversity of WRKY proteins within the cotton genome.

To further investigate the biological function of *WRKYs*, we identified *cis*-acting elements in the 5′-upstream regions of 2000 bp (Supplementary Table S3). A total of 12 different functions of *cis*-elements were identified, and these *cis*-acting elements related to stress responses were found abundantly in the promoter of *WRKYs* (Figure 3). The *cis*-acting elements can be divided into three categories: hormone-responsive sites (auxin-responsive element, gibberellin-responsive element, and MeJA-responsive element), transcription factor binding sites (MYB binding site, MYBHv1 binding site, and WRKY binding site), and growth and development sites (MYB binding site involved in drought inducibility, light responsiveness, flavonoid biosynthesis, low-temperature-induced responses, and defense- and stress-induced responses) (Figure 3).

2.4. *Duplication and Collinearity of GaWRKYs, GrWRKYs, GhWRKYs, and GbWRKYs*

The expansion pattern of *WRKYs* was elucidated through the construction of a duplication circos plot. In the diploid genomes of Gossypium raimondii and *Gossypium arboretum*, 88 and 102 *WRKYs*, respectively, were identified as originating from whole-genome duplication (WGD) or segmental duplication events (Figure 4A,B). Ga05G0631, Ga08G2219, Gorai.001G037800, Gorai.004G219400, and Gorai.009G062400 were tandem duplications and were distributed on chromosomes A05, A08, D01, D04, and D09, respectively (Figure 4A,B). Moreover, 17 and 14 WRKY genes were dispersed within the At or Dt sub-genomes (Supplementary Table S4). In *Gossypium hirsutum*, a substantial proportion (97.66%) of *WRKYs* were attributed to WGD or segmental duplication events. Among these, the WRKY genes Gh_A08G214800, Gh_D05G062100, and Gh_D08G210400 underwent tandem duplication, while Gh_D04G011700 and Gh_D07G055500 were dispersed, with respective locations on chromosomes A08, D04, D05, D07, and D08 (Figure 4C) (Supplementary Table S4). A total of 213 *WRKYs* had undergone whole-genome duplication (WGD) or segmental duplication events in *G. barbadense*. Gbar_A08G020300 (ChrA08), Gbar_D08G021260 (ChrD08), Gbar_A03G013230 (ChrA03), and Gbar_A11G020420 (ChrA11) appeared as tandem duplications or dispersion (Figure 4D) (Supplementary Table S4).

The hybridization event leading to the evolution of *Gossypium hirsutum* and *Gossypium barbadense* involved *Gossypium arboreum* (an A-genome species) and *Gossypium raimondii* (a D-genome species). A syntenic map was generated to investigate the evolutionary connections of WRKY genes between *G. hirsutum* and three other species (Supplementary Table S5). Through MCScan analysis, 571, 621, and 1044 duplicated gene pairs were identified between *G. hirsutum* and *G. arboreum*, *G. hirsutum* and *G. raimondii*, and *G. hirsutum* and *G. barbadense*, respectively (Figure 5). Notably, in *G. arboreum* and *G. raimondii*, the highest number of collinear relationships occurred on ChrA11 (85) and ChrD07 (98), with ChrA09 and ChrD05 exhibiting the fewest collinear relationships at 17 and 21, respectively (Figure 5A,B). Additionally, *G. barbadense* displayed 87, 75, 73, 65, and 78 collinear relationships on chromosomes A05, A11, D05, D07, and D11 among the 1044 gene pairs (Figure 5C). In short, the aforementioned results indicate an uneven distribution of collinear relationships across chromosomes, suggesting occurrences of deletion and duplication events within the WRKY gene family.

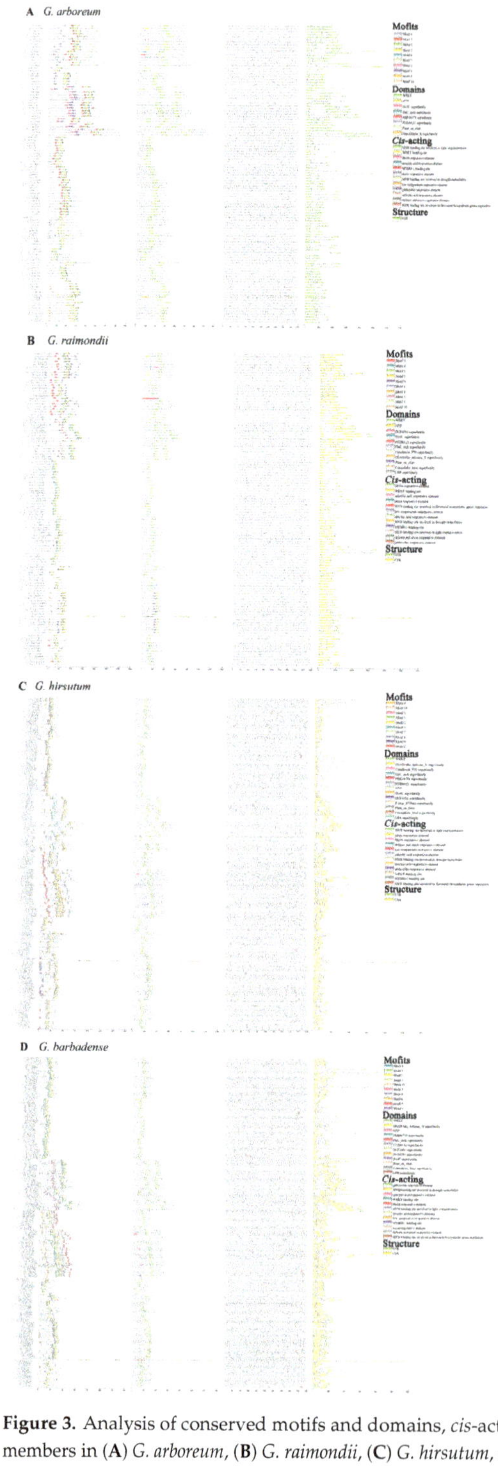

Figure 3. Analysis of conserved motifs and domains, *cis*-acting elements, and structures of WRKY members in (**A**) *G. arboreum*, (**B**) *G. raimondii*, (**C**) *G. hirsutum*, and (**D**) *G. barbadense*. The identification elements are represented by distinct colored boxes. The black lines of the gene structure indicate non-conserved regions.

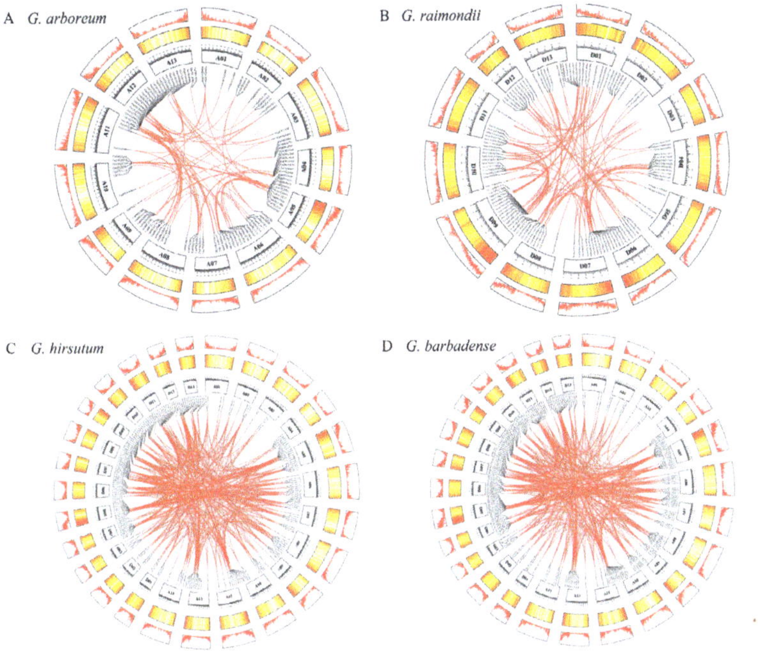

Figure 4. Duplicated WRKY genes based on the collinearity of all chromosomes in (**A**) *G. arboreum*, (**B**) *G. raimondii*, (**C**) *G. hirsutum*, and (**D**) *G. barbadense*. The number of genes is presented by a heatmap and a linear map, of which the red presents regions of high gene density, and yellow indicates a low-density region. The WRKY gene pairs with a syntenic relationship are linked by red lines, and the scale on the boxes above is in megabases (Mb).

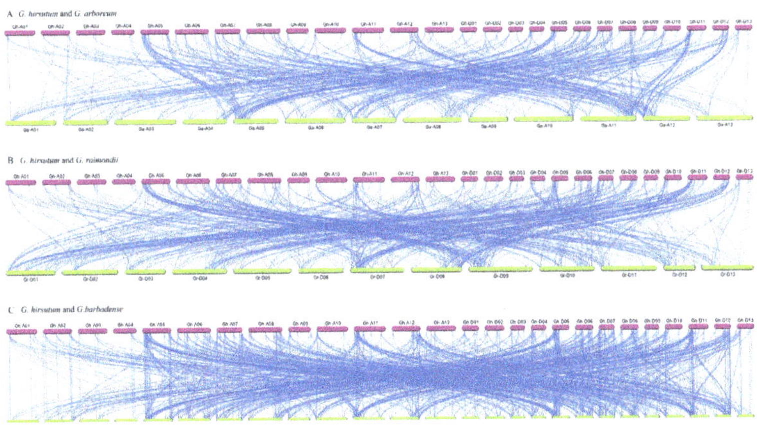

Figure 5. Synteny analysis of WRKY genes. Orthologous relationships between (**A**) *G. hirsutum* and *G. arboretum*, (**B**) *G. hirsutum* and *G. raimondii*, and (**C**) *G. hirsutum* and *G. barbadense* were investigated. Blue lines highlight duplicated WRKY gene pairs, while the gray lines in the background indicate all collinear relationships.

2.5. Expression Profiling and qRT-PCR Verification of GhWRKY Responses to Salt and Drought Stress

Transcriptome analysis of *Gossypium hirsutum* revealed distinct expression patterns of *GhWRKYs* under salt and drought stress conditions. Notably, following 3 h of salt treatment and 3 to 6 h of PEG treatment, a cluster of WRKY genes, such as Gh_A05G156700.1, Gh_D03G026500.1, Gh_A05G368400.1, and Gh_A06G109400.1, displayed elevated expression levels. Moreover, Gh_A05G156700.1, Gh_D02G067800.1, Gh_A08G149000.1, Gh_D08G210300.1, and Gh_D08G191400.1 exhibited peak expression levels after 1 and 6 h of salt treatment or 1 h of PEG treatment. Additionally, Gh_A08G031700.1, Gh_A09G013200.1, and Gh_D03G050200.1 reached their highest expression levels after 12 h of salt treatment or 12 h of PEG treatment. These findings indicate that a subset of *GhWRKYs* are responsive to salt and drought stress in *G. hirsutum* (Figure 6).

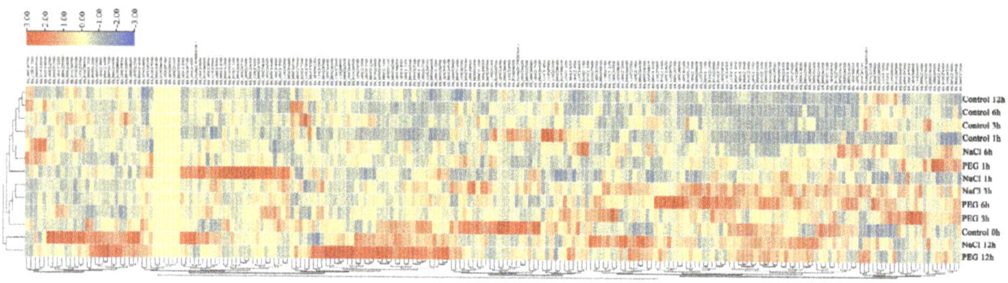

Figure 6. A cluster heatmap of expression patterns of *GhWRKYs* in response to NaCl and PEG treatment. Each line represents the expression of WRKY genes in different treatments, and the expression values in the row scale were normalized. The color scale varies from red to blue, indicating the high or low expression of each WRKY gene.

Next, qRT-PCR was employed to assess the expression dynamics of selected *GhWRKYs* in response to PEG and NaCl treatments. As expected, Gh_A08G031700.1, Gh_D02G067800.1, and Gh_A05G156700.1 exhibited distinct expression patterns in the presence of PEG and NaCl solutions (Figure 7). Specifically, Gh_A08G031700.1 demonstrated sensitivity to both PEG and NaCl treatments, with upregulation observed at 3 and 12 h under PEG treatment, and at 1, 12, and 24 h under NaCl treatment ($p < 0.05$) (Figure 7A). Gh_D02G067800.1 responded primarily to salt stress, showing a significant increase in expression at 6 and 12 h under NaCl treatment ($p < 0.05$) (Figure 7B). Moreover, Gh_A05G156700.1 displayed enhanced expression levels at 1, 3, and 6 h following PEG treatment ($p < 0.05$), and similarly exhibited upregulation after NaCl treatment at 3, 6, and 12 h (Figure 7C). Thus, Gh_A05G156700.1 was chosen for subsequent functional validation studies under salt and drought stress conditions.

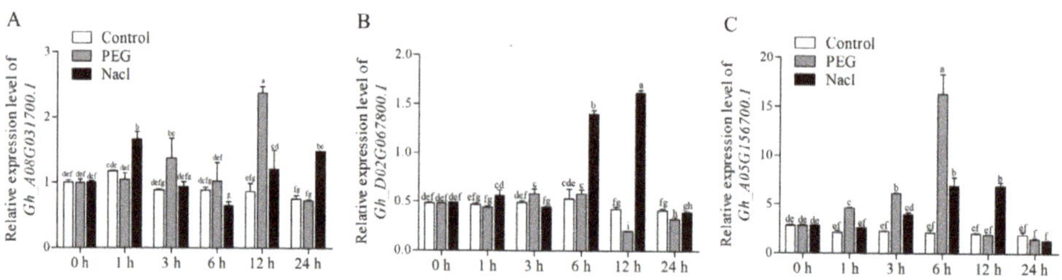

Figure 7. The expression levels of 3 WRKY genes in the leaves of *G. hirsutum* seedlings under PEG and NaCl stress. Gh_A08G037100.1 expression in the control group was set to 100% at 0 h. (**A–C**) represent the relative expression level of Gh_A08G031700.1, Gh_D02G067800.1, and Gh_A05G156700.1, respectively. Data represent the means ± SE from three independent experiments. The error bar represents the standard error of the mean, and the lowercase letter above the bar indicates a significant difference ($p < 0.05$).

2.6. *GhWRKY31* Improved the Tolerance of Transgenic *Arabidopsis* to Drought and Salt Stress

qRT-PCR analysis confirmed the upregulation of *GhWRKY31* (Gh_A05G156700.1) in response to both drought and salt stress. Subsequently, a study was conducted to explore the role of *GhWRKY31* by assessing the drought and salt stress tolerance of transgenic *Arabidopsis* plants overexpressing *GhWRKY31* following homozygous molecular characterization (Supplementary Figure S1). Both seed germination and root length in WT *Arabidopsis* were significantly suppressed by mannitol and NaCl treatments. The germination rates of WT plants were notably reduced to 83% under 100 mM, 76% under 200 mM, and 58% under 300 mM mannitol treatment (Figure 8A,B), and were suppressed to 56% and 36% under 100 mM and 150 mM NaCl treatment, respectively (Figure 8E,F). Meanwhile, the root length of WT was also inhibited under 100 mM (2.84 cm), 200 mM (2.23 cm), and 300 mM (1.62 cm) mannitol (Figure 8C,D), and was suppressed to 2.24 and 1.84 cm under 50 mM and 100 mM salt conditions, respectively (Figure 8G,H). On the contrary, the germination rates and root length of *GhWRKY31* OE lines were significantly higher than those of WT. The germination rates were nearly 100%, 100%, and 90% under 100 mM, 200 mM, and 300 mM mannitol. Under 50 mM, 100 mM, and 150 mM NaCl solution, the germination rates of *GhWRKY31* OE lines were almost up to 100% (Figure 8B,F). In addition, the root length of OE lines was 3.36 cm, 3.14 cm, 2.33 cm, 3.24 cm, and 2.37 cm under 100 mM, 200 mM, 300 mM mannitol, 50 mM, and 100 mM NaCl treatments, respectively. These measurements were significantly longer than those of WT ($p < 0.05$) (Figure 8D,H). Hence, the heterologous expression of *GhWRKY31* in *Arabidopsis* significantly improved drought and salt tolerance.

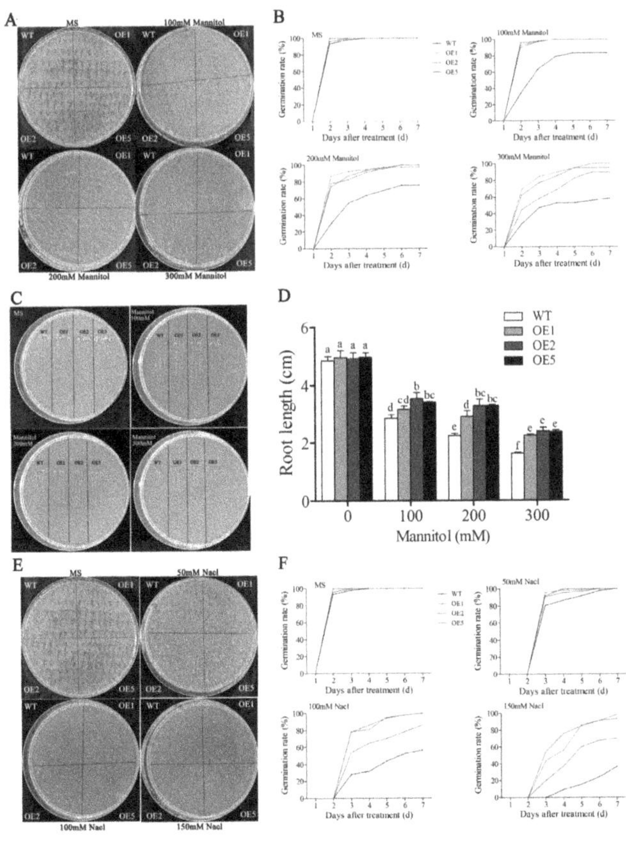

Figure 8. *Cont.*

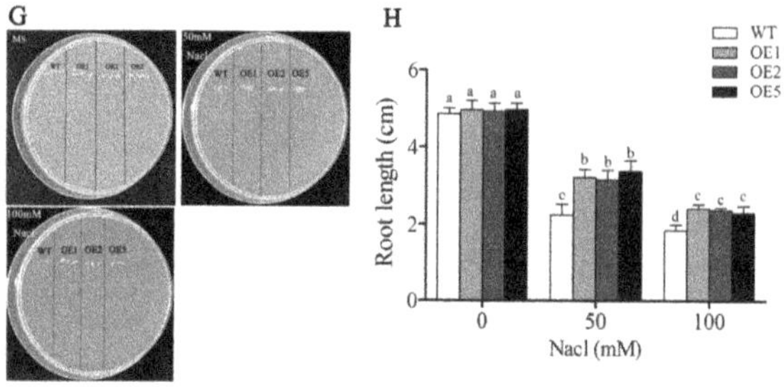

Figure 8. The germination rates and root length of *GhWRKY31* OE lines and WT under mannitol and salt conditions. (**A**,**E**) Phenotypic comparison of seedlings grown on 1/2 MS with 0 mM, 100 mM, 200 mM, and 300 mM mannitol or 0 mM, 50 mM, 100 mM, and 150 mM NaCl after 7 days. (**B**,**F**) Germination rates of seedlings grown under the conditions described in (**A**,**E**). (**C**,**D**,**G**,**H**) Phenotypic comparison and root length of seedlings grown on 1/2 MS with 0 mM, 100 mM, 200 mM, and 300 mM mannitol or 0 mM, 50 mM, and 100 mM NaCl after 7 days. Data represent the means ± SE from three independent experiments. The lowercase letters above the bar indicate the significant difference ($p < 0.05$).

2.7. VIGS of GhWRKY31 Reduced Drought and Salt Tolerance in G. hirsutum

To further elucidate the function of *GhWRKY31* in *G. hirsutum*, VIGS was employed to decrease the transcription level of *GhWRKY31*. qRT-PCR was used to evaluate the silencing efficiency of *GhWRKY31*. The expression level of *GhWRKY31* was reduced by approximately 75% in pYL156: GhWRKY31 plants (Supplementary Figure S2). As expected, no stress-related phenotype was observed in the seedlings of 'TM1+pYL156: 00' and 'TM1+pYL156: *GhWRKY31*' under water conditions. Nevertheless, the leaves of 'TM1+pYL156: *GhWRKY31*' seedlings exhibited shrinkage and yellowing characteristics compared with 'TM1+pYL156: 00' (empty vector seedlings) under 200 mM NaCl treatment (Figure 9A). Meanwhile, after a 14-day water-deficit treatment, the leaves of 'TM1+pYL156: 00' showed a healthier phenotype compared to 'TM1+pYL156: *GhWRKY31*' seedlings. The latter exhibited symptoms such as shrinkage, rolling, wilting, and death (Figure 9A). Additionally, the ABA and proline contents accumulated less in 'TM1+pYL156: *GhWRKY31*' seedlings than in the control group seedlings under drought and salt stress. Meanwhile, MDA accumulation was higher in 'TM1+pYL156: *GhWRKY31*' seedlings compared to the seedlings in the control group. Moreover, the activities of peroxidase (POD) and superoxide dismutase (SOD) were higher in plants in the control group than in 'TM1+pYL156: *GhWRKY31*' plants under drought and salt stress (Figure 9B–F).

2.8. GhWRKY31 Regulates the Expression of Salt- and Drought-Induced Genes

The *GhWRKY31*-silenced cotton seedlings exhibited heightened sensitivity to drought and salt stress. To elucidate the target genes of *GhWRKY31* in response to drought and salt in cotton, we conducted qRT-PCR analysis to determine whether *GhWRKY31* is essential for the expression of ABA-, drought-, and salt-induced genes. The expression levels of *GhRD29*, *GhNAC4*, *GhABF1*, *GhABF2*, *GhDREB2*, *GhP5CS*, and *GhSOS1* were induced in the control group under drought and NaCl stress. However, silencing *GhWRKY31* resulted in a decrease in the induction of the seven genes under drought and salt stress. Specifically, the expression levels of *GhABF1*, *GhABF2*, *GhP5CS*, and *GhSOS1* were suppressed to levels lower than those observed in the control group (Figure 10A).

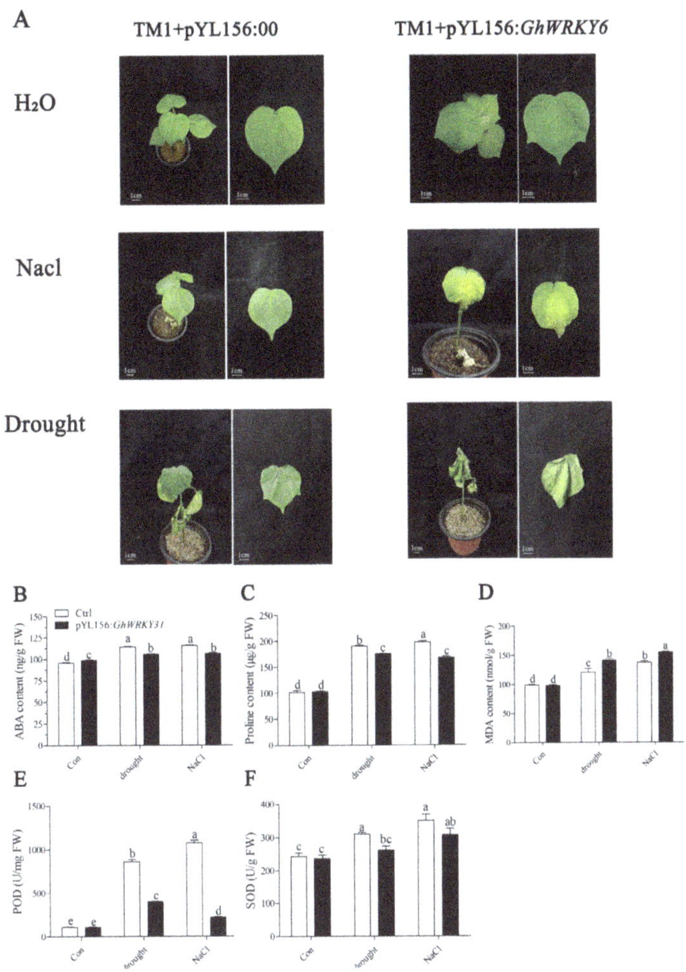

Figure 9. *GhWRKY31*-VIGS cotton seedlings exhibit increased sensitivity to drought and salt stress. (**A**) Leaf phenotypes showed shrinkage, yellowing, wilting, and death under water deficit conditions and 200 mM NaCl treatment. (**B**) ABA, (**C**) proline, (**D**) MDA content, (**E**) POD, and (**F**) SOD activity under water deficit conditions and 200 mM NaCl treatment. Data represent the means ± SE from three independent experiments. The lowercase letter above the bar indicates the significant difference ($p < 0.05$).

Further analysis focused on the identification of the W box (TTGACC/T) motif, which is crucial for the specific DNA binding of WRKY family members in the promoter regions of the aforementioned seven genes. We found that one, two, two, one, and three W boxes (TTGACC) were located in the promoter regions of *GhP5CS*, *GhABF1*, *GhRD29*, *GhABF2*, and *GhDREB2*, respectively (Figure 10B). Subsequently, molecular docking studies were performed using HDOCK v1.1 and PyMOL 2.5.0 software to explore potential interaction sites between the GhWRKY31 protein and the W box motifs of these five genes. The confidence scores for the interactions of GhWRKY31 with *GhP5CS*, *GhABF1*, *GhRD29*, *GhABF2*, and *GhDREB2* were determined as 0.8611, 0.9525/0.9050, 0.7619/0.8815, 0.8930, and 0.8576/0.8654/0.9492, respectively. The results indicated the formation of stable complexes between the WRKY domain of GhWRKY31 and the adjacent W box sequences, sustained by robust hydrogen bond interactions (Figure 10C).

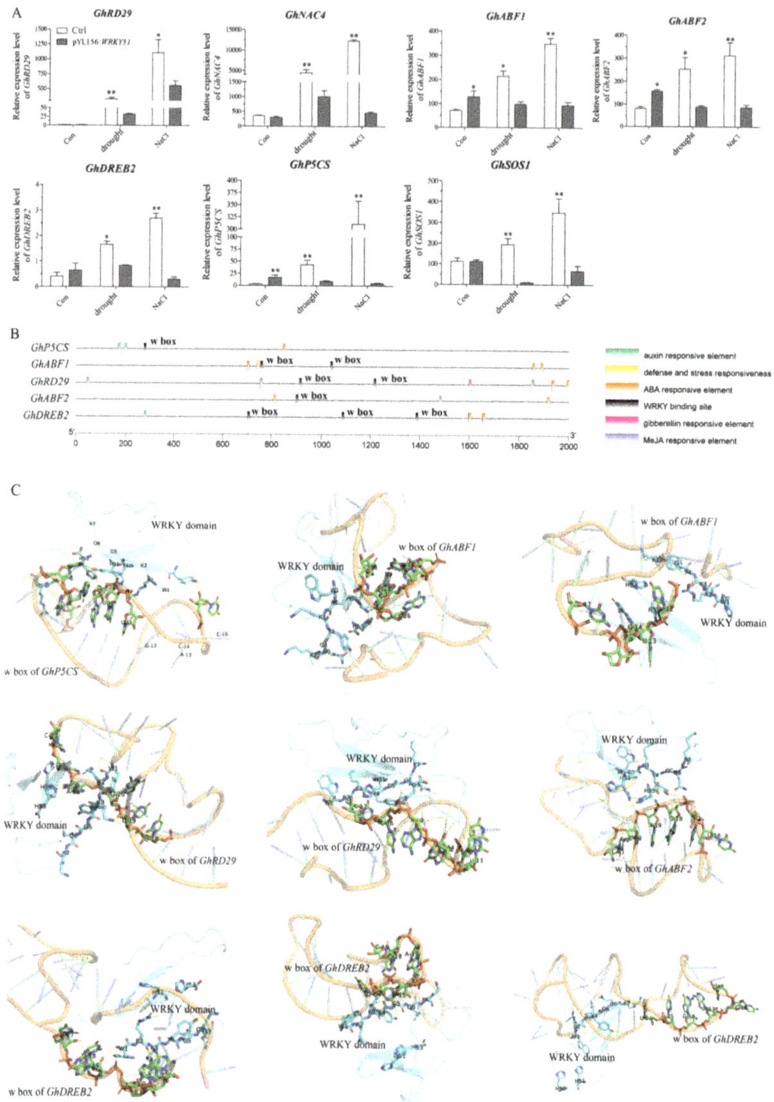

Figure 10. The gene expression levels induced by salt and drought were regulated by *GhWRKY31* in *G. hirsutum* leaves. (**A**) Silencing of *GhWRKY31* inhibits salt- and drought-induced gene expression. The data are shown as the mean ± SD from three independent biological replicates. (**, $p < 0.01$; *, $p < 0.05$; Student's *t*-test). (**B**) The *cis*-acting elements are located 2000bp upstream of the *GhP5CS*, *GhABF1*, *GhRD29*, *GhABF2*, and *GhDREB2* promoters. (**C**) The 3D structure of molecular docking for the binding of the GhWRKY31 protein and the W boxes of *GhP5CS*, *GhABF1*, *GhRD29*, *GhABF2*, and *GhDREB2*. The yellow dashed line represents hydrogen bonding interactions.

2.9. *GhWRKY31 Binds to the Promoter Regions of GhABF1, GhDREB2, and GhRD29*

The Yeast one-hybrid (Y1H) assay was employed to further investigate the binding affinity of the GhWRKY31 protein to *GhP5CS*, *GhABF1*, *GhABF2*, *GhDREB2*, and *GhRD29*.

Firstly, we confirmed that 100 ng/mL of AbA could inhibit the self-activation of pAbAi-bait. The results showed that the transformation yeast containing the combination of GhWRKY31 with the W box (TTGACC/T) of *GhABF1*, *GhDREB2*, *GhRD29*, *GhP5CS*, and *GhABF2* grew on SD/-Leu medium. The GhWRKY31 protein specifically bound to the fragment that contained the core TTGACC/T motif of *GhABF1*, *GhDREB2*, and *GhRD29* in the SD/-Leu+AbA (100 ng/mL) medium. These findings support the conclusion that GhWRKY31 directly binds to the promoter regions of *GhABF1*, *GhDREB2*, and *GhRD29* (Figure 11).

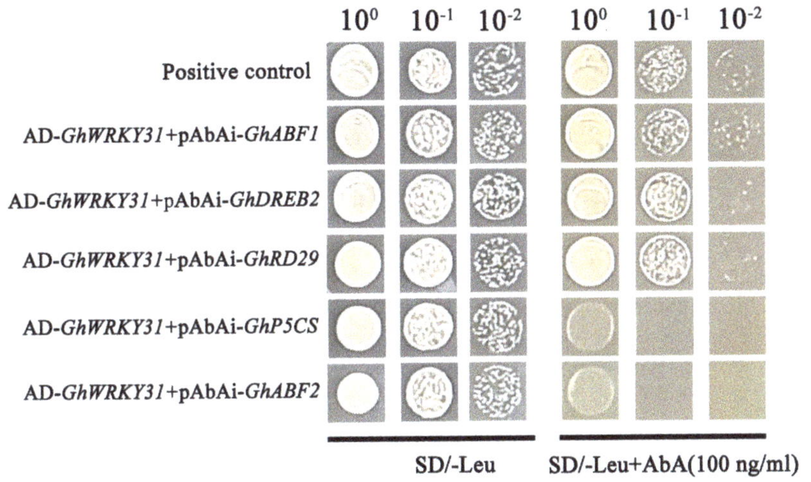

Figure 11. Y1H assay of GhWRKY31 with *GhABF1*, *GhDREB2*, *GhRD29*, *GhP5CS*, and *GhABF2*. The promoters of *GhABF1*, *GhDREB2*, *GhRD29*, *GhP5CS*, and *GhABF2*, which contain the putative TTGACC/T transformation (W box), were constructed in the pAbAi vector. The ORF of *GhWRKY31* was constructed in the pGADT7 vector. Yeast cells were diluted with distilled water (10^0 to 10^{-2}) and cultured on SD/-Leu medium supplemented with 100 ng/mL of Aureobasidin A (AbA).

3. Discussion

WRKY TFs are ubiquitously present throughout the plant kingdom, representing a remarkably conserved protein family. Currently, genome-wide studies of the WRKY gene family have been extensively conducted [14,15], revealing their crucial involvement in responding to various abiotic stressors [8,35,36]. Cotton, one of the most important economic crops, has remained relatively scarcely researched with regard to its WRKY gene family. Hence, the study investigates the evolution and function of WRKY genes in cotton based on analysis of genome-wide duplication, heterogenous expression in *Arabidopsis*, VIGS in *G. hirsutum* 'TM1', molecular docking, and Y1H.

In this study, 112 *GaWRKYs*, 119 *GrWRKYs*, 217 *GhWRKYs*, and 222 *GbWRKYs* were identified. Since cotton underwent hybridization and polyploidization 1.5 Mya, the number of WRKY genes in tetraploid cotton has increased to be ~2-fold greater than that of diploid cotton [37]. Next, the 670 *WRKYs* were divided into seven subgroups (Figure 1) based purely on phylogenetic data [38] and were unevenly distributed among different subfamilies. The analysis of chromosomal positioning showed the absence of WRKY genes on chrD01, chrD09, and chrD10, and the acquisition of WRKY genes on chrD05, chrD06, and chrD11 during the formation of tetraploid cotton. These findings provide valuable insights into the evolutionary dynamics of the WRKY gene family in cotton.

To further elucidate the evolutionary relationships among WRKY TFs in cotton, an analysis of the conserved motifs and domains of WRKY genes was conducted. Each WRKY gene typically exhibited one or two conserved WRKYGQK domains and a distinctive zinc-finger structure at the C-terminus, comprising consecutive conserved motifs (Figure 3). The WRKY domain was predominantly situated at the central region of the protein sequences, with consistent motifs and domains observed within the same subgroup across the four cotton species analyzed. These findings suggest a high degree of conservation among WRKY genes throughout cotton evolution. Nonetheless, sequence similarity in regions outside the motifs and domains of WRKY genes was comparatively lower, indicating complexity and diversity in the evolutionary trajectory of cotton WRKY genes (Figure 3). *Cis*-acting elements located in the promoter region are recognized for their crucial role in gene expression regulation, offering insights into gene functionality. The *cis*-acting elements of *WRKYs* encompass phytohormone response elements, development and stress-related elements, as well as transcription factor binding sites (Figure 3). These elements are likely to influence hormonal responses, abiotic stress responses, and interactions with transcription factors. Notably, similar *cis*-acting elements have been documented in other plant species such as *Vitis vinifera* [39], *Calohypnum plumiforme* [40], and *Chrysanthemum lavandulifolium* [41]. Furthermore, the structural analysis revealed significant variations in the proportions of UTRs and CDSs among the different cotton species, potentially attributed to homologous recombination events resulting from the artificial domestication of cotton [30,42].

In general, WGD, segmental duplication, tandem duplication, and transposon-induced duplication represent primary mechanisms capable of modifying the function, evolution, and configuration of TFs, leading to the emergence of novel subfamilies [43–45]. Our study revealed that the frequencies of WGD or segmental duplication events exceeded those of tandem duplication, underscoring the fact that the amplification and evolution of WRKY genes are predominantly driven by WGD or segmental duplication, with tandem duplication playing a secondary role (Figure 4). Likewise, these evolutionary events were also found in mung bean [10], wheat [15], and cherry [46], with WGD and segmental duplication events exerting a primary effect. In subsequent investigations into the evolutionary mechanisms of WRKY genes, synteny analysis revealed numerous collinear WRKY gene pairs between *G. hirsutum* and three other species (Figure 5). This conservation may be attributed to the stability in gene number and arrangement during the 1.5 million years of hybridization, polyploidy, and evolutionary processes in cotton. Nonetheless, partial WRKY genes were lost during evolution, potentially as a result of the artificial domestication process spanning 8000 years. The retention of these preserved WRKY genes is believed to significantly enhance cotton's survival and adaptability, as well as the quality and length of its fibers [30,33].

Currently, WRKY TFs have been identified to play a crucial role in the regulation of plant responses to drought and salt stress [20,23–25]. Our investigations indicated that a cluster of WRKY genes exhibited differential expression in response to PEG or NaCl treatment (Figure 6), and the *GhWRKY31* emerged as a potential candidate gene associated with salt and drought stress response in *Gossypium hirsutum* (Figure 7). Subsequently, to delve deeper into the function of *GhWRKY31*, the Super1300: *WRKY31* vector was engineered, leading to the generation of homozygous *GhWRKY31* transgenic *Arabidopsis* lines. Assessment of phenotypic traits indicated that under stress conditions, the germination rate and root length of the WT were notably inferior to those of the *GhWRKY31* OE lines, suggesting that *GhWRKY31* was found to confer dual resistance to salt and drought stress in *Arabidopsis* (Figure 8). Similarly, the overexpression of *GhWRKY39-1* in *Nicotiana benthamiana* not only heightened salt stress tolerance but also conferred enhanced resistance to bacterial pathogen infection [47]. Transgenic tobacco overexpressing *GhWRKY25* exhibited improved seedling tolerance to salt stress while displaying decreased resistance to mannitol-induced osmotic and drought stress [48]. Furthermore, numerous WRKY genes have been documented to actively respond to osmotic, drought, and salt stress in various plant species. For example, the overexpression of *MbWRKY5* [49], *MfWRKY40* [50], *CmWRKY10* [51], and *TaWRKY93* [52] in *Arabidopsis* or tobacco led to heightened resistance to osmotic and high-salinity stress compared to the WT, whereas overexpression of *ZmWRKY17* [53], *CdWRKY50* [54], and

VvWRKY50 [55] resulted in susceptibility under PEG, mannitol, or NaCl treatment. These outcomes underscore the pivotal role of heterologous WRKY gene expression in diverse abiotic stress responses in *Arabidopsis* and tobacco.

To enhance the understanding of *GhWRKY31*'s function, drought and salt tolerance assessments were conducted in *G. hirsutum* utilizing VIGS technology. The leaves of *GhWRKY31*-silenced cotton seedlings exhibited heightened sensitivity to water-deficit and NaCl conditions. Reduced levels of ABA and proline content, coupled with elevated MDA accumulation, indicated decreased resistance to drought and salt stress in VIGS cotton plants. Analysis of POD and SOD activities confirmed that the WT plants possessed superior ROS-scavenging capacity compared to *GhWRKY31*-silenced cotton seedlings (Figure 9). Notably, *ABF1/2*, *DREB2*, and *RD29* were identified as key players in ABA-dependent or ABA-independent responses to drought and salt stress, exerting a positive regulatory function under drought and NaCl conditions [56,57]. The induction of *P5CS*, a pivotal enzyme in proline biosynthesis, in response to drought and high salt levels was observed [58]. Our investigation revealed suppressed expression levels of *GhABF1*, *GhABF2*, *GhDREB2*, *GhRD29*, and *GhP5CS* in *GhWRKY31*-VIGS cotton leaves. Similarly, silencing of *GhWRKY46* [59] and *XsWRKY20* [60] via virus-induced gene silencing resulted in increased sensitivity to drought or salt stress, evidenced by weakened physiological phenotypes, heightened MDA content, diminished proline accumulation, and notable inhibition of stress-related gene expression levels, such as *ABI3*, *ABF2*, *DREB1*, *DREB2*, *RD22*, *LEA5*, and *P5CS*, in *WRKY*-silenced seedlings. Furthermore, molecular docking analysis illustrated the formation of stable complexes through multiple hydrogen bonds between the WRKYGQK domain of GhWRKY31 and the W boxes of *GhABF1*, *GhDREB2*, and *GhRD29* (Figure 10). Additional research has corroborated that the WRKYGQK domain of *SlWRKY3/4*, *CcWRKY1/51/70*, and *HvWRKY46* can establish hydrogen bonds with the W box of stress-related genes, exhibiting diverse bonding strengths among these members of the WRKY subfamily [61–63]. Additionally, Y1H analysis confirmed that *GhABF1*, *GhDREB2*, and *GhRD29* directly interact with the GhWRKY31 protein (Figure 11). Therefore, GhWRKY31, serving as a positive regulator in response to drought and salinity stress, has the potential to confer salt and drought resistance in *Arabidopsis* and cotton through the upregulation of *GhABF1*, *GhRD29*, and *GhDREB2* (Figure 12). Collectively, these findings not only support the advancement of research on WRKY genes implicated in stress resilience in cotton but also establish a theoretical framework for plant cultivation in arid and saline environments.

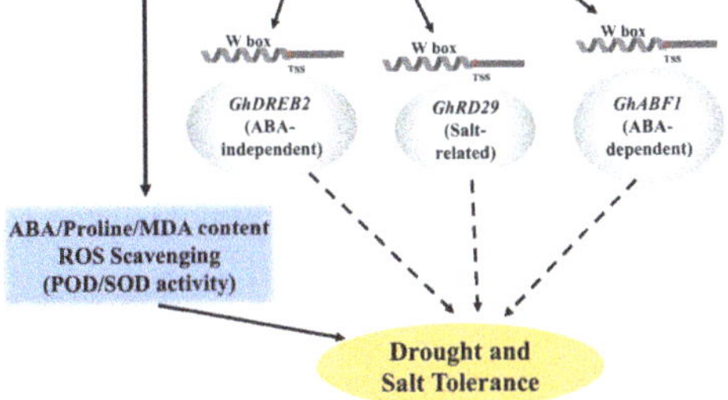

Figure 12. A working model of the role of the *GhWRKY31* module in drought and salt stress responses in cotton.

4. Materials and Methods

4.1. Identification of WRKY Family Members

Four cotton genome assembly files (FASTA format) and genome annotation files (GFF3 format), including *G. arboreum* (CRI version, strain SXY1) [30], *G. raimondii* (JGI version, strain Ulbr.) [31], *G. hirsutum* (CRI version, strain Tm-1) [32], and *G. barbadense* (HAU version, strain 3–79) [33], were downloaded from Cotton FGD [64]. The Hidden Markov Model (HMM) file (PF03106) was downloaded from the Pfam database (http://pfam.xfam.org/ (accessed on 22 April 2023)) [65]. HMMER 3.0 [66] was used to screen the potential WRKY proteins, and the key parameters were set as default (1×10^{-5}). Next, WRKY proteins were manually screened using SMART (http://smart.emblheidelberg.de/ (accessed on 2 June 2023)) and NCBI CDD (https://www.ncbi.nlm.nih.gov/Structure/bwrpsb/bwrpsb.cgi/ (accessed on 3 June 2023)). Finally, non-WRKY domains and incorrect and repetitive family members were deleted.

4.2. Multiple Sequence Alignment and Phylogenetic Tree Construction

The full-length amino acid sequences of WRKY proteins were aligned using the ClustalW program. Based on the alignments provided, a maximum likelihood tree was constructed using the MEGA 7.0 program (http://www.megasoftware.net/ (accessed on 19 June 2023)) [67], and the bootstrap test was carried out with 1000 iterations. Finally, the phylogenetic tree was plotted using interactive tree of life v5.0 (iTOL) (https://itol.embl.de/ accessed on 25 June 2023)) [68].

4.3. Chromosomal Locations, Gene Structure, Conserved Motifs and Domains, and Cis-Acting Elements of WRKY Proteins

To map the chromosomal distribution of WRKY genes in 4 cotton species, the above reference genomes and annotation files, as well as WRKY protein IDs, were incorporated into the gene location visualization toolkit of TBtools [69]. For the analysis of WRKY gene structures, we extracted information on WRKY gene structures using reference genomes and annotation files. We then visualized the WRKY gene structures using the gene structure toolkit of TBtools. The conserved motifs of WRKY proteins were analyzed using the MEME database (http://meme-suite.org/ (accessed on 5 July 2023)), and the conserved domains were obtained through the NCBI CD-Search (https://www.ncbi.nlm.nih.gov/Structure/cdd/wrpsb.cgi (accessed on 8 July 2023)). The acquired data were then plotted using TBtools. To investigate the *cis*-acting elements of *WRKY* promoters, the 5′-upstream regions of 2000 bp were downloaded from Cotton FGD. Subsequently, the sequences were analyzed using the PlantCARE database (http://bioinformatics.psb.ugent.be/webtools/plantcare/html/ (accessed on 21 July 2023)) [70] and visualized with TBtools.

4.4. Duplication and Collinearity Analysis of WRKY Proteins

The MCScan program [71] was used to detect gene pairs with a BLASTp search (e-value < 10^{-5}). Next, the chromosome length file, gene density file, and WRKY ID highlighting file were created from the reference genomes and annotation files, respectively. The prepared files were separately placed into the multiple collinearity scanning toolkit, dual synteny plotter toolkit, and advance circos toolkit of TBtools [72] for analysis of gene collinearity relationships and duplication events among the *WRKYs* in four cotton species.

4.5. Cotton Materials and Stress Treatments

Upland cotton *G. hirsutum* (Tm-1) was obtained from Cotton Research of the Chinese Academy of Agricultural Sciences (Anyang, Henan Province, China). The cotton seeds were sterilized with 3% H_2O_2 for 12 h and then washed with distilled water. Subsequently, the seedlings were grown in a greenhouse at 28 °C with a 16 h light/8 h dark photoperiod until the second true leaf expanded.

For the validation of expression levels for *GhWRKYs* under drought and salt stress treatments, cotton seedlings were irrigated with 20% (*w*/*v*) PEG6000 (drought-mimicking) and 200 mM NaCl solution, respectively. Seedlings with water were used as the control group. All leaves were collected at 0, 1, 3, 6, 12, and 24 h and stored at −80 °C for further experiments.

4.6. Transcriptome Analysis and qRT-PCR Verification of WRKY Genes

The RNA-seq raw data of PEG- and NaCl-treated *G. hirsutum* were downloaded from NCBI (https://www.ncbi.nlm.nih.gov/bioproject/?term=PRJNA490626 (accessed on 3 September 2023)) [73]. Firstly, we downloaded the raw sequencing data. Following the removal of adapters by Fastq and Trim Galore, the sequencing reads were aligned to the genome of *G. hirsutum* using STAR [74]. Next, RSEM [75] was employed to obtain the expression quantification (FPKM value) of *GhWRKYs*. The FPKM values of *WRKY* genes were log2-transformed and plotted using the TBtools heatmap.

The total RNA of *G. hirsutum* was extracted using a FastPure Universal Plant Total RNA Isolation Kit (Vazyme, Nanjing, China). The first-strand cDNA was synthesized using a TaKaRa kit (TaKaRa, Japan). RT-qPCR was performed using the Universal SYBR qPCR Master Mix kit (Vazyme, Nanjing, China), with 1 μL of cDNA template, 0.5 μL each of forward and reverse primers (at a working concentration of 10 μM), 5 μL of SYBR qPCR master mix, and 3 μL of nuclease-free water. The primers were designed using Primer Premier 5.0 (Supplementary Table S6). *GhActin7* was used as an internal control. A total volume of 10 μL was carried out in the Light Cycler® 96 fluorescence quantitative PCR instrument (ABI7500; Applied Biosystems, America). The expression levels of WRKY genes were calculated by the $2^{-\Delta\Delta Ct}$ method [76].

4.7. Heterologous Overexpression and Stress Tolerance Assay in Arabidopsis

The wild-type (WT) *Arabidopsis* (Ecotype Col-0) was used as the receptor for *GhWRKY31* genetic transformation. The seeds of WT *Arabidopsis* were surface-sterilized with 5% sodium hypochlorite and washed with sterile water 5 times. These seeds were then stored at 4 °C for 24 h. Next, the seeds were evenly sown on 1/2 MS solid mediums and cultured in a greenhouse (16 h light/8 h dark cycle, 22 °C) for 7 days. Next, these seedlings were replanted in a 3:1 mixture of vermiculite and nutrient soil.

The *GhWRKY31* CDS was inserted into the *XbaI*/*KpnI* restriction enzyme sites of the Super1300 plasmid. The Super1300:*GhWRKY31* vector was transformed into *Agrobacterium tumefaciens* strain GV3101, and full-flowering *Arabidopsis* seedlings were used for genetic transformation by the floral dip method [77]. The *GhWRKY31* overexpression (OE) lines were selected using hygromycin. Here, the 4-week-old OE lines of *GhWRKY31* were identified by RT-qPCR. We finally obtained 5 independent OE lines of *GhWRKY31* and named them OE1, 2, 3, 4, and 5.

The seeds of WT and *GhWRKY31* OE lines of *Arabidopsis* were evenly planted on 1/2 MS solid media containing different concentrations of mannitol (0, 100, 200, and 300 mM) and NaCl (0, 50, 100, and 150 mM). The germination rate was recorded for 7 consecutive days using a magnifier. To measure root length, seedlings were initially grown upright on fresh 1/2 MS solid medium for 3 days, and then transferred to 1/2 MS solid medium supplemented with mannitol (0, 100, 200, and 300 mM) and NaCl (0, 50, and 100 mM) for a period of 5 to 7 days.

4.8. Virus-Induced Gene Silencing (VIGS), Stress Treatments, and Determination of Biochemical Indexes

As a previous study described [78], the CDS of *GhWRKY31* was amplified from *G. hirsutum* using RT-PCR. The CDS of *GhWRKY31* was inserted into the pYL156 vector to construct a pYL156: *GhWRKY31* fusion vector. Subsequently, the pYL156: 00 (empty vector) and pYL156: *GhWRKY31* vectors were severally transformed into *A. tumefaciens* strain GV3101. The bacterial fluid of recombinant GV3101 was used to infect the cotyledons of *G. hirsutum* seedlings through injection. The leaves were collected for RNA extraction and to detect interference efficiency using qRT-PCR.

For the drought and salt tolerance assays, seedlings of 'TM1'+pYL156: 00 (empty vector injection) and 'TM1'+pYL156: *GhWRKY31* (*GhWRKY31* injection) were treated for 14 days with water, water-deficit conditions, and 200 mM NaCl solution. Additionally, the cotton leaves were crushed and added to the cold alcohol extract. The ABA was dissolved from the plant cells into the extraction solution by stirring at a low temperature. A centrifuge was used to separate the plant residues and cell fragments suspended in the alcohol extract. The ABA concentrate was obtained by transferring the alcohol extract into a new centrifuge tube. Subsequently, ABA levels (ng/g. FW) were detected using an ABA Elisa kit (SIONBESTBIO, YX-E21782, Shanghai, China). Following the crushing of cotton leaves, proline and MDA extracts were added separately and the mixture was subjected to low-temperature stirring to ensure full contact. A centrifuge was employed to isolate plant residues and cell fragments that were suspended in the extract. Analysis of proline (μg/g. FW) (Solarbio, BC0290, Beijing, China) and MDA (nmol/g. FW) (Solarbio, BC0025, Beijing) contents was carried out using dedicated kits for each compound.

To determine SOD (U/g. FW) and POD (U/mg. FW) activity, cotton leaves were finely crushed and mixed with a test tube containing phosphoric acid buffer and polyethylene glycol /EDTA. The mixture was stirred under ice bath conditions to ensure complete contact between the sample and the extract, facilitating the release of the POD and SOD enzymes from the cells. The resulting extract was then centrifuged and transferred to a new centrifuge tube for activity determination using a POD kit (Solarbio, BC0095, Beijing, China) and a SOD kit (Solarbio, BC0175, Beijing, China).

4.9. Molecular Docking Simulation

The interaction between the GhWRKY31 protein and the W box of stress-related genes was investigated using HDOCK v1.1 software. The nucleotide sequence of stress-related genes and the amino acid sequence of GhWRKY31 were introduced into receptors and ligands modules in HDOCK (http://hdock.phys.hust.edu.cn/ (accessed on 6 January 2024)) [79]. The output interaction model files were imported into PyMOL 2.5.0. The center of the docking boxes, which were based on the position of the crystal ligand, were constructed minutely. Next, the atoms for polarity docking were selected, and the docking relationship was plotted. In addition, the confidence score (CS) indicates the likelihood of binding between two molecules ($CS = 1.0/[1.0 + e^{0.02\times(\text{docking score}+150)}]$). The two molecules would be very likely to bind if CS > 0.7.

4.10. Yeast One-Hybrid (Y1H) Assay

The Y1H assay was performed following the same methodology as a previous study [80]. The full-length sequence of *GhWRKY31* was cloned into the pGADT7 vector between the *EcoRI* and *BamHI* sites. The recombinant plasmid was co-transformed into yeast Y1HGold with pAbAi-*GhP5CS*, pAbAi-*GhABF1*, pAbAi-*GhABF2*, pAbAi-*GhDREB2*, and pAbAi-*GhRD29*. pGADT-53 was used as a positive control, and all transformed candidates were grown on SD/-Ura/-Leu medium with 0 or 100 ng/mL of Aureobasidin A (AbA) for 3–5 days.

4.11. Statistical Analysis

The data were statistically analyzed using SPSS 10.0 software and plotted by GraphPad Prism 5.0. There were 3 biological replicates for each experiment, and the data were presented as means ± SD of three independent experiments. Experimental data were analyzed by Student's *t*-test, and the bars with different letters indicate significant differences ($p < 0.05$).

Supplementary Materials: The following supporting information can be downloaded at: https://www.mdpi.com/article/10.3390/plants13131814/s1, Table S1. WRKY genes in *G. arboreum*, *G. raimondii*, *G. hirsutum*, and *G. barbadense*. Table S2. WRKY genes containing 2 conserved motifs in *G. arboreum*, *G. raimondii*, *G. hirsutum*, and *G. barbadense*. Table S3. *cis*-acting elements in the promoter regions of WRKY genes in *G. arboreum*, *G. raimondii*, *G. hirsutum*, and *G. barbadense*. Table

S4. Dispersed, segmental, tendam and proximal duplications of WRKY genes in *G. arboreum*, *G. raimondii*, *G. hirsutum*, and *G. barbadense*. Table S5. Synteny analysis of WRKY genes in *G. arboreum*, *G. raimondii*, *G. hirsutum*, and *G. barbadense*. Table. S6 The primers used in the study. Figure S1 qRT-PCR identification of *GhWRKY31* transgenic *Arabidopsis* lines. The two-week-old seedlings were used to perform the molecular identification of *GhWRKY31* transgenic *Arabidopsis* plant and lines. WT: wild type; OE1, 2, 3, 4, and 5: *GhWRKY31* transgenic *Arabidopsis* lines of T_3 generation. The lower letter above the bar indicates the significant difference ($p < 0.05$). Figure S2 qRT-PCR identification of silencing efficiency in cotton seedings. The data represent the means ± SE from three independent experiments. Independent t-tests indicated that there was significant difference among the 'TM1', 'TM1 + pYL156:00' and 'TM1+pYL156:*GhWRKY31*'. The lower letter above the bar indicates the significant difference ($p < 0.05$).

Author Contributions: Conceptualization, T.D. and J.S.; methodology, T.D.; software, T.D.; validation, T.D. and J.S.; formal analysis, T.D., H.L. and Y.D.; investigation, T.D., H.L. and Y.D.; resources, T.D. and Y.W.; data curation, T.D.; writing—original draft preparation, T.D. and Y.W.; writing—review and editing, P.C. and H.D.; visualization, T.D. and J.S.; supervision, H.D.; project administration, H.D.; funding acquisition, H.D. All authors have read and agreed to the published version of the manuscript.

Funding: This research was supported by the financial aid of the Program for Innovative Research Team (in Science and Technology), University of Henan Province (No. 23IRTSTHN022); Science and Technology R&D Program of Henan Province (No. 222301420097).

Data Availability Statement: The data that support the findings of this study are available from the corresponding authors upon reasonable request.

Acknowledgments: The study was supported by "The High-Performance Computing Center of Henan Normal University". The cotton seeds were kindly provided by Xi Wei from the Institute of Cotton Research, Chinese Academy of Agricultural Sciences.

Conflicts of Interest: The authors declare no conflicts of interest.

References

1. Gao, Y.; Lu, Y.; Wu, M.; Liang, E.; Li, Y.; Zhang, D.; Yin, Z.; Ren, X.; Dai, Y.; Deng, D.; et al. Ability to remove Na and retain K correlates with salt tolerance in two maize inbred lines seedlings. *Front. Plant Sci.* **2016**, *7*, 1716. [CrossRef] [PubMed]
2. Peck, S.; Mittler, R. Plant signaling in biotic and abiotic stress. *J. Exp. Bot.* **2020**, *71*, 1649–1651. [CrossRef] [PubMed]
3. Zandalinas, S.I.; Fichman, Y.; Devireddy, A.R.; Sengupta, S.; Azad, R.K.; Mittler, R. Systemic signaling during abiotic stress combination in plants. *Proc. Natl. Acad. Sci. USA* **2020**, *17*, 13810–13820. [CrossRef] [PubMed]
4. Devireddy, A.R.; Zandalinas, S.I.; Fichman, Y.; Mittler, R. Integration of reactive oxygen species and hormone signaling during abiotic stress. *Plant J.* **2021**, *105*, 459–476. [CrossRef] [PubMed]
5. Harrison Day, B.L.; Carins-Murphy, M.R.; Brodribb, T.J. Reproductive water supply is prioritized during drought in tomato. *Plant Cell Environ.* **2022**, *45*, 69–79. [CrossRef] [PubMed]
6. Agarwal, P.; Baranwal, V.K.; Khurana, P. Genome-wide analysis of bZIP transcription factors in wheat and functional characterization of a TabZIP under abiotic stress. *Sci. Rep.* **2019**, *9*, 4608. [CrossRef] [PubMed]
7. Li, B.; Liu, R.; Liu, J.; Zhang, H.; Tian, Y.; Chen, T.; Li, J.; Jiao, F.; Jia, T.; Li, Y.; et al. ZmMYB56 regulates stomatal closure and drought tolerance in maize seedlings through the transcriptional regulation of *ZmTOM7*. *New Crops* **2024**, *1*, 100012. [CrossRef]
8. Li, W.; Pang, S.; Lu, Z.; Jin, B. Function and mechanism of WRKY transcription factors in abiotic stress responses of plants. *Plants* **2020**, *9*, 1515. [CrossRef] [PubMed]
9. Liu, F.; Xi, M.; Liu, T.; Wu, X.; Ju, L.; Wang, D. The central role of transcription factors in bridging biotic and abiotic stress responses for plants' resilience. *New Crops* **2024**, *1*, 100005. [CrossRef]
10. Tariq, R.; Hussain, A.; Tariq, A.; Khalid, M.H.B.; Khan, I.; Basim, H.; Ingvarsson, P.K. Genome-wide analyses of the mung bean NAC gene family reveals orthologs, co-expression networking and expression profiling under abiotic and biotic stresses. *BMC Plant Biol.* **2022**, *22*, 343. [CrossRef]
11. Wang, Y.; Zhu, R.; Shi, M.; Huang, Q.; Zhang, S.; Kai, G.; Guo, S. Genome-Wide identification and comparative analysis of WRKY transcription factors related to momilactone biosynthesis in *Calohypnum plumiforme*. *Front. Ecol. Evol.* **2022**, *9*, 809729. [CrossRef]
12. Zhang, Q.; Zhang, J.; Wei, H.; Fu, X.; Ma, L.; Lu, J.; Wang, H.; Yu, S. Genome-wide identification of NF-YA gene family in cotton and the positive role of *GhNF-YA10* and *GhNF-YA23* in salt tolerance. *Int. J. Biol. Macromol.* **2020**, *165*, 2103–2115. [CrossRef] [PubMed]
13. Zhu, J.; Wei, X.; Yin, C.; Zhou, H.; Yan, J.; He, W.; Yan, J.; Li, H. *ZmEREB57* regulates OPDA synthesis and enhances salt stress tolerance through two distinct signalling pathways in *Zea mays*. *Plant Cell Environ.* **2023**, *46*, 2867–2883. [CrossRef] [PubMed]

14. Wei, Y.; Jin, J.; Liang, D.; Gao, J.; Li, J.; Xie, Q.; Lu, C.; Yang, F.; Zhu, G. Genome-wide identification of *Cymbidium sinense* WRKY gene family and the importance of its Group III members in response to abiotic stress. *Front. Plant Sci.* **2022**, *13*, 969010. [CrossRef] [PubMed]
15. Ye, H.; Qiao, L.; Guo, H.; Guo, L.; Ren, F.; Bai, J.; Wang, Y. Genome-Wide identification of wheat WRKY gene family reveals that *TaWRKY75-A* is referred to drought and salt resistances. *Front. Plant Sci.* **2021**, *12*, 663118. [CrossRef] [PubMed]
16. Bakshi, M.; Oelmüller, R. WRKY transcription factors. *Plant Signal. Behav.* **2015**, *9*, e27700. [CrossRef] [PubMed]
17. Chen, Y.; Li, L.; Xu, Q.; Kong, Y.; Wang, H.; Wu, W. The *WRKY6* transcription factor modulates PHOSPHATE1 expression in response to low Pi stress in *Arabidopsis*. *Plant Cell* **2009**, *21*, 3554–3566. [CrossRef] [PubMed]
18. Rinerson, C.I.; Rabara, R.C.; Tripathi, P.; Shen, Q.J.; Rushton, P.J. The evolution of WRKY transcription factors. *BMC Plant Biol.* **2015**, *15*, 66. [CrossRef] [PubMed]
19. Ding, Z.; Yan, J.; Xu, X.; Yu, D.; Li, G.; Zhang, S.; Zheng, S. Transcription factor *WRKY46* regulates osmotic stress responses and stomatal movement independently in *Arabidopsis*. *Plant J.* **2014**, *79*, 13–27. [CrossRef]
20. Ren, X.; Chen, Z.; Liu, Y.; Zhang, H.; Zhang, M.; Liu, Q.; Hong, X.; Zhu, J.; Gong, Z. ABO3, a WRKY transcription factor, mediates plant responses to abscisic acid and drought tolerance in *Arabidopsis*. *Plant J.* **2010**, *63*, 417–429. [CrossRef]
21. Li, S.; Fu, Q.; Chen, L.; Huang, W.; Yu, D. *Arabidopsis thaliana WRKY25*, *WRKY26*, and *WRKY33* coordinate induction of plant thermotolerance. *Planta* **2011**, *233*, 1237–1252. [CrossRef] [PubMed]
22. Ma, J.; Gao, X.; Liu, Q.; Shao, Y.; Zhang, D.; Jiang, L.; Li, C. Overexpression of *TaWRKY146* increases drought tolerance through inducing stomatal closure in *Arabidopsis thaliana*. *Front. Plant Sci.* **2017**, *8*, 2036. [CrossRef] [PubMed]
23. Wang, Y.; Shu, Z.; Wang, W.; Jiang, X.; Li, D.; Pan, J.; Li, X. *CsWRKY2*, a novel WRKY gene from *Camellia sinensis*, is involved in cold and drought stress responses. *Biol. Plant.* **2016**, *60*, 443–451. [CrossRef]
24. He, L.; Wu, Y.; Zhao, Q.; Wang, B.; Liu, Q.; Zhang, L. Chrysanthemum *DgWRKY2* gene enhances tolerance to salt stress in transgenic Chrysanthemum. *Intern. J. Mol. Sci.* **2018**, *19*, 2062. [CrossRef] [PubMed]
25. Liu, Q.; Zhong, M.; Li, S.; Pan, Y.; Jiang, B.; Jia, Y.; Zhang, H. Overexpression of a chrysanthemum transcription factor gene, *DgWRKY3*, in tobacco enhances tolerance to salt stress. *Plant Physiol. Biochem.* **2013**, *69*, 27–33. [CrossRef] [PubMed]
26. Wang, K.; Bai, Z.; Liang, Q.; Liu, Q.; Zhang, L.; Pan, Y.; Liu, G.; Jiang, B.; Zhang, F.; Jia, Y. Transcriptome analysis of chrysanthemum (*Dendranthema grandiflorum*) in response to low temperature stress. *BMC Genom.* **2018**, *19*, 319. [CrossRef] [PubMed]
27. Jiang, J.; Ma, S.; Ye, N.; Jiang, M.; Cao, J.; Zhang, J. WRKY transcription factors in plant responses to stresses. *J. Integr. Plant Biol.* **2017**, *59*, 86–101. [CrossRef] [PubMed]
28. Wendel, J.F.; Brubaker, C.L.; Seelanan, T. The origin and evolution of Gossypium. Springer, Dordrecht, Netherlands, 2010.
29. Zhang, H.; Li, Y.; Wang, B.; Chee, P. Recent advances in cotton genomics. *Int. J. Plant Genom.* **2008**, *2008*, 742304. [CrossRef] [PubMed]
30. Du, X.; Huang, G.; He, S.; Yang, Z.; Sun, G.; Ma, X.; Li, N.; Zhang, X.; Sun, J.; Liu, M.; et al. Resequencing of 243 diploid cotton accessions based on an updated A genome identifies the genetic basis of key agronomic traits. *Nat. Genet.* **2018**, *50*, 796–802. [CrossRef]
31. Paterson, A.H.; Wendel, J.F.; Gundlach, H.; Guo, H.; Jenkins, J.; Jin, D.; Llewellyn, D.; Showmaker, K.C.; Shu, S.; Udall, J.; et al. Repeated polyploidization of *Gossypium* genomes and the evolution of spinnable cotton fibres. *Nature* **2012**, *492*, 423–427. [CrossRef]
32. Yang, Z.; Ge, X.; Yang, Z.; Qin, W.; Sun, G.; Wang, Z.; Li, Z.; Liu, J.; Wu, J.; Wang, Y.; et al. Extensive intraspecific gene order and gene structural variations in upland cotton cultivars. *Nat. Commun.* **2019**, *10*, 2989. [CrossRef] [PubMed]
33. Wang, M.; Tu, L.; Yuan, D.; Zhu, D.; Shen, C.; Li, J.; Liu, F.; Pei, L.; Wang, P.; Zhao, G.; et al. Reference genome sequences of two cultivated allotetraploid cottons, *Gossypium hirsutum* and *Gossypium barbadense*. *Nat. Genet.* **2019**, *51*, 224–229. [CrossRef] [PubMed]
34. Abdelraheem, A.; Esmaeili, N.; O'Connell, M.; Zhang, J. Progress and perspective on drought and salt stress tolerance in cotton. *Ind. Crops Prod.* **2019**, *130*, 118–129. [CrossRef]
35. Chen, S.; Cao, H.; Huang, B.; Zheng, X.; Liang, K.; Wang, G.; Sun, X. The WRKY10-VQ8 module safely and effectively regulates rice thermotolerance. *Plant Cell Environ.* **2020**, *45*, 2126–2144. [CrossRef] [PubMed]
36. Ma, L.; Li, X.; Zhang, J.; Yi, D.; Li, F.; Wen, H.; Liu, W.; Wang, X. *MsWRKY33* increases alfalfa (*Medicago sativa* L.) salt stress tolerance through altering the ROS scavenger via activating *MsERF5* transcription. *Plant Cell Environ.* **2023**, *46*, 3887–3901. [CrossRef] [PubMed]
37. Chen, Z.; Sreedasyam, A.; Ando, A.; Song, Q.; De Santiago, L.M.; Hulse-Kemp, A.M.; Ding, M.; Ye, W.; Kirkbride, R.C.; Jenkins, J.; et al. Genomic diversifications of five Gossypium allopolyploid species and their impact on cotton improvement. *Nat. Genet.* **2020**, *52*, 525–533. [CrossRef] [PubMed]
38. Rushton, P.J.; Bokowiec, M.T.; Han, S.; Zhang, H.; Brannock, J.F.; Chen, X.; Laudeman, T.W.; Timko, M.P. Tobacco transcription factors: Novel insights into transcriptional regulation in the Solanaceae. *Plant Physiol.* **2008**, *147*, 280–295. [CrossRef] [PubMed]
39. Huang, T.; Yang, J.; Yu, D.; Han, X.; Wang, X. Bioinformatics analysis of WRKY transcription factors in grape and their potential roles prediction in sugar and abscisic acid signaling pathway. *J. Plant Biochem. Biot.* **2021**, *30*, 67–80. [CrossRef]
40. Wang, Z.; Zhao, X.; Ren, Z.; Abou-Elwafa, S.F.; Pu, X.; Zhu, Y.; Dou, D.; Su, H.; Cheng, H.; Liu, Z.; et al. *ZmERF21* directly regulates hormone signaling and stress-responsive gene expression to influence drought tolerance in maize seedlings. *Plant Cell Environ.* **2022**, *45*, 312–328. [CrossRef]

41. Muhammad, A.K.; Kang, D.; Wu, Y.; Wang, Y.; Ai, P.; Wang, Z. Characterization of WRKY gene family in Whole-Genome and exploration of flowering improvement genes in *Chrysanthemum lavandulifolium*. *Front. Plant Sci.* **2022**, *26*, 861193. [CrossRef]
42. Huang, G.; Wu, Z.; Percy, R.G.; Bai, M.; Li, Y.; Frelichowski, J.E.; Hu, J.; Wang, K.; Yu, J.; Zhu, Y. Genome sequence of *Gossypium herbaceum* and genome updates of *Gossypium arboreum* and *Gossypium hirsutum* provide insights into cotton A-genome evolution. *Nat. Genet.* **2020**, *52*, 516–524. [CrossRef] [PubMed]
43. De Smet, R.; Sabaghian, E.; Li, Z.; Saeys, Y.; Van de Peer, Y. Coordinated functional divergence of genes after genome duplication in *Arabidopsis thaliana*. *Plant Cell* **2017**, *29*, 2786–2800. [CrossRef] [PubMed]
44. Lv, Q.; Li, W.; Sun, Z.; Ouyang, N.; Jing, X.; He, Q.; Wu, J.; Zheng, J.; Zheng, J.; Tang, S.; et al. Resequencing of 1,143 indica rice accessions reveals important genetic variations and different heterosis patterns. *Nat. Commun.* **2020**, *11*, 4778. [CrossRef] [PubMed]
45. Zhou, Y.; Zhao, X.; Li, Y.; Xu, J.; Bi, A.; Kang, L.; Xu, D.; Chen, H.; Wang, Y.; Wang, Y.; et al. Triticum population sequencing provides insights into wheat adaptation. *Nat. Genet.* **2020**, *52*, 1412–1422. [CrossRef] [PubMed]
46. Ji, X.; Zhang, M.; Wang, D.; Li, Z.; Lang, S.; Song, X. Genome-wide identification of WD40 superfamily in *Cerasus humilis* and functional characteristics of *ChTTG1*. *Int. J. Biol. Macromol.* **2023**, *225*, 376–388. [CrossRef] [PubMed]
47. Shi, W.; Hao, L.; Li, J.; Liu, D.; Guo, X.; Li, H. The *Gossypium hirsutum* WRKY gene *GhWRKY39-1* promotes pathogen infection defense responses and mediates salt stress tolerance in transgenic *Nicotiana benthamiana*. *Plant Cell* **2014**, *33*, 483–498. [CrossRef] [PubMed]
48. Liu, X.; Song, Y.; Xing, F.; Wang, N.; Wen, F.; Zhu, C. *GhWRKY25*, a group I WRKY gene from cotton, confers differential tolerance to abiotic and biotic stresses in transgenic *Nicotiana benthamiana*. *Protoplasma* **2016**, *253*, 1265–8121. [CrossRef]
49. Han, D.; Hou, Y.; Wang, Y.; Ni, B.; Li, Z.; Yang, G. Overexpression of a *Malus baccata* WRKY transcription factor gene (*MbWRKY5*) increases drought and salt tolerance in transgenic tobacco. *Can. J. Plant Sci.* **2019**, *99*, 2. [CrossRef]
50. Huang, Z.; Wang, J.; Li, Y.; Song, L.; Chen, D.; Liu, L.; Jiang, C. A WRKY Protein, MfWRKY40, of resurrection plant *Myrothamnus flabellifolia* plays a positive role in regulating tolerance to drought and salinity stresses of *Arabidopsis*. *Int. J. Mol. Sci.* **2022**, *23*, 8145. [CrossRef]
51. Jaffar, M.A.; Song, A.; Faheem, M.; Chen, S.; Jiang, J.; Liu, C.; Fan, Q.; Chen, F. Involvement of *CmWRKY10* in drought tolerance of *Chrysanthemum* through the ABA-signaling pathway. *Int. J. Mol. Sci.* **2016**, *17*, 693. [CrossRef]
52. Qin, Y.; Tian, Y.; Liu, X. A wheat salinity-induced WRKY transcription factor *TaWRKY93* confers multiple abiotic stress tolerance in *Arabidopsis thaliana*. *Biochem. Biophys. Res. Commun.* **2015**, *464*, 428–433. [CrossRef] [PubMed]
53. Cai, R.; Dai, W.; Zhang, C.; Wang, Y.; Wu, M.; Zhao, Y.; Ma, Q.; Xiang, Y.; Cheng, B. The maize WRKY transcription factor *ZmWRKY17* negatively regulates salt stress tolerance in transgenic *Arabidopsis* plants. *Planta* **2017**, *246*, 1215–1231. [CrossRef] [PubMed]
54. Huang, X.; Amee, M.; Chen, L. Bermudagrass *CdWRKY50* gene negatively regulates plants' response to salt stress. *Environ. Exp. Bot.* **2021**, *188*, 104513. [CrossRef]
55. Zhang, L.; Zhang, R.; Ye, X.; Zheng, X.; Tan, B.; Wang, W.; Li, Z.; Li, J.; Cheng, J.; Feng, J. Overexpressing *VvWRKY18* from grapevine reduces the drought tolerance in *Arabidopsis* by increasing leaf stomatal density. *J. Plant. Physiol.* **2022**, *275*, 153741. [CrossRef] [PubMed]
56. Liu, W.; Thapa, P.; Park, S.W. *RD29A* and *RD29B* rearrange genetic and epigenetic markers in priming systemic defense responses against drought and salinity. *Plant Sci.* **2023**, *337*, 111895. [CrossRef] [PubMed]
57. Yoshida, T.; Mogami, J.; Yamaguchi-Shinozaki, K. ABA-dependent and ABA-independent signaling in response to osmotic stress in plants. *Curr. Opin. Plant Biol.* **2014**, *21*, 133–139. [CrossRef] [PubMed]
58. Székelyg, G.; Ábrahám, E.; Cseplo, A.; Rigó, G.; Zsigmond, L.; Csiszar, J.; Ayaydin, F.; Strizhov, N.; Jasik, J.; Schmelzer, E. Duplicated *P5CS* genes of *Arabidopsis* play distinct roles in stress regulation and developmental control of proline biosynthesis. *Plant J.* **2008**, *53*, 11–28. [CrossRef] [PubMed]
59. Li, Y.; Chen, H.; Li, S.; Yang, C.; Ding, Q.; Song, C.; Wang, D. *GhWRKY46* from upland cotton positively regulates the drought and salt stress responses in plant. *Environ. Exp. Bot.* **2021**, *186*, 104438. [CrossRef]
60. Xiong, C.; Zhao, S.; Yu, X.; Sun, Y.; Li, H.; Ruan, C.; Li, J. Yellowhorn drought-induced transcription factor *XsWRKY20* acts as a positive regulator in drought stress through ROS homeostasis and ABA signaling pathway. *Plant Physiol. Biochem.* **2020**, *155*, 187–195. [CrossRef]
61. Aamir, M.; Singh, V.K.; Meena, M.; Upadhyay, R.S.; Gupta, V.K.; Singh, S. Structural and functional insights into *WRKY3* and *WRKY4* transcription factors to unravel the WRKY-DNA (W-Box) complex interaction in tomato (*Solanum lycopersicum* L.). A computational approach. *Front. Plant Sci.* **2017**, *8*, 819. [CrossRef]
62. Pandey, B.; Grover, A.; Sharma, P. Molecular dynamics simulations revealed structural differences among WRKY domain-DNA interaction in barley (*Hordeum vulgare*). *BMC Genom.* **2018**, *19*, 132. [CrossRef] [PubMed]
63. Singh, A.; Sharma, A.K.; Singh, N.K.; Sonah, H.; Deshmukh, R.; Sharma, T.R. Understanding the effect of structural diversity in WRKY transcription factors on DNA binding efficiency through molecular dynamics simulation. *Biology* **2019**, *8*, 83. [CrossRef]
64. Zhu, T.; Liang, C.; Meng, Z.; Sun, G.; Meng, Z.; Guo, S.; Zhang, R. CottonFGD: An integrated functional genomics database for cotton. *BMC Plant Biol.* **2017**, *17*, 101. [CrossRef] [PubMed]
65. Finn, R.D.; Bateman, A.; Clements, J.; Coggill, P.; Eberhardt, R.Y.; Eddy, S.R.; Heger, A.; Hetherington, K.; Holm, L.; Mistry, J.; et al. PFAM: The protein families database. *Nucleic Acids Res.* **2014**, *42*, 222–230. [CrossRef] [PubMed]

66. Mistry, J.; Finn, R.D.; Eddy, S.R.; Bateman, A.; Punta, M. Challenges in homology search: HMMER3 and convergent evolution of coiled-coil regions. *Nucleic Acids Res.* **2013**, *41*, e121. [CrossRef] [PubMed]
67. Kumar, S.; Stecher, G.; Tamura, K. MEGA7: Molecular evolutionary genetics analysis version 7.0 for bigger datasets. *Mol. Biol. Evol.* **2016**, *33*, 1870–1874. [CrossRef]
68. Letunic, I.; Bork, P. Interactive Tree of Life (iTOL) v5: An online tool for phylogenetic tree display and annotation. *Nucleic Acids Res.* **2021**, *49*, 293–296. [CrossRef] [PubMed]
69. Chen, C.; Chen, H.; Zhang, Y.; Thomas, H.R.; Frank, M.H.; He, Y.; Xia, R. TBtools: An integrative toolkit developed for interactive analyses of big biological data. *Mol. Plant* **2020**, *13*, 1194–1202. [CrossRef]
70. Lescot, M.; Déhais, P.; Thijs, G.; Marchal, K.; Moreau, Y.; Van de Peer, Y.; Rouzé, P.; Rombauts, S. PlantCARE, a database of plant cis-acting regulatory elements and a portal to tools for in silico analysis of promoter sequences. *Nucleic Acids Res.* **2002**, *30*, 325–327. [CrossRef]
71. Tang, H.; Bowers, J.E.; Wang, X.; Ming, R.; Alam, M.; Paterson, A.H. Synteny and collinearity in plant genomes. *Science* **2008**, *320*, 486–488. [CrossRef]
72. Chen, C.; Wu, Y.; Xia, R. A painless way to customize Circos plot: From data preparation to visualization using TBtools. *iMeta* **2022**, *1*, e35. [CrossRef] [PubMed]
73. Hu, Y.; Chen, J.; Fang, L.; Zhang, Z.; Ma, W.; Niu, Y.; Ju, L.; Deng, J.; Zhao, T.; Lian, J.; et al. *Gossypium barbadense* and *Gossypium hirsutum* genomes provide insights into the origin and evolution of allotetraploid cotton. *Nat. Genet.* **2019**, *51*, 739–748. [CrossRef]
74. Dobin, A.; Davis, C.A.; Schlesinger, F.; Drenkow, J.; Zaleski, C.; Jha, S.; Batut, P.; Chaisson, M.; Gingeras, T.R. STAR: Ultrafast universal RNA-seq aligner. *Bioinformatics* **2013**, *29*, 15–21. [CrossRef] [PubMed]
75. Li, B.; Dewey, C.N. RSEM: Accurate transcript quantification from RNA-Seq data with or without a reference genome. *BMC Bioinform.* **2011**, *12*, 323. [CrossRef]
76. Livak, K.J.; Schmittgen, T.D. Analysis of relative gene expression data using real-time quantitative PCR and the 2(−Delta Delta C(T)) method. *Methods* **2001**, *25*, 402–408. [CrossRef] [PubMed]
77. Clough, S.J.; Bent, A.F. Floral dip: A simplified method for Agrobacterium-mediated transformation of *Arabidopsis thaliana*. *Plant J.* **1998**, *16*, 735–743. [CrossRef]
78. Wei, X.; Geng, M.; Yuan, J.; Zhan, J.; Liu, L.; Chen, Y.; Wang, Y.; Qin, W.; Duan, H.; Zhao, H.; et al. GhRCD1 promotes cotton tolerance to cadmium by regulating the GhbHLH12–GhMYB44–GhHMA1 transcriptional cascade. *Plant Biotechnol. J.* **2024**, *22*, 1777–1796. [CrossRef]
79. Yan, Y.; Tao, H.; He, J.; Huang, S.Y. The HDOCK server for integrated protein-protein docking. *Nat. Protoc.* **2020**, *15*, 1829–1852. [CrossRef]
80. Yu, M.; Liu, J.; Du, B.; Zhang, M.; Wang, A.; Zhang, L. NAC transcription factor *PwNAC11* activates ERD1 by interaction with ABF3 and DREB2a to enhance drought tolerance in transgenic *Arabidopsis*. *Int. J. Mol. Sci.* **2021**, *22*, 6952. [CrossRef]

plants MDPI

Review

Salt Tolerance in Soybeans: Focus on Screening Methods and Genetics

Rong-Xia Guan [1,*,†], Xiao-Yang Guo [1,†], Yue Qu [2], Zheng-Wei Zhang [1], Li-Gao Bao [3], Rui-Yun Ye [4], Ru-Zhen Chang [1] and Li-Juan Qiu [1,*]

1 The National Key Facility for Crop Gene Resources and Genetic Improvement (NFCRI), Key Lab of Soybean Biology, Ministry of Agriculture, State Key Laboratory of Crop Gene Resources and Breeding, Institute of Crop Sciences, Chinese Academy of Agricultural Sciences, Beijing 100081, China; guoxiaoyanggxy@163.com (X.-Y.G.); 13121233799@163.com (Z.-W.Z.); rx_guan@sina.com (R.-Z.C.)

2 Australian Research Council Centre of Excellence in Plant Energy Biology, School of Agriculture, Food and Wine, Waite Research Institute, University of Adelaide, Glen Osmond, SA 5064, Australia; yue.qu@adelaide.edu.au

3 Agriculture and Animal Husbandry Technology Promotion Center of Inner Mongolia Autonomous Region, Hohhot 010018, China; nmgnmyjs@163.com

4 The Economic Development Center of China State Farm, Beijing 100122, China; guochan2669@163.com

* Correspondence: guanrongxia@caas.cn (R.-X.G.); qiulijuan@caas.cn (L.-J.Q.)

† These authors contributed equally to this work.

Abstract: Salinity greatly affects the production of soybeans in arid and semi-arid lands around the world. The responses of soybeans to salt stress at germination, emergence, and other seedling stages have been evaluated in multitudes of studies over the past decades. Considerable salt-tolerant accessions have been identified. The association between salt tolerance responses during early and later growth stages may not be as significant as expected. Genetic analysis has confirmed that salt tolerance is distinctly tied to specific soybean developmental stages. Our understanding of salt tolerance mechanisms in soybeans is increasing due to the identification of key salt tolerance genes. In this review, we focus on the methods of soybean salt tolerance screening, progress in forward genetics, potential mechanisms involved in salt tolerance, and the importance of translating laboratory findings into field experiments via marker-assisted pyramiding or genetic engineering approaches, and ultimately developing salt-tolerant soybean varieties that produce high and stable yields. Progress has been made in the past decades, and new technologies will help mine novel salt tolerance genes and translate the mechanism of salt tolerance into new varieties via effective routes.

Keywords: soybean; salinity stress; salinity tolerance; ion homeostasis; gene identification

1. Introduction

Soil salinity is due to the accumulation of soluble salts, including chlorides and sulfates of sodium, and more than 3% of farmlands are seriously threatened by salinization [1,2]. Salinity usually inhibits crop growth through the osmotic effect, which reduces a plant's water take up or ion toxic effect, which inhibits enzyme activity [3]. The extent and severity of the effect of saline soils on crop production is predicted to worsen because of factors such as global warming and inadequate drainage of irrigated land [4–6]. The need to expand agriculture into marginal lands, coupled with increasing global food requirements due to increasing population sizes, requires the development of crops that can achieve higher yields in soils with higher salt contents [7,8]. Salt stress reduces the yield of most crops [4,9–11]. The exploration of salt-tolerant crop plants that can withstand high salinity is considered one of the most effective biological strategies to cope with this problem and sustain food production [12]. Among the wide range of salt concentrations in saline soils, it was found that moderately saline soil had a minimal effect on reducing the soybean seed yield of salt-tolerant varieties, indicating the promising potential for using these varieties in saline fields [13].

As a source of vegetable protein and oil worldwide, soybean (*Glycine max* L. Merrill) plays important roles in human nutrition, animal feed, and oilseed production [14]. It was reported that extracts of soil solutions with conductivity (ECe) values of 5 dS m^{-1} or greater affect soybean germination and later developmental stages [15,16]. Soil salinity values of 7.3 and 9.6 dS m^{-1} caused complete stand loss of salt-sensitive and intermediate salt-tolerant soybeans, respectively. In contrast, the stands of salt-tolerant varieties were not appreciably decreased with salt, indicating inheritance control over salt tolerance [13]. Significant progress has been made in the evaluation and genetic analysis of soybean salt tolerance during the past few decades. In this review, we focus on the advances in genetics and possible mechanisms of salinity tolerance in soybeans at different growth stages. Suggestions are made for future studies that aim to improve salt tolerance via the ontogeny of soybeans.

2. Identification of Salt-Tolerant Accessions in Soybean

2.1. Salt Tolerance Evaluation at Soybean Germination Stage

Many germination tests have illustrated the negative effects of salt exposure on soybean germination. High salt concentrations decrease the soybean germination rate and inhibit the growth of radicle and lateral roots [17]. A study conducted in 1983 provided genetic resources on soybean salt tolerance at the germination stage and identified salt-tolerant soybean landraces and varieties under controlled conditions. The researchers used the relative injury index as a criterium to score the salt tolerance of soybean accessions from level 1 (a relative injury index of 0–20%) to level 5 (90.1–100%) at high salinity (a 1.6% NaCl solution) and were able to select tolerant soybean genotypes with scores of lower than 3 (a relative injury index of lower than 65%) [18]. In the Seventh Five-Year Plan of China, a total of 10,128 soybean accessions were evaluated for salt tolerance at the germination stage, and the results showed that 9.1% of genotypes were tolerant (based on relative injury index levels 1–3) [19]. Parameters, such as germination percentage (G%), tissue water content (TWC), and root length (TL), were considered useful indicators for the selection of salt-tolerant soybean in the germination stage while exposed to 150 mM and 200 mM NaCl stress [20]. The relative germination index (ST-GI) and relative germination rate (ST-GR) were significantly positively correlated with each other under 150 mM NaCl stress and could be used as indexes of salt tolerance at the germination stage [21].

2.2. Evaluation of Salt-Tolerant Soybean Accessions at Emergence Stage

Salt tolerance at the emergence stage is likely to be more important than in the germination stage because germinated seeds under certain saline conditions, like germination on salinized filter paper, may not break through the soil crust when the soil surface is hard [22]. Salinity decreased the emergence rates of soybeans grown in soil with less than 1.0% salt (a dry soil base), despite all soybeans achieving maximum germination rates. Differences in relative emergence rates were observed between the different soybean varieties grown under higher salt contents, which might be due to the effect of chloride ions [13,23]. At a salt concentration of 330 mM, the soybean cv. Williams attained a high germination rate (81%); however, seedling growth declined to 5%, even when exposed to a lower 220 mM NaCl stress [23]. This suggests that soybean seeds can survive under saline conditions during the germination stage, but the stress can be fatal at the emergence stage. Only one soybean genotype was used in this experiment; therefore, a variety of soybean accessions should be used to evaluate the correlations between genotypic differences at the germination and emergence stages. Variations in salt tolerance were reported in near-isogenic lines (NILs) that differ in maturity in the background of the soybean cultivars 'Lee' and 'Essex', indicating that specific genes in the NILs of maturity groups IV or VI may related to salt tolerance [24]. Recently, two flowering-related loci, *E2* and *J*, were found to be related to soybean seedling salt tolerance. Knockout of *E2* generated soybean lines with enhanced salt tolerance and shortened maturity [25], and loss of the function of *J* reduced salt tolerance and prolonged the maturity of soybean [26], indicating that flowering earlier might help soybeans avoid salinity toxicity. However, the effects of maturity-related genes on germination under salinity stress remain unknown. In fields salinized with sodium chloride and calcium chloride salts, soybean emergence was significantly reduced when the ECe reached 11 dS m^{-1} [24]. In order to overcome the spatial heterogeneity in saline field trials, Liu et al. (2020) developed a method using vermiculite as a culture substrate treated with a salt solution after soybean sowing, whereby soybeans grown under 150 mM NaCl showed a different salt tolerance. The salt tolerance index (SI) calculated using the relative growth of seedlings was found to be significantly related to salt tolerance and was used for the identification of salt-tolerant soybean accessions [27].

2.3. Salt Tolerance of Soybean at Seedling Stage

The screening of salt tolerance at the different stages of soybean growth has been conducted for more than half a century (Table 1). Much of the research has focused on soybean seedling stages because salt tolerance increases with the progression of plant age [18]. Much effort has been devoted to developing rapid, visual methods for the selection of salt-tolerant accessions. One such effort categorized soybeans as "includers" (salt-sensitive) or "excluders" (salt-tolerant). Soybeans grown under saline conditions accumulate more chloride ions in the stems and leaves, which causes severe leaf necrosis and even mortality, and were thus assigned as "includer" soybeans. Those soybeans that showed no necrosis and only a moderate growth reduction were assigned as "excluder" soybeans [28]. Three Cl excluders and four includers were grown hydroponically with 0, 40, 80, 120, and 160 mM NaCl for seven days. Includers and excluders showed the greatest differences at the 120 mM NaCl stress level, where the average leaf Na^{+} and Cl^{-} contents were 2.64 and 1.96 times higher for includers than excluders, respectively. Thus, the addition of 120 mM NaCl in the hydroponic system was the most effective concentration for screening salt-tolerant soybean genotypes without chemical analysis [29]. A method using sandy soil in plastic containers (named the PC method) was compared with the hydroponic method by exposing 14 soybean genotypes to salt at the V2–V3 stages. The leaf scorch scores and leaf chloride contents were comparable for both methods and salt damage appeared approximately four days sooner in the PC method; thus, the cheaper and less labor-intensive PC method was considered the better one that could be adopted by soybean breeders [30]. Using two Cl includers and two excluders as materials, a pot assay of soybean exposed to 120 mM NaCl over two weeks (for 2 h in a salt solution each day) was recently conducted to establish a leaf scorch scale (LSS) to visually rate the level of salt tolerance of plants. The scale ranges from one (healthy dark green leaves with no chlorosis) to nine (necrotic leaves). The results indicate that the LSS was positively correlated with Cl^{-} contents in leaves (r = 0.87–0.88; $p < 0.001$), which suggests that the LSS and Cl^{-} content may be used as criteria to identify tolerant genotypes [31]. In a simple screening method using vermiculite as a substrate, 200 mM NaCl was added to the tray every two days over five days (totaling 600 mM NaCl) after soybean unifoliate leaves were fully expanded, and a significant negative relationship of the leaf chlorophyll content (SPAD value) with the leaf Na^{+} content and salt tolerance was observed eight days after the final NaCl solution application [32]. This simple method was effectively used in salt tolerance gene mapping and functional analysis [33–35]. In a pot assay, a 25 mM NaCl solution was added to the soil every alternate day until a total concentration of 150 mM NaCl was added to each pot. A total of 170 soybean accessions were assigned to four groups: tolerant, moderately tolerant, moderately sensitive, and sensitive accessions. The relative total dry weight (DW) of 30-day-old seedlings, followed by the relative shoot and petiole DW, were considered as the more important discriminatory variables, while the relative root DW was considered as a secondary variable used to segregate accessions into the four groups [36]. This is in agreement with a previous study that reported the shoot growth of soybeans was more affected than the root growth under saline conditions [37]. In contrast, Lee et al.

(2008) [30] observed that the shoot dry weight was less affected than the root dry weight. The conflicting results may be attributed to the experimental methods and mediums as well as genotypes used in each study. Overall, the comparisons suggest that relative dry weight is an unreliable indicator to assess salt tolerance in plants, and screening methods based on leaf injury have been widely used for salt tolerance evaluation at the seedling stage. To better understand the genetic control of salt tolerance, efforts should be spent on developing effective screening methods that mimic saline conditions in the field. Moreover, precautions must be taken to avoid common issues in large-scale evaluations of the salt tolerance of soybeans in the field when irrigating large fields with saline water [13,19]. For example, soil salinity can greatly vary from the head- to the tail-ends of furrow-irrigated fields and negatively affect the precision of evaluating plant salt tolerance [24,38].

Table 1. Evaluation and identification of salt-tolerant accessions of soybeans at different developmental stages.

Stage	Stress Condition	Indicators	Total Accessions	No. of Tolerant Accessions	Proportion of Tolerant Accessions	Reference
Germination	1.6% NaCl	Salt damage index	10,128	924	9.1%	Shao et al. (1993) [19]
Germination	2.0% NaCl	Relative salt damage rate	760	141	18.5%	Li et al. (1996) [39]
Germination	150 and 200 mM NaCl	GR	10	3	30.0%	Shelke et al. (2017) [20]
Germination	150 mM NaCl	IR, GR, and GI	191	/	0.0%	Kan et al. (2015) [40]
Germination	1.2% NaCl	Relative salt damage rate	793	117	14.8%	Jiang et al. (2012) [41]
Emergence	Saline soil: 3.1–13.7 dS m^{-1}	Relative seedling emergence rate	6	2	33.3%	Abel and MacKenzie (1964) [13]
Emergence	Saline soil: 3–6 dS m^{-1}	GR	7	3	42.9%	Wang et al. (1999) [24]
Emergence	150 mM NaCl	SI	27	10	37.0%	Liu et al. (2020) [27]
Seedling	5.0–10.2 dS m^{-1}	Cl^- content in leaves	6	4	66.7%	Abel and MacKenzie (1964) [13]
Seedling	Saline water (EC = 15–17 dS m^{-1})	Green loss grade (1–5)	10,128	457	4.5%	Shao et al. (1993) [19]
Seedling	25 mM to 150 mM NaCl	SDW	170	18	10.6%	Mannan et al. (2010) [36]
Seedling	120 mM NaCl	Leaf damage	7	3	42.9%	Valencia et al. (2008) [29]
Seedling	100 mM NaCl	Leaf damage	14	5	35.7%	Lee et al. (2008) [30]
Seedling	200 mM NaCl	Leaf damage	8	4	50.0%	Jiang et al. (2013) [32]
Seedling	21 ± 3 dS m^{-1}	Salt damage index	793	41	5.17%	Jiang et al. (2012) [41]
Seedling	120 mM NaCl	Leaf damage	98	36	36.7%	Ledesma et al. (2016) [31]
Emergence	200 mM NaCl	SI	27	12	44.4%	Liu et al. (2020) [27]
Whole period	Saltwater irrigation	Degree of salt damage	2000	7	0.4%	Shao et al. (1986) [18]
Whole period	Saline soil	Relative salt tolerance index	793	35	4.41%	Jiang et al. (2012) [41]

2.4. Salt Tolerance Identification in Later Soybean Growing Period

Salt stress affects many agronomic traits of soybean, especially yield, mainly via the reduction in the branch number, pod number, grain weight, and 100-grain weight, which ultimately leads to yield reduction [42]. Screening soybeans throughout the whole growing period in a saline field is difficult, especially for a large number of accessions, due to salinity heterogeneity and uncontrolled environments. Saline water irrigation in field or pot experiments was used for the identification of salt tolerance at later soybean growing stages. Soybean accessions were planted in saline fields and irrigated with saline water (EC = 20 to 24 dS m^{-1}) at the flowering and podding stages, and seven tolerant varieties like Wenfeng 7, Jindou 33, and Tiefeng 8 were identified [18]. Using salt-tolerant varieties (Wenfeng 7, Zhongye 1, Tiefeng 8, and Zhonghuang 10) as controls, a two-year salt tolerance evaluation of 280 soybean varieties was conducted in saline fields during the whole growth period of soybean, the biomass and grain weight of each variety were investigated at the maturity stage, the relative salt tolerance index was calculated, and only 35 soybean varieties showed high tolerance in both years, in which only 3 and 11 were tolerant at the germination and seedling stages, respectively [41]. In a potting experiment, soybean varieties were treated with 80 mM NaCl at the V3, R2, R4, and R6 stages, respectively. Biomass and pod weight were greatly affected in both tolerant and sensitive genotypes when salt was applied at the R6 stage, indicating that R6 is one of the most sensitive stages to salinity stress [43].

3. Genetics of Salinity Tolerance in Soybeans

3.1. Genetic Control of Salt Tolerance in Soybeans

Variations in salt tolerance exist in soybeans at different developmental stages, indicating potential tolerance gene resources in soybean accessions. To understand the genetic architecture of salt tolerance and improve selection efficiency, linkage and association analyses were used to identify loci controlling salt tolerance at the germination stage. Indices (the relative imbibition ratio (IR); the relative ratio of germination index (GI]); and the relative germination rate (GR)) representing possible mechanisms were used, and 11 QTLs located on chromosomes 2, 7, 8, 10, 17, and 18 underlying the complex traits were identified in an RIL population, NJRIKY, developed from a cross between Kefeng 1 and Nannong 1138-2 [21]. In the association analysis of natural soybean populations, 11 SNP- and 22 SSR-trait associations were identified [21,40]. Although several candidate genes on chromosomes 8, 9, and 18 were verified in response to salt stress, no consistent loci were identified in bi-parental segregating and natural populations [21]. The significant correlation between different indices, such as the GI and GR, and the co-association with related markers indicated that effective indices like the ratio of the germination rate under salt conditions to the germination rate under no-salt conditions (ST–GR) can be used for future experiments. The advantage of using the GR over the GI is that to obtain the GI, we need to manually count the number of germinated seeds in each Petri dish daily throughout the experiment, while to obtain the GR, we only need to count the number of germinated seeds once, at the end of the experiment. In an association mapping study of salt-tolerance-related markers in the emergence stage, the salt tolerance index (STI) based on the phenotypes of root length (LR), fresh or dry root weight (FWR; DWR), the biomass of seedlings (BS), and the length of hypocotyls (LH) were used as indices, and 19 QTLs were detected on various chromosomes, with only 2 related to LR-STI. Additional results showed that epistatic interactions between QTLs related to FWR-STI had strong effects ($r^2 > 5\%$) [44]. The minor effects and fewer conserved QTLs in different populations indicated that salt tolerance at the germination and emergence stages are likely controlled by quantitative loci.

In order to investigate the potential genetic control of salt tolerance, crosses were made using soybean chloride includers and excluders and 30-day-old seedlings of segregating populations grown in fields that were furrow-irrigated with saline water (an equal mixture of NaCl and $CaCl_2$). The F_2 populations of the includers × excluders were segregated in ratios of three non-necrotic plants (low chloride content) to one necrotic plant (very high chloride content), indicating the likelihood of a single gene governing the inheritance of salt tolerance. The gene symbols *Ncl* and *ncl* were proposed for the dominant (excluder) and recessive (includer) alleles, respectively. The phenotype of individuals was primarily sorted according to the level of leaf necrosis, while only 10 individuals were determined by the chloride concentration [28]. More than two decades later, three salt-tolerant and three salt-sensitive soybean cultivars were used to make crosses and evaluated for salt tolerance inheritance. Populations were planted in saline fields and irrigated with saline water (EC = 4.2 to 21 dS m^{-1}, depending on current drought or non-drought conditions). The resulting segregation of phenotypes indicated that salt tolerance in all of the three tolerant cultivars was governed by a dominant gene [45]. In the $F_{2:3}$ population derived from the cross of Peking (salt-sensitive) and wild soybean NY36-87 (salt-tolerant), the ratio of salt-tolerant to separation to salt-sensitive families was consistent with 1:2:1, indicating that the seedling salt tolerance of the NY36-87 wild soybean is controlled by a dominant single gene [46].

An $F_{2:5}$ population from a cross of the salt-tolerant cultivar S-100 and the salt-sensitive cultivar Tokyo, for the first time, was used for salt tolerance QTL mapping. The heritability of salt tolerance was 0.85 and 0.48 in the field and greenhouse environments, respectively. A major salt tolerance QTL was mapped on the soybean linkage group N (LG N, Chr. 03) in a 3.6 cM interval between SSR markers Sat_091 and Satt237 based on the phenotypes of plants grown in a field, greenhouse, and combined environments of the two [47]. This major QTL was confirmed in several salt-tolerant soybean varieties, such as Nannong

1138-2, Tiefeng 8, Jidou 12, Fiskeby III, and FT-Abyora [33,48–51]. The genetic and mapping results have been described in more detail in recently published reviews [15,52–55]. The leaf sodium (LSC) and leaf chloride (LCC) contents of the $F_{2:3}$ population derived from Williams 82 and the tolerant soybean Fiskeby III were used as physiological traits in QTL mapping. For the LCC, only one genomic region with a high R^2 (58.9%) was identified on Chr. 03, where the major salt tolerance QTL was located. While for the LSC, except for the locus on Chr. 03, another dominant gene (a positive allele in the sensitive parent Williams 82) was located on Chr. 13, which explained 11.5% of the observed total variation, and no significant epistatic interactions were detected between these two loci [51]. Genome-wide association mapping based on leaf chloride concentration and SPAD showed SNPs on Chr. 02, 03, 14, 16, and 20, and all were significantly associated with both traits. These results suggest novel genes are involved in soybean salt tolerance at the seedling stage and the potential application of these SNP markers in the evaluation of accessions and breeding selection [56]. Recently, two major QTLs associated with the LSS and CCR were identified in Williams82 × PI483460B RIL populations. *qSalt_Gm03*, associated with the CCR and LSS, was located in the same region as the known salt tolerance gene *GmCHX1*. Another new locus, *qSalt_Gm18*, significantly associated with the LSS, was mapped on Chr. 18. The salt tolerance alleles of the two loci were both from PI483460B [57].

Salt tolerance gene mapping of wild soybeans has also obtained vital progress in addition to that of cultivated soybeans. In F_2 populations derived from soybean cultivars and salt-tolerant wild soybeans, a major salt tolerance QTL was mapped on Chr. 03 [46,58], indicating that the same QTL or major gene was present on Chr. 03 in wild and cultivated soybeans (Figure 1). An F_2 population of PI483463 × S-100 was used to determine the allelic relationship of wild accession PI483463 and cultivar S-100. The population was segregated as 15 (tolerant):1 (sensitive), indicating that the gene in wild soybean was different from that in S-100, and the gene was assigned as *Ncl2* in PI483463 [59]. However, the salt tolerance QTL in PI483463 was mapped within a 658 kb region on Chr. 03 using an RIL population derived from PI483463 and Hutcheson [60]. Because different sensitive parents were used in these two studies and PI483463 showed a higher tolerance than S-100 after 30 days of salt stress, it is difficult to rule out the possibility that PI483463 has a different salt tolerance gene [59,60]. A new salt tolerance locus on Chr. 18, named *GmSALT18*, was identified in an $F_{2:3}$ population derived from the salt-sensitive variety Peking and the salt-tolerant wild soybean NY36-87 [46]. These wild soybeans should be further investigated to clone novel salt tolerance genes.

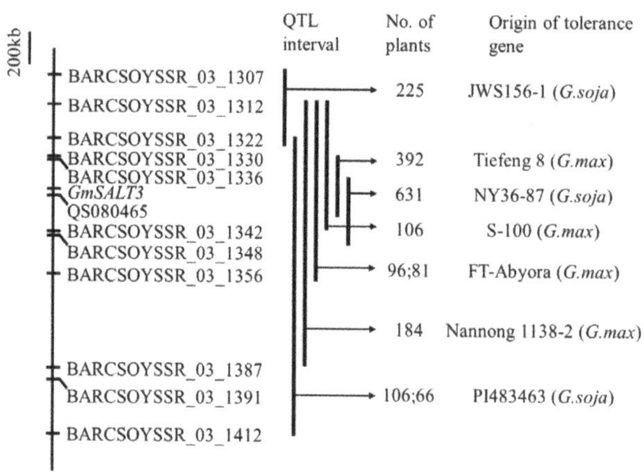

Figure 1. Schematic of salt tolerance QTL/gene mapping interval in Chr. 03 at the seedling stage. Bold vertical lines signify the QTL interval in each study, and the origin of the salt tolerance gene and the number of individuals in the segregating populations are also marked. All markers are *Glyma.Wm82.a2* version [34,46–49,58,60].

3.2. Candidate Gene Contributes to Salt Tolerance in Soybean

A major salt tolerance locus *qST-8* related to salt tolerance at the soybean germination stage was mapped onto Chr. 08 using QTL mapping in the RIL population and GWAS in the natural population. *Glyma.08g102000*, which belongs to the CDF (cation diffusion facilitator) family, was found to be the candidate gene of *GmCDF1*. Hairy root transformation experiments showed that the gene negatively regulated soybean salt tolerance by maintaining K^+–Na^+ homeostasis in shoots under salt stress. Haplotype analysis showed that two SNPs were significantly associated with salt tolerance, and Hap2 was more salt-tolerant than Hap1 (Table 2) [61].

The major salt tolerance QTL locus on Chr. 03 was conserved in both cultivated and wild soybeans. The isolation of the dominant gene has been the focus of extensive research efforts. By re-sequencing 96 RI lines derived from the salt-tolerant wild soybean W05 and the sensitive cultivar C08, a bin map was constructed, and a salt-related QTL was mapped in a 388 kb genomic region that overlapped with a previously mapped *Ncl* locus on Chr. 03. A root-specific expressed cation/H^+ exchanger gene *Glyma03g32900* was identified as the candidate gene, which was named as *GmCHX1*. The expression of *GmCHX1* in hairy roots leads to a higher fresh root weight than the control. Moreover, transgenic tobacco BY-2 cells showed higher survival rates under 100 mM NaCl treatment, confirming the salt tolerance function of this candidate gene was from wild soybean (Table 2) [62]. A map-based cloning strategy was conducted for fine mapping the salt tolerance gene *GmSALT3* (a salt-tolerance-associated gene on chromosome 3) in cultivated soybean Tiefeng 8, and only *Glyma03g32900*, an endoplasmic-reticulum-localized gene, was predicted to be present in a 17.5 kb candidate region according to the reference genome Williams 82 (Table 2) [34]. Salt-tolerant wild soybeans were used for the identification of a novel salt tolerance gene, and a 7 bp InDel in the promoter region of *Glyma.11G149900* (*GsERD15B*) was found to be associated with salt tolerance. Genetic transformation proved that a Hap2-type promoter enhanced hairy root growth under salt stress, and 87.5% (42 of 48) of tolerant soybeans belong to Hap2. The average STR (salt tolerance rating) of Hap1 is 4.20, which is significantly higher than that of Hap2 (1.64) (Table 2) [63].

In addition to forward genetics, a series of genes encoding ion transporters and transcriptional factors were cloned from soybean via homologous cloning and functionally evaluated in *Arabidopsis*, tobacco, or soybean [64–70]. Functionally verified salt tolerance genes in soybeans have been carefully summarized in a recent review [71].

Table 2. Salt tolerance genes identified in soybean and related molecular markers.

Tolerance Gene	Associated Markers	Salt Tolerance	Reference
GmSALT3	Pro-Ins, H2-Ins, H3-MboII, H4-NlaIII, and H5-Del	Seedling stage	Guan et al. (2021) [72]
	Tn-I, I-S, TGCT-D, and C-I	Seedling stage	Lee et al. (2018) [73]
GmCHX1	—	Seedling stage	Qi et al. (2014) [62]
	M1, M2, M3, M4, and M5	Seedling stage	Patil et al. (2016) [74]
GmCDF1	—	Germination stage Seedling stage	Zhang et al. (2019) [61]
GsERD15B	dCAPS-*GsERD15B*-promoter	Seedling stage	Jin et al. (2021) [63]

3.3. Dissecting Salt Tolerance Mechanisms in Soybean

Mechanisms related to the salt stress response, including signaling, osmotic stress, and ionic homeostasis, have been reviewed in detail [71]. In this review, we focus on salt tolerance genes cloned using forward genetics, because these genes are likely to be more valuable in marker-assisted selection breeding. However, this does not preclude that other genes may also contribute to salt tolerance breeding.

Independent studies have cloned the major salt tolerance gene *GmSALT3/GmCHX1/GmNcl*, which regulates salt tolerance at the soybean seedling stage [34,62,75]. GmSALT3 is an ER-localized protein regulating ions transport to shoots in a root-dependent manner [34]. The salt tolerance gene *Ncl* can reduce Na^+, K^+, and Cl^- accumulation in soybean leaves under salt stress and function like cation–chloride cotransporter (CCC) genes [75]. Recently, it was proved in a heterologous system that the GmSALT3 protein contributed Na^+, K^+, and Cl^- transport in *Xenopus laevis* oocytes. Detailed analysis of three sets of salt-tolerant NILs (NIL-*GmSALT3*) and salt-sensitive NILs (NIL-*Gmsalt3*) showed that *GmSALT3* mediates Na^+ and Cl^- exclusion from shoots via net xylem loading or phloem re-translocation, although the exact molecular mechanism requires further study [35,76]. Transcriptomic analysis has been used to unravel the molecular mechanisms of *GmSALT3*, which suggests GmSALT3 might help to detoxify ROS toxicity through the flavonoid biosynthesis pathway [77]. Recently, it was reported that the membrane-bound NAC with trans-membrane motif1-like (NTL) transcription factor GmNTL1 can bind to the promoter of *GmSALT3/GmCHX1/GmNcl* to promote soybean salt tolerance by activating gene transcription [78], while the other gene, *GmERD15B*, might promote soybean salt tolerance via the up-regulation of stress-related genes, including *GmbZIP1*, *GmP5CS*, *GmCAT4*, and *GmSOS1* [63].

Salinity inhibits seed germination in plants by altering different growth processes including the imbibition of water, enzyme activities, and hormonal balance [3]. *GmCDF1* is the only candidate gene related to salt tolerance at the germination stage, the function of which was proved in soybean hairy roots. Overexpression of *GmCDF1* could decrease soybean salt tolerance by affecting K^+–Na^+ homeostasis in soybean roots and shoots [61]. The mechanism through which *GmCDF1* regulates soybean germination under salt stress needs further exploration.

4. Salt-Tolerant Accessions and Molecular Markers in Soybean Breeding for Saline Soils

4.1. Major Genes Involved in Soybean Salt Tolerance Provide Molecular Approach to Tolerant Soybean Screening and Breeding

An RAPD marker tightly linked to the salt tolerance gene was identified in the soybean cultivar Jindou 33 and used for the identification of tolerant accessions [79,80]. Subsequent sequencing of specific fragments of the RAPD marker in soybeans showed that the sequence is part of *Glyma03g32920.1*, from which a sequence-characterized amplified region (SCAR) marker was developed for the fine mapping of the salt tolerance gene [33]. Two SSR markers, Sat_091 and Satt237, were suggested to be useful for salt-tolerant soybean breeding due to the tight linkage of SSR markers with major QTLs on linkage group N and the association of marker alleles with salt tolerance in soybean descendants [47]. With the cloning of the major salt tolerance gene *GmSALT3/GmCHX1/GmNcl* [34,62,75], variations in the gene promoter and coding regions were identified in diverse soybean accessions [72,74]. Taking those haplotypes observed in more than ten soybean accessions as the main haplotypes, there are seven main haplotypes of *GmSALT3/GmCHX1/GmNcl* reported in soybean (Figure 2). Haplotype H1/Hn/SV-1 is the only conserved tolerant allele that has a functional domain. H2/HTn/SV-2 is a salt-sensitive haplotype with a 3.78 kb Ty1/copia retrotransposon insertion in the third exon; H5-1/Hd-2 is a sensitive haplotype with a 4 bp deletion in exon 2; and H5-2/Hd-3 has the same 4 bp deletion as that of H5-1/Hd-2, with an additional C > G variation in exon 3. H3 and H4 are two sensitive haplotypes only observed in Chinese soybean accessions [34,72]. SV-3 is a sensitive haplotype with a ~180 bp deletion in exon 3, and this allele was only reported by Patil et al. [74]. SNP assays and PCR-based markers were developed according to the variations in *GmSALT3/GmCHX1* and showed the precise identification (>90%) of salt-tolerant accessions, providing functional markers for targeted breeding [72–74]. A new variation in the promoter region of *GmCHX1* was proved to be a conditional gene-expression-related allele that existed in four salt-tolerant lines. It provides a new allele for salt tolerance breeding [81]. A 7 bp Indel in the promoter region of an early responsive to dehydration 15B (*GsERD15B*) gene was found to be related to the salt tolerance rating in wild soybean (*G. soja*), and a dCAPS marker was designed to distinguish the two alleles [63]. It remains unknown what the variation of *GsERD15B* in cultivated soybeans is and how it can be used in soybean breeding. Ten haplotypes of *GmCDF1* were detected that control salt tolerance at the germination stage, and haplotype Hap2 was more tolerant than Hap1 [61], while no further molecular markers related to this gene were reported.

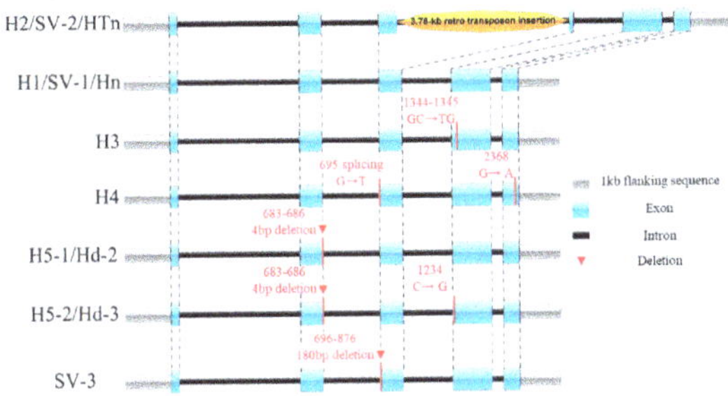

Haplotype	Phenotype	Haplotype	Phenotype	Haplotype	Phenotype
H2	Sensitive	SV-2	Sensitive	HTn	Sensitive
H1	Tolerant	SV-1	Tolerant	Hn	Tolerant
H3	Sensitive				
H4	Sensitive				
H5-1	Sensitive			Hd-2	Sensitive
H5-2	Sensitive			Hd-3	Sensitive
		SV-3	Sensitive		

Figure 2. Main structure and coding variations of *GmSALT3* observed in soybean accessions. Blue boxes indicate exons, the black bars represent introns, gray boxes represent the 1 kb flanking sequence of *GmSALT3*, dotted lines indicate the exon position, orange block represents the 3.78 kb retro transposon insertion, and key variants are listed with red letters (only those haplotypes observed in more than 10 accessions were counted) [72–74].

4.2. Creation of Salt-Tolerant Soybean

The identification of salt tolerance loci is likely to contribute to the development of salt-tolerant soybean varieties. To confirm the function of known tolerance genes like *GmSALT3*, near-isogenic lines harboring *GmSALT3* (NIL-T) or *Gmsalt3* (NIL-S) were created using a marker-assisted strategy. Each pair of NILs contained 95.6–99.3% genetic similarity and were used to elucidate gene function in salinized soil. No yield penalty was observed for *GmSALT3* under normal field conditions, and a significantly higher 100-seed weight and total plant seed weight were found in NIL-T lines in salinized fields [35]. Under salt stress, NILs with the salt tolerance allele showed a yield decrease of less than 29.5%, whereas NILs with the salt sensitivity allele experienced a more pronounced yield reduction of 44.0–55.8%, indicating the presence of the salt tolerance gene contributed to sustainable soybean production in saline fields [75]. Commercial soybean cultivars containing *GmSALT3*, such as Zhonghuang 30 and Zhonghuang 13, which have been approved to be salt-tolerant, are potential resources for the breeding of salt-tolerant soybeans [72].

In addition to the introgression of *GmSALT3* into soybeans via marker-assisted selection, new salt-tolerant lines have been created using transgenic approaches. Overexpression lines with the transcription factor *GmSIN1* showed rapid emergence and higher yields compared with the salt-tolerant variety Wei6823 under saline conditions [64]. Soybean seedlings with overexpression of nuclear factor Y C subunit *GmNF-YC14* had higher biomass than the wild type under salt stress [70]. Transgenic soybean lines overexpressing a class B heat shock factor *HSFB2b* had higher survival rates than wild-type Jack after 7 d of 300 mM NaCl treatment, and the variations in the promoter of *HSFB2b* may be useful for breeding tolerant soybeans [68]. By introducing nuclear factor Y subunit *GmNFYA* into soybean Jack, the plant height and survival rate were greatly improved under 300 mM NaCl stress [69].

5. Perspective

The salt tolerance of soybeans is the result of contributions from genetic loci involved in different developmental stages. Full-seed germination is the initial step for plants to achieve greater yields in saline fields, especially where the timing of sowing usually depends on rainfall events. Studies have shown that the genetic control of salt tolerance varies at different growth stages in soybeans [13,35,56]. Despite these studies, there is still a gap in knowledge on the genetic responses and underlying mechanisms involved in salt tolerance between the germination and emergence stages. This may be partly due to difficulties in conducting these types of studies, which are time-consuming and labor-intensive. They require screening a large number of accessions for germination and seedling vigor to determine maximum levels of salinity stress on plants and are inherently more difficult under more unpredictable field conditions compared with laboratory conditions [82]. Therefore, it is necessary to develop ways to reduce these complexities and difficulties by developing, for example, feasible selection indicators that can bridge the differences between lab and field environments and high-throughput screening technologies to more accurately and precisely phenotype large numbers of samples, especially when measuring plants grown in control conditions could be avoided. WinRoots is a system recently developed for soybean phenomics study, which made RGB (red–green–blue) images of the roots and shoots canopy phenotype easily collected [83]. Identifying traits related to salt stress responses at particular developmental stages using thermal sensors and RGB imaging will lead to the identification of major QTLs on trait variation that can be applied to breeding. The salt tolerance genes that have been genetically characterized have created opportunities to develop salt-tolerant soybean varieties via marker-assisted selection (MAS) using tightly linked or functional molecular markers.

Salt stress causes a reduction in growth because energetic resources must be allocated away from photosynthetic processes to accommodate the need for osmotic adjustment [4]. The introduction of salt tolerance genes into crops, for example, *Nax2* in wheat and *GmSATL3* in soybean, only mitigates the losses in yield due to salt stress rather than restoring the full yield achievable in non-saline fields due to the energy used for osmotic adjustment [35,51,84]. Therefore, knowing how to more effectively use salt tolerance genes or their regulators in soybean breeding via genetic engineering, i.e., transgenic manipulation or genomic editing, to improve yield under both saline and non-saline conditions depends on our advances in understanding the underlying mechanisms and warrants further exploration. 'Omics' approaches may be useful to determine the roles of known genes and identify the pathways and essential genes involved in salt tolerance (Figure 3).

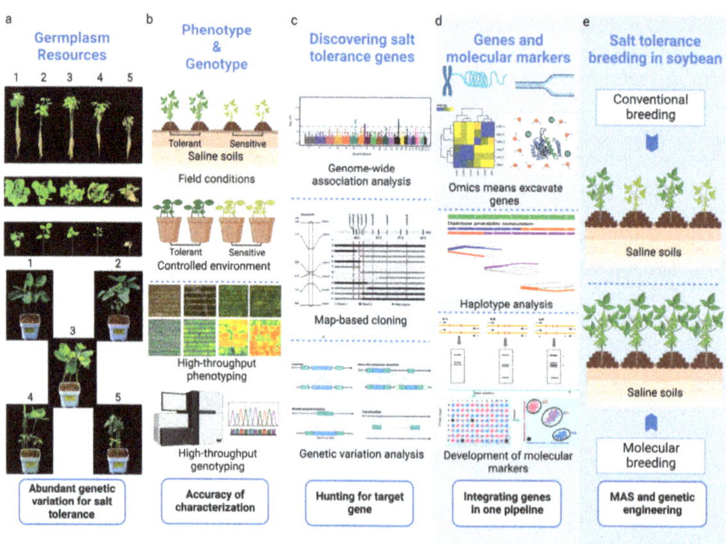

Figure 3. Breeding salt-tolerant soybeans. (**a**) Abundant variations in salt tolerance exist in soybean accessions. Numbers from 1 (the most tolerant) to 5 (the most sensitive) indicate salt tolerance level.

(**b**) Characterizing soybean accessions with high-throughput phenotyping methods under conditions that mimic realistic saline field conditions and cost-effective genotyping approaches. (**c**) Identifying target genes and functional variants for salt-tolerance-related traits. (**d**) Integration of single-nucleotide variations and 'omics' at gene level. (**e**) With knowledge of key gene variations and related pathways, MAS and other biological approaches will become the most effective way for breeding.

Although we have evaluated some genetic loci, as salt tolerance is a very complex physiological process, the following challenges remain: (1) approaches that can be easily used to screen the phenotypes of soybean accessions; (2) major gene(s) contributing to specific developmental stages without a yield penalty under both saline and non-saline conditions; and (3) effective ways for trait stacking to obtain salt-tolerant soybeans throughout the whole growing period.

Author Contributions: R.-X.G., R.-Z.C. and L.-J.Q. designed the overall outline. X.-Y.G., Y.Q., Z.-W.Z., L.-G.B. and R.-Y.Y. reviewed the literature and collated data. X.-Y.G. drew the graphs and charts. R.-X.G., X.-Y.G. and L.-J.Q. wrote the manuscript. All authors have read and agreed to the published version of the manuscript.

Funding: This research was funded by the Natural Science Foundation of China (31830066) and the National Key R&D Program of China (2022YFD1201702).

Data Availability Statement: Not applicable.

Conflicts of Interest: The authors declare no conflicts of interest.

References

1. FAO. Saline Soils and Their Management. Food and Agriculture Organization of the United Nations. 2016. Available online: http://www.fao.org/3/x5871e/x5871e04.htm (accessed on 20 December 2023).
2. FAO. Salt-Affected Soils. Food and Agriculture Organization of the United Nations. 2021. Available online: https://www.fao.org/soils-portal/soil-management/management-of-some-problem-soils/salt-affected-soils/more-information-on-salt-affect (accessed on 20 December 2023).
3. Munns, R. Genes and salt tolerance: Bringing them together. *New Phytol.* **2005**, *167*, 645–663. [CrossRef]
4. Singh, A. Soil salinization management for sustainable development: A review. *J. Environ. Manag.* **2021**, *277*, 111383. [CrossRef]
5. Mukhopadhyay, R.; Sarkar, B.; Jat, H.S.; Sharma, P.C.; Bolan, N.S. Soil salinity under climate change: Challenges for sustainable agriculture and food security. *J. Environ. Manag.* **2021**, *280*, 111736. [CrossRef]
6. Hassani, A.; Azapagic, A.; Shokri, N. Global predictions of primary soil salinization under changing climate in the 21st century. *Nat. Commun.* **2021**, *12*, 6663. [CrossRef]
7. FAO. Can Technology Deliver on the Yield Challenge to 2050? Expert Meeting on How to Feed the World in 2050. Food and Agriculture Organization of the United Nations. 2010. Available online: http://www.fao.org/fileadmin/templates/wsfs/docs/Issues_papers/HLEF2050_Technology.pdf (accessed on 20 December 2023).
8. Zelm, E.V.; Zhang, Y.; Testerink, C. Salt tolerance mechanisms of plants. *Annu. Rev. Plant Biol.* **2020**, *71*, 403–433. [CrossRef]
9. Srivastava, S.P.; Bhandari, T.M.S.; Yadav, C.R.; Joshi, M.; Erskine, W. Boron deficiency in Lentil: Yield loss and geographic distribution in a germplasm collection. *Plant Soil* **2000**, *219*, 147–151. [CrossRef]
10. Wichelns, D.; Qadir, M. Achieving sustainable irrigation requires effective management of salts, soil salinity, and shallow groundwater. *Agric. Water Manag.* **2015**, *157*, 31–38. [CrossRef]
11. Chang, X.; Gao, Z.; Wang, S.; Chen, H. Modelling long-term soil salinity dynamics using SaltMod in Hetao Irrigation District, China. *Comput. Electron. Agric.* **2019**, *156*, 447–458. [CrossRef]
12. Epstein, E.; Norlyn, J.D.; Rush, D.W.; Kingsbury, R.W.; Kelly, D.B.; Gunningham, G.A.; Wrona, A.F. Saline culture of crops: A genetic approach. *Science* **1980**, *210*, 399–404. [CrossRef]
13. Abel, G.H.; MacKenzie, A.J. Salt tolerance of soybean varieties (*Glycin max* L. Merrill) during germination and later growth. *Crop. Sci.* **1964**, *4*, 157–161. [CrossRef]
14. American Soybean Association. *SoyStats: A Reference Guide to Important Soybean Facts & Figures*; American Soybean Association: St. Louis, MO, USA, 2021.
15. Phang, T.H.; Shao, G.; Lam, H.M. Salt tolerance in soybean. *J. Integr. Plant Biol.* **2008**, *50*, 1196–1212. [CrossRef]
16. Mass, E.V.; Hoffman, G.J. Crop salt tolerance—current assessment. *J. Irrig. Drain. Div.* **1977**, *103*, 115–134. [CrossRef]
17. Shao, G.; Wan, C.; Li, S. Preliminary study on the physiology of soybean tolerance to salt stress at germinating stage. *Crops* **1994**, *6*, 25–27, (In Chinese with Chinese Abstract).
18. Shao, G.; Song, J.; Liu, H. Identification of salt tolerance in soybean germplasm resources. *Sci. Agric. Sin.* **1986**, *6*, 30–35. (In Chinese with English Abstract)

19. Shao, G.; Chang, R.; Chen, Y. Advances of salt tolerance in soybean. *Soybean Sci.* **1993**, *3*, 244–248. (In Chinese with Chinese Abstract)
20. Shelke, D.B.; Pandey, M.; Nikalje, G.C.; Zaware, B.N.; Suprasanna, P.; Nikam, T.D. Salt responsive physiological, photosynthetic and biochemical attributes at early seedling stage for screening soybean genotypes. *Plant Physiol. Biochem.* **2017**, *118*, 519–528. [CrossRef]
21. Kan, G.; Ning, L.; Li, Y.; Hu, Z.; Zhang, W.; He, X.; Yu, D. Identification of novel loci for salt stress at the seed germination stage in soybean. *Breed. Sci.* **2016**, *66*, 530–541. [CrossRef]
22. Hollington, P.A. Technological breakthrough in screening/breeding wheat varieties for salt tolerance. In Proceedings of the National Conference on 'Salinity Management in Agriculture', Karnal, India, 2–5 December 1998.
23. Hosseini, M.K.; Powell, A.A.; Bingham, I.J. Comparison of the seed germination and early seedling growth of soybean in saline conditions. *Seed Sci. Res.* **2002**, *12*, 165–172. [CrossRef]
24. Wang, D.; Shannon, M.C. Emergence and seedling growth of soybean cultivars and maturity groups under salinity. *Plant Soil* **1999**, *214*, 117–124. [CrossRef]
25. Dong, L.; Hou, Z.; Li, H.; Li, Z.; Fang, C.; Kong, L.; Li, Y.; Du, H.; Li, T.; Wang, L.; et al. Agronomical selection on loss-of-function of *GIGANTEA* simultaneously facilitates soybean salt tolerance and early maturity. *J. Integr. Plant Biol.* **2022**, *64*, 1866–1882. [CrossRef]
26. Cheng, Q.; Gan, Z.; Wang, Y.; Lu, S.; Hou, Z.; Li, H.; Xiang, H.; Liu, B.; Kong, F.; Dong, L. The soybean gene *J* contributes to salt stress tolerance by up-regulating salt-responsive genes. *Front. Plant Sci.* **2020**, *11*, 272. [CrossRef]
27. Liu, X.; Chang, R.; Guan, R.; Qiu, L. Establishment of screening method for salt tolerant soybean at emergence stage and screening of tolerant germplasm. *Acta Agron. Sin.* **2020**, *46*, 1–8. (In Chinese with English Abstract)
28. Abel, G.H. Inheritance of the capacity for chloride inclusion and chloride exclusion by soybeans. *Crop. Sci.* **1969**, *9*, 697–698. [CrossRef]
29. Valencia, R.; Chen, P.; Ishibashi, T.; Conatser, M. A rapid and effective method for screening salt tolerance in soybean. *Crop. Sci.* **2008**, *48*, 1773–1779. [CrossRef]
30. Lee, J.D.; Smothers, S.L.; Dunn, D.; Villagarcia, M.; Shumway, C.R.; Carter, T.E.; Shannon, J.G. Evaluation of a simple method to screen soybean genotypes for salt tolerance. *Crop. Sci.* **2008**, *48*, 2194–2200. [CrossRef]
31. Ledesma, F.; Lopez, C.; Ortiz, D.; Chen, P.; Korth, K.L.; Ishibashi, T.; Zeng, A.; Orazaly, M.; Florez-Palacios, L. A simple greenhouse method for screening salt tolerance in soybean. *Crop. Sci.* **2016**, *56*, 585–594. [CrossRef]
32. Jiang, J.; Guan, R.; Guo, Y.; Chang, R.; Qiu, L. Simple evaluation method of tolerance to salt at seedling stage in soybean. *Acta Agron. Sin.* **2013**, *39*, 1248–1256, (In Chinese with English Abstract). [CrossRef]
33. Guan, R.; Chen, J.; Jiang, J.; Liu, G.; Liu, Y.; Tian, L.; Yu, L.; Chang, R.; Qiu, L. Mapping and validation of a dominant salt tolerance gene in the cultivated soybean (*Glycine max*) variety Tiefeng 8. *Crop. J.* **2014**, *2*, 358–365. [CrossRef]
34. Guan, R.; Qu, Y.; Guo, Y.; Yu, L.; Liu, Y.; Jiang, J.; Chen, J.; Ren, Y.; Liu, G.; Tian, L.; et al. Salinity tolerance in soybean is modulated by natural variation in *GmSALT3*. *Plant J.* **2014**, *80*, 937–950. [CrossRef]
35. Liu, Y.; Yu, L.; Qu, Y.; Chen, J.; Liu, X.; Hong, H.; Liu, Z.; Chang, R.; Gilliham, M.; Qiu, L.; et al. GmSALT3, which confers improved soybean salt tolerance in the field, increases leaf Cl^- exclusion prior to Na^+ exclusion but does not improve early vigor under salinity. *Front. Plant Sci.* **2016**, *7*, 1485. [CrossRef]
36. Mannan, M.A.; Karim, M.A.; Khaliq, Q.A.; Haque, M.M.; Mian, M.A.K.; Ahmed, J.U. Assessment of genetic divergence in salt tolerance of soybean (*Glycine max* L.) genotypes. *J. Crop. Sci. Biot.* **2010**, *13*, 33–38. [CrossRef]
37. Essa, T.A. Effect of salinity stress on growth and nutrient composition of three soybean (*Glycine max* L. Merrill) cultivars. *J. Agron. Crop. Sci.* **2002**, *188*, 86–93. [CrossRef]
38. Rhoades, J.D.; Lesch, S.M.; Burch, S.L.; Letey, J.; LeMert, R.D.; Shouse, P.J.; Oster, J.D.; O'Halloran, T. Salt distribution in cracking soils and salt pick up by runoff waters. *J. Irrig. Drain. Eng.* **1997**, *123*, 323–328. [CrossRef]
39. Li, X.; Chen, W.; Li, Z. Identification of salt tolerance of soybean germplasm resources in Shandong province. *Acta Agron. Sin.* **1996**, *4*, 11–13. (In Chinese with Chinese Abstract)
40. Kan, G.; Zhang, W.; Yang, W.; Ma, D.; Zhang, D.; Hao, D.; Hu, Z.; Yu, D. Association mapping of soybean seed germination under salt stress. *Mol. Genet. Genom.* **2015**, *290*, 2147–2162. [CrossRef] [PubMed]
41. Jiang, Q.; Hu, Z.; Zhang, H.; Wang, M.; Tang, J.; Ni, Z.; Jiang, F. Identification and study on salt tolerance of soybean germplasm resources. *J. Plant Genet. Resour.* **2012**, *13*, 726–732. (In Chinese with Chinese Abstract)
42. Chang, R.; Chen, Y.; Shao, G.; Wan, C. Effect of salt on agronomic traits and grain quality of soybean. *Soybean Sci.* **1994**, *2*, 101–105. (In Chinese with English Abstract)
43. Zhang, P.; Xu, C.; Xu, K.; Zhang, Z.; Li, D.; Ji, P.; Feng, Y. A rapid method for the identification of salt tolerance in soybean varieties and the study of salt tolerance in different periods. *China J. Oil Crop. Sci.* **2013**, *35*, 572–578. (In Chinese with Chinese Abstract)
44. Zhang, W.; Niu, Y.; Bu, S.; Li, M.; Feng, J.; Zhang, J.; Yang, S.; Odinga, M.M.; Wei, S.; Liu, X.; et al. Epistatic association mapping for alkaline and salinity tolerance traits in the soybean germination stage. *PLoS ONE* **2014**, *9*, e84750. [CrossRef]
45. Shao, G.; Chang, R.; Chen, Y.; Yan, S. Study on inheritance of salt tolerance in soybean. *Acta Agron. Sin.* **1994**, *20*, 721–726. (In Chinese with English Abstract)
46. Guo, X.; Jiang, J.; Liu, Y.; Yu, L.; Chang, R.; Guan, R.; Qiu, L. Identification of a novel salt tolerance-related locus in wild soybean (*Glycine soja* Sieb. & Zucc.). *Front. Plant Sci.* **2021**, *12*, 791175. [PubMed]

47. Lee, G.J.; Carter, T.E., Jr.; Villagarcia, M.R.; Li, Z.; Zhou, X.; Gibbs, M.O.; Boerma, H.R. A major QTL conditioning salt tolerance in S-100 soybean and descendent cultivars. *Theor. Appl. Genet.* **2004**, *109*, 1610–1619. [CrossRef] [PubMed]
48. Chen, H.; Cui, S.; Fu, S.; Gai, J.; Yu, D. Identification of quantitative trait loci associated with salt tolerance during seedling growth in soybean (*Glycine max* L.). *Aust. J. Agric. Res.* **2008**, *59*, 1086–1091. [CrossRef]
49. Hamwieh, A.; Tuyen, D.D.; Cong, H.; Benitez, E.R.; Takahashi, R.; Xu, D.H. Identification and validation of a major QTL for salt tolerance in soybean. *Euphytica* **2011**, *179*, 451–459. [CrossRef]
50. Shi, X.; Yan, L.; Yang, C.; Yan, W.; Moseley, D.O.; Wang, T.; Liu, B.; Di, R.; Chen, P.; Zhang, M. Identification of a major quantitative trait locus underlying salt tolerance in 'Jidou 12' soybean cultivar. *BMC Res. Notes* **2018**, *11*, 95. [CrossRef]
51. Do, T.D.; Vuong, T.D.; Dunn, D.; Smothers, S.; Patil, G.; Yungbluth, D.C.; Chen, P.; Scaboo, A.; Xu, D.; Carter, T.E.; et al. Mapping and confirmation of loci for salt tolerance in a novel soybean germplasm, Fiskeby III. *Theor. Appl. Genet.* **2018**, *131*, 513–524. [CrossRef]
52. Chen, H.; Ye, H.; Do, T.D.; Zhou, J.; Valliyodan, B.; Shannon, G.J.; Chen, P.; Chen, X.; Nguyen, H.T. Advances in genetics and breeding of salt tolerance in soybean. *Salin. Responses Toler. Plants* **2018**, *2*, 217–237.
53. Chen, H.; Liu, X.; Zhang, H.; Yuan, X.; Gu, H.; Cui, X.; Chen, X. Advances in salinity tolerance of soybean: Genetic diversity, heredity, and gene identification contribution to improving salinity tolerance. *J. Integr. Agric.* **2018**, *17*, 2215–2221. [CrossRef]
54. Tian, Y.; Gao, F.; Cao, P.; Gao, C. Recent developments of salt tolerance gene in soybean. *Soybean Sci.* **2018**, *37*, 629–636. (In Chinese with English Abstract)
55. Zhang, W.; Liao, X.; Yu, D.; Kan, G. A review of salt tolerance in soybean (*Glycine max* (L.) Merrill). *Soils Crops* **2018**, *7*, 284–292. (In Chinese with English Abstract)
56. Zeng, A.; Chen, P.; Korth, K.; Hancock, F.; Pereira, A.; Brye, K.; Wu, C.; Shi, A. Genome-wide association study (GWAS) of salt tolerance in worldwide soybean germplasm lines. *Mol. Breed.* **2017**, *37*, 30. [CrossRef]
57. Li, Y.; Ye, H.; Vuong, T.D.; Zhou, L.; Do, T.D.; Chhapekar, S.S.; Zhao, W.; Li, B.; Jin, T.; Gu, J.; et al. A novel natural variation in the promoter of *GmCHX1* regulates the conditional gene expression to improve salt tolerance in soybean. *J. Exp. Bot.* **2023**, erad404. [CrossRef] [PubMed]
58. Hamwieh, A.; Xu, D. Conserved tolerance quantitative trait locus (QTL) in wild and cultivated soybeans. *Breed. Sci.* **2008**, *58*, 355–359. [CrossRef]
59. Lee, J.D.; Shannon, J.G.; Vuong, T.D.; Nguyen, H.T. Inheritance of salt tolerance in wild soybean (*Glycine soja* Sieb. and Zucc.) germplasm PI483463. *J. Hered.* **2009**, *100*, 798–801. [CrossRef] [PubMed]
60. Ha, B.K.; Vuong, T.D.; Velusamy, V.; Nguyen, H.T.; Shannon, J.G.; Lee, J.D. Genetic mapping of quantitative trait loci conditioning salt tolerance in wild soybean (*Glycine soja*) PI 483463. *Euphytica* **2013**, *193*, 79–88. [CrossRef]
61. Zhang, W.; Liao, X.; Cui, Y.; Ma, W.; Zhang, X.; Du, H.; Ma, Y.; Ning, L.; Wang, H.; Huang, F.; et al. A cation diffusion facilitator, *GmCDF1*, negatively regulates salt tolerance in soybean. *PLoS Genet.* **2019**, *15*, e1007798. [CrossRef]
62. Qi, X.; Li, M.; Xie, M.; Liu, X.; Ni, M.; Shao, G.; Song, C.; Yim, A.K.Y.; Tao, Y.; Wong, F.L.; et al. Identification of a novel salt tolerance gene in wild soybean by whole-genome sequencing. *Nat. Commun.* **2014**, *5*, 4340. [CrossRef] [PubMed]
63. Jin, T.; Sun, Y.; Shan, Z.; He, J.; Wang, N.; Gai, J.; Li, Y. Natural variation in the promoter of *GsERD15B* affects salt tolerance in soybean. *Plant Biotechnol. J.* **2021**, *19*, 1155–1169. [CrossRef]
64. Li, S.; Wang, N.; Ji, D.; Zhang, W.; Wang, Y.; Yu, Y.; Zhao, S.; Lyu, M.; You, J.; Zhang, Y.; et al. A GmSIN1/GmNCED3s/GmRbohBs feed-forward loop acts as a signal amplifier that regulates root growth in soybean exposed to salt stress. *Plant Cell* **2019**, *31*, 2107–2130. [CrossRef]
65. Nguyen, Q.H.; Vu, L.T.K.; Nguyen, L.T.N.; Pham, N.T.T.; Nguyen, Y.T.H.; Le, S.V.; Chu, M.H. Overexpression of the *GmDREB6* gene enhances proline accumulation and salt tolerance in genetically modified soybean plants. *Sci. Rep.* **2019**, *9*, 19663. [CrossRef]
66. Zhao, M.; Yin, L.; Ma, J.; Zheng, J.; Wang, Y.; Lan, J.; Fu, J.; Chen, M.; Xu, Z.; Ma, Y. The roles of *GmERF135* in improving salt tolerance and decreasing ABA sensitivity in soybean. *Front. Plant Sci.* **2019**, *10*, 940. [CrossRef]
67. Jia, Q.; Sun, S.; Kong, D.; Song, J.; Wu, L.; Yan, Z.; Zuo, L.; Yang, Y.; Liang, K.; Lin, W.; et al. Ectopic expression of *Gs5PTase8*, a soybean inositol polyphosphate 5-phosphatase, enhances salt tolerance in plants. *Int. J. Mol. Sci.* **2020**, *21*, 1023. [CrossRef]
68. Bian, X.; Li, W.; Niu, C.; Wei, W.; Hu, Y.; Han, J.; Lu, X.; Tao, J.; Jin, M.; Qin, H.; et al. A class B heat shock factor selected for during soybean domestication contributes to salt tolerance by promoting flavonoid biosynthesis. *New Phytol.* **2020**, *225*, 268–283. [CrossRef]
69. Lu, L.; Wei, W.; Tao, J.; Lu, X.; Bian, X.; Hu, Y.; Cheng, T.; Yin, C.; Zhang, W.; Chen, S.; et al. Nuclear factor Y subunit *GmNFYA* competes with GmHDA13 for interaction with GmFVE to positively regulate salt tolerance in soybean. *Plant Biotechnol. J.* **2021**, *19*, 2362–2379. [CrossRef]
70. Yu, T.; Liu, Y.; Fu, J.; Ma, J.; Fang, Z.; Chen, J.; Zhang, L.; Lu, Z.; Zhou, Y.; Chen, M.; et al. The NF-Y-PYR module integrates the abscisic acid signal pathway to regulate plant stress tolerance. *Plant Biotechnol. J.* **2021**, *19*, 2589–2605. [CrossRef]
71. Leung, H.S.; Chan, L.Y.; Law, C.H.; Li, M.W.; Lam, H.M. Twenty years of mining salt tolerance genes in soybean. *Mol. Breed.* **2023**, *43*, 45. [CrossRef]
72. Guan, R.; Yu, L.; Liu, X.; Li, M.; Chang, R.; Gilliham, M.; Qiu, L. Selection of the salt tolerance gene *GmSALT3* during six decades of soybean breeding in China. *Front. Plant. Sci.* **2021**, *12*, 794241. [CrossRef]

73. Lee, S.; Kim, J.; Sundaramoorthy, J.; Park, G.T.; Lee, J.D.; Kim, J.H.; Chung, G.; Seo, H.S.; Song, J.T. Identification of *GmSALT3* haplotypes and development of molecular markers based on their diversity associated with salt tolerance in soybean. *Mol. Breed.* **2018**, *38*, 86. [CrossRef]
74. Patil, G.; Do, T.; Vuong, T.D.; Valliyodan, B.; Lee, J.D.; Chaudhary, J.; Shannon, J.G.; Nguyen, H.T. Genomic-assisted haplotype analysis and the development of high-throughput SNP markers for salinity tolerance in soybean. *Sci. Rep.* **2016**, *6*, 19199. [CrossRef]
75. Do, T.D.; Chen, H.; Vu, H.T.; Hamwieh, A.; Yamada, T.; Sato, T.; Yan, Y.; Cong, H.; Shono, M.; Suenaga, K.; et al. *Ncl* synchronously regulates Na^+, K^+, and Cl^- in soybean and greatly increases the grain yield in saline field conditions. *Sci. Rep.* **2016**, *6*, 19147. [CrossRef]
76. Qu, Y.; Guan, R.; Bose, J.; Henderson, S.W.; Wege, S.; Qiu, L.; Gilliham, M. Soybean CHX-type ion transport protein GmSALT3 confers leaf Na^+ exclusion via a root derived mechanism, and Cl^- exclusion via a shoot derived process. *Plant Cell Environ.* **2021**, *44*, 856–869. [CrossRef] [PubMed]
77. Qu, Y.; Guan, R.; Yu, L.; Berkowitz, O.; David, R.; Whelan, J.; Ford, M.; Wege, S.; Qiu, L.; Gilliham, M. Enhanced reactive oxygen detoxification occurs in salt-stressed soybean roots expressing *GmSALT3*. *Physiol. Plant* **2022**, *174*, e13709. [CrossRef] [PubMed]
78. Zhang, W.; Zhi, W.; Qiao, H.; Huang, J.; Li, S.; Lu, Q.; Wang, N.; Li, Q.; Zhou, Q.; Sun, J.; et al. H_2O_2-dependent oxidation of the transcription factor GmNTL1 promotes salt tolerance in soybean. *Plant Cell* **2023**, *36*, 112–135. [CrossRef] [PubMed]
79. Guo, B.; Qiu, L.; Shao, G.; Chang, R.; Liu, L.; Xu, Z.; Li, X.; Sun, J. Tagging salt tolerant gene using PCR markers in soybean. *Sci. Agric. Sin.* **2000**, *33*, 10–16, (In Chinese with English Abstract).
80. Guo, B.; Qiu, L.; Shao, G.; Xu, Z. Molecular marker assisted identification and utilization of salt tolerant soybean germplasm. *Soybean Sci.* **2002**, *1*, 56–61, (In Chinese with English Abstract).
81. Feng, C.; Gao, H.; Zhou, Y.; Jing, Y.; Li, S.; Yan, Z.; Xu, K.; Zhou, F.; Zhang, W.; Yang, X.; et al. Unfolding molecular switches for salt stress resilience in soybean: Recent advances and prospects for salt-tolerant smart plant production. *Front. Plant Sci.* **2023**, *14*, 1162014. [CrossRef]
82. Munns, R.; James, R.A.; Gilliham, M.; Flowers, T.J.; Colmer, T.D. Tissue tolerance: An essential but elusive trait for salt-tolerant crops. *Funct. Plant Biol.* **2016**, *43*, 1103–1113. [CrossRef]
83. Zhang, Y.; Zhang, W.; Cao, Q.; Zheng, X.; Yang, J.; Xue, T.; Sun, W.; Du, X.; Wang, L.; Wang, J.; et al. WinRoots: A high-throughput cultivation and phenotyping system for plant phenomics studies under soil stress. *Front. Plant Sci.* **2022**, *12*, 794020. [CrossRef]
84. Munns, R.; James, R.A.; Xu, B.; Athman, A.; Conn, S.J.; Jordans, C.; Byrt, C.S.; Hare, R.A.; Tyerman, S.D.; Tester, M.; et al. Wheat grain yield on saline soils is improved by an ancestral Na^+ transporter gene. *Nat. Biotechnol.* **2012**, *30*, 360–364. [CrossRef]

plants

MDPI

Article

Exploring Genetics by Environment Interactions in Some Rice Genotypes across Varied Environmental Conditions

Mohamed I. Ghazy [1], Mohamed Abdelrahman [1], Roshdy Y. El-Agoury [1], Tamer M. El-hefnawy [1], Sabry A. EL-Naem [1], Elhousini M. Daher [1] and Medhat Rehan [2,3,*]

1 Rice Research and Training Department, Field Crops Research Institute, Agricultural Research Center, Kafrelsheikh 33717, Egypt; m_ghazy2050@yahoo.com (M.I.G.); maabdelrahman@arc.sci.eg or abdelrahman.rrtc@gmail.com (M.A.); roshdyrrtc@gmail.com (R.Y.E.-A.); tito.307@gmail.com (T.M.E.-h.); sabryelnaem@gmail.com (S.A.E.-N.); elhousani.daher@yahoo.com (E.M.D.)
2 Department of Plant Production and Protection, College of Agriculture and Veterinary Medicine, Qassim University, Buraydah 51452, Saudi Arabia
3 Department of Genetics, Faculty of Agriculture, Kafrelsheikh University, Kafr El-Sheikh 33516, Egypt
* Correspondence: m.rehan@qu.edu.sa or medhat.rehan@agr.kfs.edu.eg

Abstract: Rice production faces challenges related to diverse climate change processes. Heat stress combined with low humidity, water scarcity, and salinity are the foremost threats in its cultivation. The present investigation aimed at identifying the most resilient rice genotypes with yield stability to cope with the current waves of climate change. A total of 34 rice genotypes were exposed to multilocation trials. These locations had different environmental conditions, mainly normal, heat stress with low humidity, and salinity-affected soils. The genotypes were assessed for their yield stability under these conditions. The newly developed metan package of R-studio was employed to perform additive main effects and multiplicative interactions modelling and genotype-by-environment modelling. The results indicated that there were highly significant differences among the tested genotypes and environments. The main effects of the environments accounted for the largest portion of the total yield sum of squared deviations, while different sets of genotypes showed good performance in different environments. AMMI1 and GGE biplots confirmed that Giza179 was the highest-yielding genotype, whereas Giza178 was considered the most-adopted and highest-yielding genotype across environments. These findings were further confirmed by the which–won–where analysis, which explained that Giza178 has the greatest adaptability to the different climatic conditions under study. While Giza179 was the best under normal environments, N22 recorded the uppermost values under heat stress coupled with low humidity, and GZ1968-S-5-4 manifested superior performance regarding salinity-affected soils. Giza 177 was implicated regarding harsh environments. The mean vs. stability-based rankings indicated that the highest-ranked genotypes were Giza179 > Giza178 > IET1444 > IR65600-77 > GZ1968-S-5-4 > N22 > IR11L236 > IR12G3213. Among them, Giza178, IR65600-77, and IR12G3213 were the most stable genotypes. Furthermore, these results were confirmed by cluster-analysis-based stability indices. A significant and positive correlation was detected between the overall yield under all the environments with panicle length, number of panicles per plant, and thousand grain weight. Our study sheds light on the notion that the Indica/Japonica and Indica types have greater stability potential over the Japonica ones, as well as the potential utilization of genotypes with wide adaptability, stability, and high yield, such as Giza178, in the breeding programs for climate change resilience in rice.

Keywords: multilocation; climate change; heat stress; GGE biplot; AMMI analysis; yield stability; genotypes ranking; sustainability; biodiversity

1. Introduction

Rice (*Oryza sativa* L.) plays a vital role as a staple component in the diets of significant portions of the global population. As the world's population is projected to reach 10 billion by 2058 [1], the importance of rice production becomes even more critical. Approximately 755 million tons of paddy rice are produced annually from about 162 million hectares of land [2]. However, in order to ensure food security and combat poverty, there is a pressing need for a substantial increase, exceeding 60%, in high-quality rice production [3,4].

To meet this escalating demand, programs of rice breeding have implemented various strategies to enhance yield potential and develop high-yielding rice varieties. These strategies have often focused on utilizing specific germplasms tailored to a single environmental condition, such as submergence, upland, lowland, salinity, or drought. However, the ever-growing threat of climate change poses new challenges, emphasizing the importance of stability and adaptability in rice cultivars across different environmental conditions. Therefore, it is crucial to model genotype-by-environment interactions (GEI), quantify genotypic resilience, and assess the stability of the parental genotypes that are frequently used as donor varieties. Such an approach is necessary to increase the efficiency in genotype selection and determine the adaptability of the genotype under multienvironment trials (METs) [3,5–9]. Parental selection for crosses might consider high adaptation (genotype capability to positively adapt to surrounding stressors) and yield stability (genotype capacity to react in relation to the environment's yield prospective) across various environments (places, seasons, or both). Taking these considerations into account, parent selection is also critical for reproduction tasks looking for a greater area of protection, particularly in places with diverse environmental and soil conditions [10]. Wide adaptability and stability under different environments are the most important characteristics that should be imbedded as criteria for varietal selection in breeding programs. Rice genotypes with widespread adaptability could be recommended as elite parents in rice breeding programs to improve general adaptability in various climates, particularly with the unpredictability of climate change events and the potential occurrence of extreme stressors simultaneously or successively. Several statistical approaches have been created to enhance the precision of genotype x environment interactions and to aid understanding the stability as well as adaptability of tested genotypes [3,11].

Grain yield is considered the most reliable determinant of genotypic performance, mainly determined by the additive effects of genotype (G), environment (E), and the non-additive effect of GEI [12]. Evaluating different rice varieties across diverse rice production locations provides essential insights into their performance and resilience to the inherent environmental factors unique to each location. Furthermore, this approach allows for tailored genotype selection tuned to distinct natural environments [13]. Additionally, such multilocation preliminary yield trials facilitate the selection of promising varieties by demonstrating their yield potential and stability across diverse environments [14]. So long as the environment differs in terms of climatic conditions, the varieties that exhibit stability across these environments demonstrate greater resilience to the effects of climate change.

Several statistical methods have been developed to measure GEI, including regression coefficient [15], sum of squared deviations due to regression [16], and additive main effects and multiplicative interaction (AMMI) [17]. Another method, known as G + G × E (genotype + genotype × environment, GGE) polygon view, incorporates a graphical approach (GGE biplot), which characterizes the mega-environment, ranks varieties, and identifies the best genotypes in each environment (which–won–where) [18,19].

In recent a development, a novel R package named "metan" has been specifically designed to analyze multienvironment trials (METs) [20]. This package offers a workflow-based procedure that encompasses sequential functions for assessing commonly used parametric and nonparametric stability statistics [21]. The metan package provides a comprehensive set of tools for MET data management, manipulating, analyzing, and visualizing the MET data. It has been successfully employed to quantify yield stability in various crops, including rice [14], lentil [22], wheat [23], soybean [24,25], sugarcane [26], and others [27–31].

In the current investigation, the yield performance of different rice genotypes was evaluated at different rice growing environments. These genotypes are commonly employed as donors for different breeding purposes, such as salinity, water shortage, and disease resistance. These genotypes are tested under environments where heat stress, low humidity, and salinity stress are naturally inherited in these locations. This study aims to conduct a comprehensive multilocation evaluation of yield performance, identifying genotypes with both high yield potential and stability in diverse locations.

2. Results

2.1. GY Combined AMMI Analysis of Variance and Genotypic Variability

To cope with climate change, 34 genotypes were assessed at four locations during two seasons of study (2021 and 2022). The obtained results indicated highly significant differences among the genotypes across the different environments under study (Figure 1A,B and Table 1). The AMMI model analysis of variance for the 34 genotypes across the four locations throughout the two seasons exhibited a significant influence of all the factors analyzed on the genotypes' yield performance (Table 1). The main effects of environments accounted for a substantial portion, as much as 71.29%, of the total sum of the squared deviations for the genotypes' yield, while the proportion that accounted for G and GEI was 14.77 and 13.94, respectively. Accordingly, 29.32% of the grain yield variability could be attributed to identifying genotypes with narrow adaptability.

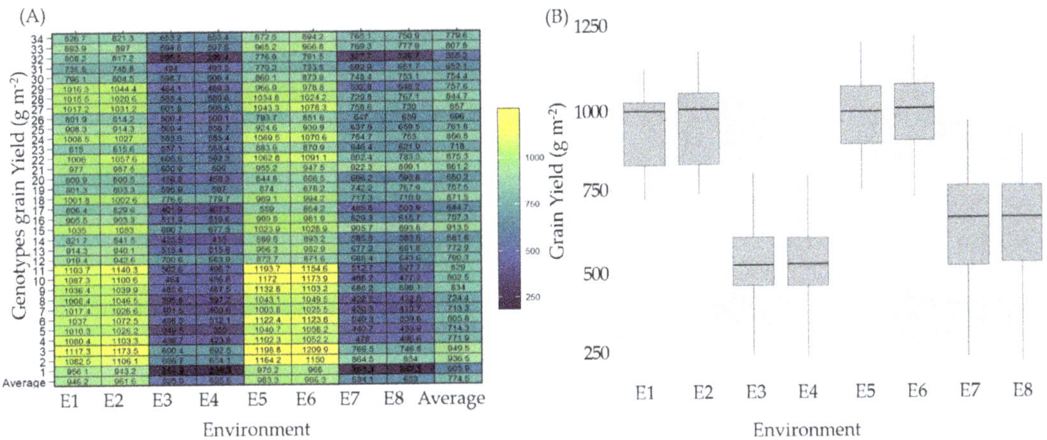

Figure 1. Genotypic performance across the different locations. (**A**) Genotypes' grain yield means across the 8 different environments. (**B**) Box plot of grain yield for the 8 different environments explaining the differences among the 4 locations.

The variability among the genotypes is supported by the differences between the mean yield values of the genotypes under study as they ranged from 238.5 g m^{-2} for Giza177 (G1) in 2022 at the Alexandria location to 1209 g m^{-2} for Giza179 (G3) in 2022 at the Gemmiza location (Figure 1A, Supplementary file: Table S3). Furthermore, the average yield production throughout the locations was also highly changeable. The average yield at Alexandria revealed the lowest values (525.9 g m^{-2} and 525.5 g m^{-2} for the two seasons, respectively), while the Gemmiza site recorded the maximum values (987.7 and 990.4 g m^{-2}, respectively) for the two seasons of study (Figure 1B).

2.2. GEI-Structure-Based Additive Main Effect and Multiplicative Interaction Model

The decomposition of the effect in the interaction-related multiplicative variance of the GEI was depicted by AMMI analysis and yielded four significant principal components of interaction (IPCAs, Table 1). These four IPCAs explained 99.5 percent of the total GEI effects, contributing 88.9%, 7.9%, 2.2%, and 0.5% for IPCA1, IPCA 2, IPCA3, and IPCA4, respectively. Meanwhile, IPCA5, IPCA6, and IPCA7 displayed nonsignificant impacts, and their ratio amounted to 0.5. IPCA1 and IPCA2 together gathered about 96.9%, explaining that both IPCAs are the best predictive model and sufficient for explaining the GEI. These results further indicate the extensive interaction of multivariate datasets with the PCA output, subsequently extracting and understanding the patterns regarding GEI that existed in the yield performance of the 34 genotypes under the different environments.

Table 1. AMMI analysis of variance model and GEI decomposition.

Source	Df	Sum Sq	Mean Sq	F Value	Pr (>F)	Proportion	Accumulated
ENV	7	32,277,639	4,611,091	5554.827	0		
REP(ENV)	16	13,281.69	830.1054	1.703935	0.042241		
GEN	33	6,685,134	202,579.8	415.83	0		
GEN:ENV	231	6,311,279	27,321.56	56.08221	8.8×10^{-275}		
PC1	39	5,612,970	143,922.3	295.43	0	88.9	88.9
PC2	37	499,545.4	13,501.23	27.71	0	7.9	96.9
PC3	35	141,380.5	4039.443	8.29	0	2.2	99.1
PC4	33	28,625.39	867.4361	1.78	0.0054	0.5	99.5
PC5	31	15,441.88	498.1253	1.02	0.439	0.2	99.8
PC6	29	10,612.79	365.9583	0.75	0.8262	0.2	100
PC7	27	2703	100.1111	0.21	1	0	100
Residuals	528	257,225.6	487.1697				
Total	1046	51,855,839	49,575.37				

ENV: environments; REP: replicates; GEN: genotypes; PC: principal components; Df: degree of freedom; Sq: squares; and Pr: probability.

2.3. *Detection of Stable and High-Yielding Genotypes via AMMIs Model Biplot*

To identify high-yielding and stable genotypes across the environments, the AMMI1 biplot was generated. The AMMI1 biplot plots the means of the genotypes and environments against their PCA1 (Figure 2A). The AMMI1 biplot revealed that those environments located at Alexandria (E3 and E4) and Kharga oasis (E7 and E8) produced the lowest yields across the genotypes. Similarly, Sakha103 (G32) showed the minimized grain yield average across the different environments presented in the current research (Figure 2A and Supplementary file: Table S3). Furthermore, the AMMI1 biplot exhibited that Giza179 (G3) had the uppermost yielding values, whereas Giza178 (G2) was considered the most widely adopted and highest-yielding genotype across the environments. Based on the genotypes' performance with PC1 estimates, a value close to zero means that the genotype has general adaptability under the different environments. Morobereccan (G17) and Nerica 7 (G16) were the most stable genotypes, owing the least PC1 values of 0.22 and 0.3, respectively (Figure 2A and Supplementary file: Table S4).

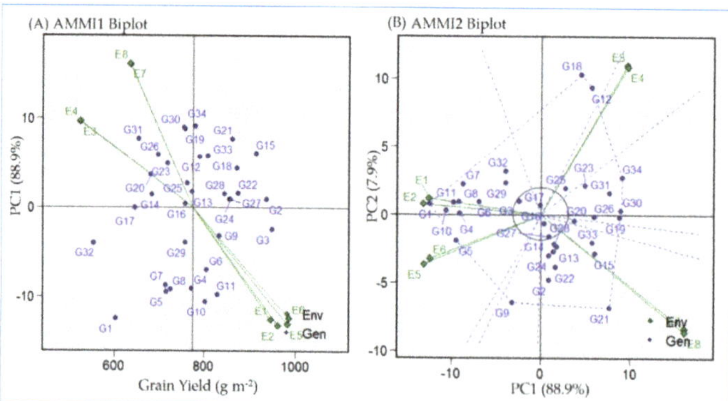

Figure 2. (**A**) The "AMMI1" biplot displays the main effect (GY) and IPC1 effect values explaining the relationship among tested genotypes and environments. (**B**) The "AMMI2" biplot displays the main

axes of G+GEI effect (IPCA1 and IPCA2) values for the tested genotypes and environments. The tested genotypes are 34 (G1:G34 in blue color) grown in four locations in the two consecutive years, 2021 and 2022 (E1 and E2 = Sakha; E3 and E4 = Alexandria; E5 and E6 = Gemmiza; E7 and E8 = Kharga oasis).

The AMMI2 biplot pinpointed the genotype stability performance regarding yield based on IPCA1 and IPCA2 values (Figure 2B). The genotypes and environments with lower IPCA1 and IPCA2 values that are plotted in close proximity to the plot origin are considered the most stable ones. This explains the reduced interaction ability of genotypes and their adaptability regarding the different environments under study. Based on the AMMI2 biplot, Nerica 7 (G16), Morobereccan (G17), IR11L236 (G27), and Giza179 (G3) are demonstrated to be considerable genotypes. Both the AMMI1 and AMMI2 biplots indicated the instability of Giza177 (G1) and Vandana (G30) for the different environments in the present research.

2.4. Which–Won–Where Approach Based on GGE Biplot for Detecting the Best-Performing Genotypes

The GGE biplot polygon view based on the which–won–where structure of a MET approach is the simplest and most efficient method for detecting the genotype and its environmental interaction (Figure 3A). It is used for interpreting GEI and detecting superior genotypes across different environments. In the current investigation, the biplot polygon view showed the tested environments in two different sectors. The environments of the locations Gemmiza (E5 and E6) and Sakha (E1 and E2) were located in the same sector, while the other two locations (Alexandria (E3 and E4) and Kharga oasis (E7 and E8)) were separated in another sector. The genotypes at the corner of each section of those environments had the highest yield for the corresponding environments. The genotypes joined by the polygon are the farthest from the origin and called vertex genotypes. Those vertex genotypes located in the same sector or close to a specific environment are considered the best genotypes for this environment. Accordingly, Giza179 (G3) is the best-performing genotype in Sakha and Gemmiza, followed by Sakha Super 300 (G11). Likewise, N22 (G21) followed by IET1444 (G15) are considered the most suitable genotypes for environments E7 and E8 of Kharga oasis. IET1444 (G15) and Giza178 (G2) had stable performance for the environments located in their sectors. Furthermore, Giza178 (G2) exists in the middle of all the environments; consequently, it is stable across all the study environments. In contrast, Giza177 (G1) is situated on the opposite side of E3, E4, E7, and E8, which reflects the minimum appropriateness for these environments. Similarly, Dular (G31) exists on the opposite side of E1, E2, E5, and E6, reflecting the same situation.

2.5. Grain Yield Versus Weighted Average of Absolute Score Stability Index Biplot

Various stability statistics were estimated and presented in the Supplementary Materials (Table S5). Based on GY, which considers the main breeder selection criteria, genotypes Giza179, Giza178, IET1444, IR65600-77, GZ1968-S-5-4, and N22 achieved the highest rank in this regard. The weighted average of absolute scores (WAASB) was also used to better identify the most adapted genotypes across the different environments based on the mean GY and stability. The GY× WAAS biplot displayed the distribution of the tested rice genotypes and environments based on the genotypes' GY mean and WAASB values, as presented in Figure 3B.

The first quadrant I, contains those genotypes that are low yielding and unstable across all tested environments. Among these genotypes Giza177 (G1), G4, G5, G7, G8, G19, G30, and G31. These genotypes are less desirable as they showed lower grain yield and inconsistent performance compared to the mean of overall grain yield. In addition, this quadrant has low GY environments E3, E4, E7 and E8. Accordingly, the genotypes and environments located in this quadrant have the largest response to GEI. The second section (quadrant II), contains the environments E1, E2, E5 and E6, coupled with the genotypes that have a GY above average with a high GEI response, such as G6, G10, G11, G34, and G21. These genotypes are less reliable as they have high yield under normal conditions, but didn't display consistently under harsh environments. GY × WAAS biplot also presented genotypes that showed relative stable performance across the evaluated environments. These genotypes were existed in the quadrants III and IV. G13, G14, G16, G17, G20, G23, G26, G29 and G32 presented minimum yield but stable performance across the tested environments (Figure 3B). Those genotypes are more suitable for harsh environments. At the same time, G2, G3, G9, G12, G15, G18, G22, G24, G27, G28 and G33 located in quadrant IV. This quadrant contains the genotypes showing high yield performance without being influenced significantly by specific environmental conditions. These genotypes are desirable for broad adaption ability.

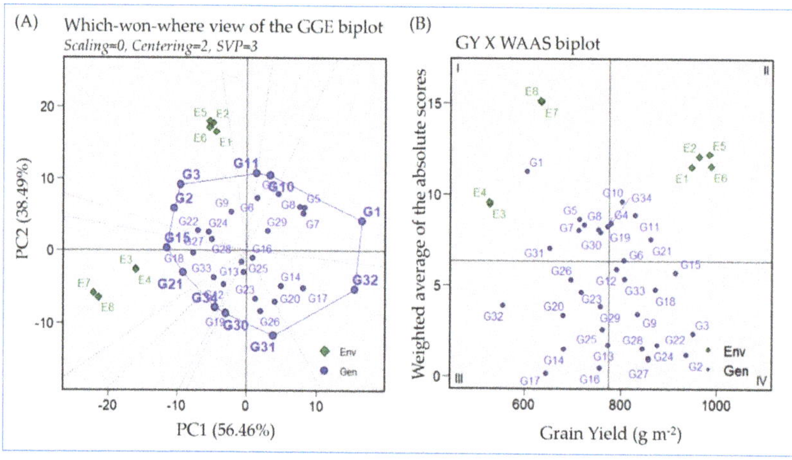

Figure 3. (**A**) The "which–won–where" polygon biplot displays the winning genotypes at each environment. (**B**) The "GY vs. WAAS" biplot displays the most-adopted genotypes across the tested environments. The tested genotypes are 34 (G1:G34 in blue color) grown in four locations in the two consecutive years, 2021 and 2022 (E1 and E2 = Sakha; E3 and E4 = Alexandria; E5, and E6 = Gemmiza; E7 and E8 = Kharga oasis).

2.6. GGE Biplot—Means Versus Stability Model and Ranking of Rice Genotypes' Performance

The genotypes' mean performance versus stability biplot provides a visual tool for discriminating the tested genotypes (Figure 4A,B). This biplot presents the two PCs (1 and 2), which, for their additive percentage, explain the G + GE effects, respectively. The single-arrowed line in the biplot (Figure 4A) is the average environmental average (AEA) and points towards higher mean performance across the tested genotypes. Regarding the AEA, the average environment corresponding to the average values of the two PCs is pinpointed by the arrowhead in Figure 4A and further circled in Figure 4B. The genotypes located in the circle are considered to be the best genotypes. The perpendicular line to the AEA is called the average ordinate environment (AOE), and the intersection is the point that represents both the average mean performance and high stability. Other perpendicular lines linking the genotypes to the AEA explain the stability of the genotype. The closeness of genotypes to the AEA explains their stability across environments. However, using the ranking of the biplot, the ideal genotype is Giza178 (G2), being in the center of the circle. Giza178 had a high yield and the best adaptability among the other genotypes under consideration. Giza179 (G3) also exhibited high yielding ability. Giza 177 (G1) manifested a decrease in yielding ability with decreased stability compared to the other genotypes under study. The ranking of the genotypes from the worst to the best could be followed using the AEA direction. The yield average reduced in Sakha103 (G32) to the lowest values across the environments, followed by Giza177 (G1), Morberekan (G17), Dular (G31), IRAT112 (G20), and Nerica 9 (G14), following the AEA direction until Giza179 (G3). The uppermost-ranked genotypes were assigned for Giza179 (G3) > Giza178 (G2) > IET1444 (G15) > IR65600-77 (G22) > GZ1968-S-5-4 (G18) > N22 (G21) > IR11L236 (G27) > IR12G3213 (G24).

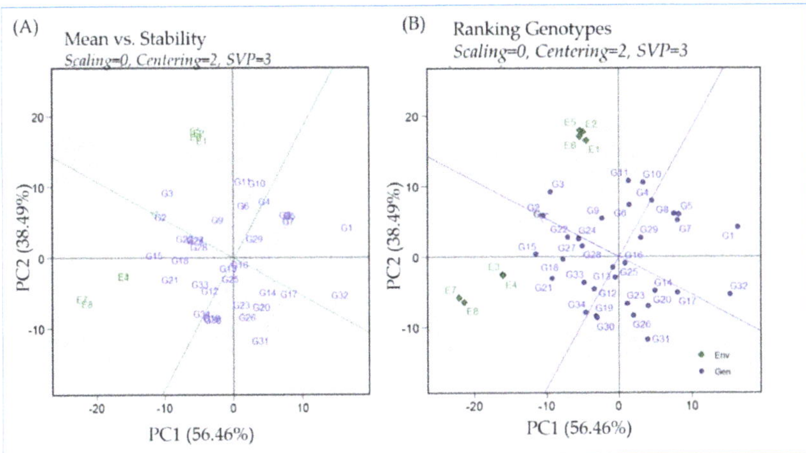

Figure 4. (**A**) The "mean versus stability" model describing the interaction effect of the tested rice genotypes evaluated across eight environments. (**B**) The "ranking genotypes" model of biplot to assess the ideal genotype. The tested genotypes are 34 (G1:G34 in blue color) grown in four locations in the two consecutive years, 2021 and 2022 (E1 and E2 = Sakha; E3 and E4 = Alexandria; E5 and E6 = Gemmiza; E7 and E8 = Kharga oasis).

2.7. Cluster-Analysis-Based Stability Indices

Several stability indices were measured based on the genotypes' yield records in the environments under the present constructed study. Stability indices such as environmental variance, mean variance component, GE variance component, joint regression analysis, Tai's stability statistics, coefficient of variance, superiority index, and AMMI-based stability statistics are normally used to rank the genotypes based on their estimated values. These indices are Shukla_R, Wi_g_R, Wi_f_R, Wi_u_R, Ecoval_R, Sij_R, Pi_a_R, Pi_f_R, Pi_u_R, Gai_R, S1_R, S2_R, S3_R, S6_R, N1_R; Supplementary Table S5. The indices' values corresponding to the tested genotypes were manipulated based on the squared Euclidean distance to conduct hierarchical cluster analysis via Ward's method (Figure 5). This method was used to group the genotypes with the same stability in the same cluster. Based on this fact, the tested rice genotypes were grouped in two main clusters (CL), resembling the main grouping pattern. Each CL was divided into two subclusters (SCL). The first SCL colored with red contains genotypes that had similar stability measures. Those genotypes (Giza179, Giza178, IR65600-77, IR11L236, IR12G3213, and IR12G3222) have the highest overall mean performance across the different environments. These genotypes are among the top-ranked ones based on mean vs. stability ranking. Furthermore, IET1444 was clustered with N22, and they are the top-performing genotypes under Kharga oasis environments. GZ1968-S-5-4 is the top-performing genotype under Alexandria conditions, whereas IET1444 is considered among the top-four-performing genotypes. For the Gemmiza location, Sakha104 and Sakha 107 are among the top-performing genotypes, whereas Sakha104 and Sakha109 are categorized at the same rank in the overall mean performance of the genotypes as compared to other genotypes. These results confirm the goodness of the stability indices in ranking the genotypes and their utilization in clustering the genotypes into different groups.

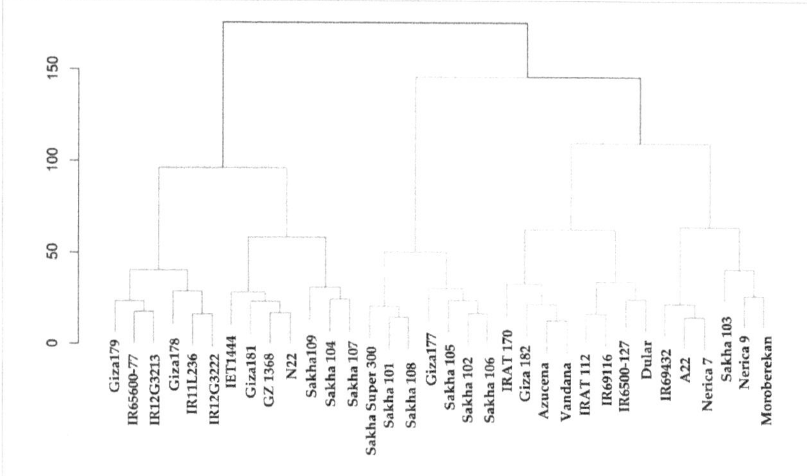

Figure 5. Hierarchical dendrogram classifying the 34 rice genotypes based on their ranks for GY and stability statistics conducted via Ward's method.

The Sakha Super 300, Sakha101, and Sakha 108 genotypes showed top performance in the Sakha and Gemmiza locations. However, their rank deteriorated in the unfavorable environments of Alexandria and Kharga oasis as compared to the other tested genotypes. Giza177, Sakha105, Sakha102, and Saka 106 were grouped together in the middle subcluster, indicating their average performance under favorable conditions and low performance under unfavorable conditions. At the same time, IRAT17, Giza182, Azucena, and Vandana were grouped together and provided a slightly more stable performance in unfavorable conditions and an unstable performance under favorable conditions when compared to the other genotypes. Moreover, IRAT112, IR69116, IR6500-127, and Dular in the same subcluster had average performance under harsh environments and low performance under favored environments.

The Sakha 103, Nerica 9, and Moroberekan genotypes grouped together in the purple subcluster and displayed a reduction in their performance with regard to the overall performance under the different environments. In the last subcluster, IR69432, A22, and Nerica7 located together due to their average ranking across the different environments.

2.8. Correlation Coefficient Analysis

The correlations among the studied yield-related characteristics and yield were analyzed under the different environments (Figure 6). All the evaluated characteristics have a significant correlation with yield. A significant and positive correlation was detected between the overall yield under all the environments with panicle length (r= 0.825), number of panicles per plant (r = 0.904), and thousand grain weight (r = 0.491). In contrast, the total yield performance negatively correlated with the genotypes' sterility percentage under all the environments. Interestingly, panicle length showed a highly significant and positive correlation with yield under harsh conditions in contrast to the negative ones under favored environments. Furthermore, sterility was negatively correlated with all the measured characteristics. In the density plots presented in the diagonal, it is noted that E1, E2, E6, E8, and E2 have the highest peaks compared to the others for yield, panicle length, number of panicles per plant, thousand grain weight, and sterility percentages, respectively.

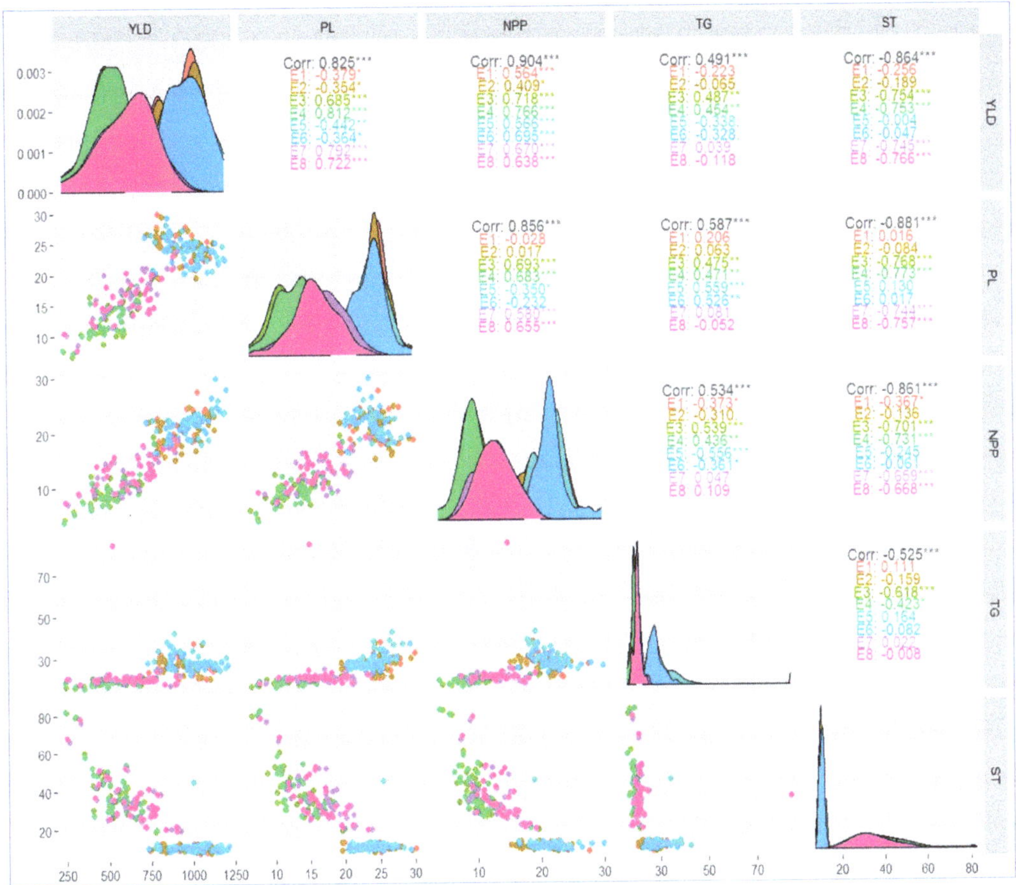

Figure 6. Spearman correlation coefficients for the yield, and other studied yield-related traits. Scatterplots of each trait's pair of numeric variables are situated in the left part of the figure. Variable distribution is drawn on the diagonal. YLD: yield; PL: panicle length (cm); NPP: number of panicles per plant; TG: thousand grain weight (g); and ST: sterility (%). These traits were recorded for the tested genotypes grown in the four locations for the two consecutive years, 2021 and 2022 (E1 and E2 = Sakha; E3 and E4 = Alexandria; E5 and E6 = Gemmiza; E7 and E8 = Kharga oasis). * $p \leq 0.05$, ** $p \leq 0.01$, *** $p \leq 0.001$.

3. Discussion

Breeding for yield stability across different environments is becoming the most popular program with the current waves of climatic changes. Annually, there are major fluctuations in climate records, especially with regard to water scarcity, temperature, and humidity. Those stressors are considered among the main ones affecting rice production [10]. Identifying the most suitable genotypes for a specific climatic location involves an important step of starting a breeding program for such an environment. AMMI analysis of variance of 34 rice genotypes over eight environments for grain yield is provided in Table 1. Significant differences were detected among E, G, and GEI. This means that the genotypes showed different behavior in their environments. This enables the breeder to justify the selection of genotypes based on the magnitude of interaction with the environment [32,33]. Environments accounted for the highest sum of squares, 70.68%, explaining the highest differences among them. This indicates that different sets of genotypes appeared to be high-yielding in different environments. Accordingly, selection of the genotypes based on the environment is effective while considering genotype performance and GEI. The high effect of environments was mainly due to the unique environmental features for each location. These variability among the environmental conditions may have activated some yield-enhancing genes in different genotypes, which developed a significant GEI and resulted in high yield variability (Figure 1A,B). While investigating the genotypic yield stability across environments, the research reported by Barona et al. 2021, Dewi et al. 2014, Tariku et al. 2013, and Zewdu et al. 2020 [34–37] also concluded that the environmental effect recorded the highest impact. Meanwhile, other investigations reported genotypes' effects to be the most effective impacts. In these investigations, the differences between environments were mainly due to one effect, such as identifying the most stable rice genotype across different fertilizer levels [14].

Apparently, our investigation confirms that high variability exists between the tested environments. Two locations had favored environments, Sakha and Gemmiza (Supplementary Table S6), where the environmental conditions and soil properties are suitable for good rice growth. Contrarily, Kharga oasis and Sobahya had unfavored environments. Kharga oasis has high temperature records during the growing seasons together with low humidity, while the Alexandria site suffered from poor soil conditions. These conditions were reflected in the genotypes' yield performance across environments (Figure 1A,B). Consequently, the rankings of the tested genotypes, based on yield performance, were subject to changes in specific locations. Giza179, Sakha Super 300, Sakha 108, and Giza178 developed the highest grain yield under the favored conditions of Sakha and Gemmiza. Meanwhile, GZ1968-S-5-4, IRAT170, IET1444, and Giza179 were the best at the Alexandria location despite affecting the soil with salinity. Moreover, at the unfavored environment of Kharga oasis, with heat stress accompanied by low humidity conditions, genotypes N22, IET144, Giza178, and IR65600-77 recorded the best values and the highest ranking. These findings demonstrate the influence of GEI on the performance of the tested genotypes. GEI reduced the performance of the evaluated genotypes by affecting their yield under unfavored conditions. This output is supported by the findings of others [38,39].

To quantify the magnitude of the GEI in the current investigation, the GEI principal components (PCs) were assessed. GEI was partitioned into seven PCs (Table 1). The first two significant IPCs (PC 1 and PC 2) accounted for approximately 96.9% of the interaction between the E and G. This indicates that these two PCAs were sufficient to present the GEI complex patterns. Consequently, all the interaction information could be speculated by plotting these two PCs. Our findings provide proofing for the results of several earlier investigations obtained by others that support the idea that the first two IPCs of the AMMI model are the most important during quantifying GEI [14,34–37].

The AMMI model provides an effective analytical procedure to understand the GEI through a graphical biplot tool to clearly identify the ideal genotypes [21,40,41]. Among these are AMMI1 and two biplots. AMMI1 revealed that Giza 178 is the best genotype with regard to stability and mean performance across environments (Figure 2A). Genotypes with small interaction with the environment have IPCA1 values close to zero [5,21]. Consequently, Giza178 showed wider adaption to the environments (normal, heat stress, and salinity stress) under the current research, and high yielding performance. Contrarily, Giza177 showed small adaptability to the tested environments. This is mainly due to lacking specific genes that could provide the genotype resistance to the harsh environments of Kharga oasis and Sobahya. The AMMI2 biplot summarizes information based on the first two PCs (PC1 and PC2 in GEI). Genotypes with low IPCA1 and IPCA2 values have low interaction with the environments [38–40,42–44]. These genotypes are centered near to the origin of the AMMI2 biplot. Among these genotypes, Giza179 had high yielding performance and had general adaptability to the conditions under study. Giza179 is an Egyptian rice cultivar bred for climate change resistance [9,45].

Evaluating the genotypes across multilocation trials provide the fundamentals regarding variation. The GGE biplot clarifies the best-performing genotypes across all the environments [42,46]. The polygon view of the GGE biplot (which–won–where) illustrated the winning genotypes for each environment under study (Figure 3A). This biplot view considers the most efficient and simplest method for characterizing the genotypes and their interaction with the environment [5,14]. In this biplot, the genotypes with the longest distance from the biplot origin in the same direction or close to one or more from the environment are considered the best genotypes for those environments. Giza179 is the superior-performing genotype in the Sakha and Gemmiza environments, followed by Sakha Super 300. These environments are considered ideal for rice crop growth in Egypt. Meanwhile, in a harsh environment where heat stress is excited in Kharga oasis, N22 and IET1444 were preferable. This finding is supported by N22 and IET1444 harboring specific genes for heat and water shortage stress tolerance [43,44,47–49]. Furthermore, IET1444 and Giza178 were located in the middle between the eight studied environments regarding their stable performance [5], while Giza177 was inferior for the environments located in Kharga oasis and Sobahya. This was mainly due to the harsh environments coupled with heat and salinity stress. Giza177 is a popular Egyptian variety with high grain quality characteristics but is sensitive to heat and salinity stress [47,50].

We further calculated the stability index of WAAS to consider the sum of the absolute values of the genotypes' IPCAs [39,43]. Then, a GY × WAAS biplot was generated to better clarify the superior-performing genotypes. In terms of WAAS values, Moroberekan represents the minimal and desirable WAAS value. Moroberekan is a drought-blast-resistant genotype but with poor yield potential under favored conditions [47,51]. These features provided the genotype with stability performance across the different environments tested in the current study. However, as the biplot considers both GY and WAAS values for the genotypes, Giza178 and Giza179 are the superior ones (Figure 3B). These genotypes had high yield and minimum estimates for WAAS stability index. The GY × WAAS biplot further divided the tested genotypes into four different categories (i.e., high-yielding stable, high-yielding not stable, low-yielding stable, and low-yielding unstable genotypes). This classification provides a clear vision for rice breeders to predict the genotypic performance of each genotype under study prior to integrating them in the different programs.

It is essential to assess the average stability together with the genotype's performance through the means versus stability model of a GGE biplot (Figure 4A,B). This model is capable of detecting high-ranking genotypes with great stability performance based on AEC decisions [23,52]. Giza179 is considered the highest, whereas the ideal genotype was Giza178 via being high-yielding and more stable than Giza179. In contrast, Giza177 had low stability with sensitivity to harsh environments under study, which decreased its overall yield performance. Based on AEC ranking, the lowest-ranked genotypes are Sakha103, Giza177, Moroberekan, Dular, IRAT112, and Nerica 9, while the top-ranked genotypes are Giza179 (G3) > Giza178 (G2) > IET1444 (G15) > IR65600-77 (G22) > GZ1968-S-5-4 (G18) > N22 (G21) > IR11L236 (G27) > IR12G3213 (G24). Among them, Giza178, IR65600-77, and IR12G3213 were the most stable ones. These genotypes could be utilized in breeding programs while having high yielding ability under unfavored environments and also maintaining good performance under normal conditions.

To further identify the groups of the genotypes based on the different stability indices, cluster analysis was conducted. It is worth mentioning that, based on this analysis, the genotypes with the highest overall mean performance across the different environments were clustered in a single SCL. These genotypes are Giza179, Giza178, IR65600-77, IR11L3213, IR12G3213, and IR12G3222, which are the top-ranked ones based on mean vs. stability ranking (Figure 5).

Correlation coefficients indicated that number of panicles per plant exhibited the highest positive correlation with grain yield under all the environments. However, the overall correlation coefficient between panicle length and grain yield records was significantly positive, and the coefficient was positive under unfavored environments and negative under favored environments (Figure 6). This indicated the importance of panicle length as a characteristic for increasing yield under harsh environments. This positive relationship suggests that it is possible to improve yield by breeding cultivars with higher panicle length. These findings are supported by those obtained by Laza et al. [49,52], who highlighted the importance of panicle size as a promising characteristic for enhancing cultivars' yield potential.

Considering our results, Giza179 and Giza178 exhibited high performance. Giza178 was the ideal genotype and had the widest adaptability. GZ1968-S-5-4 displayed the best genotype performance under Alexandria environments, where salinity stress affected the soil. At the same time, N22 presented the best values in Kharga oasis since high temperature coupled with low humidity are the common environmental conditions at this location.

4. Materials and Methods

4.1. Plant Materials

In the present research, a total of 34 rice genotypes were carefully selected. These genotypes could be classified into two sets. The first set consisted of elite Egyptian rice cultivars, specifically selected for their adaptability within the Egyptian rice production systems. These cultivars have been extensively cultivated and grown throughout Egypt, making them representative of the local rice landscape. The second set comprised international donor genotypes, known for their tolerance to salinity and water shortage stresses (Table 2). These donor genotypes were incorporated into the breeding program to enhance the resilience of the Egyptian elite cultivars against these specific constraints.

Table 2. List of genotypes used in the study.

No.	Genotype	Origin	Type
1	Giza 177	Egypt	Japonica
2	Giza 178	Egypt	Indica/Japonica
3	Giza 179	Egypt	Indica/Japonica
4	Sakha 101	Egypt	Japonica
5	Sakha 102	Egypt	Japonica
6	Sakha 104	Egypt	Japonica
7	Sakha 105	Egypt	Japonica
8	Sakha 106	Egypt	Japonica
9	Sakha 107	Egypt	Japonica
10	Sakha 108	Egypt	Japonica
11	Sakha Super 300	Egypt	Japonica
12	IRAT 170	Ivory Cost	Tropical japonica
13	A22	Srilanka	Indica
14	Nerica 9	Ivory cost	Indica
15	IET 1444	Inbdia	Indica
16	Nerica 7	Ivory Cost	Indica
17	Moroberekan	Guinea	Japonica
18	GZ 1368-S-5-4	Egypt	Indica
19	Azucena	Philippine	Japonica
20	IRAT 112	Ivory Cost	Indica
21	N22	India	Aus
22	IR65600-77	Philippines	Indica
23	IR69116	Philippines	Indica
24	IR12G3213	Philippines	Indica
25	IR69432	Philippines	Indica
26	IR6500-127	Philippines	Indica
27	IR11L236	Philippines	Indica
28	IR12G3222	Philippines	Indica
29	Sakha 109	Egypt	Japonica
30	Vandana	India	Indica
31	Dular	India	Indica
32	Sakha 103	Egypt	Japonica
33	Giza 181	Egypt	Indica
34	Giza 182	Egypt	Indica

4.2. Experimental Locations and Climatic Conditions

The research trials were conducted across four distinct locations in Egypt, as presented in Table 3.

Table 3. List of four locations used in the present study.

No.	Location (Governorate)	Altitude–Latitude
1	Sakha (Kafr El-Sheikh)	31.09° N and 30.9° E
2	Gemmiza (ElGharbya)	30.88° N and 31.05° E
3	Sobahya (Alexandria)	31.2° N and 29.9° E
4	Kharga oasis (New Valley)	25.4° N and 30.5° E

The trials were carried out over two consecutive growing seasons, 2021 and 2022. The temperature data for the four locations over the two years were acquired from https://weather.com/ (accessed on 3 December 2022), while the humidity data were extracted from the NASA POWER project for agricultural needs (https://power.larc.nasa.gov/ (accessed on 1 June 2023)).

Each of the four locations exhibits unique climatic conditions and soil properties. Kharga location, for instance, experiences high temperatures accompanied by very low humidity during the two growing seasons (Figures 7 and 8, Supplementary Table S1). The climatic conditions varied between the two seasons (Figures 1 and 2). Alexandria (Sobahya) location recorded the lowest maximum temperature and the highest humidity among the four locations, whereas the minimum temperature was exhibited at Gemmiza location during the two cropping seasons.

The physical and chemical analyses of the 8 environments' soils are presented in Table S2. The soils of the Gemmiza and Sakha locations are clay for the two seasons, while in Alexandria location was sandy clay loam. In contrast, Kharga oasis included loam sandy and clay sandy for 2021 and 2022 seasons, respectively. The Alexandria location soil analysis indicated the existence of high salt stress since Na^+ was 59.10 and 58.94, whereas Cl^- amounted to 74.5 and 70.78 for the two seasons, 2021 and 2022, respectively.

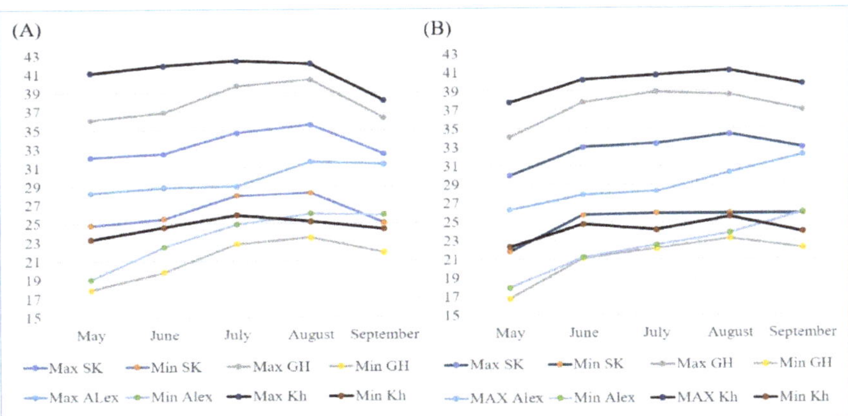

Figure 7. The monthly average maximum and minimum temperature (°C) at four locations during 2021 (**A**) and 2022 (**B**) rice seasons.

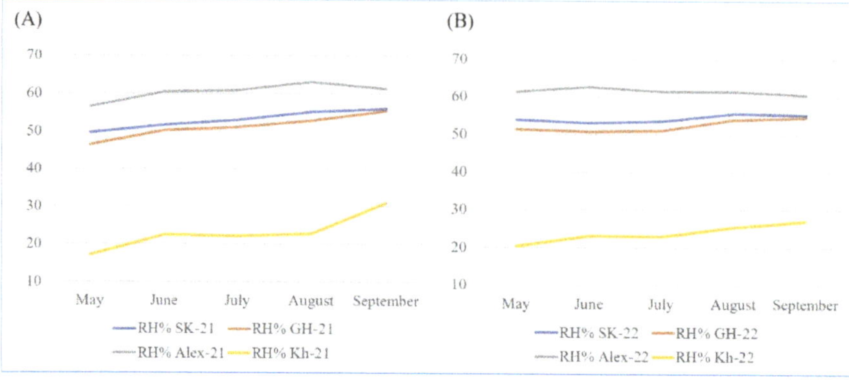

Figure 8. The monthly average of relative humidity (%) at the four locations during 2021 (**A**) and 2022 (**B**) rice seasons.

4.3. Experimental Design and Data Recording

Randomized complete block design (RCBD) with three replications was adopted at each of the four different locations during the two successive seasons. The experiment consisted of total eight environments, representing the combination of the four locations and the two years. Each genotype's seeds were planted in seedbed in May and transplanted after 25 days. The experimental plots were structured with five rows 2 m long with planting space of 20 cm × 20 cm. This layout was standardized for each genotype within every environment. To maintain the uniformity of all factors, except for the varying climatic conditions specific to each environment, standard agronomical practices such as field preparation, fertilizer application, and weed control were implemented. At the harvest stage, the plots were harvested at harvest stage and grain yield (g m^{-1}) was recorded for each experimental plot after adjusting the moisture content. Furthermore, the yield-related characteristics, such as number of panicles per plant, panicle length (cm), thousand grain weight (g), and sterility (%), were studied.

4.4. Statistical Analysis

The statical packages available in R (R Core Team [53]) version 4.2.3. were used to conduct multivariate procedures for stability analysis according to AMMI and GGE. The "metan" package [20] was applied for AMMI analysis, whereas AMMI Model merged ANOVA and PCA techniques. The package was also employed for GGE biplot analysis [54], stability statistical analysis, and weighted average of absolute scores (WAAS) [43], while GGE-Biplot-GUI package was functioned to support the GGE-biplot-based analysis. Additionally, hierarchical cluster analysis was conducted via "Nbclust" package [55]. The AMMI analysis followed the below mathematical equation: $Y_{ge} = \mu + \alpha_g + \beta_e + \Sigma_n \lambda_n \gamma_{gn} \delta_{en} + \rho_{ge}$, where Y_{ge} represents the grain yield for a particular genotype (g) in a given environment (e), μ is the genotype grand mean, α_g is the deviation of genotype performance from the mean, β_e stands for the deviation of environment from the mean, λ_n signifies the singular value of n component, γ_{gn} indicates the value of eigenvector for genotype (g), and δ_{en} is the value of eigenvector for e and $\rho_{ge,}$ which is the remaining residual [55], while the GGE biplot model mathematical equation was $P_{ij} = (y_{ij} - \mu - \delta_j)/\lambda_j = (\beta_i + \epsilon_{ij})/\lambda_j$, where P_{ij} is the matrix for genotype i and environment j, μ is the genotype grand mean, δ_j is the column (environment) main effect, λ_j is an evaluating factor, β_i is the row (genotype) main effect, and ϵ_{ij} stands for GEI, and y_{ij} is a two-way table for G and E [56]. The Spearman correlation coefficient was calculated among the grain yield, the studied yield-related characteristics using the GGaly R software package.

5. Conclusions

Different sets of genotypes showed high yield performance under different environments. While the environments had the highest portion of yield sum of squared deviations, the results showed that the selection should be conducted based on genotypes' performance with respect to GEI. Furthermore, the results of AMMI, GGE biplot, and which–won–where showed that Giza178 had less response to the differences among the tested environments. This indicates that, compared to the tested genotypes, it has wide adaptability and capability as a climate-change-resilient genotype.

Supplementary Materials: The following supporting information can be downloaded at https://www.mdpi.com/article/10.3390/plants13010074/s1, Table S1: The monthly maximum and minimum temperature (°C) as well as relative humidity (%) at four locations during 2021 and 2022 rice seasons; Table S2: Physical and chemical properties of experimental field soils at four locations in the rice growing seasons; Table S3: Genotypes mean performance under the eight different environments; Table S4: Genotypes and environments overall mean performance and their PC1; Table S5: Genotypes rankings based on yield performance and several stability indices; Table S6: Favorable and unfavorable classification of the environments.

Author Contributions: M.A. and M.I.G. conceptualized the study. M.I.G., M.A., R.Y.E.-A., T.M.E.-h., S.A.E.-N. and E.M.D. conducted field experiments. M.I.G., R.Y.E.-A., T.M.E.-h., S.A.E.-N. and E.M.D. worked on the availability of weather records and soil analysis. M.A. and M.I.G. conducted the statistical analysis and wrote the first manuscript draft. M.R. revised the final version. All authors have read and agreed to the published version of the manuscript.

Funding: This research was supported by RRTC, Field Crops Research Institute, ARC, Egypt. The researchers would like to thank the Deanship of Scientific Research, Qassim University for funding publication of this project.

Data Availability Statement: The data supporting the findings of the current study are available within the paper and within its Supplementary Materials published online.

Acknowledgments: The researchers would like to thank the Deanship of Scientific Research, Qassim University for funding publication of this project.

Conflicts of Interest: The authors declare no conflicts of interest.

References

1. Chamie, J. Population Growth Diversity Continuing in the Twenty-first Century. In *Population Levels, Trends, and Differentials: More Important Population Matters*; Springer: Berlin/Heidelberg, Germany, 2023; pp. 179–182.
2. Shaheen, S.M.; Antoniadis, V.; Shahid, M.; Yang, Y.; Abdelrahman, H.; Zhang, T.; Hassan, N.E.; Bibi, I.; Niazi, N.K.; Younis, S.A. Sustainable applications of rice feedstock in agro-environmental and construction sectors: A global perspective. *Renew. Sustain. Energy Rev.* **2022**, *153*, 111791. [CrossRef]
3. Lee, S.Y.; Lee, H.-S.; Lee, C.-M.; Ha, S.-K.; Park, H.-M.; Lee, S.-M.; Kwon, Y.; Jeung, J.-U.; Mo, Y. Multi-Environment Trials and Stability Analysis for Yield-Related Traits of Commercial Rice Cultivars. *Agriculture* **2023**, *13*, 256. [CrossRef]
4. Abdelrahman, M.; Zhao, K. Genome editing and rice grain quality. In *The Future of Rice Demand: Quality Beyond Productivity*; Springer: Cham, Switzerland, 2020; pp. 395–422.
5. Chairi, F.; Aparicio, N.; Serret, M.D.; Araus, J.L. Breeding effects on the genotype× environment interaction for yield of durum wheat grown after the Green Revolution: The case of Spain. *Crop J.* **2020**, *8*, 623–634. [CrossRef]
6. Khan, M.M.H.; Rafii, M.Y.; Ramlee, S.I.; Jusoh, M.; Al Mamun, M. AMMI and GGE biplot analysis for yield performance and stability assessment of selected Bambara groundnut (*Vigna subterranea* L. Verdc.) genotypes under the multi-environmental trials (METs). *Sci. Rep.* **2021**, *11*, 22791. [CrossRef] [PubMed]
7. Omrani, A.; Omrani, S.; Khodarahmi, M.; Shojaei, S.H.; Illés, Á.; Bojtor, C.; Mousavi, S.M.N.; Nagy, J. Evaluation of grain yield stability in some selected wheat genotypes using AMMI and GGE biplot methods. *Agronomy* **2022**, *12*, 1130. [CrossRef]
8. Yue, H.; Gauch, H.G.; Wei, J.; Xie, J.; Chen, S.; Peng, H.; Bu, J.; Jiang, X. Genotype by environment interaction analysis for grain yield and yield components of summer maize hybrids across the huanghuaihai region in China. *Agriculture* **2022**, *12*, 602. [CrossRef]
9. Mehana, M.; Abdelrahman, M.; Emadeldin, Y.; Rohila, J.S.; Karthikeyan, R. Impact of genetic improvements of rice on its water use and effects of climate variability in Egypt. *Agriculture* **2021**, *11*, 865. [CrossRef]
10. Anputhas, M.; Samita, S.; Abeysiriwardena, D. Stability and adaptability analysis of rice cultivars using environment-centered yield in two-way ANOVA model. *Commun. Biometry Crop Sci.* **2011**, *6*, 80–86.

11. Sandhu, N.; Yadaw, R.B.; Chaudhary, B.; Prasai, H.; Iftekharuddaula, K.; Venkateshwarlu, C.; Annamalai, A.; Xangsayasane, P.; Battan, K.R.; Ram, M. Evaluating the performance of rice genotypes for improving yield and adaptability under direct seeded aerobic cultivation conditions. *Front. Plant Sci.* **2019**, *10*, 159. [CrossRef]
12. Bocianowski, J.; Tratwal, A.; Nowosad, K. Genotype by environment interaction for main winter triticale varieties characteristics at two levels of technology using additive main effects and multiplicative interaction model. *Euphytica* **2021**, *217*, 26. [CrossRef]
13. Banterng, P.; Patanothai, A.; Pannangpetch, K.; Jogloy, S.; Hoogenboom, G. Yield stability evaluation of peanut lines: A comparison of an experimental versus a simulation approach. *Field Crops Res.* **2006**, *96*, 168–175. [CrossRef]
14. Abdelrahman, M.; Alharbi, K.; El-Denary, M.E.; Abd El-Megeed, T.; Naeem, E.-S.; Monir, S.; Al-Shaye, N.A.; Ammar, M.H.; Attia, K.; Dora, S.A. Detection of Superior Rice Genotypes and Yield Stability under Different Nitrogen Levels Using AMMI Model and Stability Statistics. *Plants* **2022**, *11*, 2775. [CrossRef] [PubMed]
15. Finlay, K.; Wilkinson, G. The analysis of adaptation in a plant-breeding programme. *Aust. J. Agric. Res.* **1963**, *14*, 742–754. [CrossRef]
16. Eberhart, S.t.; Russell, W. Stability parameters for comparing varieties 1. *Crop Sci.* **1966**, *6*, 36–40. [CrossRef]
17. Gauch, H.G.; Zobel, R.W. Predictive and postdictive success of statistical analyses of yield trials. *Theor. Appl. Genet.* **1988**, *76*, 1–10. [CrossRef] [PubMed]
18. Yan, W. Singular-value partitioning in biplot analysis of multienvironment trial data. *Agron. J.* **2002**, *94*, 990–996.
19. Yan, W.; Hunt, L.A.; Sheng, Q.; Szlavnics, Z. Cultivar evaluation and mega-environment investigation based on the GGE biplot. *Crop Sci.* **2000**, *40*, 597–605. [CrossRef]
20. Olivoto, T.; Lúcio, A.D.C. metan: An R package for multi-environment trial analysis. *Methods Ecol. Evol.* **2020**, *11*, 783–789. [CrossRef]
21. Pour-Aboughadareh, A.; Khalili, M.; Poczai, P.; Olivoto, T. Stability indices to deciphering the genotype-by-environment interaction (GEI) effect: An applicable review for use in plant breeding programs. *Plants* **2022**, *11*, 414. [CrossRef]
22. Hossain, M.A.; Sarker, U.; Azam, M.G.; Kobir, M.S.; Roychowdhury, R.; Ercisli, S.; Ali, D.; Oba, S.; Golokhvast, K.S. Integrating BLUP, AMMI, and GGE Models to Explore GE Interactions for Adaptability and Stability of Winter Lentils (Lens culinaris Medik.). *Plants* **2023**, *12*, 2079. [CrossRef]
23. Tanin, M.J.; Sharma, A.; Saini, D.K.; Singh, S.; Kashyap, L.; Srivastava, P.; Mavi, G.; Kaur, S.; Kumar, V.; Kumar, V. Ascertaining yield and grain protein content stability in wheat genotypes having the Gpc-B1 gene using univariate, multivariate, and correlation analysis. *Front. Genet.* **2022**, *13*, 1001904. [CrossRef] [PubMed]
24. Artur, C.R.R.; Moiana, L.D.; Maleia, M.P.; Valentini, G.; Sumbuleiro, A.D.; Marcos, M.A. Evaluation of the grain yield performance of 5 soybean genotypes in Mozambique using the GGE Biplot method. *Afr. J. Biotechnol.* **2023**, *22*, 61–70.
25. Rani, R.; Raza, G.; Ashfaq, H.; Rizwan, M.; Shimelis, H.; Tung, M.H.; Arif, M. Analysis of genotype× environment interactions for agronomic traits of soybean (*Glycine max* [L.] Merr.) using association mapping. *Front. Genet.* **2023**, *13*, 1090994. [CrossRef] [PubMed]
26. Kumar, R.; Dhansu, P.; Kulshreshtha, N.; Meena, M.R.; Kumaraswamy, M.H.; Appunu, C.; Chhabra, M.L.; Pandey, S.K. Identification of Salinity Tolerant Stable Sugarcane Cultivars Using AMMI, GGE and Some Other Stability Parameters under Multi Environments of Salinity Stress. *Sustainability* **2023**, *15*, 1119. [CrossRef]
27. Anisha, A.; Rajappa, P.; Parashuram, P.; Hemalatha, V.; Dhanyashree, R.; Tonapi, V.A.; Sujatha, K.; Girish, G.; Madhusudhana, R. Selection of post-rainy sorghum landraces combining multi-traits mean performance and stability. *Euphytica* **2022**, *218*, 176. [CrossRef]
28. Rubiales Olmedo, D.; Moral, A.; Flores Gil, F. Agronomic Performance of Broomrape Resistant and Susceptible Faba Bean Accession. *Agronomy* **2022**, *12*, 1421. [CrossRef]
29. Hussain, T.; Akram, Z.; Shabbir, G.; Manaf, A.; Ahmed, M. Identification of drought tolerant Chickpea genotypes through multi trait stability index. *Saudi J. Biol. Sci.* **2021**, *28*, 6818–6828. [CrossRef]
30. Rossi, A.; Clemente, C.; Tavarini, S.; Angelini, L.G. Variety and Sowing Date Affect Seed Yield and Chemical Composition of Linseed Grown under Organic Production System in a Semiarid Mediterranean Environment. *Agronomy* **2022**, *13*, 45. [CrossRef]
31. Krishna, K.S.; Chaudhary, R.K.; Kumar, M. Deciphering the Genotype× Environment Interaction for Identification of Superior Genotypes of Mango (*Mangifera indica* L.) using Ammi Stability Measures. *Int. J. Environ. Clim. Chang.* **2022**, *12*, 348–358. [CrossRef]
32. Adil, N.; WANI, S.H.; RAFIQEE, S.; Mehrajuddin, S.; SOFI, N.R.; SHIKARI, A.B.; HUSSAIN, A.; MOHIDDIN, F.; JEHANGIR, I.A.; KHAN, G.H. Deciphering genotype× environment interaction by AMMI and GGE biplot analysis among elite wheat (*Triticum aestivum* L.) genotypes of himalayan region. *Ekin J. Crop Breed. Genet.* **2022**, *8*, 41–52.
33. Erdemci, I. Investigation of genotype× environment interaction in chickpea genotypes using AMMI and GGE biplot analysis. *Turk. J. Field Crops* **2018**, *23*, 20–26. [CrossRef]
34. Barona, M.A.; Díaz, R.S.; Suárez, R.R. Adaptability and grain yield stability of rice hybrids and varieties in Venezuela. *Bioagro* **2021**, *33*, 181–190. [CrossRef]
35. Dewi, A.K.; Chozin, M.A.; Triwidodo, H.; Aswidinnoor, H. Genotype× environment interaction, and stability analysis in lowland rice promising genotypes. *Int. J. Agron. Agric. Res.* **2014**, *5*, 74–84.
36. Tariku, S.; Lakew, T.; Bitew, M.; Asfaw, M. Genotype by environment interaction and grain yield stability analysis of rice (*Oryza sativa* L.) genotypes evaluated in north western Ethiopia. *Net J. Agric. Sci.* **2013**, *1*, 10–16.

37. Zewdu, Z.; Abebe, T.; Mitiku, T.; Worede, F.; Dessie, A.; Berie, A.; Atnaf, M. Performance evaluation and yield stability of upland rice (*Oryza sativa* L.) varieties in Ethiopia. *Cogent Food Agric.* **2020**, *6*, 1842679. [CrossRef]
38. Liang, S.; Ren, G.; Liu, J.; Zhao, X.; Zhou, M.; McNeil, D.; Ye, G. Genotype-by-environment interaction is important for grain yield in irrigated lowland rice. *Field Crops Res.* **2015**, *180*, 90–99. [CrossRef]
39. Singh, G.; Khanna, R.; Kaur, R.; Kaur, K.; Kaur, R.; Sharma, N.; Mangat, G.S. Performance under multi-environment trial for quantitative traits of rice (*Oryza sativa* L.) genotypes in North-West India (Punjab). *Ecol. Genet. Genom.* **2023**, *28*, 100190. [CrossRef]
40. Gauch Jr, H.G. A simple protocol for AMMI analysis of yield trials. *Crop Sci.* **2013**, *53*, 1860–1869. [CrossRef]
41. Gauch Jr, H.G.; Piepho, H.P.; Annicchiarico, P. Statistical analysis of yield trials by AMMI and GGE: Further considerations. *Crop Sci.* **2008**, *48*, 866–889. [CrossRef]
42. Daba, S.D.; Kiszonas, A.M.; McGee, R.J. Selecting High-Performing and Stable Pea Genotypes in Multi-Environmental Trial (MET): Applying AMMI, GGE-Biplot, and BLUP Procedures. *Plants* **2023**, *12*, 2343. [CrossRef]
43. Olivoto, T.; Lúcio, A.D.; da Silva, J.A.; Marchioro, V.S.; de Souza, V.Q.; Jost, E. Mean performance and stability in multi-environment trials I: Combining features of AMMI and BLUP techniques. *Agron. J.* **2019**, *111*, 2949–2960. [CrossRef]
44. Pour-Aboughadareh, A.; Barati, A.; Gholipoor, A.; Zali, H.; Marzooghian, A.; Koohkan, S.A.; Shahbazi-Homonloo, K.; Hou-seinpour, A. Deciphering genotype-by-environment interaction in barley genotypes using different adaptability and stability methods. *J. Crop Sci. Biotechnol.* **2023**, *26*, 547–562. [CrossRef]
45. Abo-Yousef, M.; Sedeek, S.; EL-Rafaee, I.; Hammoud, S.; EL-Abd, A.; El-Malkey, M.; EL-Namaky, R.; Ammar, M.; Abdelkhalik, A.; Zayed, B. Giza 179 Egyptian rice variety: As a new, early, high-yielding, tolerant to saline, and climate change challenge. *Egypt. J. Agric. Res.* **2023**, *101*, 567–588.
46. Shrestha, J.; Subedi, S.; Acharya, R.; Sharma, S.; Subedi, M. Grain yield stability of maize (*Zea mays* L.) hybrids using ammi model and GGE biplot analysis. *SAARC J. Agric.* **2021**, *19*, 107–121. [CrossRef]
47. Abdelaal, K.; Mazrou, Y.; Mohamed, A.; Ghazy, M.; Barakat, M.; Hafez, Y.; Gaballah, M. The different responses of rice genotypes to heat stress associated with morphological, chlorophyll and yield characteristics. *Not. Bot. Horti Agrobot. Cluj-Napoca* **2021**, *49*, 12550. [CrossRef]
48. Gaballah, M.M.; Ghoneim, A.M.; Rehman, H.U.; Shehab, M.M.; Ghazy, M.I.; El-Iraqi, A.S.; Mohamed, A.E.; Waqas, M.; Shamsudin, N.A.A.; Chen, Y. Evaluation of morpho-physiological traits in rice genotypes for adaptation under irrigated and water-limited environments. *Agronomy* **2022**, *12*, 1868. [CrossRef]
49. Jagadish, S.; Muthurajan, R.; Rang, Z.W.; Malo, R.; Heuer, S.; Bennett, J.; Craufurd, P.Q. Spikelet proteomic response to combined water deficit and heat stress in rice (*Oryza sativa* cv. N22). *Rice* **2011**, *4*, 1–11. [CrossRef]
50. Shehab, M.; Iovene, M.; Ciancio, A.; Colagiero, M.; Finetti-Sialer, M. Transcriptome analysis provides novel insights into salt stress response in two egyptian rice varieties with different tolerance levels. *Rice Sci.* **2022**, *29*, 499–502. [CrossRef]
51. Grondin, A.; Dixit, S.; Torres, R.; Venkateshwarlu, C.; Rogers, E.; Mitchell-Olds, T.; Benfey, P.N.; Kumar, A.; Henry, A. Physiological mechanisms contributing to the QTL qDTY 3.2 effects on improved performance of rice Moroberekan x Swarna BC 2 F 3: 4 lines under drought. *Rice* **2018**, *11*, 1–17. [CrossRef]
52. Laza, M.R.C.; Peng, S.; Akita, S.; Saka, H. Effect of panicle size on grain yield of IRRI-released indica rice cultivars in the wet season. *Plant Prod. Sci.* **2004**, *7*, 271–276. [CrossRef]
53. R Development Core Team. *R: A Language and Environment for Statistical Computing*; R Development Core Team: Vienna, Austria, 2010.
54. Yan, W.; Kang, M.S. *GGE biplot analysis: A graphical tool for breeders, geneticists, and agronomists*; CRC press: Boca Raton, FL, USA, 2002.
55. Charrad, M.; Ghazzali, N.; Boiteau, V.; Niknafs, A. NbClust: An R package for determining the relevant number of clusters in a data set. *J. Stat. Softw.* **2014**, *61*, 1–36. [CrossRef]
56. Yan, W.; Tinker, N.A. Biplot analysis of multi-environment trial data: Principles and applications. *Can. J. Plant Sci.* **2006**, *86*, 623–645. [CrossRef]

plants MDPI

Article

Melatonin-Mediated Enhancement of Photosynthetic Capacity and Photoprotection Improves Salt Tolerance in Wheat

Di Yan [1,†], Jiajie Wang [1,†], Zhenzong Lu [1], Rui Liu [1], Yue Hong [1], Baocai Su [1], Ye Wang [1,2], Zhen Peng [1,2], Chunxin Yu [1,2], Yuerong Gao [1,2], Ziyan Liu [1,2], Zhaoshi Xu [3], Liusheng Duan [1,2,4] and Runzhi Li [1,2,*]

[1] College of Plant Science and Technology, Beijing University of Agriculture, Beijing 102206, China; di.yan@bua.edu.cn (D.Y.); jiajie.wang@bua.edu.cn (J.W.); zhenzong.lu@bua.edu.cn (Z.L.); rui.liu@bua.edu.cn (R.L.); yue.hong@bua.edu.cn (Y.H.); baocai.su@bua.edu.cn (B.S.); ye.wang@bua.edu.cn (Y.W.); zhen.peng@bua.edu.cn (Z.P.); chunxin.yu@bua.edu.cn (C.Y.); yuerong.gao@bua.edu.cn (Y.G.); ziyan.liu@bua.edu.cn (Z.L.); duanlsh@cau.edu.cn (L.D.)

[2] Beijing Key Laboratory for Agricultural Application and New Technique, Beijing 102206, China

[3] Institute of Crop Sciences, Chinese Academy of Agricultural Sciences, Beijing 100081, China; xuzhaoshi@caas.cn

[4] College of Agronomy and Biotechnology, China Agricultural University, Beijing 100193, China

[*] Correspondence: lirunzhi7639@163.com

[†] These authors contributed equally to this work.

Abstract: The role of melatonin in plant growth and response to environmental stress has been widely demonstrated. However, the physiological and molecular regulation of salt tolerance in wheat seedlings by melatonin remains unclear. In this study, we investigated changes in phenotype, physiology, photosynthetic parameters, and transcript levels in wheat seedlings to reveal the role of melatonin in the regulation of salt tolerance in wheat. The results indicate that the application of exogenous melatonin significantly alleviates growth inhibition, reactive oxygen species accumulation, and membrane oxidative damage induced by salt stress in wheat. Additionally, exogenous melatonin increased antioxidant enzyme activity and regulated photosynthetic gas exchange. Transcriptomic data showed a significant up-regulation of genes encoding light-harvesting chlorophyll protein complex proteins in photosynthesis and genes related to chlorophyll and carotenoid biosynthesis under the influence of melatonin. These results suggest that exogenous melatonin improves salt tolerance in wheat seedlings by enhancing the antioxidant, photoprotective, and photosynthesis activities.

Keywords: wheat (*Triticum aestivum* L); salt stress; melatonin; RNA-seq; photosynthesis

1. Introduction

Soil salinization is a global issue resulting from both global climate change and inappropriate agricultural production, including incorrect farming techniques and excessive fertilizer application [1,2]. More than 20% of the world's arable land has been threatened by salt stress, which is severely reducing food production [3,4], threatening sustainable agricultural development and global food security [5–8]. Salt stress causes osmotic stress, ionic toxicity, and oxidative damage to plants, which in turn inhibits seed germination, development, flowering and fruiting, and ultimately leads to reduced crop yields [9–12]. Salt-mediated osmotic stress reduces the ability of plants to draw nutrients and water from the soil, leading to growth stagnation [1,13]. Ionic toxicity caused by high Na^+ and Cl^- concentrations disrupts cellular homeostasis, increases membrane peroxidation, and inhibits photosynthesis in plants [14,15]. In addition, salt stress induces the accumulation of reactive oxygen species (ROS), such as hydrogen peroxide (H_2O_2) and superoxide anion (O_2^-), resulting in oxidative damage and inhibition of plant growth and development [16,17].

Plants have developed a complex mechanism against salt stress, including regulating endogenous hormone levels, ion homeostasis, osmotic potential, and activity of antioxidant enzymes [18–20]. Plants control ion uptake and transport by the salt overly sensitive (SOS) signaling pathway, which can maintain Na^+/K^+ in cells by exporting excess Na^+ when triggered by Ca^{2+} in the cytoplasm [21–25]. Moreover, plants remove excess Na^+ via Na^+/H^+ antiporters located at plasma and vesicle membranes [17,26,27]. Plant cells can increase the osmotic potential by accumulating osmoregulatory substances, such as soluble sugars, proline, inorganic ions, betaine, and polyols, thereby increasing salt tolerance [28]. ROS, which play an essential role in cell signaling, will be increased dramatically in plants when subjected to adversity stress and cause oxidative damage to cellular structures [9,29]. The ROS scavenging system mediated by antioxidant enzymes, including ascorbate peroxidase (APX), superoxide dismutase (SOD), peroxidase (POD), and catalase (CAT), can reduce the extent of cellular damage by scavenging the excessive accumulation of ROS, which is an important way to reduce oxidative damage in plants [7,30].

Melatonin (N-acetyl-5-methoxy tryptamine, MT) is a famous animal hormone involved in many bio-processes that include sleep, mood, day–night rhythms, seasonal reproductive physiology, temperature homeostasis, sexual behavior, antioxidant activity, and immune enhancement [31,32]. However, it is not a hormone unique to animals but is commonly found in plants [33]. Since 1995, when three laboratories simultaneously reported MT in plants, studies have detected and quantified MT in roots, shoots, leaves, fruits, and seeds of a considerable variety of plants [34–36]. To this day, MT is understood as a universal regulator of plant growth and development processes and various stress responses [31,37]. For example, MT maintains Na^+/K^+ homeostasis under salt stress by increasing root H^+-pump activity and the sensitivity of Na^+/K^+ transporter proteins to ROS and reactive nitrogen species (RNS) [38]. MT has been reported to increase seed germination and the root/stem ratio in cucumber under water stress [39]. Furthermore, MT was reported to reduce chlorophyll degradation, delay leaf senescence, improve antioxidant enzyme activity, increase osmoregulatory substances, and maintain hydrogen peroxide homeostasis in plants under abiotic stress [19,40–42]. Melatonin increased chlorophyll content, assimilation rate (A), stomatal conductance (gs), Internal CO_2 (Ci), and transpiration rate (Tr) and also increased photosystem II (PSII) activity, PSII response center protein D1, Lhcb1, and Lhcb2 in rice, maize, and tomato under various stresses [43–45].

Wheat (*Triticum aestivum* L.), the second largest food crop in the world, suffers from the salinization of arable land, which seriously affects yields [4]. Therefore, how to improve the salt tolerance of wheat is gradually becoming an agricultural research hotspot. Although some studies have demonstrated the protective effect of applied exogenous melatonin on plants under salt stress conditions, MT-mediated salinity tolerance mechanisms in wheat seedlings have been studied less. Therefore, it is important to investigate the physiological and molecular mechanisms of wheat seedlings impacted by MT under salt stress, particularly the influence of melatonin on wheat seedling photosynthesis. In this study, wheat biomass, antioxidant enzyme activity, photosynthesis, and transcriptome sequencing were measured for different treatments. The results show that exogenous MT was involved in redox metabolism, alteration of photosynthetic parameters, and biosynthesis of photosynthetic proteins during wheat growth under salt stress, which improved the salt tolerance of wheat at the seedling stage.

2. Results

2.1. Phenotypes of Wheat Seedlings in Different Treatments under Salt Stress

After 20 days of growth under salt stress conditions, the phenotypes of wheat seedlings had significant differences (Figure 1). Compared to the seedlings grown under CK (normal condition without MT) and CKM (normal condition with MT) conditions, the plant height and root length of seedlings of S (salt stress without MT) and SM (salt stress whit MT) groups were significantly decreased, and the wilting and yellowing of the first leaves were more serious. After MT application, the plant height and leaf length of SM treatment seedlings significantly increased by 13.66% and 12.70%, respectively, compared to those of untreated seedlings under salt stress, while the leaf tip wilt index significantly decreased by 72.53% (Figure 1A–D). Meanwhile, the application of exogenous MT led to a 46.2% and 47.97% increase in fresh weight and a 20.3% and 45.91% increase in dry weight of aboveground and belowground seedlings, respectively, under salt stress (Figure 1E,F). Notedly, MT had no significant effect on primary root length (Figure 1C), but significantly increased the length and number of lateral roots (Figure 1B). Overall, our results show that melatonin improved the growth of wheat under salt stress but had no significant effect on wheat growth under normal conditions.

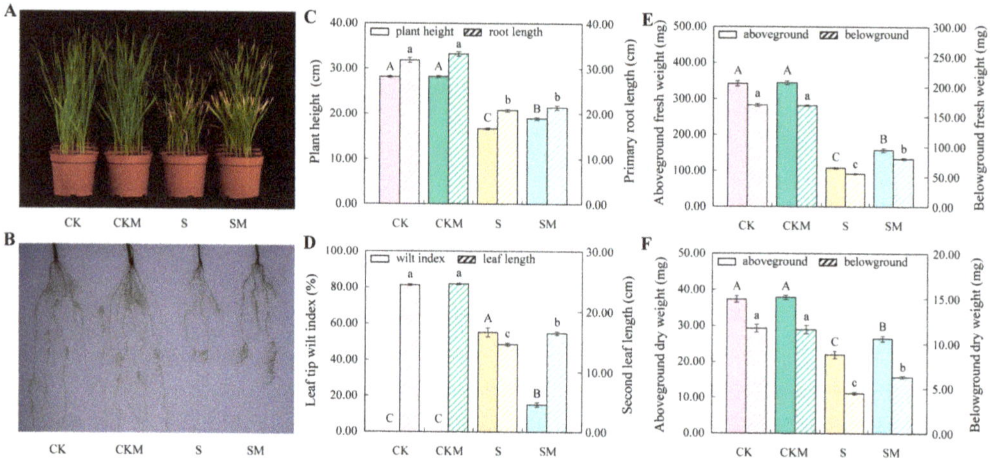

Figure 1. The effect of different conditions (CK, CKM, S, and SM) on the growth of wheat seedlings. (**A**) Phenotypes; (**B**) primary root growth; (**C**) plant height and root length; (**D**) leaf tip wilt index and second leaf length; (**E**) above- and belowground fresh weight; (**F**) dry weight. Values are the averages of replicates ± deviation (SD, n = 3). Different letters indicate significant differences according to the Duncan test ($p < 0.05$). Uppercase letters indicate comparisons between indicators on the left and lowercase letters indicate comparisons between indicators on the right.

2.2. *Measuring the Antioxidant and Osmoregulatory Abilities of MT*

To reveal the antioxidant capacity of MT and its effect on osmoregulation, we determined the REL, MDA content, osmoregulatory substance content, ROS content, and antioxidant enzyme activities of wheat seedlings under different treatments. The results of measuring REL, MDA, proline, and soluble sugar contents of wheat seedlings show a significant increase in both leaves and roots under salt stress, indicating that salt stress damaged the membrane structure of wheat seedlings and induced the accumulation of osmoregulatory substances (Figure 2A–D). However, exogenous MT significantly reduced REL and MDA content in seedlings, while increased proline content in leaves and soluble sugar content in the root system under salt stress were detected, but no significant effect on seedlings under normal conditions (Figure 2A–D), which suggested that MT reduced salt stress-induced oxidative damage and alleviated osmotic stress by increasing osmoregulatory substances.

Under stress conditions, the level of ROS in plants increases dramatically [46]. In this study, exogenous MT had no significant effect on H_2O_2 and O_2^- in leaves and roots under normal conditions, but the size and number of brown and blue spots on the leaves and roots were significantly reduced compared to those in untreated seedlings under salt stress, and the results of H_2O_2 and O_2^- content measurements were consistent with the staining results (Figure 2E–H). These results indicate that MT can effectively reduce the accumulation of ROS in wheat seedling tissues under salt stress.

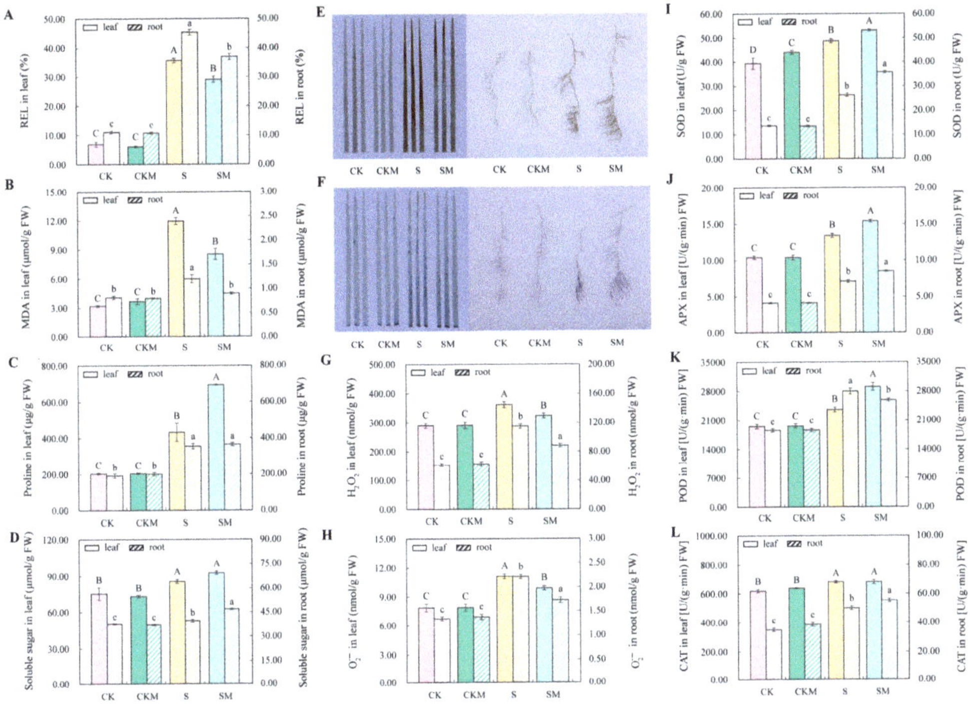

Figure 2. Effect of exogenous MT on physiological indicators of wheat seedlings under salt stress. (**A**) The rate of REL in leaves and roots; (**B**) MDA content in leaves and roots; (**C**) proline content in leaves and roots; (**D**) soluble sugar content in leaves and roots; (**E**) histochemical staining of H_2O_2 levels in leaves and roots—H_2O_2 was brown-spotted; (**F**) histochemical staining of O_2^- levels in leaves and roots—O_2^- was blue spotted; (**G**) H_2O_2 content in leaves and roots; (**H**) O_2^- content in leaves and roots; (**I**) SOD activities in leaves and roots; (**J**) APX activities in leaves and roots; (**K**) POD activities in leaves and roots; (**L**) CAT activities in leaves and roots. Different letters indicate significant differences according to the Duncan test ($p < 0.05$). Uppercase letters indicate comparisons between indicators on the left and lowercase letters indicate comparisons between indicators on the right.

The results of antioxidant enzyme (SOD, APX, POD, and CAT) activities assay show that the antioxidant enzyme activities of wheat seedlings applied with exogenous MT did not change significantly under normal conditions, but increased significantly under salt stress (Figure 2I–L). The activities of SOD, APX, and POD in leaves and the activities of SOD, APX, and CAT in the root system were significantly increased (Figure 2I–L).

2.3. *Photosynthetic Parameters and Chlorophyll Content of Wheat Seedlings under Different Treatments*

The assimilation rate of seedlings was significantly decreased under salt stress compared to normal conditions (Figure 3A). However, exogenous melatonin increased the assimilation rate of seedlings, which was significantly increased by 69.70% at 600 $\mu mol \cdot m^2 \cdot s^{-1}$ in the SM group compared with the S group (Figure 3A). Compared with normal conditions, Ci was increased by salt stress but decreased after the application of MT (Figure 3B). At different light intensities, the gs and Tr of wheat seedlings in the SM group increased by an average of 9.7% and 12.87%, respectively, compared to the S group (Figure 3C,D). The WUE of wheat increased with increasing light intensity (Figure 3E). Salt-stressed plants showed a significant decrease in the WUE of seedlings, while exogenous MT significantly attenuated the effect of salt stress on WUE (Figure 3E). Salt stress increased the chlorophyll content of wheat, while exogenous MT treatment further increased the chlorophyll content of wheat under salt stress (Figure 3F). These results suggest that MT enhanced the photosynthesis of wheat seedlings under salt stress by improving the assimilation rate and WUE, which promoted the seedlings' growth.

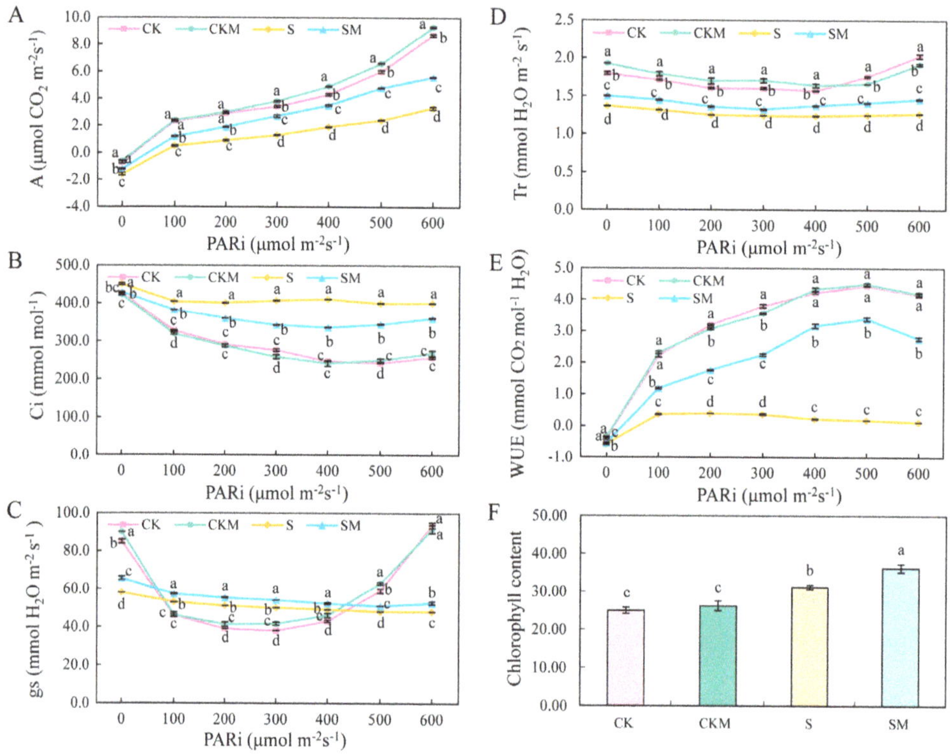

Figure 3. Effect of exogenous MT on photosynthetic parameters of wheat seedlings. (**A**) Assimilation rate (**A**); (**B**) internal CO_2 (Ci); (**C**) stomatal Conductance (gs); (**D**) transmission rate (Tr); (**E**) water use efficiency (WUE); (**F**) chlorophyll content. Different letters indicate significant differences according to the Duncan test ($p < 0.05$).

2.4. Comprehensive Transcriptome Analysis of S and SM at Different Times

To investigate the molecular mechanism of exogenous MT to improve salt tolerance in wheat seedlings, we performed RNA-seq on the seedlings of the S group and SM group at 24, 36, 48, and 60 h after MT application. Each sample obtained an average of 7.85 GB of data, with an average of Q30 > 91.26% (Table S2). To verify the reliability of RNA-seq, six genes were randomly selected for qRT-PCR analysis, and the results show that their expression patterns were consistent with the RNA-seq data (Figure S1), indicating that the RNA-seq data are reliable and can be used for further analysis. The results of principal component analysis (PCA) and cluster analysis show that samples from different time points of the same treatment were clustered together, while samples from different treatments had significant dispersion, indicating that MT had a significant effect on transcript levels in seedlings from all four different time points under salt stress (Figure 4A,B).

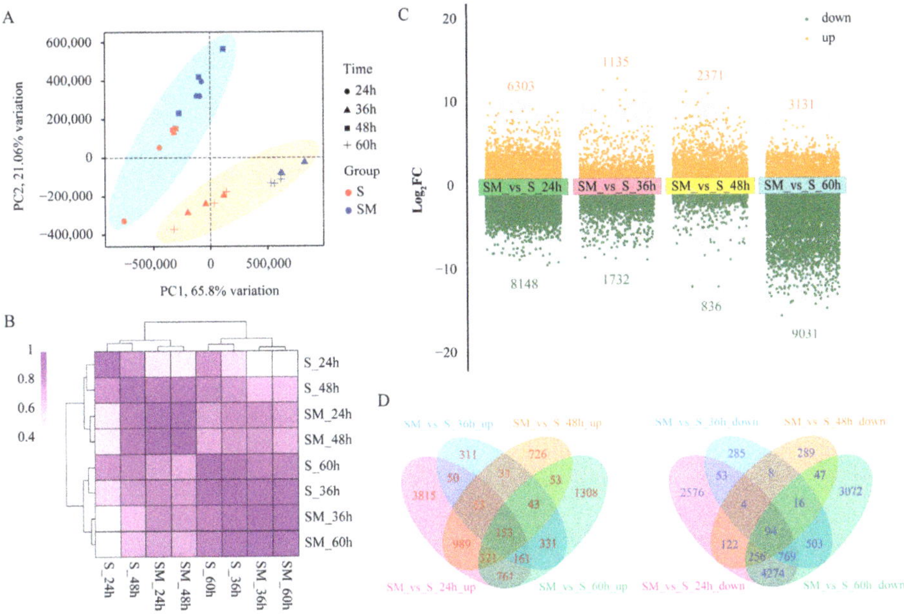

Figure 4. The effect of MT on transcript levels in wheat seedlings from all four different time points under salt stress. (**A**) Principal component analysis (PCA); (**B**) heat map of clustering among treatments; (**C**) DEGs for SM-vs-S at different time points; (**D**) Venn diagram of DEGs.

A total of 14,451 (6303 up-regulated, 8148 down-regulated), 2867 (1135 up-regulated, 1732 down-regulated), 3207 (2371 up-regulated, 836 down-regulated), and 12,162 (3131 up-regulated, 9031 down-regulated) differentially expressed genes (DEGs) were identified in SM treatments compared to S treatments at 24, 36, 48, and 60 h, respectively (Figure 4C). The Venn diagram showed that a total of 153 up-regulated overlapping DEGs and 94 down-regulated overlapping DEGs were identified at all different time points, indicating that these genes were involved in the MT-mediated regulation of salt tolerance in wheat seedlings both day and night (Figure 4D).

2.5. Functional and Pathway Annotation of DEGs

GO enrichment analysis was performed to annotate the DEGs of seedlings between SM-vs-S comparisons (Figure 5A, Tables S3 and S4). The results show that the up-regulated DEGs were mainly enriched in "photosynthesis, light harvesting in photosystem I (GO:0009768)", "sucrose biosynthetic process (GO:0005986)", "response to cytokinin (GO:0009735)", "chlorophyll biosynthetic process (GO:0015995)", and "carbon fixation (GO:0015977)" (Figure 5A, Table S3). The down-regulated DEGs in GO terms were mainly enriched in the "chlorophyll catabolic process (GO:0015996)", "lipid oxidation (GO:0034440)", "auxin catabolic process (GO:0009852)", and "sodium ion import across plasma membrane (GO:0098719)" (Figure 5A, Table S4).

To gain insight into the biological processes affected by exogenous MT, we performed KEGG pathway analysis (Figure 5B, Tables S5 and S6). Among them, the up-regulated DEGs mainly involved "photosynthesis-antenna proteins (ko00196)", "porphyrin and chlorophyll metabolism (ko00860)", and "carotenoid biosynthesis (ko00906)" pathways (Figure 5B, Table S5). The down-regulated DEGs were mainly enriched in "porphyrin and chlorophyll metabolism (ko00860)" and "arachidonic acid metabolism (ko00590)" (Figure 5B, Table S6). The results of GO and KEGG reveal that MT enhanced the response of wheat to salt stress mainly by regulating the expression of genes involved in photosynthesis, phytohormone regulation, and energy metabolism.

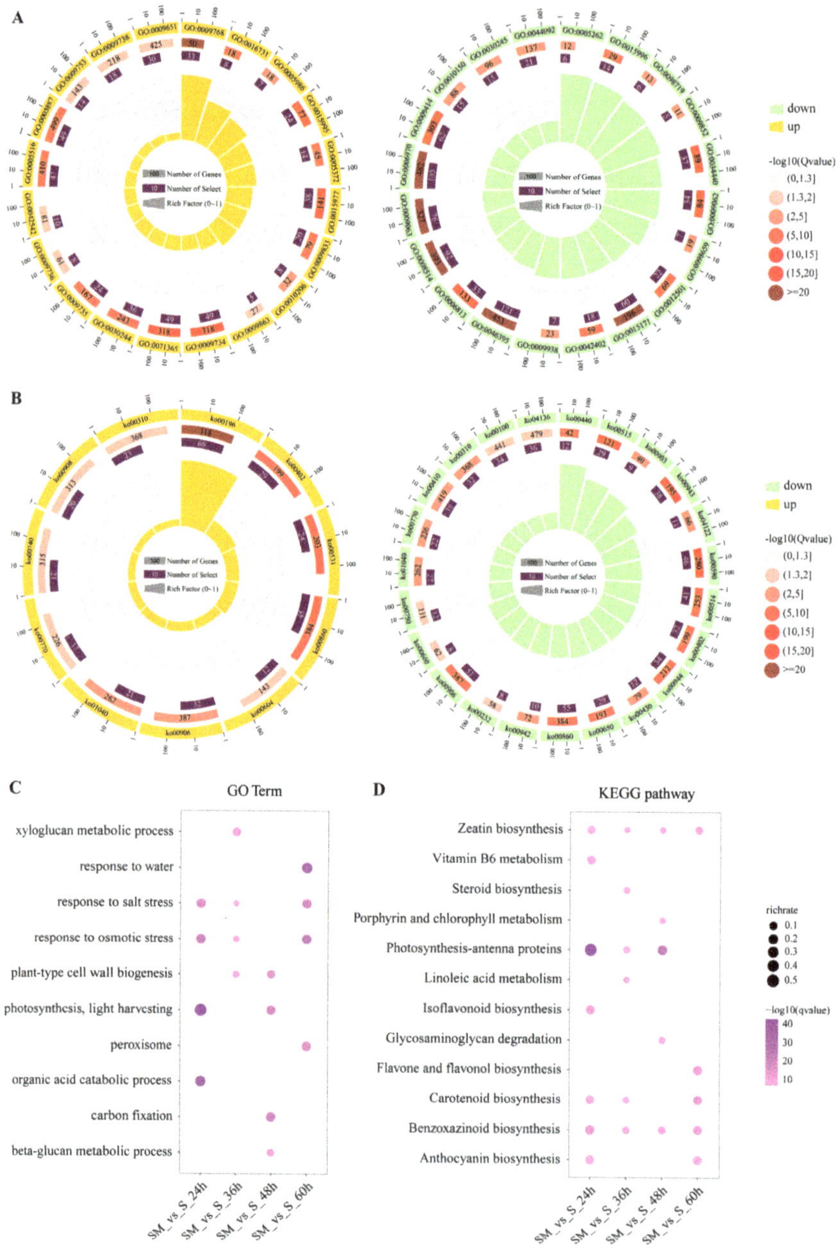

Figure 5. Functional and pathway annotation of DEGs in SM-vs-S comparison. (**A**) Gene ontology (GO) enrichment analysis of DEGs; (**B**) Kyoto Encyclopedia of Genes and Genomes (KEGG) enrichment analysis of DEGs; (**C**) GO enrichment analysis; (**D**) KEGG enrichment analysis of DEGs in different time comparison groups of wheat seedling leaves SM-vs-S.

Then, we analyzed the function and pathway of DEGs identified at different time points, respectively. The GO results show that MT-induced DEGs were mainly enriched in stress-responsive terms like "response to osmotic stress" and "response to salt stress" and in terms related to cell structure like "xyloglucan metabolic process" and "plant-type cell wall biogenesis" at 24 h (Figure 5A). In addition, DEGs were mainly enriched in "carbon fixation" and "photosynthesis, light harvesting" which are related to photosynthesis at 48 h, and in "response to water" and "peroxisome" at 60 h (Figure 5A). The results of KEGG enrichment analysis showed that DEGs were enriched to the "benzoxazinoid biosynthesis" and "zeatin biosynthesis" pathways at all times. Photosynthesis-antenna proteins were enriched at 24, 36, and 48 h. In addition, carotenoid biosynthesis was enriched at 24, 36, and 60 h (Figure 5B). Overall, MT improved salt tolerance in wheat by regulating the expression of genes in photosynthesis and antioxidant-related pathways.

2.6. Expression Pattern Clustering of the DEGs

To investigate the expression pattern of DEGs, we used the Mfuzz clustering method to perform a time series analysis of all DEGs. The result show that all DEGs were clustered into eight clusters according to their different expression patterns (Figure 6). Among them, C1, C2, C4, and C6, which clustered 2874, 1437, 1270, and 1256 DEGs, respectively, were considered to have a great contribution to improved salt tolerance in wheat. The DEGs clustered in C1 and C2 were consistently down-regulated by MT induction, while those clustered in C4 and C6 were consistently up-regulated by MT (Figure 6). GO enrichment analysis showed that the DEGs in C1 were mainly involved in "programmed cell death", "organic anion transmembrane transporter activity", and "sodium ion transport", while the DEGs in C2 were mainly enriched in "response to water", "carboxylic acid catabolic process", and "lipid oxidation". These results suggest that MT had a positive effect on the maintenance of cellular osmotic potential and protected cell structure by inhibiting lipid oxidation and controlling ion transportation. The DEGs in the C6 were mainly associated with "photosynthesis, light harvesting in photosystem I", "carbon fixation", and "circadian rhythm". The DEGs in C4 were mainly related to "solute: cation symporter activity", "water channel activity", and "auxin-activated signaling pathway". These results suggest that MT increased biomass by enhancing carbon fixation and photosynthesis in wheat seedlings under salt stress, and enhanced the response of wheat to salt stress by increasing the auxin-activated signaling pathway (Figure 6).

2.7. Transcription Factor and Co-Expression Network Analysis in S and SM Treatments

To investigate the regulation of transcription factors in wheat seedlings by MT under salt stress, the statistical analysis of transcription factors and network analysis were performed. A larger number of transcription factors (TFs) genes in 48 TF families were differentially regulated in wheat under salt stress (Figure 7A). The gene families with the highest number of differentially expressed TFs included the AP2/EREBP, bHLH, MYB, NAC, and WRKY families with 96, 97, 140, 110, and 81 DEGs, respectively (Figure 7A). Next, a co-expression network was created to identify key TFs associated with MT to improve salt tolerance in seedlings. Notably, *TraesCS4A02G017700 (CDF1)* and *TraesCS5A02G311300 (CBF14)* were at the center of the network and may be key TFs encoding genes induced by MT to regulate salt tolerance in wheat seedlings (Figure 7B).

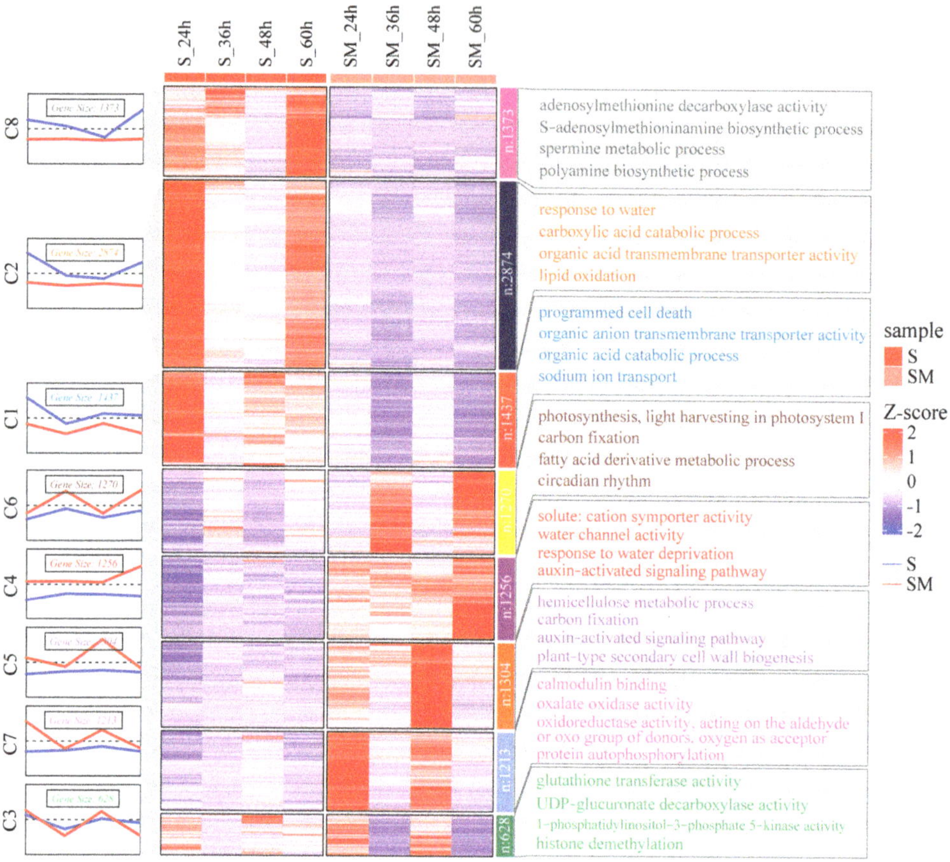

Figure 6. Heat map of expression pattern clustering and GO enrichment analysis of DEGs in SM-vs-S comparison. The folded line graph represented the Mfuzz clustered expression pattern of DEGs.

2.8. *Effect of Exogenous MT on LHC*

Notably, the results of KEGG enrichment analysis show that photosynthetic antenna protein-related genes were significantly up-regulated by MT induction. The LHC, also known as the light-harvesting antenna complex, is responsible for collecting light energy for transmission to the photosynthetic reaction center, including LHCI and LHCII (Figure 8). In this study, 17 DEGs of five binding proteins in LHCI were found to be up-regulated at all times (Figure 8). A total of 55 DEGs encoding LHCII-related proteins, of which 40 Lhcb1 encoding genes, were up-regulated at 24, 36, and 48 h, while 15 DEGs encoding Lhcb2-6 were up-regulated at all times (Figure 8). The up-regulation of LHC-related genes indicates that MT improves photosynthesis in seedlings by increasing the leaves' light-harvesting ability under salt stress.

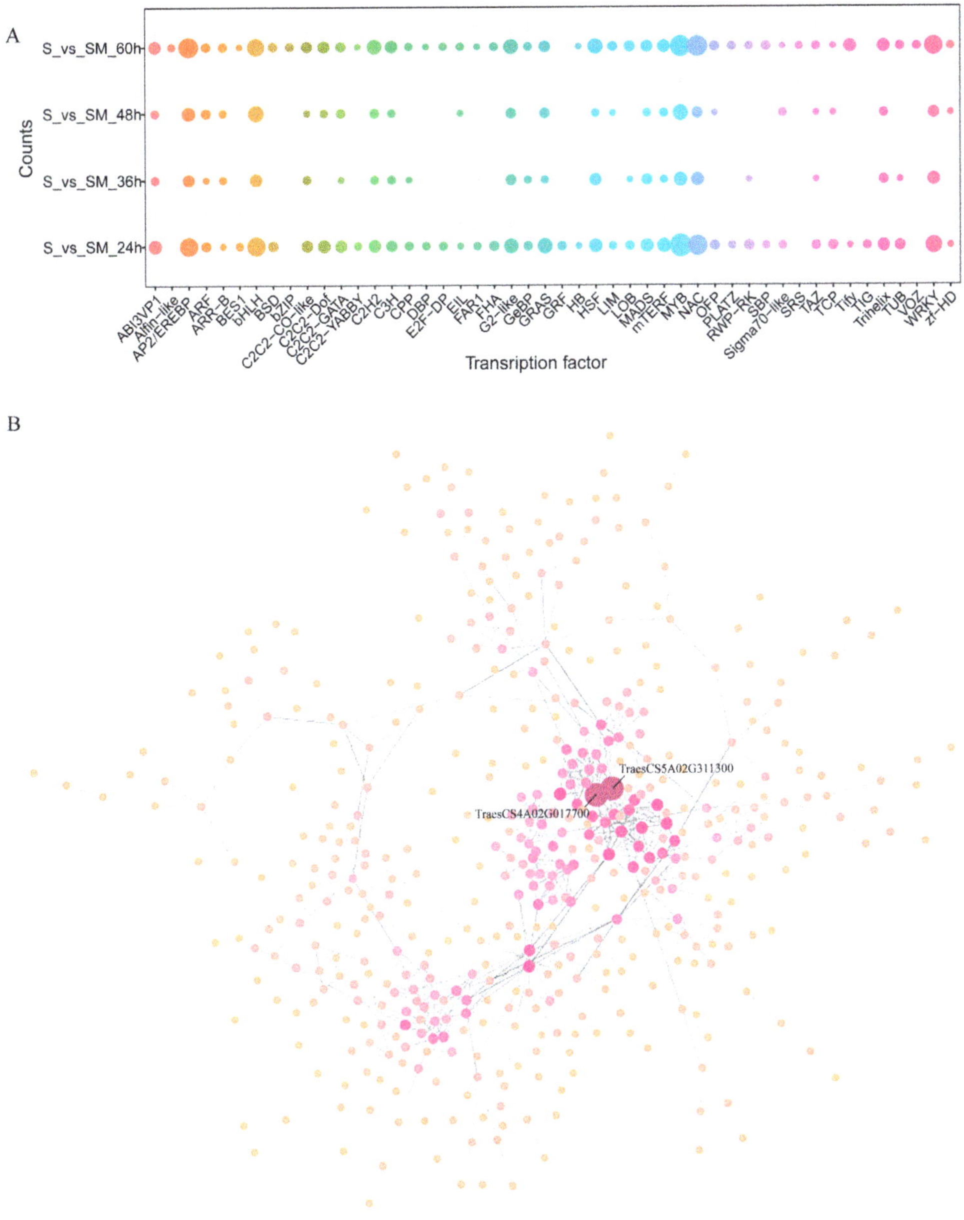

Figure 7. Transcription factor and co-expression network analysis in S and SM treatments. (**A**) Classifications of TFs in the SM-vs-S comparisons at different times; (**B**) Co-expression network of DEGs.

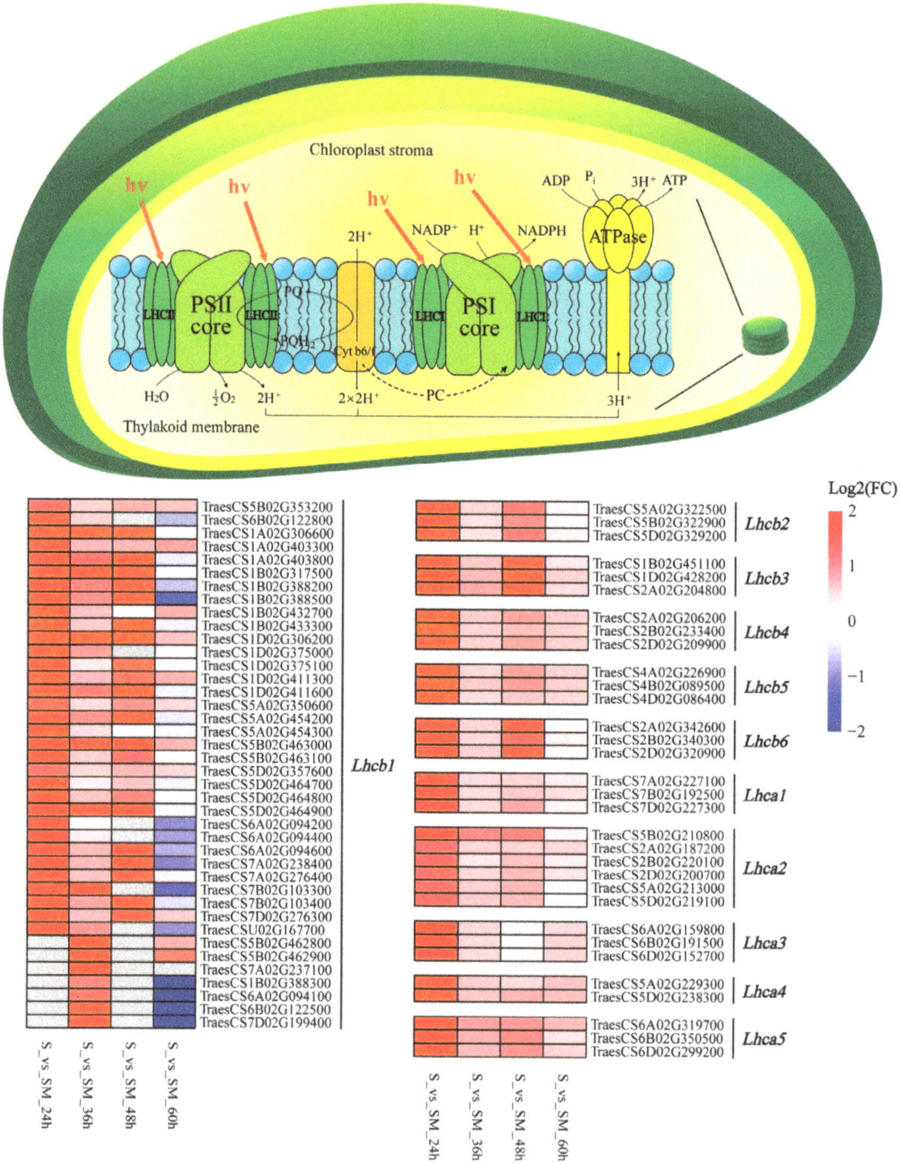

Figure 8. Photosynthesis light reaction process. PSI: Photosystem I; PSII: Photosystem II; LHCI: Light-harvesting chlorophyll protein complex I; LHCII: Light-harvesting chlorophyl protein complex II; PQ: Plastoquinone; PQH_2: Plastoquinol-1; Cyt b6/f: Cytochrome b6/f complex; PC: Plastocyanin; ATPase: ATP Synthase.

2.9. Effect of Exogenous MT on Photosynthetic Pigments

Chlorophylls and carotenoids are involved in the uptake, transfer, and conversion of light energy in photosynthesis. A total of 100 DEGs in the "porphyrin and chlorophyll metabolism" pathway were regulated by MT (Figures 9 and S2). Among them, *EARS*, *HemA*, and *HemF* were up-regulated by MT under salt stress (Figures 9 and S2). MT induced the up-regulation of genes involved in chlorophyll biosynthesis, such as *chlD*, *chlM*, *por*, *DVR*, *HCAR*, and *chlP* (Figures 9 and S2). However, MT down-regulated the expression of *HemC*, *HemH*, *HO*, *SGR*, *NOL*, *PAO*, and *PPD* in seedlings under salt stress (Figures 9 and S2). These results indicate that MT improved chlorophyll biosynthesis and inhibited chlorophyll catabolism in seedlings.

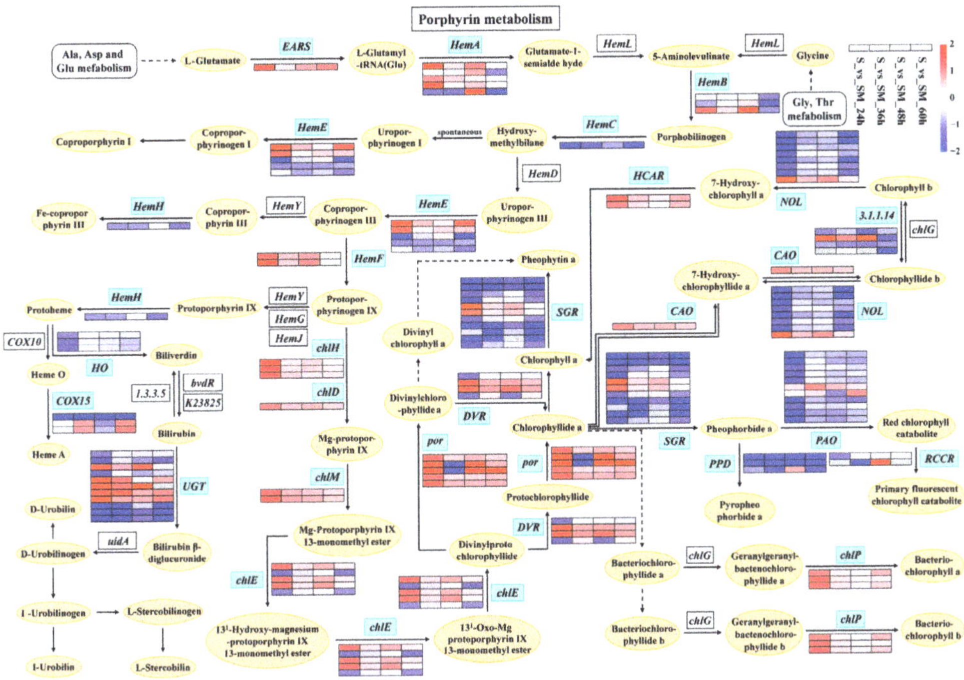

Figure 9. Porphyrin and chlorophyll metabolism pathway. *por*: protochlorophyllide reductase; *HemF*: coproporphyrinogen III oxidase; *HemE*: uroporphyrinogen decarboxylase; *HemB*: porphobilinogen synthase; *HemH*: protoporphyrin/coproporphyrin ferrochelatase; *EARS*: glutamyl-tRNA synthetase; *COX15*: heme a synthase; *HemA*: glutamyl-tRNA reductase; *chlH*: magnesium chelatase subunit H; *chlD*: magnesium chelatase subunit D; *chlM*: magnesium–protoporphyrin O-methyltransferase; *chlE*: magnesium–protoporphyrin IX monomethyl ester (oxidative) cyclase; *3.1.1.14*: chlorophyllase; *chlP*: geranylgeranyl diphosphate; *PAO*: pheophorbide a oxygenase; *UGT*: glucuronosyltransferase; *PPD*: pheophorbidase; *RCCR*: red chlorophyll catabolite reductase; *CAO*: chlorophyllide a oxygenase; *NOL*: chlorophyll(ide) b reductase; *HemC*: hydroxymethylbilane synthase; *HCAR*: 7-hydroxymethyl chlorophyll a reductase; *DVR*: divinyl chlorophyllide a 8-vinyl-reductase; *HO*: heme oxygenase (biliverdin-producing, ferredoxin); *SGR*: magnesium dechelatase.

A total of 83 DEGs in the carotenoid biosynthesis pathway were identified (Figures 10 and S3). Among them, *crtL2*, *crtH*, *ABA1*, and *NPQ1* were up-regulated by MT under salt stress, whereas *CYP707A*, *AOG*, *Z-ISO*, and *crtZ* were down-regulated (Figures 10 and S3). In the lutein biosynthesis process, the synthesis of ε-carotene was significantly up-regulated, while the synthesis of the lutein was significantly inhibited (Figure 10).

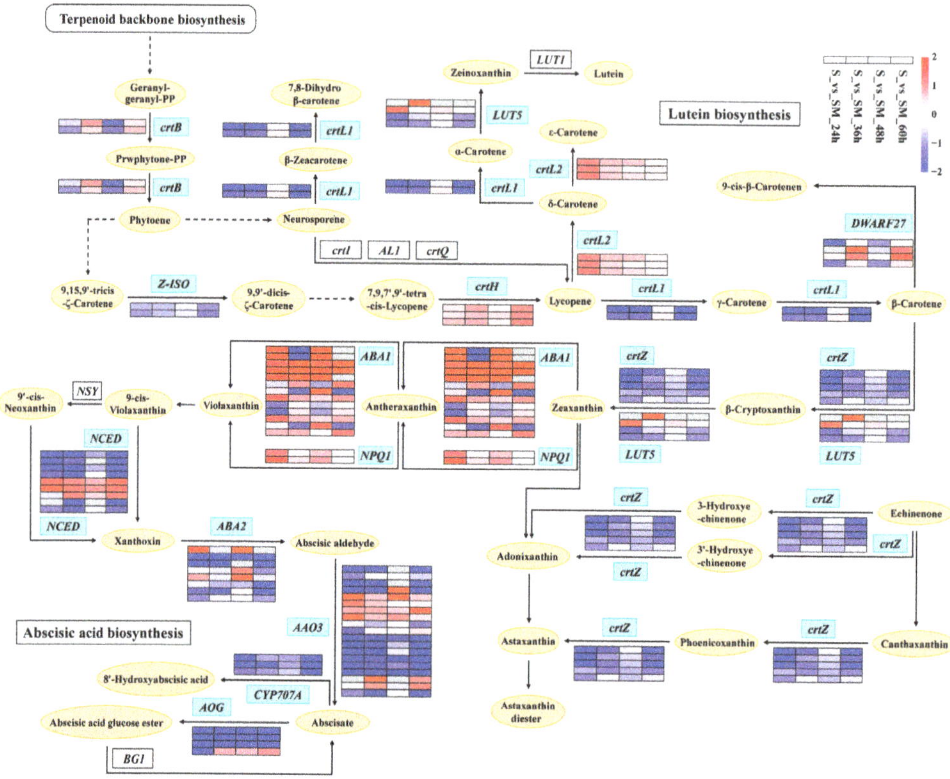

Figure 10. Carotenoid biosynthesis pathway. *crtB*: 15-cis-phytoene synthas; *crtL1*: lycopene beta-cyclase; *crtL2*: lycopene epsilon-cyclase; *crtH*: prolycopene isomerase; *ABA1*: zeaxanthin epoxidase; *NPQ1*: violaxanthin de-epoxidase; *NCED*: 9-cis-epoxycarotenoid dioxygenase; *ABA2*: xanthoxin dehydrogenase; *AAO3*: abscisic-aldehyde oxidase; *CYP707A*: (+)-abscisic acid 8′-hydroxylase; *AOG*: abscisate beta-glucosyltransferase; *Z-ISO*: zeta-carotene isomerase; *crtZ*: beta-carotene 3-hydroxylase; *LUT5*: beta-ring hydroxylase; *DWARF27*: beta-carotene isomerase.

3. Discussion

3.1. Exogenous MT Alleviates the Growth Inhibition of Wheat Seedlings under Salt Stress

It is well known that crop growth, development, and production are inhibited by abiotic stresses [47]. However, plants have evolved different physiological, biochemical, and morphological strategies to cope with salt stress [48]. Growth inhibition is the most evident characteristic of plants exposed to environmental stresses [49]. The most commonly reported function of MT is to promote plant growth under abiotic stresses [50]. In the present study, exogenous MT application alleviated the inhibition of wheat growth caused by salt stress and significantly increased the plant height and biomass of wheat under salt stress (Figure 1), which is consistent with the results of earlier studies on canola, wheat, and naked oat [51–53]. Under salt stress, plant water absorption is hindered due to the disruption of the osmotic potential balance in the root cells, resulting in a physiological drought in the plant [13]. Exogenous MT promoted the accumulation of osmoregulatory substances, such as proline and soluble sugar, which maintain the osmotic balance of cells (Figure 2C,D).

Ionic toxicity occurs when Na^+ concentration reaches a threshold in plant cells, and in severe cases necrotic spots are formed in the leaves [54]. In the present study, the leaf tip wilt index of wheat under salt stress was 55.27%, whereas the leaf tip wilt index of wheat applied with exogenous MT under salt stress was 15.18% (Figure 1D), indicating that MT significantly alleviated the leaf damage caused by salt stress. Meanwhile, osmotic stress and ion toxicity induced by salt stress caused the excessive accumulation of ROS in the electron transport process, which caused oxidative damage to the cytoplasmic membrane [55,56]. MT, which can scavenge H_2O_2 directly [57], is an effective antioxidant [40]. Our study indicated that exogenous MT increased the activities of antioxidant enzymes and reduced the accumulation of ROS in wheat to reduce oxidative damage induced by salt stress (Figure 2). Zhang et al. also found that MT increased the activity of antioxidant enzymes and reduced the accumulation of ROS in cucumbers under water stress [39].

3.2. Exogenous MT Alleviates the Photosynthetic Inhibition Affected by Salt Stress

Photosynthesis is the main driver of plant growth and production [58]. Plant growth has been reported to be affected by salt stress-induced reduced photosynthesis [59]. The decrease in photosynthesis under salt stress may be related to stomatal limiting factors and non-stomatal limiting factors [60]. The salt stress-induced reduction in photosynthesis is caused by the disruption of the light-harvesting chlorophyll protein complex (LHC) and Photosystem II (PSII), the inhibition of chlorophyll biosynthesis, and the repair mechanism of PSII, blockage of electron transport, reduction in enzyme activity and CO_2 supply, and promotion of stomatal closure [61–64]. Our study further demonstrated the role of MT in maintaining photosynthesis under stressful environments. Compared to the MT-untreated group, exogenous MT increased photosynthetic parameters and chlorophyll content to improve the ability of photosynthesis in wheat under stress (Figure 3). Cui et al. reported that MT increased gs and Tr in wheat under drought stress [65], illustrating that MT may influence WUE by affecting gs. Under salt stress, the assimilation rate may be influenced by the stomatal limitation factor and non-stomatal limitation factor [66]. Stomatal limiting factors include a decrease in gs, which limits the entry of CO_2 from the air into the mesophyll cells and affects the photosynthetic rate. Non-stomatal limiting factors may reduce photosynthesis by inhibiting carbon assimilation, including decreased LHC light energy uptake, reduced PS activity, and reduced enzyme activity required for carbon fixation, resulting in the underutilization of CO_2 and increase in Ci. In this study, we demonstrated that exogenous MT has a regulatory effect on non-stomatal limiting factors, exhibiting a significant increase in Ci under low gs and a significant decrease after the application of exogenous MT under salt stress (Figure 3B). These results suggest that MT may have increased the photosynthetic rate by increasing the efficiency of carbon assimilation in photosynthesis. In the study of cold-stressed wheat seedlings, the authors considered that MT enhances photosynthetic rate through its protective effects on Rubisco expression and photosynthetic pigments, thus ensuring the normal growth of wheat [67]. Zhou et al. suggested that MT could alleviate the inhibition of photosynthetic electron transport and D1 repair cycle protein synthesis under salt stress, thereby enhancing the tolerance of photosynthetic activity to salt stress in tomatoes [66].

3.3. Exogenous MT Enhances the Light Harvesting of Photosynthesis

Photosynthesis plays an important role in plant growth and development by converting light energy into chemical energy through two stages: photoreaction and carbon fixation. During the photoreaction stage, the LHC, which binds to PSII, is mainly involved in light harvesting, transfer, and photoprotection (Figure 8) [68,69]. In this study, we found that the genes encoding Lhca (Lhca1, Lhca2, Lhca3, Lhca4, Lhca5) and Lhcb (Lhcb1, Lhcb2, Lhcb3, Lhcb4, Lhcb5, Lhcb6, Lhcb7) were abundantly up-regulated by MT in response to salt stress (Figure 8). Jiang et al. found that the expression of Lhcb1 was up-regulated under cold, heat, salt, and drought stress in Apium graveolens [70]. Under osmotic stress, MT was able to up-regulate the levels of D1, Lhcb5, and Lhcb6 proteins and promote the dephosphorylation of LHCII and D1 [71]. Light is the source of energy for photosynthesis, and when the rate of energy absorption exceeds the rate of electron transfer, it leads to photooxidative damage [72]. The previous study reported that Lhcb4 plays a decisive role in LHCII and cannot be replaced by other Lhc subunits, while deletion mutants of Lhcb4 are defective in photoprotection [73]. Meanwhile, Lhcb5 (CP26) was demonstrated to be involved in non-photochemical quenching (NPQ) related to photoprotection in PSII [74]. Furthermore, *Lhca1*, *Lhca2*, *Lhca3*, and *Lhca4* encode the four antenna proteins which are responsible for connecting PSI and LHCB [75]. Our results suggest that the application of exogenous MT may enhance photosynthesis by improving the capacity of light energy capture and photoprotection in wheat leaves under salt stress.

3.4. Exogenous MT Affects the Synthesis of Chlorophylls and Carotenoids

Chlorophylls and carotenoids are two types of photosynthetic pigments in plants. It has been shown that a part of the chlorophyll converts light energy into chemical energy at the photoreaction center, while the rest of the chlorophyll and carotenoids bind to the LHC for light energy capture and transfer [76–78]. MT can mitigate the reduction in photosynthetic pigments and improve plant growth under stressful conditions [44,71]. Similar conclusions were obtained from our study. MT significantly increased the total chlorophyll content of wheat leaves under salt stress (Figure 3F). The levels of the chlorophyll synthesis precursor L-Glutamyl-tRNA were elevated due to the significant up-regulation of *EARS* (Figure 9), and genes related to the carotenoid synthesis pathway were also regulated by MT (Figure 10). Jahan et al. found that MT significantly up-regulated the expression of *protochlorophyllide oxidoreductase* (*por*), *chlorophyll a oxygenase* (*CAO*) genes, and increased chlorophyll content, in agreement with our findings [45]. In addition, violaxanthin and neoxanthin in the carotenoid synthesis pathway are also precursors for the biosynthesis of the phytohormone ABA [79]. Carotenoids, as auxiliary pigments in photosynthesis, can direct photons not absorbed by chlorophyll molecules to the reaction centers of photosynthesis [80]. Carotenoids also play an important role in photoprotection by quenching free radical triplet-state chlorophyll and singlet oxygen, or by active NPQ, or heat dissipation, before oxidative damage occurs [80–82]. These results indicate that exogenous MT improves the light-harvesting ability and photoprotective capacity of seedlings by increasing the total chlorophyll content and regulating the chlorophyll and carotenoid synthesis and metabolism, thereby improving the salt tolerance of wheat seedlings.

4. Materials and Methods

4.1. Materials and Treatments

Wheat "Zhongmai886" (ZM886) was used in this study. MT was purchased from Bio Basic Inc. (BBI, Shanghai, China) and all other chemicals were purchased from Sinopharm Chemical Reagent Beijing Co., Ltd. (Beijing, China).

Wheat seeds were sown in plastic pots (11 cm $\times$ 9 cm) containing a mixture of grass peat and vermiculite (1:1.5). To investigate the response of wheat seedlings to melatonin under salt stress, wheat seedlings were divided into control group (CK, normal conditions) and salt stress group (S, treated with 300 mM NaCl) at V2 stage. On the same evening, half of the seedlings in each group were sprayed with 5 mL of 100 μM MT in each pot of wheat leaves (CKM and SM) and the other half of the seedlings were sprayed with the same mass of distilled water (CK and S) for 5 days. The NaCl and MT concentrations were chosen in a pre-experiment. Each group of treatments had three biological replicates. The experiment was carried out in the Wisdom Agriculture Experimental Greenhouse at Beijing University of Agriculture, where were maintained at 25/20 °C (day/night), 40% relative humidity, 600 $\mu mol \cdot m^2 \cdot s^{-1}$ photosynthetically active radiation, and 12/12 h (light/dark) photoperiod.

4.2. Morphological Observation

Morphological indices were measured after 20 days of MT treatment in seedlings. The plant height, length of the first leaf (L_1), length of the wilt and yellowed part of the first leaf (L_2), and root length of wheat seedlings were measured with a straightedge, and fresh dry weight was measured with an analytical balance. Three replicates were used for each treatment.

$$\text{Leaf tip wilt index } (\%) = L_2/L_1 \times 100\%$$

4.3. Measurement of Malondialdehyde Content and Relative Electrolyte Leakage Rate (REL)

Physiological indicators were measured after 15 days of MT treatment in seedlings. All measurements were performed in three biological replicates. The content of malondialdehyde (MDA) was determined according to the method described by Hodges [83] and Diao [84] with some modifications. A total of 0.4 g of seedling leaves or roots and 8 mL 10% trichloroacetic acid (TCA) were added into a pre-cooled mortar and ground; then, they were centrifuged at 25 °C and 4000× *g* for 10 min and the supernatant was the crude extract of MDA. A total of 2 mL of the centrifuged supernatant (2 mL of distilled water for the control) was mixed with 2 mL of 0.6% thiobarbituric acid (TBA) and the MDA absorption was spectrophotometrically measured at 450, 532, and 600 nm. Relative electrolyte leakage was determined according to the method described by Dionisio [85] and Zhang [39]. A total of 1 g of fresh wheat leaves or roots were cut into pieces and placed in the test tube with 10 mL of distilled deionized water. The extracts were incubated in a constant temperature water bath at 32 °C for 2 h and the electrical conductivity (EC1) of the extracts cooled to 25 °C was measured using an electrical conductivity meter (DDS-307A, YOKE Instrument, Shanghai, China). The samples were boiled for 20 min to release all electrolytes and cooled to 25 °C to measure the boiling conductivity (EC2). The value of distilled deionized water was also measured (EC3). The REL was expressed by the following formula:

$$\text{REL} = (\text{EC1} - \text{EC3})/(\text{EC2} - \text{EC3}) \times 100\%.$$

4.4. Measurement of Proline and Soluble Sugars Contents

The proline content was determined according to the method of Ye [86]. Briefly, 1 g of fresh wheat leaves or roots was homogenized in 10 mL of 3% sulphosalicylic acid and extracted in a boiling water bath for 10 min. Then, 2 mL extract liquid was incubated in a boiling water bath for 30 min with 2 mL ice acetic acid and 2 mL acidic ninhydrin solution. After cooling, the extract was extracted with toluene and measured at 520 nm. The soluble sugar content was determined according to the method described by Shi [87], with some modifications. The samples were incubated in a boiling water bath for 30 min. The supernatant after centrifugation (25 °C, 5000× *g*, 10 min) was mixed with 0.25 mL anthrone and 2.5 mL concentrated sulfuric acid and incubated in a boiling water bath for 10 min. The absorbance at 630 nm was measured and the content was calculated according to the standard curve of the sucrose standard.

4.5. Measurement of Reactive Oxygen Species and Antioxidant Enzyme Activities

The contents of H_2O_2 and O_2^- in leaves and roots were determined according to the methods of Zhang [88] and Ke et al. [89]. Pre-cooled acetone was used to extract the H_2O_2 in the sample and detected by monitoring the absorbance of the titanium peroxide complex at 412 nm. A total of 1 mL of 65 mM phosphate buffer (pH = 7.8) was added to 0.3 g of samples, centrifuged in a centrifuge (4 °C, 5000× *g*, 15 min), and the supernatant was the O_2^- extraction. The O_2^- content of the leaves was detected by the absorbance at 530 nm of the pink azo compound formed by the hydroxylamine reaction.

Histochemical staining for H_2O_2 and O_2^- was conducted according to the methods of Thorsten [90] and Dunand et al. [91]. H_2O_2 staining was performed by immersing fresh wheat leaves in a solution containing 1 mg/mL 3,3′-diaminobenzidine (DAB, pH = 3.0) under light-proof conditions. O_2^- staining was performed by immersing fresh wheat leaves in 100 mM Nitro blue tetrazolium (NBT, dissolved in 50 mM phosphate buffer, pH = 7.5) solution. In both cases, the leaves were shaken in a shaker for 4–5 h (37 °C, 80–100 r/min). The leaves were rinsed with ethanol/lactic acid/glycerol (3:1:1; *v*/*v*) for 15 min in a 90–95 °C water bath to remove chlorophyll and any unreacted dye for observation later.

A total of 0.2 g of sample was added in a centrifuge tube with 2 mL phosphate buffer (pH = 7.8), and then centrifuged in a centrifuge (4 °C, 12,000× g, 20 min), and finally the supernatant was collected for analysis of enzyme activities. The enzymatic activity of superoxide dismutase (SOD) was measured by its inhibition of the photochemical reduction in nitroblue tetrazolium using the method of Stewart and Bewley [92]. Peroxidase (POD) activity was determined by converting guaiacol to tetra guaiacol and measuring at 470 nm using the method of Andrea [93]. Catalase (CAT) activity was measured by detecting the change in absorbance of H_2O_2 at 240 nm using the method of Patra [94]. Ascorbate peroxidase (APX) activity was determined according to the method described in Nakano [95] and combined with the method in Zhou [96]. A total of 0.3 g of the sample was added to 2 mL of 50 mM PBS (pH = 7.8), 0.2 mM EDTA, 2 mM AsA, and then centrifuged in a centrifuge (4 °C, 12,000× g, 20 min). APX activity was assessed by measuring the change in absorbance over time at 290 nm.

4.6. Measurement of Gas Exchange Parameters and Chlorophyll Content

A portable photosynthesizer (CIRAS-3, HANSHA SCIENCE AND TECHNOLOGY GROUP CO., LIMITED, China, HongKong) was used to measure gas exchange parameters, including assimilation rate (A, μmol CO_2 m^{-2} s^{-1}), internal CO_2 (Ci, mmol mol^{-1}), stomatal conductance (gs, mmol H_2O m^{-2} s^{-1}), transport rate (Tr, mmol H_2O m^{-2} s^{-1}), and water use efficiency (WUE, mmol CO_2 mol^{-1} H_2O). Measurements were performed in a greenhouse with room and leaf temperatures maintained at 25 °C, reference CO_2 at 425 μmol mol^{-1}, and relative humidity between 60 and 80%. Chlorophyll content was measured using a chlorophyll meter (SPAD-502 PLUS, KONICA MINOLTA, Japan). Each treatment was replicated three times and the middle of the second leaf of wheat was selected for measurement.

4.7. RNA-Seq Analysis and Quantitative Real-Time PCR Assays

Total RNA was isolated from wheat leaves treated with S and SM after 24, 36, 48, and 60 h by using the Plant Total RNA Kit (Beijing Zoman, Beijing, China) according to the manufacturer's protocol. Three biological replicates were set up for each treatment group at different time points. Transcriptome sequencing was performed by the Beijing Genomics Institute (BGI, Beijing, China) based on the MGISEQ-2000 platform. Clean reads were compared to genomic sequences by Bowtie2 (v2.2.5, Ben Langmead, Baltimore, MD, USA) and then gene expression was calculated for each sample using RSEM (v1.2.8, Bo Li, Madison, WI, USA) [97]. Identification of differentially expressed genes (DEGs) between SM and S treatments was performed using DESeq2 (v1.34.0, Michael I Love, Boston, Germany) [98]. *Padj* $\leq$ 0.05 and $|\log_2\text{FoldChange}| \geq 1.0$ were used as standards to recognize the DEGs. The clusterProfile (v4.2.2, Guangchuang Yu, Guangzhou, China) was used for GO (gene ontology) enrichment analysis and (Kyoto Encyclopedia of Genes and Genomes) KEGG pathway analysis. Gene expression patterns were clustered using Mfuzz (v2.54.0, Lokesh Kumar, Berlin, Germany) in R [99].

Quantitative Real-time PCR (qRT-PCR) was used to test the reliability of the transcriptome data. Specific primers of DEGs were designed using the online quantitative primer database (https://bioinfo.ut.ee/primer3-0.4.0/, accessed on 12 April 2023). The primer sequence information of *TaActin1* (internal reference gene) and six randomly selected DEGs are shown in Table S1. LightCycler 96 Instrument (F. Hoffmann-La Roche Ltd., Indianapolis, IN, USA) was applied for qRT-PCR analysis. The samples used for the experiments were consistent with RNA-seq, with three biological replicates per treatment. Relative expression was calculated by the $2^{-\Delta\Delta CT}$ method [100].

4.8. Statistical Analysis

One-way ANOVA was performed using IBM SPSS Statistics 22 software (v22.0, IBM Corp, Armonk, NY, USA) for the evaluation of significant differences, and the final results qre expressed as mean ± standard error, with $p < 0.05$ being statistically significant.

5. Conclusions

The results of this study show that MT could effectively alleviate the growth retardation of wheat seedlings induced by salt stress. In addition, MT can reduce the accumulation of reactive oxygen species induced by salt stress, increase the activity of antioxidant enzymes, and alleviate the oxidative damage induced by salt stress. In this study, we found that exogenous MT increased the assimilation rate and chlorophyll content of wheat seedlings under salt stress, and MT improved the salt tolerance of wheat by regulating the expression of genes of photosynthetic capacity, photoprotection, and antioxidant-related pathways. Taken together, the present study revealed the regulatory mechanism of exogenous MT in improving the photosynthetic capacity and salt tolerance of seedlings under salt stress through the genes related to light capture and photoprotection through the measurement of physiological and biochemical indices and transcriptome analysis, which will provide a theoretical basis for the cultivation of new salt-tolerant varieties in the future.

Supplementary Materials: The following supporting information can be downloaded at: https://www.mdpi.com/article/10.3390/plants12233984/s1.

Author Contributions: R.L. (Runzhi Li): Conceptualization, Methodology, Supervision, Writing—Reviewing and Editing. D.Y. and J.W.: Investigation, Methodology, Data curation, Writing—Original draft preparation. Z.L. (Zhenzong Lu), R.L. (Rui Liu), Y.H. and B.S.: Investigation. Z.X., L.D., Y.W., Z.P., C.Y., Y.G. and Z.L. (Ziyan Liu): Resources. All authors have read and agreed to the published version of the manuscript.

Funding: This work was supported by grants from the General Plan for Scientific Research Projects of Beijing Education Commission (KM201910020009).

Data Availability Statement: The RNA-seq dataset in this study has been uploaded to SRA database in NCBI (BioProject ID: PRJNA1034758).

Conflicts of Interest: The authors declare no conflict of interest. The funders had no role in the design of the study; in the collection, analyses, or interpretation of data; in the writing of the manuscript; or in the decision to publish the results.

References

1. Zhu, J.-K. Plant Salt Tolerance. *Trends Plant Sci.* **2001**, *6*, 66–71. [CrossRef] [PubMed]
2. Chen, L.; Lu, B.; Liu, L.; Duan, W.; Jiang, D.; Li, J.; Zhang, K.; Sun, H.; Zhang, Y.; Li, C.; et al. Melatonin Promotes Seed Germination under Salt Stress by Regulating ABA and GA3 in Cotton (*Gossypium hirsutum* L.). *Plant Physiol. Biochem.* **2021**, *162*, 506–516. [CrossRef]
3. Wang, M.; Qi, Z.; Pei, W.; Cheng, Y.; Mao, K.; Ma, F. The Apple Argonaute Gene MdAGO1 Modulates Salt Tolerance. *Environ. Exp. Bot.* **2023**, *207*, 105202. [CrossRef]
4. Wang, M.; Xia, G. The landscape of molecular mechanisms for salt tolerance in wheat. *Crop. J.* **2018**, *6*, 42–47. [CrossRef]
5. Ju, F.; Pang, J.; Sun, L.; Gu, J.; Wang, Z.; Wu, X.; Ali, S.; Wang, Y.; Zhao, W.; Wang, S.; et al. Integrative Transcriptomic, Metabolomic and Physiological Analyses Revealed the Physiological and Molecular Mechanisms by Which Potassium Regulates the Salt Tolerance of Cotton (*Gossypium hirsutum* L.) Roots. *Ind. Crops Prod.* **2023**, *193*, 116177. [CrossRef]
6. Gao, Y.; Lu, Y.; Wu, M.; Liang, E.; Li, Y.; Zhang, D.; Yin, Z.; Ren, X.; Dai, Y.; Deng, D.; et al. Ability to Remove Na+ and Retain K+ Correlates with Salt Tolerance in Two Maize Inbred Lines Seedlings. *Front. Plant Sci.* **2016**, *7*, 1716. [CrossRef] [PubMed]
7. Wang, J.; Lv, P.; Yan, D.; Zhang, Z.; Xu, X.; Wang, T.; Wang, Y.; Peng, Z.; Yu, C.; Gao, Y.; et al. Exogenous Melatonin Improves Seed Germination of Wheat (*Triticum aestivum* L.) under Salt Stress. *Int. J. Mol. Sci.* **2022**, *23*, 8436. [CrossRef]
8. Li, S.; Wang, Y.; Gao, X.; Lan, J.; Fu, B. Comparative Physiological and Transcriptome Analysis Reveal the Molecular Mechanism of Melatonin in Regulating Salt Tolerance in Alfalfa (*Medicago sativa* L.). *Front. Plant Sci.* **2022**, *13*, 919177. [CrossRef]
9. Shahzad, B.; Rehman, A.; Tanveer, M.; Wang, L.; Park, S.K.; Ali, A. Salt Stress in Brassica: Effects, Tolerance Mechanisms, and Management. *J. Plant Growth Regul.* **2022**, *41*, 781–795. [CrossRef]
10. Ismail, A.M.; Horie, T. Genomics, Physiology, and Molecular Breeding Approaches for Improving Salt Tolerance. *Annu. Rev. Plant Biol.* **2017**, *68*, 405–434. [CrossRef]
11. Chen, M.; Yang, Z.; Liu, J.; Zhu, T.; Wei, X.; Fan, H.; Wang, B. Adaptation Mechanism of Salt Excluders under Saline Conditions and Its Applications. *Int. J. Mol. Sci.* **2018**, *19*, 3668. [CrossRef]
12. Park, H.-S.; Kazerooni, E.A.; Kang, S.-M.; Al-Sadi, A.M.; Lee, I.-J. Melatonin Enhances the Tolerance and Recovery Mechanisms in *Brassica juncea* (L.) Czern. Under Saline Conditions. *Front. Plant Sci.* **2021**, *12*, 593717. [CrossRef] [PubMed]

13. Bazihizina, N.; Colmer, T.D.; Barrett-Lennard, E.G. Response to Non-Uniform Salinity in the Root Zone of the Halophyte Atriplex Nummularia: Growth, Photosynthesis, Water Relations and Tissue Ion Concentrations. *Ann. Bot.* **2009**, *104*, 737–745. [CrossRef] [PubMed]
14. Liang, W.; Ma, X.; Wan, P.; Liu, L. Plant Salt-Tolerance Mechanism: A Review. *Biochem. Biophys. Res. Commun.* **2018**, *495*, 286–291. [CrossRef] [PubMed]
15. Hao, S.; Wang, Y.; Yan, Y.; Liu, Y.; Wang, J.; Chen, S. A Review on Plant Responses to Salt Stress and Their Mechanisms of Salt Resistance. *Horticulturae* **2021**, *7*, 132. [CrossRef]
16. Hernández, J.A.; Ferrer, M.A.; Jiménez, A.; Barceló, A.R.; Sevilla, F. Antioxidant Systems and $O_2{}^{\cdot -}/H_2O_2$ Production in the Apoplast of Pea Leaves. Its Relation with Salt-Induced Necrotic Lesions in Minor Veins. *Plant Physiol.* **2001**, *127*, 817–831. [CrossRef]
17. Zhao, G.; Zhao, Y.; Yu, X.; Kiprotich, F.; Han, H.; Guan, R.; Wang, R.; Shen, W. Nitric Oxide Is Required for Melatonin-Enhanced Tolerance against Salinity Stress in Rapeseed (*Brassica napus* L.) Seedlings. *Int. J. Mol. Sci.* **2018**, *19*, 1912. [CrossRef]
18. Zhu, J.-K. Salt and Drought Stress Signal Transduction in Plants. *Annu. Rev. Plant Biol.* **2002**, *53*, 247–273. [CrossRef]
19. Kurt-Celebi, A.; Colak, N.; Torun, H.; Dosedělová, V.; Tarkowski, P.; Ayaz, F.A. Exogenous Melatonin Ameliorates Ionizing Radiation-Induced Damage by Modulating Growth, Osmotic Adjustment and Photosynthetic Capacity in Wheat Seedlings. *Plant Physiol. Biochem.* **2022**, *187*, 67–76. [CrossRef]
20. Waadt, R.; Seller, C.A.; Hsu, P.-K.; Takahashi, Y.; Munemasa, S.; Schroeder, J.I. Plant Hormone Regulation of Abiotic Stress Responses. *Nat. Rev. Mol. Cell Biol.* **2022**, *23*, 680–694. [CrossRef]
21. Li, C.; Wang, P.; Wei, Z.; Liang, D.; Liu, C.; Yin, L.; Jia, D.; Fu, M.; Ma, F. The Mitigation Effects of Exogenous Melatonin on Salinity-Induced Stress in Malus Hupehensis. *J. Pineal Res.* **2012**, *53*, 298–306. [CrossRef]
22. Parida, A.K.; Das, A.B. Salt Tolerance and Salinity Effects on Plants: A Review. *Ecotoxicol. Environ. Saf.* **2005**, *60*, 324–349. [CrossRef] [PubMed]
23. Munns, R.; Tester, M. Mechanisms of Salinity Tolerance. *Annu. Rev. Plant Biol.* **2008**, *59*, 651–681. [CrossRef]
24. Shi, H.; Ishitani, M.; Kim, C.; Zhu, J.-K. The Arabidopsis Thaliana Salt Tolerance Gene SOS1 Encodes a Putative Na+/H+ Antiporter. *Proc. Natl. Acad. Sci. USA* **2000**, *97*, 6896–6901. [CrossRef] [PubMed]
25. Quan, R.; Lin, H.; Mendoza, I.; Zhang, Y.; Cao, W.; Yang, Y.; Shang, M.; Chen, S.; Pardo, J.M.; Guo, Y. SCABP8/CBL10, a Putative Calcium Sensor, Interacts with the Protein Kinase SOS2 to Protect Arabidopsis Shoots from Salt Stress. *Plant Cell* **2007**, *19*, 1415–1431. [CrossRef] [PubMed]
26. Shi, H.; Zhu, J.-K. Regulation of Expression of the Vacuolar Na+/H+ Antiporter Gene AtNHX1 by Salt Stress and Abscisic Acid. *Plant Mol. Biol.* **2002**, *50*, 543–550. [CrossRef]
27. Blumwald, E. Sodium Transport and Salt Tolerance in Plants. *Curr. Opin. Cell Biol.* **2000**, *12*, 431–434. [CrossRef]
28. Munns, R.; Passioura, J.B.; Colmer, T.D.; Byrt, C.S. Osmotic Adjustment and Energy Limitations to Plant Growth in Saline Soil. *New Phytol.* **2020**, *225*, 1091–1096. [CrossRef]
29. Hazman, M.; Hause, B.; Eiche, E.; Nick, P.; Riemann, M. Increased Tolerance to Salt Stress in OPDA-Deficient Rice ALLENE OXIDE CYCLASE Mutants Is Linked to an Increased ROS-Scavenging Activity. *J. Exp. Bot.* **2015**, *66*, 3339–3352. [CrossRef]
30. Yu, R.; Zuo, T.; Diao, P.; Fu, J.; Fan, Y.; Wang, Y.; Zhao, Q.; Ma, X.; Lu, W.; Li, A.; et al. Melatonin Enhances Seed Germination and Seedling Growth of Medicago Sativa Under Salinity via a Putative Melatonin Receptor MsPMTR1. *Front. Plant Sci.* **2021**, *12*, 702875. [CrossRef]
31. Calvo, J.R.; González-Yanes, C.; Maldonado, M.D. The Role of Melatonin in the Cells of the Innate Immunity: A Review. *J. Pineal Res.* **2013**, *55*, 103–120. [CrossRef]
32. Galano, A.; Tan, D.X.; Reiter, R.J. Melatonin as a Natural Ally against Oxidative Stress: A Physicochemical Examination. *J. Pineal Res.* **2011**, *51*, 1–16. [CrossRef] [PubMed]
33. Kolář, J.; Macháčková, I. Melatonin in Higher Plants: Occurrence and Possible Functions. *J. Pineal Res.* **2005**, *39*, 333–341. [CrossRef] [PubMed]
34. Hattori, A.; Migitaka, H.; Iigo, M.; Itoh, M.; Yamamoto, K.; Ohtani-Kaneko, R.; Hara, M.; Suzuki, T.; Reiter, R.J. Identification of Melatonin in Plants and Its Effects on Plasma Melatonin Levels and Binding to Melatonin Receptors in Vertebrates. *Biochem. Mol. Biol. Int.* **1995**, *35*, 627–634.
35. Dubbels, R.; Reiter, R.J.; Klenke, E.; Goebel, A.; Schnakenberg, E.; Ehlers, C.; Schiwara, H.W.; Schloot, W. Melatonin in Edible Plants Identified by Radioimmunoassay and by High Performance Liquid Chromatography-Mass Spectrometry. *J. Pineal Res.* **1995**, *18*, 28–31. [CrossRef]
36. Manchester, L.C.; Tan, D.X.; Reiter, R.J.; Park, W.; Monis, K.; Qi, W. High Levels of Melatonin in the Seeds of Edible Plants: Possible Function in Germ Tissue Protection. *Life Sci.* **2000**, *67*, 3023–3029. [CrossRef] [PubMed]
37. Arnao, M.B.; Hernández-Ruiz, J. The Physiological Function of Melatonin in Plants. *Plant Signal Behav.* **2006**, *1*, 89–95. [CrossRef]
38. Yan, F.; Wei, H.; Ding, Y.; Li, W.; Chen, L.; Ding, C.; Tang, S.; Jiang, Y.; Liu, Z.; Li, G. Melatonin Enhances Na+/K+ Homeostasis in Rice Seedlings under Salt Stress through Increasing the Root H+-Pump Activity and Na+/K+ Transporters Sensitivity to ROS/RNS. *Environ. Exp. Bot.* **2021**, *182*, 104328. [CrossRef]
39. Zhang, N.; Zhao, B.; Zhang, H.-J.; Weeda, S.; Yang, C.; Yang, Z.-C.; Ren, S.; Guo, Y.-D. Melatonin Promotes Water-Stress Tolerance, Lateral Root Formation, and Seed Germination in Cucumber (*Cucumis sativus* L.). *J. Pineal Res.* **2013**, *54*, 15–23. [CrossRef]

40. Liang, C.; Zheng, G.; Li, W.; Wang, Y.; Hu, B.; Wang, H.; Wu, H.; Qian, Y.; Zhu, X.-G.; Tan, D.-X.; et al. Melatonin Delays Leaf Senescence and Enhances Salt Stress Tolerance in Rice. *J. Pineal Res.* **2015**, *59*, 91–101. [CrossRef]
41. Kołodziejczyk, I.; Kaźmierczak, A.; Posmyk, M.M. Melatonin Application Modifies Antioxidant Defense and Induces Endoreplication in Maize Seeds Exposed to Chilling Stress. *Int. J. Mol. Sci.* **2021**, *22*, 8628. [CrossRef]
42. Ni, J.; Wang, Q.; Shah, F.A.; Liu, W.; Wang, D.; Huang, S.; Fu, S.; Wu, L. Exogenous Melatonin Confers Cadmium Tolerance by Counterbalancing the Hydrogen Peroxide Homeostasis in Wheat Seedlings. *Molecules* **2018**, *23*, 799. [CrossRef] [PubMed]
43. Han, Q.-H.; Huang, B.; Ding, C.-B.; Zhang, Z.-W.; Chen, Y.-E.; Hu, C.; Zhou, L.-J.; Huang, Y.; Liao, J.-Q.; Yuan, S.; et al. Effects of Melatonin on Anti-Oxidative Systems and Photosystem II in Cold-Stressed Rice Seedlings. *Front. Plant Sci.* **2017**, *8*, 785. [CrossRef] [PubMed]
44. Chen, Y.-E.; Mao, J.-J.; Sun, L.-Q.; Huang, B.; Ding, C.-B.; Gu, Y.; Liao, J.-Q.; Hu, C.; Zhang, Z.-W.; Yuan, S.; et al. Exogenous Melatonin Enhances Salt Stress Tolerance in Maize Seedlings by Improving Antioxidant and Photosynthetic Capacity. *Physiol. Plant* **2018**, *164*, 349–363. [CrossRef] [PubMed]
45. Jahan, M.S.; Guo, S.; Baloch, A.R.; Sun, J.; Shu, S.; Wang, Y.; Ahammed, G.J.; Kabir, K.; Roy, R. Melatonin Alleviates Nickel Phytotoxicity by Improving Photosynthesis, Secondary Metabolism and Oxidative Stress Tolerance in Tomato Seedlings. *Ecotoxicol. Environ. Saf.* **2020**, *197*, 110593. [CrossRef]
46. Tan, D.-X.; Manchester, L.C.; Terron, M.P.; Flores, L.J.; Reiter, R.J. One Molecule, Many Derivatives: A Never-Ending Interaction of Melatonin with Reactive Oxygen and Nitrogen Species? *J. Pineal Res.* **2007**, *42*, 28–42. [CrossRef] [PubMed]
47. Li, H.; Li, X.; Zhang, D.; Liu, H.; Guan, K. Effects of Drought Stress on the Seed Germination and Early Seedling Growth of the Endemic Desert Plant Eremosparton Songoricum (Fabaceae). *EXCLI J.* **2013**, *12*, 89–101. [PubMed]
48. Huang, B.; Chen, Y.-E.; Zhao, Y.-Q.; Ding, C.-B.; Liao, J.-Q.; Hu, C.; Zhou, L.-J.; Zhang, Z.-W.; Yuan, S.; Yuan, M. Exogenous Melatonin Alleviates Oxidative Damages and Protects Photosystem II in Maize Seedlings Under Drought Stress. *Front Plant Sci.* **2019**, *10*, 677. [CrossRef]
49. Zhang, R.; Sun, Y.; Liu, Z.; Jin, W.; Sun, Y. Effects of Melatonin on Seedling Growth, Mineral Nutrition, and Nitrogen Metabolism in Cucumber under Nitrate Stress. *J. Pineal Res.* **2017**, *62*, e12403. [CrossRef]
50. Park, S.; Back, K. Melatonin Promotes Seminal Root Elongation and Root Growth in Transgenic Rice after Germination. *J. Pineal Res.* **2012**, *53*, 385–389. [CrossRef]
51. Javeed, H.M.R.; Ali, M.; Skalicky, M.; Nawaz, F.; Qamar, R.; Rehman, A.U.; Faheem, M.; Mubeen, M.; Iqbal, M.M.; Rahman, M.H.U.; et al. Lipoic Acid Combined with Melatonin Mitigates Oxidative Stress and Promotes Root Formation and Growth in Salt-Stressed Canola Seedlings (*Brassica napus* L.). *Molecules* **2021**, *26*, 3147. [CrossRef] [PubMed]
52. Ke, Q.; Ye, J.; Wang, B.; Ren, J.; Yin, L.; Deng, X.; Wang, S. Melatonin Mitigates Salt Stress in Wheat Seedlings by Modulating Polyamine Metabolism. *Front Plant Sci.* **2018**, *9*, 914. [CrossRef] [PubMed]
53. Gao, W.; Feng, Z.; Bai, Q.; He, J.; Wang, Y. Melatonin-Mediated Regulation of Growth and Antioxidant Capacity in Salt-Tolerant Naked Oat under Salt Stress. *Int. J. Mol. Sci.* **2019**, *20*, 1176. [CrossRef] [PubMed]
54. Yeo, A. Molecular Biology of Salt Tolerance in the Context of Whole-Plant Physiology. *J. Exp. Bot.* **1998**, *49*, 915–929. [CrossRef]
55. Ahmad, P.; Jaleel, C.A.; Salem, M.A.; Nabi, G.; Sharma, S. Roles of Enzymatic and Nonenzymatic Antioxidants in Plants during Abiotic Stress. *Crit. Rev. Biotechnol.* **2010**, *30*, 161–175. [CrossRef] [PubMed]
56. Apel, K.; Hirt, H. Reactive Oxygen Species: Metabolism, Oxidative Stress, and Signal Transduction. *Annu. Rev. Plant Biol.* **2004**, *55*, 373–399. [CrossRef] [PubMed]
57. Li, X.; Tan, D.-X.; Jiang, D.; Liu, F. Melatonin Enhances Cold Tolerance in Drought-Primed Wild-Type and Abscisic Acid-Deficient Mutant Barley. *J. Pineal Res.* **2016**, *61*, 328–339. [CrossRef]
58. Flexas, J.; Carriquí, M. Photosynthesis and Photosynthetic Efficiencies along the Terrestrial Plant's Phylogeny: Lessons for Improving Crop Photosynthesis. *Plant J.* **2020**, *101*, 964–978. [CrossRef]
59. Hichem, H.; El Naceur, A.; Mounir, D. Effects of Salt Stress on Photosynthesis, PSII Photochemistry and Thermal Energy Dissipation in Leaves of Two Corn (*Zea mays* L.) Varieties. *Photosynthetica* **2009**, *47*, 517–526. [CrossRef]
60. das Neves, J.P.C.; Ferreira, L.F.P.; Vaz, M.M.; Gazarini, L.C. Gas Exchange in the Salt Marsh Species Atriplex Portulacoides L. and *Limoniastrum monopetalum* L. in Southern Portugal. *Acta Physiol Plant* **2008**, *30*, 91–97. [CrossRef]
61. Zhang, W.; Huang, W.; Yang, Q.-Y.; Zhang, S.-B.; Hu, H. Effect of Growth Temperature on the Electron Flow for Photorespiration in Leaves of Tobacco Grown in the Field. *Physiol. Plant.* **2013**, *149*, 141–150. [CrossRef] [PubMed]
62. Ahanger, M.A.; Agarwal, R.M. Salinity Stress Induced Alterations in Antioxidant Metabolism and Nitrogen Assimilation in Wheat (*Triticum aestivum* L) as Influenced by Potassium Supplementation. *Plant Physiol. Biochem.* **2017**, *115*, 449–460. [CrossRef] [PubMed]
63. Qin, C.; Ahanger, M.A.; Zhou, J.; Ahmed, N.; Wei, C.; Yuan, S.; Ashraf, M.; Zhang, L. Beneficial Role of Acetylcholine in Chlorophyll Metabolism and Photosynthetic Gas Exchange in Nicotiana Benthamiana Seedlings under Salinity Stress. *Plant Biol.* **2020**, *22*, 357–365. [CrossRef] [PubMed]
64. Zahra, N.; Al Hinai, M.S.; Hafeez, M.B.; Rehman, A.; Wahid, A.; Siddique, K.H.M.; Farooq, M. Regulation of Photosynthesis under Salt Stress and Associated Tolerance Mechanisms. *Plant Physiol. Biochem.* **2022**, *178*, 55–69. [CrossRef] [PubMed]
65. Cui, G.; Zhao, X.; Liu, S.; Sun, F.; Zhang, C.; Xi, Y. Beneficial Effects of Melatonin in Overcoming Drought Stress in Wheat Seedlings. *Plant Physiol. Biochem.* **2017**, *118*, 138–149. [CrossRef] [PubMed]

66. Zhou, X.; Zhao, H.; Cao, K.; Hu, L.; Du, T.; Baluška, F.; Zou, Z. Beneficial Roles of Melatonin on Redox Regulation of Photosynthetic Electron Transport and Synthesis of D1 Protein in Tomato Seedlings under Salt Stress. *Front. Plant Sci.* **2016**, *7*, 1823. [CrossRef]
67. Turk, H.; Erdal, S.; Genisel, M.; Atici, O.; Demir, Y.; Yanmis, D. The Regulatory Effect of Melatonin on Physiological, Biochemical and Molecular Parameters in Cold-Stressed Wheat Seedlings. *Plant Growth Regul.* **2014**, *74*, 139–152. [CrossRef]
68. Brestic, M.; Zivcak, M.; Kunderlikova, K.; Sytar, O.; Shao, H.; Kalaji, H.M.; Allakhverdiev, S.I. Low PSI Content Limits the Photoprotection of PSI and PSII in Early Growth Stages of Chlorophyll B-Deficient Wheat Mutant Lines. *Photosynth. Res.* **2015**, *125*, 151–166. [CrossRef]
69. Horton, P.; Ruban, A. Molecular Design of the Photosystem II Light-Harvesting Antenna: Photosynthesis and Photoprotection. *J. Exp. Bot.* **2005**, *56*, 365–373. [CrossRef]
70. Jiang, Q.; Xu, Z.-S.; Wang, F.; Li, M.-Y.; Ma, J.; Xiong, A.-S. Effects of Abiotic Stresses on the Expression of Lhcb1 Gene and Photosynthesis of Oenanthe Javanica and Apium Graveolens. *Biol. Plant* **2014**, *58*, 256–264. [CrossRef]
71. Lin, S.; Song, X.-F.; Mao, H.-T.; Li, S.-Q.; Gan, J.-Y.; Yuan, M.; Zhang, Z.-W.; Yuan, S.; Zhang, H.-Y.; Su, Y.-Q.; et al. Exogenous Melatonin Improved Photosynthetic Efficiency of Photosystem II by Reversible Phosphorylation of Thylakoid Proteins in Wheat under Osmotic Stress. *Front. Plant Sci.* **2022**, *13*, 966181. [CrossRef] [PubMed]
72. Alboresi, A.; Ballottari, M.; Hienerwadel, R.; Giacometti, G.M.; Morosinotto, T. Antenna Complexes Protect Photosystem I from Photoinhibition. *BMC Plant Biol.* **2009**, *9*, 71. [CrossRef] [PubMed]
73. de Bianchi, S.; Betterle, N.; Kouril, R.; Cazzaniga, S.; Boekema, E.; Bassi, R.; Dall'Osto, L. Arabidopsis Mutants Deleted in the Light-Harvesting Protein Lhcb4 Have a Disrupted Photosystem II Macrostructure and Are Defective in Photoprotection. *Plant Cell* **2011**, *23*, 2659–2679. [CrossRef] [PubMed]
74. Dall'Osto, L.; Caffarri, S.; Bassi, R. A Mechanism of Nonphotochemical Energy Dissipation, Independent from PsbS, Revealed by a Conformational Change in the Antenna Protein CP26. *Plant Cell* **2005**, *17*, 1217–1232. [CrossRef]
75. Liang, X.; Qiao, D.; Huang, M.; Yi, X.; Bai, L.; Xu, H.; Wei, L.; Zeng, J.; Cao, Y. Identification of a Gene Encoding the Light-Harvesting Chlorophyll a/b Proteins of Photosystem I in Green Alga Dunaliella Salina. *DNA Seq.* **2008**, *19*, 137–145. [CrossRef]
76. Jansson, S. A Guide to the Lhc Genes and Their Relatives in Arabidopsis. *Trends Plant Sci.* **1999**, *4*, 236–240. [CrossRef]
77. Demmig-Adams, B.; Adams, W.W. Photoprotection in an Ecological Context: The Remarkable Complexity of Thermal Energy Dissipation. *New Phytol.* **2006**, *172*, 11–21. [CrossRef]
78. Matsubara, S.; Krause, G.H.; Seltmann, M.; Virgo, A.; Kursar, T.A.; Jahns, P.; Winter, K. Lutein Epoxide Cycle, Light Harvesting and Photoprotection in Species of the Tropical Tree Genus Inga. *Plant Cell Environ.* **2008**, *31*, 548–561. [CrossRef]
79. Olson, J.A.; Krinsky, N.I. Introduction: The Colorful, Fascinating World of the Carotenoids: Important Physiologic Modulators. *FASEB J.* **1995**, *9*, 1547–1550. [CrossRef]
80. McElroy, J.S.; Kopsell, D.A.; Sorochan, J.C.; Sams, C.E. Response of Creeping Bentgrass Carotenoid Composition to High and Low Irradiance. *Crop. Sci.* **2006**, *46*, 2606–2612. [CrossRef]
81. Frank, H.A.; Cogdell, R.J. Carotenoids in Photosynthesis. *Photochem. Photobiol.* **1996**, *63*, 257–264. [CrossRef]
82. Demmig-Adams, B.; Gilmore, A.M.; Adams, W.W. Carotenoids 3: In Vivo Function of Carotenoids in Higher Plants. *FASEB J.* **1996**, *10*, 403–412. [CrossRef]
83. Hodges, D.M.; DeLong, J.M.; Forney, C.F.; Prange, R.K. Improving the Thiobarbituric Acid-Reactive-Substances Assay for Estimating Lipid Peroxidation in Plant Tissues Containing Anthocyanin and Other Interfering Compounds. *Planta* **1999**, *207*, 604–611. [CrossRef]
84. Diao, M.; Ma, L.; Wang, J.; Cui, J.; Fu, A.; Liu, H. Selenium Promotes the Growth and Photosynthesis of Tomato Seedlings Under Salt Stress by Enhancing Chloroplast Antioxidant Defense System. *J. Plant Growth Regul.* **2014**, *33*, 671–682. [CrossRef]
85. Dionisio-Sese, M.L.; Tobita, S. Antioxidant Responses of Rice Seedlings to Salinity Stress. *Plant Sci.* **1998**, *135*, 1–9. [CrossRef]
86. Ye, T.; Shi, H.; Wang, Y.; Chan, Z. Contrasting Changes Caused by Drought and Submergence Stresses in Bermudagrass (Cynodon Dactylon). *Front. Plant Sci.* **2015**, *6*, 951. [CrossRef] [PubMed]
87. Shi, H.; Qian, Y.; Tan, D.-X.; Reiter, R.J.; He, C. Melatonin Induces the Transcripts of CBF/DREB1s and Their Involvement in Both Abiotic and Biotic Stresses in Arabidopsis. *J. Pineal Res.* **2015**, *59*, 334–342. [CrossRef] [PubMed]
88. Zheng, C.; Jiang, D.; Liu, F.; Dai, T.; Jing, Q.; Cao, W. Effects of Salt and Waterlogging Stresses and Their Combination on Leaf Photosynthesis, Chloroplast ATP Synthesis, and Antioxidant Capacity in Wheat. *Plant Sci.* **2009**, *176*, 575–582. [CrossRef]
89. Ke, D.; Sun, G.; Wang, Z. Effects of Superoxide Radicals on ACC Synthase Activity in Chilling-Stressed Etiolated Mungbean Seedlings. *Plant Growth Regul.* **2007**, *51*, 83–91. [CrossRef]
90. Jabs, T.; Dietrich, R.A.; Dangl, J.L. Initiation of Runaway Cell Death in an Arabidopsis Mutant by Extracellular Superoxide. *Science* **1996**, *273*, 1853–1856. [CrossRef]
91. Dunand, C.; Crèvecoeur, M.; Penel, C. Distribution of Superoxide and Hydrogen Peroxide in Arabidopsis Root and Their Influence on Root Development: Possible Interaction with Peroxidases. *New Phytol.* **2007**, *174*, 332–341. [CrossRef] [PubMed]
92. Stewart, R.R.; Bewley, J.D. Lipid Peroxidation Associated with Accelerated Aging of Soybean Axes. *Plant Physiol.* **1980**, *65*, 245–248. [CrossRef] [PubMed]
93. Polle, A.; Otter, T.; Seifert, F. Apoplastic Peroxidases and Lignification in Needles of Norway Spruce (*Picea abies* L.). *Plant Physiol.* **1994**, *106*, 53–60. [CrossRef] [PubMed]

94. Patra, H.K.; Kar, M.; Mishra, D. Catalase Activity in Leaves and Cotyledons During Plant Development and Senescence1)1) Part X of the Series "Studies on Leaf Senescence". Supported in Part by a Grant from the Council of Scientific and Industrial Research, Government of India to D.M. *Biochem. Und Physiol. Der Pflanz.* **1978**, *172*, 385–390. [CrossRef]
95. Nakano, Y.; Asada, K. Hydrogen Peroxide Is Scavenged by Ascorbate-Specific Peroxidase in Spinach Chloroplasts. *Plant Cell Physiol.* **1981**, *22*, 867–880. [CrossRef]
96. Zhou, J.; Wang, J.; Li, X.; Xia, X.-J.; Zhou, Y.-H.; Shi, K.; Chen, Z.; Yu, J.-Q. H2O2 Mediates the Crosstalk of Brassinosteroid and Abscisic Acid in Tomato Responses to Heat and Oxidative Stresses. *J. Exp. Bot.* **2014**, *65*, 4371–4383. [CrossRef] [PubMed]
97. Yang, W.-J.; Du, Y.-T.; Zhou, Y.-B.; Chen, J.; Xu, Z.-S.; Ma, Y.-Z.; Chen, M.; Min, D.-H. Overexpression of TaCOMT Improves Melatonin Production and Enhances Drought Tolerance in Transgenic Arabidopsis. *Int. J. Mol. Sci.* **2019**, *20*, 652. [CrossRef]
98. Love, M.I.; Huber, W.; Anders, S. Moderated Estimation of Fold Change and Dispersion for RNA-Seq Data with DESeq2. *Genome Biol* **2014**, *15*, 550. [CrossRef]
99. Wu, T.; Hu, E.; Xu, S.; Chen, M.; Guo, P.; Dai, Z.; Feng, T.; Zhou, L.; Tang, W.; Zhan, L.; et al. clusterProfiler 4.0: A Universal Enrichment Tool for Interpreting Omics Data. *Innovation* **2021**, *2*, 100141. [CrossRef]
100. Livak, K.J.; Schmittgen, T.D. Analysis of Relative Gene Expression Data Using Real-Time Quantitative PCR and the $2^{-\Delta\Delta CT}$ Method. *Methods* **2001**, *25*, 402–408. [CrossRef]

plants

MDPI

Article

Transcriptome Analysis of Native Kentucky Bluegrass (*Poa pratensis* L.) in Response to Osmotic Stress

Jinjing Cheng [1,†], Leilei Xiang [1,†], Meizhen Yang [1,†], Ying Liu [2], Luyi Pan [3], Zhenfei Guo [4,*] and Shaoyun Lu [1,*]

1 Guangdong Engineering Research Center for Grassland Science, College of Life Sciences, South China Agricultural University, Guangzhou 510642, China; chengjj@stu.scau.edu.cn (J.C.); 20201002007@stu.scau.edu.cn (L.X.); 20191002009@stu.scau.edu.cn (M.Y.)

2 College of Grassland Science, Qinhai University, Xining 210095, China; liuying_yanhong@sina.com

3 Instrumental Analysis and Research Center, South China Agricultural University, Guangzhou 510642, China; panluyi@scau.edu.cn

4 College of Grassland Science, Nanjing Agricultural University, Nanjing 210095, China

* Correspondence: zfguo@njau.edu.cn (Z.G.); turflab@scau.edu.cn (S.L.)

† These authors contributed equally to this work.

Abstract: Kentucky bluegrass (*Poa pratensis* L.) is an important cool season turfgrass species with a high cold tolerance, but it is sensitive to drought. It is valuable for the applications of Kentucky bluegrass to improve its drought tolerance. However, little is known about the underlying drought mechanism. In the present study, transcriptomic profiling in the roots and leaves of the Kentucky bluegrass cultivar 'Qinghai', in response to osmotic stress in the form of treatment with 2 h and 50 h of 25% (*v*/*v*) PEG-6000, was analyzed. The results showed that a large number of genes were significantly up-regulated or down-regulated under osmotic stress. The majority of genes were up-regulated in leaves but down-regulated in roots after 2 h and 50 h of osmotic stress, among them were 350 up-regulated DEGs and 20 down-regulated DEGs shared in both leaves and roots. GO and KEGG analysis showed that carbohydrate metabolism, polyamine and amino acid metabolism and the plant hormone signaling pathway were enriched in the leaves and roots of 'Qinghai' after osmotic stress. The genes involving in carbohydrate metabolism were up-regulated, and sucrose, trehalose and raffinose levels were consistently increased. The genes involved in polyamine and amino acid metabolism were up-regulated in leaves in response to osmotic stress and several amino acids, such as Glu, Met and Val levels were increased, while the genes involved in photosynthesis, carbon fixation and citrate cycle in leaves were down-regulated. In addition, the genes involved in plant hormone biosynthesis and signal transduction were altered in leaves after osmotic stress. This study provided promising candidate genes for studying drought mechanisms in 'Qinghai' and improving the drought tolerance of Kentucky bluegrass and drought-sensitive crops.

Keywords: Kentucky bluegrass; osmotic stress; transcriptomes analysis; carbohydrate metabolism; polyamine and amino acid metabolism; plant hormone signaling pathway

1. Introduction

Drought is the most common environmental stress that severely restricts plant growth and development [1]. Drought stress alters the expression of thousands of genes that results in biochemical, physiological and morphological changes in plants [2]. Photosynthesis, which is considered to be one of the most crucial biological processes for the survival of plants, is greatly affected by stomatal closure during drought stress [3], and as such, carbon fixation will be hindered [4]. Plant hormones are involved in plant adaptation to drought. Abscisic acid (ABA) is accumulated after plants are exposed to drought, which regulates stomata closure and downstream gene expression through ABA signaling [3]. Cytokinin (CTK) delays leaf senescence under drought stress [4]. Auxin (IAA) and gibberellin (GA) levels are decreased in response to drought, resulting in reduced growth for saving energy [5,6]. Polyamines (PAs) play a certain role in growth and development and in resisting adverse environmental factors. Levels of PAs including putrescine (Put), spermidine (Spd), and spermine (Spm) are increased in plants under drought conditions [7]. PAs regulate the antioxidant defense system to scavenge reactive oxygen species (ROSs) under stress conditions [8]. γ-Aminobutyric acid (GABA), that is produced from polyamine oxidation, is accumulated to protect plants against drought stress by increasing osmolytes and reducing oxidative damage via antioxidants [9]. Soluble sugars are one of the small molecular osmolytes induced by drought stress, among which sucrose is the main soluble sugar. They are largely accumulated under drought stress as a result of starch degradation, while the intermediates in sugar metabolism provide a carbon skeleton for amino acid synthesis [10].

Irrigation is an essential cultivation tool for maintaining turf quality. The turfgrass species or cultivar with increased drought tolerance is always a major issue in urban landscape and sports field applications [11]. Kentucky bluegrass (*Poa pratensis* L.) is one of the most important cool-season type turfgrass species. It has a high cold tolerance but is sensitive to drought. Changes in fatty acid composition and saturation levels and antioxidant enzyme activities are involved in drought tolerance in Kentucky bluegrass [12]. The drought-tolerant cultivar 'Midnight' maintains a higher net photosynthetic rate (P_n), and higher activities of ribulose-1,5-bisphosphate carboxylase (Rubisco) and glyceraldehyde phosphate dehydrogenase (GADPH) than the drought-sensitive cultivar 'Brilliant' during drought stress [13]. In addition, drought stress-induced injury to Kentucky bluegrass is associated with hormonal alteration, and the plants with higher levels of CTK and IAA and lower levels of ABA have better photosynthetic function and performance under drought stress [14]. Exogenous application of ethephon, silicate and 5-aminolevulinic acid increases drought tolerance in Kentucky bluegrass, with improved photosynthesis and an antioxidant defense system under drought stress [15–17]. An RNA-seq analysis of the leaves of the cultivar 'Midnight II' in response to PEG-6000 treatment revealed that DEGs were enriched in "plant hormone signal transduction" and the "MAPK signaling pathway". Some up-regulated DEGs included *PYL*, *JAZ* and *BSK* involved in the hormone signaling transduction of ABA, jasmonic acid (JA) and brassinosteroid (BR) [18]. An RNA-seq approach using three germplasm sources with different drought tolerances identified transcript isoforms exhibiting a shared response of all three germplasm sources to drought stress and transcript isoforms exhibiting a tolerance response, where the more drought-tolerant germplasm sources exhibited higher transcript differences compared to the drought-susceptible cultivar [19].

A native Kentucky bluegrass cultivar named 'Qinghai' with extreme cold tolerance was selected from the collections in Dari County, Qinghai Province, which is located in the alpine cold region at an attitude of 4000 m [20]. The molecular responses to drought in 'Qinghai' have not been investigated, and yet it is important for breeders to use this special gene resource to improve drought tolerance in Kentucky bluegrass by using modern biotechnological tools. The objective of this study was to investigate transcriptomic responses in 'Qinghai' to osmotic stress at early stage of osmotic treatment (2 h to 50 h) and the enriched KEGG pathways. Based on the analysis, some candidate key genes associated with drought tolerance could be selected and identified in the future.

2. Results

2.1. Global Analysis of Gene Expression Profiles and the Differentially Expressed Genes (DEGs) in Response to Osmotic Stress

Transcriptomic analysis based on deep RNA-seq was performed to understand the global gene expression profiles in 'Qinghai' in response to osmotic stress. A total of 18 cDNA samples from leaves and roots were sequenced using the Illumina. A total of 407,753,079 raw reads were obtained (Table S1). After removing the low-quality reads and adaptor sequences, 398,144,703 clean reads were obtained (Table S1). The clean reads were then de novo assembled using Trinity software (2.4.0), and a total of 1,090,844 transcripts were obtained. The average length of the transcripts was 831 bp, and the N50 length was 1049 bp (Table S2). In addition, 569,270 unigenes were obtained with an average length of 737 bp, and the N50 length was 870 bp (Table S2). Among them 365,785 (64.25%) unigenes could be matched to at least one database. 271,475 (47.68%), 116, 438 (20.45%), 121,782 (21.39%), 204,584 (35.93%), 248,701 (43.68%) and 103,921 (18.25%) unigenes were matched to the Non-Redundant Protein Sequence Database (Nr), NCBI nucleotide sequences (Nt), the Kyoto Encyclopedia of Genes and Genomes (KO), Swiss-prot, Pfam, Gene Ontology (GO) and Clusters of Orthologous Groups of proteins (KOG), respectively, while only 23,636 (4.15%) unigenes could be matched to all databases (Table S3).

A total of 7403 DEGs (5304 up-regulated and 2099 down-regulated) and 14,057 DEGs (8835 up-regulated and 5222 down-regulated) were obtained from leaves after 2 h and 50 h of osmotic treatment, respectively, while 41,863 DEGs (10,583 up-regulated, 31,280 down-regulated) and 86,725 DEGs (12,688 up-regulated, 74,037 down-regulated) were obtained from roots after 2 h and 50 h, respectively (Figure 1a). The data indicated that more genes were altered by osmotic treatment in roots than in leaves, and the majority of genes were up-regulated in leaves but down-regulated in roots. The Venn diagram shows that 3579 DEGs among the up-regulated genes in leaves were shared at 2 h and 50 h after osmotic treatment, and they showed three trend profiles, including 2753 DEGs in profile 9, 141 DEGs in profile 6 and 659 DEGs in profile 8 (Figure 1b). Among the down-regulated genes in leaves, 1543 DEGs were shared at 2 h and 50 h after osmotic treatment, and they were shown in profile 2 (1227 DEGs), profile 1 (306 DEGs) and profile 0 (3 DEGs) (Figure 1c). Among the up-regulated genes in roots, 3668 DEGs were shared at 2 h and 50 h after osmotic treatment, and they showed patterns in profile 9 (2243 DEGs), profile 6 (680 DEGs) and profile 8 (745 DEGs), respectively (Figure 1d). Among the down-regulated genes in roots 27,252 DEGs were shared at 2 h and 50 h after osmotic treatment, and they showed patterns in profile 2 (26,213 DEGs) and profile 1 (695 DEGs) (Figure 1e).

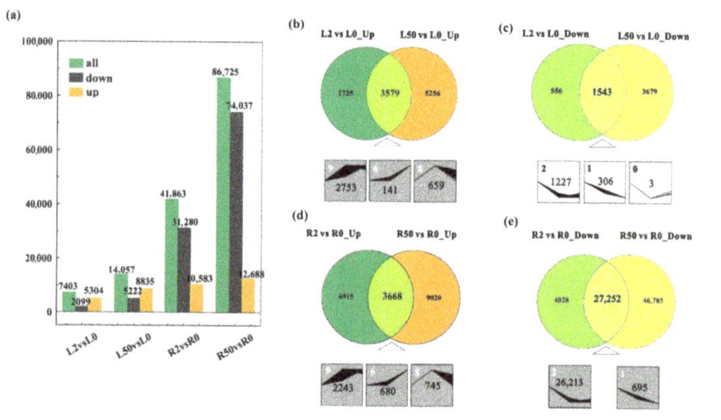

Figure 1. Global analysis of differentially expressed genes (DEGs) in leaves and roots after osmotic stress. (**a**) The number of DEGs in 'Qinghai' under osmotic stress; (**b**–**e**) the comparative analysis and trend profiles of DEGs in leaves and roots of 'Qinghai' after osmotic stress. L0, L2 and L50 indicate the genes in leaves at 0 h, 2 h or 50 h after osmotic stress. R0, R2 and R50 indicate the genes in roots at 0 h, 2 h or 50 h after osmotic stress. Expression profiles were ordered by the number of differentially expressed genes. Blocks indicate significant enrichment trends ($p \leq 0.05$). The top-left number in blocks represents the trend ID, and the different trend ID indicates different expression trends. The middle number in blocks represents the number of genes.

2.2. Analysis of Gene Ontology (GO) and KEGG Pathway Enrichment of DEGs in Leaves

The up-regulated DEGs (3579) shared at 2 h and 50 h after osmotic stress in leaves were analyzed using GO and KEGG enrichment. Based on q-value < 0.05, 208 GO terms and 18 KEGG pathways were enriched. The top 30 GO terms including 20 terms in "biological process" and ten terms in "molecular function" are listed in Figure S1a. The enriched pathways included "plant hormone signal transduction" (55 genes), "starch and sucrose metabolism" (63 genes), "galactose metabolism" (39 genes), "carotenoid biosynthesis" (21 genes), "plant-pathogen interaction" (56 genes), "phenylpropanoid biosynthesis" (32 genes), "glycerophospholipid metabolism" (31 genes), "stilbenoid, diarylheptanoid and gingerol biosynthesis" (ten genes), "diterpenoid biosynthesis" (six genes), "flavonoid biosynthesis" (nine genes), "linoleic acid metabolism" (eight genes), "arginine and proline metabolism" (21 genes), "butanoate metabolism" (12 genes), "fatty acid elongation" (ten genes), "amino sugar and nucleotide sugar metabolism" (21 genes), "beta-alanine metabolism" (14 genes), "ether lipid metabolism" (nine genes), "phenylalanine metabolism" (11 genes) (Figure 2a).

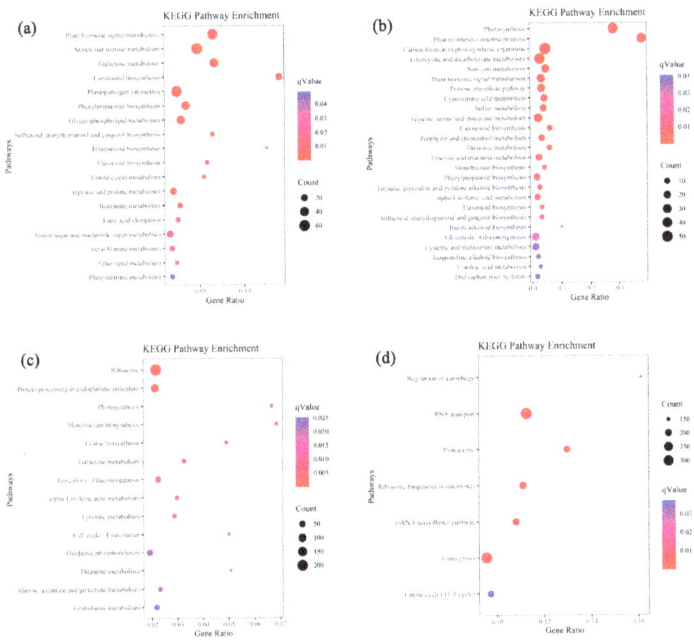

Figure 2. KEGG enrichment analysis of the up- and down-regulated DEGs in leaves and roots after osmotic stress. (**a**) The enriched KEGG pathways of the up-regulated DEGs in leaves shared at 2 h and 50 h after osmotic stress. (**b**) The enriched KEGG pathways of the down-regulated DEGs in leaves shared at 2 h and 50 h after osmotic stress. (**c**) The enriched KEGG pathways of the up-regulated DEGs in roots shared at 2 h and 50 h after osmotic stress. (**d**) The enriched KEGG pathways of the down-regulated DEGs in roots shared at 2 h and 50 h after osmotic stress. The *q*-value is represented by colors from blue to red. The size of the dark dots reflects the number of DEGs involved in each metabolism pathway.

The down-regulated DEGs (1543) shared at 2 h and 50 h after osmotic stress in leaves were analyzed using GO and KEGG enrichment. Based on *q*-value < 0.05, 286 GO terms and 26 KEGG pathways were enriched. The top 30 GO terms are listed in Figure S1b, including "biological process" (11 terms), "cellular component" (13 terms) and "molecular function" (6 terms). The enriched pathways included "photosynthesis" (46 genes), "photosynthesis-antenna proteins" (37 genes), "carbon fixation in photosynthetic organisms" (55 genes), "glyoxylate and dicarboxylate metabolism" (46 genes), "nitrogen metabolism" (24 genes), "plant hormone signal transduction" (26 genes), "pentose phosphate pathway" (24 genes), "cyanoamino acid metabolism" (17 genes), "sulfur metabolism" (17 genes), "glycine, serine and threonine metabolism" (27 genes), "carotenoid biosynthesis" (9 genes), "porphyrin and chlorophyll metabolism" (13 genes), "thiamine metabolism" (8 genes), "fructose and mannose metabolism" (15 genes), "monobactam biosynthesis" (6 genes), "phenylpropanoid biosynthesis" (15 genes), "tropane, piperidine and pyridine alkaloid biosynthesis" (seven genes), "alpha-Linolenic acid metabolism" (11 genes), "flavonoid biosynthesis" (5 genes), "stilbenoid, diarylheptanoid and gingerol biosynthesis" (five genes), "indole alkaloid biosynthesis" (two genes), "glycolysis/gluconeogenesis" (21 genes), "cysteine and methionine metabolism" (18 genes), "isoquinoline alkaloid biosynthesis" (six genes), "linoleic acid metabolism" (four genes) and "one carbon pool by folate" (seven genes) (Figure 2b).

2.3. Gene Ontology (GO) and KEGG Pathway Enrichment of DEGs in Roots

The up-regulated DEGs (3668) shared at 2 h and 50 h after osmotic stress in roots were analyzed using GO and KEGG enrichment. Based on q-value < 0.05, 401 GO terms and 14 KEGG pathways were enriched. The top 30 GO terms included "biological process" (23 terms), "cellular component" (two terms) and "molecular function" (five terms, Figure S1c). The enriched KEGG pathways included "ribosome" (221 genes), "protein processing in endoplasmic reticulum" (95 genes), "photosynthesis" (11 genes), "monobactam biosynthesis" (ten genes), "lysine biosynthesis" (11 genes), "galactose metabolism" (19 genes), "glycolysis/Gluconeogenesis" (44 genes), "alpha-linolenic acid metabolism" (19 genes), "tyrosine metabolism" (20 genes), "cell cycle-caulobacter" (eight genes), "oxidative phosphorylation" (58 genes), "thiamine metabolism" (seven genes), "alanine, aspartate and glutamate metabolism" (28 genes) and "glutathione metabolism" (32 genes, Figure 2c).

Analysis of GO and KEGG enrichment of the down-regulated DEGs (27252) shared at 2 h and 50 h after osmotic stress in roots showed that, based on q-value < 0.05, 899 GO terms and seven KEGG pathways were enriched. The top 30 GO terms included "biological process" (including ten terms), "cellular component" (including six terms) and "molecular function" (including 14 terms, Figure S1d). The enriched KEGG pathways included "regulation of autophagy" (146 genes), "RNA transport" (343 genes), "proteasome" (190 genes), "ribosome biogenesis in eukaryotes" (205 genes), "mRNA surveillance pathway" (197 genes), "endocytosis" (340 genes) and "citrate cycle (TCA cycle)" (185 genes) (Figure 2d).

2.4. DEGs Joint Analysis of the Up-Regulated and Down-Regulated in Leaves and Roots

All the up-regulated and down-regulated DEGs were analyzed using a Venn diagram. The results showed that 350 up-regulated DEGs were shared in both leaves and roots at 2 h and 50 h after osmotic treatment (Figure 3a), while 20 down-regulated DEGs were shared (Figure 3b). The up-regulated DEGs were mapped to 10 pathways (Figure 3c), including "protein processing in endoplasmic reticulum" (18 genes), "spliceosome" (16 genes), "galactose metabolism" (seven genes), "endocytosis" (15 genes), "starch and sucrose metabolism" (seven genes), "plant-pathogen interaction" (eight genes), "circadian rhythm-plant" (three genes), "glutathione metabolism" (six genes), "steroid biosynthesis" (three genes), and "indole alkaloid biosynthesis" (one gene) (Figure 3c). Three genes among twenty down-regulated DEGs could be annotated, and they were mapped to "sulfur metabolism"), "glycerophospholipid metabolism") and "cysteine and methionine metabolism" (Figure 3d).

2.5. The Genes Involving in Carbohydrate Metabolism Were Up-Regulated in Leaves and Roots in Response to Osmotic Stress

Plants accumulate sugars under drought stress. The up-regulated DEGs clustered the in sucrose and starch metabolism pathway combined with those in raffinose, trehalose and stachyose biosynthesis were further analyzed. The metabolic pathway and the key enzymes are shown in Figure 4a. The DEGs included *sucrose-phosphate synthase* (*SPS*, two genes), *sucrose synthase* (*SuSase*, six genes), *β-fructofuranosidase* (*INV*, seven genes) in sucrose biosynthesis and metabolism, *hexokinase* (*HK*, two genes), *glucan endo-1,3-beta-glucosidase 4* (*GN4*, two genes), *β-glucosidase* (*BGLX*, five genes) for fructose-6-phosphate and glucose-

6-phosphate biosynthesis. In addition, *trehalose 6-phosphate synthase* (*TPS*) and *trehalose 6-phosphate phosphatase* (*TPP*) for trehalose biosynthesis, *inositol 3-α-galactosyltransferase* (*GOLS*, eight genes), *raffinose synthases* (*RAFS*, fifteen genes) and *stachyose synthase* (*STS*, one gene) for raffinose biosynthesis were up-regulated in leaves and roots after osmotic stress. *α-Amylase* (*AMYA*) and *β-amylase* (*BMYB*, nine genes) for starch degradation and *UTP-glucose-1-phosphate uridylyltransferase* (*UGP2*), *starch synthase* (*SSS*) and *UDP-glucose 4-epimerase* (*GALE*) were also up-regulated (Figure 4b).

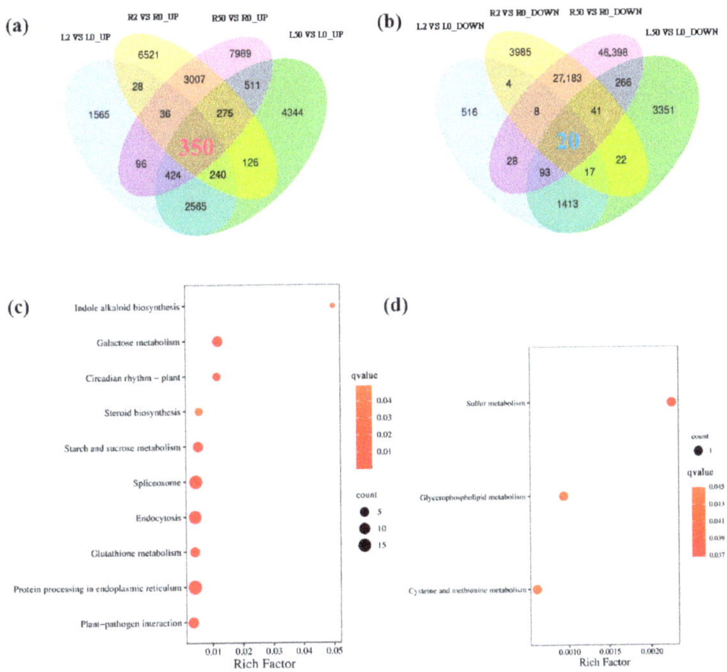

Figure 3. Comprehensive analysis of the DEGs shared in both leaves and roots. (**a**) The number of up-regulated DEGs shared in both leaves and roots after 2 h and 50 h of osmotic stress. (**b**) The number of down-regulated DEGs shared in both leaves and roots after 2 h and 50 h of osmotic stress. (**c**) The enriched KEGG pathways of the up-regulated DEGs shared in both leaves and roots after 2 h and 50 h of osmotic stress. (**d**) The enriched KEGG pathways of the down-regulated DEGs shared in both leaves and roots after 2 h and 50 h of osmotic stress. The red color digit in venn (**a**) represents the number of up-regulated DEGs share in both leaves and roots, the blue color digit in venn (**b**) represents the number of down-regulated DEGs in both leaves and roots. The *q*-value is represented by red color from light red to dark red. The size of the dark dots reflects the number of DEGs involved in each metabolism pathway.

Soluble sugars in leaves and roots in response to osmotic stress were measured. Fructose and sucrose levels were higher in leaves than in roots (Figure 4c), while trehalose and galactinol levels were higher in roots than in leaves under control condition (Figure 4d). The sucrose level was increased in leaves but not in roots after osmotic stress, while fructose and glucose levels were not altered in either leaves or roots (Figure 4c). Trehalose and raffinose levels were increased in both leaves and roots, while the galactinol level was increased in leaves but not in roots after osmotic stress (Figure 4d).

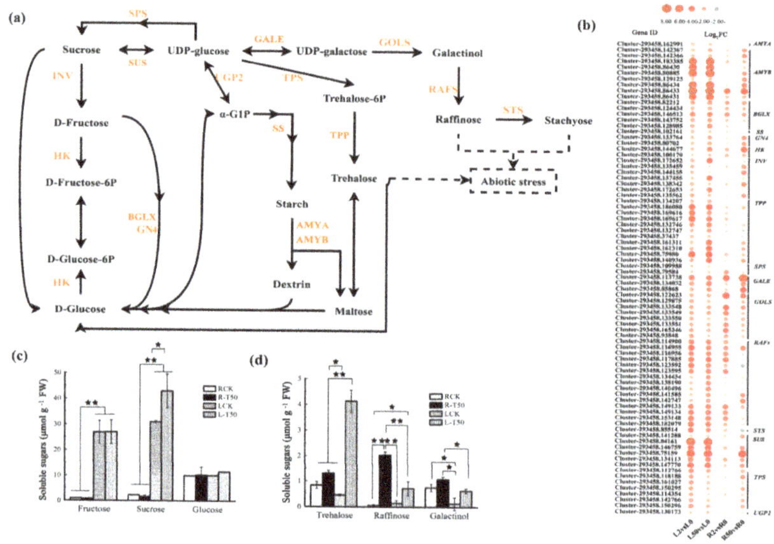

Figure 4. Analysis of DEGs involved in carbohydrate metabolism and soluble sugars in leaves and roots in response to osmotic stress. (**a**) Major metabolism pathway and key enzymes of carbohydrate metabolism. (**b**) The expression patterns of DEGs involved in carbohydrate metabolism. (**c**,**d**) Soluble sugar levels in leaves and roots in response to 50 h of osmotic stress. The dashed lines in pathway represent participation, the dashed boxes represent pathway, the solid lines in pathway represents synthesis. The color spectrum of heat map ranging from blue to red represents the $\log_2$ FC from low to high, $|\text{Log}_2\text{FC}| \geq 1$. The soluble sugars were measured after 50 h of treatment with 25% PEG. RCK and LCK indicate the control roots and leaves, respectively, while R-T50 and L-T50 indicate the samples of roots and leaves after 50 h of treatment with 25% PEG. All data are presented as means ± SE from three independent experiments. The asterisks * and ** indicate significant difference at $p < 0.05$ and 0.01, respectively.

2.6. The Genes Involving in Polyamine and Amino Acid Biosynthesis and Metabolism Were Up-Regulated in Leaves in Response to Osmotic Stress

S-adenosylmethionine decarboxylase (SpeD) and spermidine synthase (SpeE) are key enzymes for polyamine biosynthesis, while polyamine oxidase and aldehyde dehydrogenase (ALDH7A1) catalyze oxidation of polyamines to produce α-aminobutyricacid (GABA). *SpeD*, *SpeE* (three genes), *PAO2*, *3*, *4* (six genes) and *ALDH7A1* (two genes) were up-regulated after 2 h and 50 h of osmotic stress (Figure 5), indicating that polyamine biosynthesis and metabolism were involved in the response to osmotic stress.

Thirty-one genes involved in amino acid biosynthesis and metabolism were up-regulated in leaves in response to osmotic stress. *δ-1-pyrroline-5-carboxylate synthetase* (*P5CS*, seven genes) and *pyrroline-5-carboxylate reductase* (*P5CR*, two genes) involved in proline biosynthesis were up-regulated after 2 h and 50 h of osmotic stress. *Branched-chain amino acid aminotransferase* (*BCAT*, ten genes) involved in GABA biosynthesis, *chorismate mutase* (*CM*, three genes), *anthranilate synthase component II* (*ASII*, two genes), *indole-3-glycerol phosphate synthase* (*IGPS*), *anthranilate phosphoribosyltransferase* (*AnPRT*) and *phosphoribosylanthranilate isomerase* (*TRP*) involved in tryptophan biosynthesis, *4-hydroxy-tetrahydrodipicolinate synthase* (*DHDPS*, four genes) involved in lysine biosynthesis, *D-3-phosphoglycerate dehydrogenase* (*PGDH*), *L-3-cyanoalanine synthase/cysteine synthase* (*ATCYSC1*) and *cystathionine β-synthase* (*CBS*) involved in cysteine biosynthesis, *5-methyltetrahydropteroyltriglutamate-homocysteine methyltransferase* (*MHSM*) and *S-adenosylmethionine synthetase* (*SAMS*) involved in methionine biosynthesis were up-regulated after 50 h of osmotic stress (Figure 6).

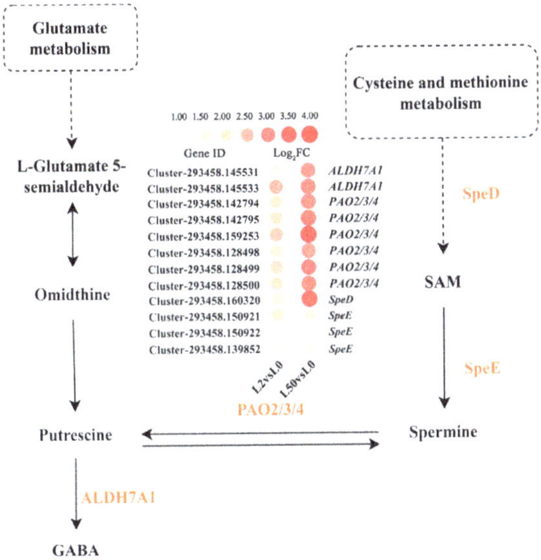

Figure 5. Analysis of DEGs involved in polyamine biosynthesis and metabolism under osmotic stress. The color spectrum ranging from yellow to red represents $\log_2$FC values from low to high, $|\mathrm{Log_2FC}| \geq 1$. The dashed lines represent source, the dashed boxes represent pathway, the solid lines represent synthesis.

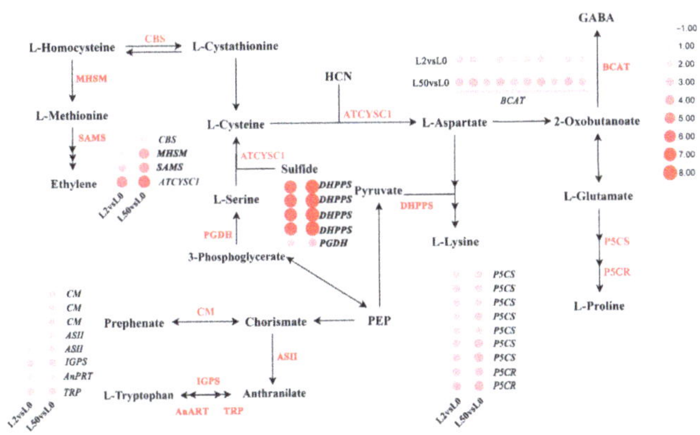

Figure 6. Analysis of DEGs involved in amino acid biosynthesis and metabolism under osmotic stress. Red color represents up-regulated genes. The color spectrum ranging from white to red represents $\log_2$FC values from low to high, $|\mathrm{Log2FC}| \geq 1$.

Free amino acids in leaves were detected in response to osmotic stress. Except Cys, Orn, Asp, Ser, Lys, Tyr and Ile, the levels of Glu, Gly, Met, Thr, Ala, Arg, His, Trp, Leu, Pro, Phe and Val in leaves were increased after 50 h of osmotic stress. The Glu, Met and Val were maintained at high levels under both control and osmotic stress conditions and were increased by osmotic stress. Pro, Val and Phe levels showed approximately 20, 36, and 33-fold increases after osmotic stress, respectively (Table 1). In addition, some of the amino acid derivatives, except Tau, Cysthi, β-AiBA and 1-Mehis, were significantly increased after osmotic stress, among them, the GABA level was increased by 43.6-fold (Table 1).

Table 1. Amino acids and their derivative levels in leaves in response to osmotic stress.

Amino Acid and Derivative	Control (μg g^{-1} DW)	Osmotic Stress (μg g^{-1} DW)
Glutamate (Glu)	642.88 ± 77.95	1142.38 ± 15.92 **
Ornithine (Orn)	0.17 ± 0.016	0.74 ± 0.34
Glycine (Gly)	0.57 ± 0.06	1.98 ± 0.47 *
Methionine (Met)	0.35 ± 0.03	2.20 ± 0.13 **
Threonine (Thr)	8.99 ± 2.47	26.50 ± 6.39 *
Asparagine (Asp)	9.61 ± 0.76	33.23 ± 10.70
Serine (Ser)	9.06 ± 2.49	40.53 ± 17.01
Alanine (Ala)	4.33 ± 0.35	21.29 ± 5.68 *
Arginine (Arg)	0.54 ± 0.19	6.22 ± 2.19 *
Lysine (Lys)	0.52 ± 0.17	9.23 ± 3.77
Tyrosine (Tyr)	0.55 ± 0.37	9.92 ± 4.44
Histidine (His)	1.17 ± 0.16	15.54 ± 5.19 *
Tryptophan (Trp)	1.04 ± 0.78	15.25 ± 5.53 *
l-isoleucine (Ile)	1.14 ± 0.36	21.32 ± 5.00
Leucine (Leu)	0.34 ± 0.14	14.68 ± 6.41 *
Proline (Pro)	5.03	102.41 ± 40.42 *
Phenylalanine (Phe)	0.92 ± 0.39	30.21 ± 10.76 *
Valine (Val)	0.94 ± 0.40	33.48 ± 12.98 **
Cysteine (Cys)	1.27 ± 0.26	1.30 ± 0.42
Taurine (Tau)	0.01 ± 0.00	0.35 ± 0.29
α-aminobutyricacid (GABA)	1.09 ± 0.44	47.60 ± 17.95 *
Cysthionine (Cysthi)	0.08 ± 0.02	0.03 ± 0.00
α-aminoadipic acid (α-AAA)	1.40 ± 0.09	13.68 ± 4.43 *
β-aminoisobutyric acid (β-AiBA)	1.14 ± 0.99	1.81 ± 1.14
P-Serine (P-Ser)	0.99 ± 0.26	2.95 ± 0.21 *
β-Alanine (β-Ala)	0.18 ± 0.05	1.71 ± 0.60 *
1-methylhistidine (1-Mehis)	0.39 ± 0.06	3.66 ± 2.44
α-aminobutyricacid (α-ABA)	0.57 ± 0.06	4.62 ± 1.14 *
Total	696.27	1573.99

Means of three independent samples and standard errors are presented. The asterisks * and ** indicate significant difference at $p < 0.05$ and 0.01, respectively.

2.7. The DEGs Involving in Photosynthesis and Carbon Fixation in Leaves Were Down-Regulated in Response to Osmotic Stress

Among the down-regulated DEGs in leaves, eighty-four genes were clustered in antenna protein (Figure 7a) and photosynthesis (Figure 7b) pathways. They included photosystem I (19 genes), photosystem II (13 genes), photosynthesis electron transport (12 genes), F-type ATPase (two genes) and the light-harvesting chlorophyll protein complex (38 genes), which were down-regulated after 2 h and 50 h of osmotic stress (Figure 7c).

The down-regulated DEGs in glyoxylate and dicarboxylate metabolism pathway were further analyzed. The pathway is shown in Figure 8b, and *ribulose-bisphosphate carboxylase small chain* (*Rubisco*) (seven genes), *phosphoglycerate kinase* (*PGK*) (two genes), *glyceraldehyde-3-phosphate dehydrogenase* (*GAPA*) (eight genes), *fructose-bisphosphate aldolase class I* (*ALDO*) (eight genes), *fructose-1,6-bisphosphatase I* (*FBPase*) (four genes) and *sedoheptulose-1,7-bisphosphatase* (*SBPase*) (two genes) were down-regulated after 2 h and 50 h of osmotic stress (Figure 8b).

2.8. The Expression of Genes Involving in Plant Hormone Biosynthesis and Signal Transduction Were Altered in Leaves in Response to Osmotic Stress

ABA is accumulated under osmotic stress to improve osmotic tolerance. Sixteen DEGs in the ABA biosynthesis pathway were up-regulated after osmotic stress, including three *beta-carotene 3-hydroxylase* (*CrtZ*), nine *9-cis-epoxycarotenoid dioxygenase* (*NCED*), one *xanthoxin dehydrogenase* (*ABA2*) and *abscisic-aldehyde oxidase 3* (*AAO3*), and two *abscisic acid 8'-hydroxylase* (*CYP707A*), but *Violaxanthin deepoxidase* (*VDE*) was down-regulated (Figure 9a). The altered expression of the above genes was consistent with ABA accumulation in plants under osmotic stress. In addition, forty-three genes involved in ABA signal transduction were up-regulated, including eight *ABA-dependent kinases SNF1-regulated protein kinase 2* (*SnRK2s*), twenty-eight *ABA negative regulated genes protein phosphatase 2C* (*PP2C*) and seven *ABA responsive element binding factor* (*ABF*), while four ABA receptor encoding genes *Pyrabactin Resistance 1-like* (*PYL*) were down-regulated (Figure 9b).

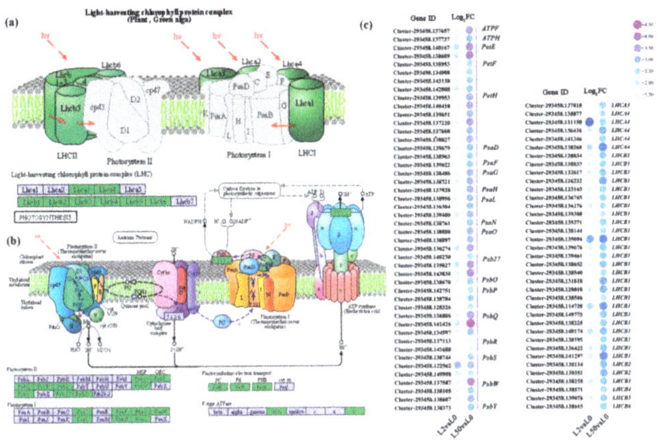

Figure 7. Analysis of DEGs involved in photosynthesis pathway in leaves. (**a**,**b**) Major pathway of photosynthesis and key antenna protein. (**c**) The expression patterns of DEGs involved in photosynthesis pathway. The green rectangles in pathway represent down-regulated DEGs, while the blue rectangles in pathway represent insignificant changes genes. The color spectrum of heat map represents log_2FC values from high to low, $|Log_2FC| \geq 1$. Schemes were retrieved from KEGG (ko00195, ko00196). Shapes and arrows follow the KEGG representation standards (https://www.kegg.jp/kegg/, accessed on 13 November 2023), except for color codes.

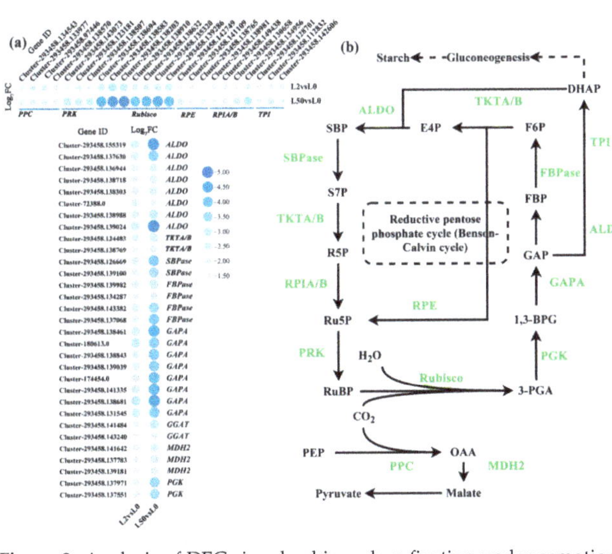

Figure 8. Analysis of DEGs involved in carbon fixation under osmotic stress. (**a**) The expression patterns of DEGs involved in carbon fixation; (**b**) major metabolism pathway and key enzymes of carbon fixation. The green digits represent down-regulated DEGs. Schemes were retrieved from KEGG (ko00710). The dashed lines represent participation, the dashed boxes represent pathway, the solid lines represent synthesis. The color spectrum of heat map ranging from white to blue represents Log_2FC values from high to low, $|Log_2FC| \geq 1$.

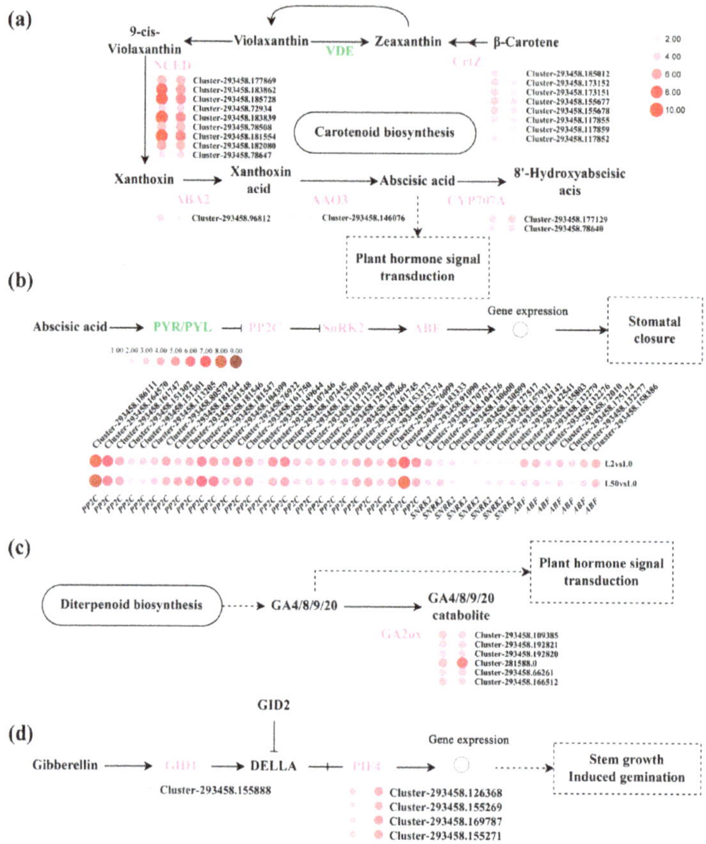

Figure 9. Analysis of DEGs involved in ABA and GA biosynthesis and signal transduction in leaves under osmotic stress. (**a**,**b**) The key enzymes and expression patterns of DGEs in the ABA biosynthesis and signal transduction. (**c**,**d**) The key enzymes and expression patterns of DGEs in the GA biosynthesis and signal transduction. The green-labeled genes represent down-regulated DGEs, while the pink-labeled genes represent up-regulated DGEs. The color spectrum of heat map ranging from white to red represents Log_2FC values from low to high, $|Log_2FC| \geq 1$. Schemes were adapted from KEGG (ko00906, ko00904, ko04075). The dashed lines represent participation, the dashed boxes represent pathway, the solid lines in (**a**,**c**) represent synthesis, the solid lines in (**b**,**d**) represent signal transduction.

Seven genes involved in gibberellins biosynthesis and the signal transduction pathway were altered after drought stress, including six *gibberellin 2-oxidase* (*GA2ox*) genes which were down-regulated and *gibberellin receptor* (*GID1*), *phytochrome-interacting factor 4* (*PIF*) which were up-regulated (Figure 9c,d). Twelve DEGs involved in the auxin signal transduction pathway were down-regulated, including three *auxin influx carriers* (*AUX1*), three *auxin-responsive proteins* (*Aux/IAA*) and two *small auxins up RNAs* (*SAUR*) (Figure S2).

2.9. The Genes Involving in Citrate Cycle in Roots Were Down-Regulated in Response to Osmotic Stress

The TCA cycle is an important aerobic pathway for the oxidation of carbohydrates and fatty acids. The cycle starts with acetyl-CoA and goes back to oxaloacetate at the end of the cycle. *Pyruvate dehydrogenase E1 component alpha subunit* (*PDHA/B/C/D*, 20 genes) expression was down-regulated. *Phosphoenolpyruvate carboxykinase ATP* (*PCKA*, ten genes) in the gluconeogenesis pathway was dramatically down-regulated (Figure 10a,b). A large number of DEGs involving in citrate cycle were down-regulated (Figure 10b,c), including *ATP citrate lyase* (*ACLY*, 11 genes), *citrate synthase* (*CS*, 14 genes), *aconitate hydratase* (*ACO*, 12 genes) and *isocitrate dehydrogenase* (*IDH1/3*, 26 genes) for the first carbon oxidation, and *2-oxoglutarate dehydrogenase E1 component* (*OGDH*, 14 genes), *2-oxoglutarate dehydrogenase E2 component* (*SUCB*, six genes), *succinyl-CoA synthetase alpha subunit* (*LSC1/2*, 19 genes), *succinate dehydrogenase (ubiquinone) flavoprotein subunit* (*SDHA/B/C*, 14 genes), *class II fumarate hydratase* (*FumA/B/C*, eight genes), *malate dehydrogenases* (*MDH1*, 20 genes).

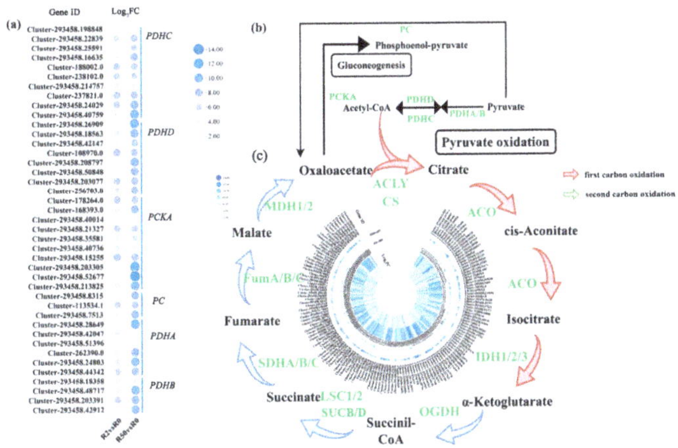

Figure 10. Analysis of DEGs involved in citrate cycle pathway in roots under osmotic stress. (**a**) The expression patterns of the DEGs involved in pyruvate oxidation and gluconeogenesis. (**b**) Major pathway and key enzymes of tyrosine biosynthesis and citrate cycle. (**c**) The key enzymes and expression patterns of the DEGs in TCA. The green-labeled genes represent down-regulated genes. The color spectrum of heat map ranging from white to blue represents log_2FC values from high to low, $|Log_2FC| \geq 1$.

2.10. Verification of Several Differentially Expressed Genes in Resposne to Osmotic Stress

Ten DEGs in leaves and roots were randomly selected for validation with qRT-PCR (Figures 11 and 12). Ten DEGs were differentially expressed among roots and leaves after osmotic stress. Among them, *ABF*, *SUS*, *AMYB*, *INV*, *HK*, *TPP* and *RAFS* expression was up-regulated, while *GAPA*, *PRK* and *Rubisco* expression was down-regulated in leaves after osmotic stress for 2 h and 50 h. *AMYB*, *INV*, *SUS*, *GAPA*, *SnRK2*, *ABF*, *PRK*, *RAFS*, *Rubisco* and *HK* expression in roots was significantly up-regulated after osmotic stress. The relative expression of these genes in leaves or roots were consistent with the RNA-seq results.

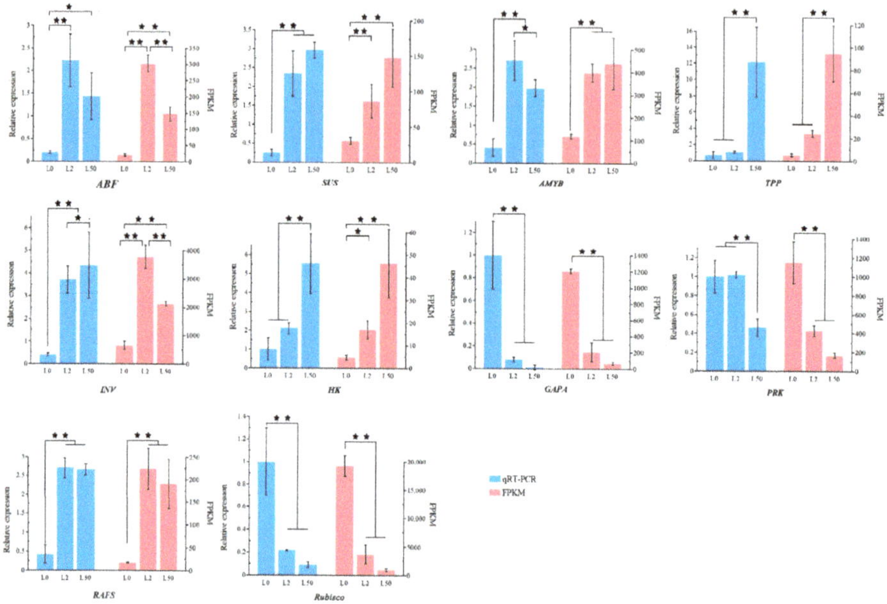

Figure 11. Validation of ten DEGs profiles in leaves using qRT-PCR. All data are presented as means ± SE from three independent experimental replicates. The asterisks * and ** above the column indicate significant difference at $p < 0.05$ and 0.01, respectively.

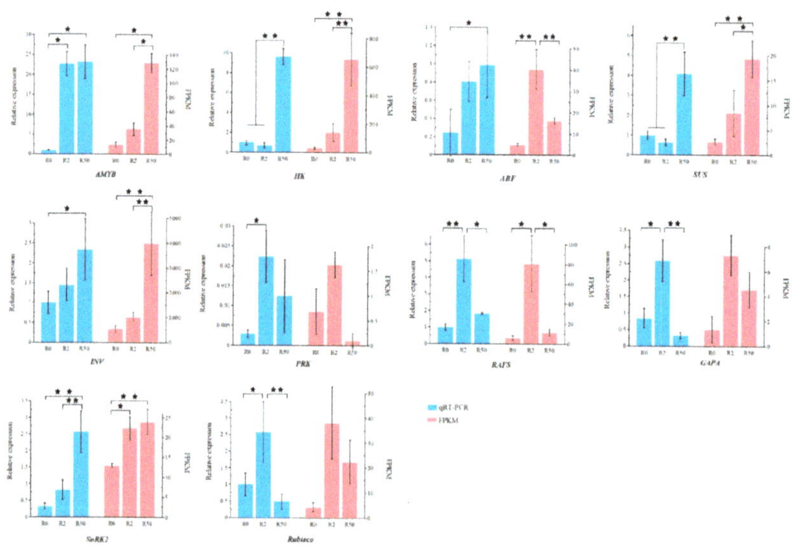

Figure 12. Validation of the ten DEG profiles in roots by qRT-PCR. All data are presented as means± SE from three independent experimental replicates. The asterisks * and ** above the column indicate significant difference at $p < 0.05$ and 0.01, respectively.

3. Discussion

Plants respond to osmotic stress by changing their morphology and physical and biochemical properties, resulting from alterations in the expression of numerous genes.

The transcriptomic profiling of 'Qinghai' in response to osmotic stress was analyzed in the present study. The results showed that 3579 genes were continuously up-regulated and 1543 genes were continuously down-regulated in leaves after 2 h and 50 h of osmotic stress, while 3668 genes were continuously up-regulated and 27,252 genes were continuously down-regulated in roots after 2 h and 50 h of osmotic stress. Some differentially enriched pathways were obtained in leaves and roots of 'Qinghai' after osmotic stress. Consistent with work on Kentucky bluegrass [19], the stress-related pathways, such as carbohydrate metabolism, polyamine and amino acid metabolism and plant hormone signaling pathway, were also differentially enriched in our research under osmotic stress. Through comparative analysis, 350 up-regulated genes and 20 down-regulated genes shared in both leaves and roots after 2 h and 50 h of osmotic stress were identified. Most of these DEGs are involved in some key biological processes, suggesting these DEGs are crucial in 'Qinghai' coping with osmotic stress.

The down-regulated genes in leaves were enriched in the Calvin–Benson cycle, photorespiration, the photosynthetic electron transport chain and antenna proteins. Photosynthesis is the first process affected by drought stress, and photosynthetic proteins have been reported to be the most affected proteins under osmotic stress [21]. The concentration of PSII and PSI proteins as well as *Lhcb* have been shown to decrease in water stress [22]. It is likely that a large number of photosynthesis genes were down-regulated expression, suggesting photosynthesis was severely restricted under osmotic stress. In addition, *ribulose-bisphosphate carboxylase small chain* (*Rubisco*), *phosphoglycerate kinase* (*PGK*), *glyceraldehyde-3-phosphate dehydrogenase* (*GAPA*), *fructose-bisphosphate aldolase class I* (*ALDO*), *fructose-1,6-bisphosphatase I* (*FBPase*) and *sedoheptulose-1,7-bisphosphatase* (*SBPase*) in carbon fixation were also down-regulated, and carbohydrate synthesis was reduced in 'Qinghai' under osmotic stress.

The TCA cycle is not only a bridge connecting carbohydrate, amino acid, lipid and protein metabolism, but also an engine to generate energy and reduce the power needed to drive metabolism [23]. The TCA cycle is also one of the most important protection systems for plants under abiotic stress [24]. Altered levels of L-asparagine and citric acid in the TCA cycle is associated with the difference in drought resistance among the soybeans [25]. Drought-resistant sorghum coped with drought stress through promoting the TCA cycle to improve sphingolipid biosynthesis [26]. In our analysis, the expression of *CS, ACO, IDH, OGDH, SUC, SDH, FUM* and *MDH* in the TCA pathway were down-regulated, suggesting that the TCA cycle in 'Qinghai' was inhibited under osmotic stress.

Carbohydrate content will directly affect many physiological processes in plants, such as photosynthesis and the glycolytic pathway [27]. Starch biosynthesis and accumulation were reported to be significantly reduced after drought stress [10]. *α-Amylase* (*AMYA*) and *β-amylase*, used for starch metabolism, were up-regulated. The degradation of starch will provide a substrate for the synthesis of other soluble sugars in response to osmotic stress. High levels of soluble sugar could improve plant resistance. SUS and SPS are key enzymes in the sucrose biosynthesis pathway, and previous research showed that SPS activity increased in wheat under drought stress [28]. INV and HK function promote the hydrolysis of sucrose into glucose. The key genes, *GOLS*, *RAF*s, *STS* and *TPS*, were involve in the trehalose and raffinose biosynthesis and were significantly up-regulated in *C. pilosula* roots after drought stress [29]. The alteration of the expression of key genes in sugar pathways are closely related to plant drought tolerance. These DEGs involved in sucrose, trehalose and raffinose biosynthesis and metabolism were up-regulated in leaves and roots of 'Qinghai', which was consistent with the increased sucrose, trehalose and raffinose concentrations after osmotic stress, while fructose and glucose levels were not altered in either leaves or roots (Figure 2c,d). In addition, the concentration of trehalose was dramatically increased in leaves of 'Qinghai' compared with the control after osmotic stress due to the constant expression of the active *TPS*s and *TPP*s in leaves, while raffinose accumulates mainly due to the high expression of *RAFS* in roots. The results were similar to many previous studies showing that oligosaccharides play an essential role in osmotic stress [30,31].

Polyamines participate in abiotic stress, and the accumulation of polyamines enhances resistance to abiotic stress in plants [32]. Spermidine can improve photosynthetic capacity and participate in hormone signal transmission under stress [33]. Overexpression of *speE* increased spermidine accumulation and enhanced tolerance to multiple environmental stresses in Arabidopsis [34]. In addition, the increased activity of *PAOs* in the backconversion pathway that catalyzes spermine and spermidine to putrescine is also involved in abiotic stress [35]. *SpeD*, *SpeE* and *PAO2, 3, 4* and *ALDH7A1* (two genes) expression was up-regulated in 'Qinghai' after 2 h and 50 h of osmotic stress, indicating that polyamine biosynthesis and metabolism were involved in the response to osmotic stress, and polyamine accumulation was associated with the strong drought tolerance of 'Qinghai'. GABA is also an important molecule and it participates in plant protection, promoting ethylene (ETH) production and affecting plant growth [36]. *ALDH7A1* expression was up-regulated, and amino acid content increased in the metabolic pathway in 'Qinghai' after osmotic stress, indicating that GABA may be also important for resisting osmotic stress in 'Qinghai'.

Phenylalanine catalyzes the oxidation of glutamate–gamma–semialdehyde into glutamate with the reduction of NAD (+) into NADH, and is involved in osmotic regulation. Glutamate plays multiple roles in abiotic stresses, such as salt, cold, heat and osmotic [37,38]. The large number of genes in the arginine and proline metabolism pathways were up-regulated after osmotic stress in osmotic-tolerant plants [39]. P5CS and P5CR are key enzymes in the proline synthesis pathway, catalyzing proline synthesis from glutamate. These observations are consistent with our analysis. Thirty-one genes involved in amino acid biosynthesis and metabolism were up-regulated in leaves in response to osmotic stress, including P5CS and P5CR. It is likely that the levels of 18 amino acids increased after osmotic stress. Among them, Pro and Phe levels showed 20 and 33.5-fold increase after osmotic stress, indicating increased proline and phenylalanine content is important for enhancing osmotic tolerance in 'Qinghai'.

ABA is a defensive phytohormone and regulates stomatal closure, gene expression and the accumulation of osmotic protectants [40]. β-carotene is the precursor of ABA biosynthesis. It is hydrolyzed to produce zeaxanthin catalyzed by CrtZ, while zeaxanthin is further oxidized to produce violaxanthin. Violaxanthin is converted to ABA via several steps catalyzed by NCED, ABA2 and AAO. *CrtZ*, *ABA2*, *NCED*, *AAO3* and *CYP707A* expression was up-regulated in 'Qinghai' after osmotic stress. The accumulated ABA activates downstream signaling components to mediate signal cross-talking with other pathways [41]. The abscisic acid receptors PYR/PYLs are a positive factor combining with ABA to inhibit PP2C by phosphorylation in ABA signal transduction. The phosphorylated PP2C release the SnRK2, while SnRK2 further regulate ABF by phosphorylation to induce related genes expression. The expression of most ZmPP2Cs were dramatically induced by multiple stresses in Maize (drought, salt, and ABA) [42]. Overexpression of ABF could improve osmotic tolerance in plants [43]. In our study, the expression of *PP2C*, *SnRK2s* and *ABF* were up-regulated, while four *PYL* genes were down-regulated after osmotic stress. There may be a complex balancing mechanism of osmotic stress in 'Qinghai'.

Gibberellins are an antagonist of ABA in the regulation of drought tolerance [44]. Gibberellins (GA) promote plant growth, but negatively regulate drought tolerance [45]. GA2ox catalyzes the oxidation of Gas, and positively regulates drought tolerance [46]. Gibberellin receptor gene GID1 was reported to regulate stomatal development. The *gid1* mutant showed impaired biosynthesis of endogenous GA under drought stress [47]. Transcription factor *PIF4* was reported to regulate auxin biosynthesis and is involved in stress response [48]. Consistent with these observations, *GA2ox*, *GID1* and *PIF4* expression was also up-regulated to synergistically regulate plant development and response to osmotic stress.

4. Materials and Methods

4.1. Plant Growth and Osmotic Stress

The Kentucky bluegrass 'Qinghai' (*Poa pratensis* cv. Qinghai) were seeded in plastic pots containing a mixture of peat and vermiculite and grown in greenhouse at 25 °C under nature light. Two-month-old seedlings were washed carefully and incubated in 1/2 Hoagland nutrient solution for one week before osmotic treatment. The seedlings were transferred to 1/2 Hoagland nutrient solution containing 25% (*v*/*v*) PEG-6000, while 1/2 Hoagland nutrient solution was used as the control. Leaves and roots were harvested after 0 h, 2 h and 50 h of stress treatment. RNA was isolated for transcriptome sequencing analysis. Free amino acids and soluble sugars were measured after 50 h of stress treatment.

4.2. RNA Isolation and RNA-Seq Analysis

Total RNA was extracted from 0.5 g roots and leaves using IZOL reagent (Invitrogen, Carlsbad, CA, USA) according to the manufacturer's instructions [26], three RNA samples at each time point were used to construct a cDNA library of repeats using Gene Denovo Biotechnology Co. (Guangzhou, China) as described by Li et al. (2022) [26] for sequencing using Illumina HiSeq TM 2500. The RNA-seq data were deposited in the sequence read archive (SRA) of the NCBI database (accession number: PRJNA1025311). The FASTQ formatted raw sequence reads were pre-processed through in-house perl scripts by removing reads containing an adapter and those with more than 5% "N" base, and low-quality reads (length > 50% of the bases at *p*-value $\leq$ 5) were removed to obtain high-quality clean data. The cleaned reads were performed de novo transcriptome assembly using Trinity (v2.90) with default settings [49].

All assembled unigenes were subjected to alignment and annotation using the following databases: Non-Redundancy Protein (NR) database (http://www.ncbi.nlm.nih.gov/, accessed on 2 June 2021), Swiss-Prot database (http://www.expasy.ch/sprot/, accessed on 2 June 2021), Pfam database (http://pfam.xfam.org/, accessed on 2 June 2021), Cluster of Orthologous Groups of proteins (COG) database (http://www.ncbi.nlm.nih.gov/COG, accessed on 2 June 2021), Gene Ontology (GO) database (http://www.geneontology.org, accessed on 2 June 2021) and the Kyoto Encyclopedia of Genes and Genomes (KEGG) database (http://www.genome.jp/kegg/, accessed on 11 November 2023) [50]. The BLASTx algorithm (v2.2.28+) was employed with an E-value threshold of $\leq 10^{-5}$ for database searches. The best BLAST hit was used to determine the sequence orientation of the unigenes

4.3. In-Depth Analysis of Differential Gene Expression

The reference sequence used for mapping the clean reads of each sample was generated from transcriptome sequences assembled using Trinity. Mapping was performed using RSEM (v1.3.1) (http://deweylab.biostat.wisc.edu/rsem/, accessed on 2 June 2021) (bowtie 2, the parameter is mismatch 0). The expression levels of individual unigenes were quantified using the metric of transcripts per kilobase of exon model per million mapped reads (TPM). Differential gene expression analysis was conducted using DESeq (v1.20.0) software; the unigenes were determined as differentially expressed genes (DEGs) with false discovery rate (FDR) $\leq$ 0.05 or absolute $\log_2$ (foldchange (FC)) value $\geq$ 1, FC = FPKM (treat)/FPKM (control), FC > 2 or FC < 0.5. Enrichment analysis for differentially expressed genes in KEGG pathways and GO terms was performed using the R package "ClusterProfile" [51], with a *q*-value $\leq$ 0.05 to identify significantly enriched GO terms and KEGG pathways.

4.4. Trend Analysis

To understand the expression patterns of DEGs, trend analysis was used to cluster genes with similar expression patterns in different time samples. Clustering of the DEGs was performed by using version 1.3.13 of Short Time-series Expression Miner (STEM) software [52], and clustered profiles were considered significant when *p* values were $\leq$0.05.

4.5. Measurements of Free Amino Acids

Fresh leaves (0.2 g) were powdered and soaked in 3 mL of 20 mM hydrochloric acid for measurements of free amino acids as previously described [53]. The mixture experienced 30 min of ultrasound vibration and was centrifuged at 12,000 rpm for 10 min. Then, 1 mL of the supernatant was mixed with 1 mL of 5% sulfosalicylic acid and placed at room temperature for 1 h and centrifuged at 12,000 rpm for 10 min. The supernatant was filtered with a 0.22 μm membrane and the free amino acids were measured using automatic amino acid analyzer Hitachi L-8900.

4.6. Measurement of Soluble Sugars

The extraction of soluble sugars was based on a previously method [54]. Fresh leaves (0.5 g) were harvested and dried at 80 °C in a baking oven. The dried leaves were powdered and soaked in 5 mL of 80% (*v/v*) ethanol in a water bath at 80 °C for 30 min. The mixture was cooled at room temperature and centrifuged at 14,000 rpm for 20 min. The supernatant was combined after 3 repeated extractions and evaporated in a water bath at 80 °C. Then, the samples were dissolved in 1 mL of ultrapure water and filtered with a 0.22 μm membrane before testing. The extracts were separated on acetonitrile-water and water with a Carbohydrate column (4.6 mm × 250 mm, 5-Micron; Agilent Technologies Inc., Santa Clara, CA, USA) and a Hi-Plex Ca column (4 mm × 250 mm, 8 μm; Agilent Technologies Inc.), respectively, and were analyzed using the Agilent Technologies 1260 Infinity II with RID detector. Sucrose, raffinose, trehalose, glucose, galactinol and fructose concentrations were calculated based on the standard curve for each sugar and calibrated with the recovery of the whole analysis procedure.

4.7. Quantitative Real-Time PCR (qRT-PCR) Analysis

Total RNA was extracted from the roots and leaves of the Kentucky bluegrass using TRIZOL reagent (Invitrogen, CA, USA) according to the manufacturer's instructions. The cDNA was synthesized using the PrimeScript™ II 1st Strand cDNA Synthesis Kit (Solarbio, Beijing, China) according to the manufacturer's instructions. The qRT-PCR was performed using SYBR® Premix Ex Taq™ II kit (Takara Biomedical Technology, Dalian, China) and a Light Cycl®96 Real-Time PCR system (Roche Life Science, Shanghai, China) according to the manufacturer's instructions. The reaction conditions were 94 °C for 5 min, followed by 40 cycles of 95 °C for 15 s and 60 °C for 1 min. The *PpActin* gene was used as the internal control. The relative expression level of gene was calculated using the $2^{-\Delta\Delta Ct}$ method [55]. The specific primers were listed in Supplementary Table S4. Three independent technical repeats and three biological replicates were performed.

5. Conclusions

Transcriptome analysis of Kentucky bluegrass cultivar 'Qinghai' showed that the majority of genes were up-regulated in leaves but down-regulated in roots in response to osmotic stress. The genes involved in stress-related pathway, such as carbohydrate metabolism, polyamine and amino acid metabolism and plant hormone signaling pathways were altered. The levels of sucrose, trehalose and raffinose, as well as amino acids, such as Glu, Val, Met and Pro, were increased after osmotic stress. Overall, our study contributes to a systematic understanding of changes in DEGs and critical metabolism in 'Qinghai' after osmotic stress. This study provides a theoretical basis for studying drought mechanisms in 'Qinghai'.

Supplementary Materials: The following supporting information can be downloaded at: https://www.mdpi.com/article/10.3390/plants12233971/s1, Table S1. Data quality summary of samples. Table S2. Summary statistics of the common vetch transcriptome assemblies. Table S3. Number of functional comments. Table S4. Specific primers of quantitative real-time PCR (qRT-PCR). Figure S1. GO enrichment analysis of the up- and down-regulated DEGs in leaves and roots. Figure S2: Analysis of DEGs involved in signal transduction of plant hormones (IAA, CTK, ETH, BR, JA) in leaves in response to osmotic stress.

Author Contributions: Conceptualization, Z.G. and S.L.; methodology, J.C.; software, J.C.; validation, J.C., L.X. and M.Y.; formal analysis, J.C.; investigation, Y.L.; resources, L.P.; data curation, J.C., L.X. and M.Y.; writing—original draft preparation, J.C., L.X. and M.Y.; writing—review and editing, Z.G. and S.L.; visualization, S.L.; supervision, Z.G. and S.L.; project administration, S.L.; funding acquisition, S.L. All authors have read and agreed to the published version of the manuscript.

Funding: This research was funded by the National Natural Science Foundation of China (U21A20241).

Data Availability Statement: Raw Illumina sequence data were deposited in the National Center for Biotechnology Information (NCBI) and be accessed in the sequence read archive (SRA) database (https://www.ncbi.nlm.nih.gov/sra, accessed on 7 October 2023). The accession number is PRJNA1025311 (https://www.ncbi.nlm.nih.gov/bioproject/PRJNA1025311, accessed on 7 October 2023), which includes 18 accession items (SAMN37714455-SAMN37714472). All data generated or analyzed during this study are included in this published article and it supplementary information files.

Conflicts of Interest: The authors declare no conflict of interest.

References

1. Gupta, A.; Rico-Medina, A.; Cao-Delgado, A.I. The physiology of plant responses to drought. *Science* **2020**, *368*, 266–269. [CrossRef]
2. Todaka, D.; Shinozaki, K.; Yamaguchi-Shinozaki, K. Recent advances in the dissection of drought-stress regulatory networks and strategies for development of drought-tolerant transgenic rice plants. *Front. Plant Sci.* **2015**, *6*, 84. [CrossRef]
3. Feller, U. Drought stress and carbon assimilation in a warming climate: Reversible and irreversible impacts. *J. Plant Physiol.* **2016**, *203*, 84–94. [CrossRef]
4. Kaur, H.; Manna, M.; Thakur, T.; Gautam, V.; Salvi, P. Imperative role of sugar signaling and transport during drought stress responses in plants. *Physiol. Plant* **2021**, *171*, 833–848. [CrossRef]
5. Mukherjee, A.; Dwivedi, S.; Bhagavatula, L.; Datta, S. Integration of light and ABA signaling pathways to combat drought stress in plants. *Plant Cell Rep.* **2023**, *42*, 829–841. [CrossRef] [PubMed]
6. Peleg, Z.; Reguera, M.; Tumimbang, E.; Walia, H.; Blumwald, E. Cytokinin-mediated source/sink modifications improve drought tolerance and increase grain yield in rice under water-stress. *Plant Biotechnol. J.* **2011**, *9*, 747–758. [CrossRef] [PubMed]
7. Wang, C.; Yang, A.; Yin, H.; Zhang, J. Influence of water stress on endogenous hormone contents and cell damage of maize seedlings. *J. Integr. Plant Biol.* **2008**, *50*, 8. [CrossRef]
8. Zhang, S.W.; Li, C.H.; Cao, J.; Zhang, Y.C.; Zhang, S.Q.; Xia, Y.F.; Sun, D.Y.; Sun, Y. Altered architecture and enhanced drought tolerance in rice via the down-regulation of Indole-3-Acetic Acid by TLD1/OsGH3.13 activation. *Plant Physiol.* **2009**, *151*, 1889–1901. [CrossRef] [PubMed]
9. Minocha, R.; Majumdar, R.; Minocha, S.C. Polyamines and abiotic stress in plants. a complex relationship. *Front. Plant Sci.* **2014**, *5*, 175. [CrossRef]
10. Liu, M.; Chen, J.; Guo, Z.; Lu, S. Differential responses of polyamines and antioxidants to drought in a centipedegrass mutant in comparison to its wild type plants. *Front. Plant Sci.* **2017**, *8*, 792. [CrossRef]
11. Zhao, Q.; Ma, Y.; Huang, X.; Song, L.; Li, N.; Qiao, M.; Li, T.; Hai, D.; Cheng, Y. GABA application enhances drought stress tolerance in wheat seedlings (*Triticum aestivum* L.). *Plants* **2023**, *12*, 2495. [CrossRef] [PubMed]
12. Du, Y.; Zhao, Q.; Chen, L.; Yao, X.; Zhang, W.; Zhang, B.; Xie, F. Effect of drought stress on sugar metabolism in leaves and roots of soybean seedlings. *Plant Physiol. Biochem.* **2020**, *146*, 1–12. [CrossRef] [PubMed]
13. Braun, R.C.; Bremer, D.J.; Ebdon, J.S.; Fry, J.D.; Patton, A.J. Review of cool-season turfgrass water use and requirements. II. responses to drought stress. *Crop. Sci.* **2022**, *62*, 1685–1701. [CrossRef]
14. Xu, L.; Han, L.; Huang, B. Membrane fatty acid composition and saturation levels associated with leaf dehydration tolerance and post-drought rehydration in Kentucky Bluegrass. *Crop. Sci.* **2011**, *51*, 273–281. [CrossRef]
15. Xu, L.X.; Yu, J.J.; Han, L.B.; Huang, B.R. Photosynthetic enzyme activities and gene expression associated with drought tolerance and post-drought recovery in Kentucky bluegrass. *Environ. Exp. Bot.* **2013**, *89*, 28–35. [CrossRef]
16. Zhang, X.; Goatley, M.; Wu, W.; Ervin, E.; Shang, C. Drought-induced injury is associated with hormonal alteration in Kentucky bluegrass. *Plant Signal. Behav.* **2019**, *14*, e1651607. [CrossRef]
17. Niu, K.; Ma, X.; Liang, G.; Ma, H.; Jia, Z.; Liu, W.; Yu, Q. 5-Aminolevulinic acid modulates antioxidant defense systems and mitigates drought-induced damage in Kentucky bluegrass seedlings. *Protoplasma* **2017**, *254*, 2038–2094. [CrossRef]
18. Saud, S.; Fahad, S.; Chen, Y.J.; Ihsan, M.Z.; Hammad, H.M.; Nasim, W.; Amanullah, J.; Arif, M.; Alharby, H. Effects of nitrogen supply on water stress and recovery mechanisms in Kentucky Bluegrass plants. *Front. Plant Sci.* **2017**, *8*, 983. [CrossRef]
19. Zhang, J.; Gao, Y.; Xu, L.; Han, L. Transcriptome analysis of Kentucky bluegrass subject to drought and ethephon treatment. *PLoS ONE* **2021**, *16*, e261472. [CrossRef]

20. Chen, Y.; Chen, Y.; Shi, Z.; Jin, Y.; Sun, H.; Xie, F.; Zhang, L. Biosynthesis and signal transduction of ABA, JA, and BRs in response to drought stress of Kentucky Bluegrass. *Int. J. Mol. Sci.* **2019**, *20*, 1289. [CrossRef]
21. Sun, H.; Shi, Q.; Liu, N.Y.; Zhang, S.B.; Huang, W. Drought stress delays photosynthetic induction and accelerates photoinhibition under short-term fluctuating light in tomato. *Plant Physiol. Biochem.* **2023**, *196*, 152–161. [CrossRef] [PubMed]
22. Dalal, V.K.; Tripathy, B.C. Water-stress induced downsizing of light-harvesting antenna complex protects developing rice seedlings from photo-oxidative damage. *Sci. Rep.* **2018**, *8*, 5955. [CrossRef]
23. Wang, L.; Fu, J.; Li, M.; Fragner, L.; Weckwerth, W.; Yang, P. Metabolomic and proteomic profiles reveal the dynamics of primary metabolism during seed development of Lotus (*Nelumbo nucifera*). *Front. Plant Sci.* **2016**, *7*, 750. [CrossRef] [PubMed]
24. Gupta, R.; Gupta, N. Tricarboxylic Acid Cycle. In *Fundamentals of Bacterial Physiology and Metabolism*, 12th ed.; Springer: Singapore, 2021; pp. 327–346.
25. Wang, X.; Wang, X.; Li, X.; Dong, S. Physiology and metabonomics reveal differences in drought resistance among soybean varieties. *Bot. Stud.* **2022**, *63*, 8. [CrossRef] [PubMed]
26. Li, Y.; Tan, B.; Wang, D.; Mu, Y.; Li, G.; Zhang, Z.; Pan, Y.; Zhu, L. Proteomic analysis revealed different molecular mechanisms of response to PEG stress in drought-sensitive and drought-resistant sorghums. *Int. J. Mol. Sci.* **2022**, *23*, 13297. [CrossRef] [PubMed]
27. Guo, J.; Qu, L.; Hu, Y.; Lu, W.; Lu, D. Proteomics reveals the effects of drought stress on the kernel development and starch formation of waxy maize. *BMC Plant Biol.* **2021**, *21*, 434. [CrossRef]
28. Dinakar, C.; Djilianov, D.; Bartels, D. Photosynthesis in desiccation tolerant plants: Energy metabolism and antioxidative stress defense. *Plant Sci.* **2012**, *182*, 29–41. [CrossRef]
29. Liang, Y.; Wei, G.; Ning, K.; Li, M.; Zhang, G.; Luo, L.; Zhao, G.; Wei, J.; Liu, Y.; Dong, L.; et al. Increase in carbohydrate content and variation in microbiome are related to the drought tolerance of Codonopsis pilosula. *Plant Physiol. Biochem.* **2021**, *165*, 19–35. [CrossRef]
30. Han, B.; Fu, L.; Zhang, D.; He, X.; Chen, Q.; Peng, M.; Zhang, J. Interspecies and intraspecies analysis of trehalose contents and the biosynthesis pathway gene family reveals crucial roles of trehalose in osmotic-stress tolerance in Cassava. *Int. J. Mol. Sci.* **2016**, *17*, 1077. [CrossRef]
31. Echeverria, A.; Larrainzar, E.; Li, W.; Watanabe, Y.; Sato, M.; Tran, C.D.; Moler, J.A.; Hirai, M.Y.; Sawada, Y.; Tran, L.P.; et al. Medicago sativa and Medicago truncatula show contrasting root metabolic responses to drought. *Front. Plant Sci.* **2021**, *12*, 652143. [CrossRef]
32. Alcázar, R.; Bueno, M.; Tiburcio, A.M. Polyamines: Small Amines with Large Effects on Plant Abiotic Stress Tolerance. *Cells* **2020**, *9*, 2373. [CrossRef]
33. Cuevas, J.C.; López-Cobollo, R.; Alcázar, R.; Zarza, X.; Koncz, C.; Altabella, T.; Salinas, J.; Tiburcio, A.F.; Ferrando, A. Putrescine as a signal to modulate the indispensable ABA increase under cold stress. *Plant Signal. Behav.* **2009**, *4*, 219–220. [CrossRef] [PubMed]
34. Huang, W.; Ma, X.; Wang, Q.; Gao, Y.; Xue, Y.; Niu, X.; Yu, G.; Liu, Y. Significant improvement of stress tolerance in tobacco plants by overexpressing a stress-responsive aldehyde dehydrogenase gene from maize (*Zea mays*). *Plant Mol. Biol.* **2008**, *68*, 451–463. [CrossRef] [PubMed]
35. Pakdel, H.; Hassani, S.; Ghotbi-Ravandi, A.; Bernard, F. Contrasting the expression pattern change of polyamine oxidase genes and photosynthetic efficiency of maize (*Zea mays* L.) genotypes under drought stress. *J. Biosci.* **2020**, *45*, 73. [CrossRef]
36. Serraj, R.; Shelp, B.J.; Sinclair, T.R. Accumulation of γ-aminobutyric acid in nodulated soybean in response to drought stress. *Physiol. Plant* **2010**, *102*, 79–86. [CrossRef] [PubMed]
37. Cheng, Y.; Zhang, X.; Sun, T.; Tian, Q.; Zhang, W.H. Glutamate receptor homolog3.4 is involved in regulation of seed germination under salt stress in Arabidopsis. *Plant Cell Physiol.* **2018**, *59*, 978–988. [CrossRef] [PubMed]
38. Li, Z.G.; Ye, X.Y.; Qiu, X.M. Glutamate signaling enhances the heat tolerance of maize seedlings by plant glutamate receptor-like channels-mediated calcium signaling. *Protoplasma* **2019**, *256*, 1165–1169. [CrossRef]
39. Hatmi, S.; Gruau, C.; Trotel-Aziz, P.; Villaume, S.; Rabenoelina, F.; Baillieul, F.; Eullaffroy, P.; Clément, C.; Ferchichi, A.; Aziz, A. Drought stress tolerance in grapevine involves activation of polyamine oxidation contributing to improved immune response and low susceptibility to Botrytis cinerea. *J. Exp. Bot.* **2015**, *66*, 775–787. [CrossRef]
40. Chen, X.; Wang, T.; Rehman, A.U.; Wang, Y.; Qi, J.; Li, Z.; Song, C.; Wang, B.; Yang, S.; Gong, Z. Arabidopsis U-box E3 ubiquitin ligase PUB11 negatively regulates drought tolerance by degrading the receptor-like protein kinases LRR1 and KIN7. *Integr. Plant Biol.* **2021**, *63*, 494–509. [CrossRef]
41. Kuromori, T.; Seo, M.; Shinozaki, K. ABA Transport and plant water stress responses. *Trends Plant Sci.* **2018**, *23*, 513–522. [CrossRef] [PubMed]
42. He, Z.; Wu, J.; Sun, X.; Dai, M. The Maize clade a PP2C phosphatases play critical roles in multiple abiotic stress responses. *Int. J. Mol. Sci.* **2019**, *20*, 3573. [CrossRef]
43. Nakashima, K.; Yamaguchi-Shinozaki, K.; Shinozaki, K. The transcriptional regulatory network in the drought response and its crosstalk in abiotic stress responses including drought, cold, and heat. *Front. Plant Sci.* **2014**, *5*, 170. [CrossRef]
44. Bushman, B.S.; Robbins, M.D.; Thorsted, K.; Robins, J.G.; Warnke, S.E.; Martin, R.; Harris-Shultz, K. Transcript responses to drought in Kentucky bluegrass (*Poa pratensis* L.) germplasm varying in their tolerance to drought stress. *Environ. Exp. Bot.* **2021**, *190*, 104571. [CrossRef]
45. Chen, H.; Li, P.; Yang, C. NAC-like gene GIBBERELLIN SUPPRESSING FACTOR regulates the gibberellin metabolic pathway in response to cold and drought stresses in Arabidopsis. *Sci. Rep.* **2019**, *9*, 19226. [CrossRef]

46. Shi, J.B.; Wang, N.; Zhou, H.; Xu, Q.H.; Yan, G.T. The role of gibberellin synthase gene GhGA2ox1 in upland cotton (*Gossypium hirsutum* L.) responses to drought and salt stress. *Biotechnol. Appl. Biochem.* **2019**, *66*, 298–308. [CrossRef] [PubMed]
47. Du, H.; Chang, Y.; Huang, F.; Xiong, L. GID1 modulates stomatal response and submergence tolerance involving abscisic acid and gibberellic acid signaling in rice. *J. Integr. Plant Biol.* **2015**, *57*, 954–968. [CrossRef] [PubMed]
48. Ueda, H.; Ito, T.; Inoue, R.; Masuda, Y.; Nagashima, Y.; Kozuka, T.; Kusaba, M. Genetic interaction among phytochrome, ethylene and abscisic acid signaling during dark-induced senescence in *Arabidopsis thaliana*. *Front. Plant Sci.* **2020**, *11*, 564. [CrossRef] [PubMed]
49. Grabherr, M.G.; Haas, B.J.; Yassour, M.; Levin, J.Z.; Thompson, D.A.; Amit, I.; Adiconis, X.; Fan, L.; Raychowdhury, R.; Zeng, Q. Full-length transcriptome assembly from RNA-Seq data without a reference genome. *Nat. Biotechnol.* **2011**, *29*, 644–652. [CrossRef]
50. Kanehisa, M.; Furumichi, M.; Sato, Y.; Kawashima, M.; Ishiguro-Watanabe, M. KEGG for taxonomy-based analysis of pathways and genomes. *Nucleic Acids Res.* **2022**, *51*, 587–592. [CrossRef]
51. Yu, G.; Wang, L.G.; Han, Y.; He, Q.Y. clusterProfiler. an R package for comparing biological themes among gene clusters. *Omics* **2012**, *16*, 284–287. [CrossRef]
52. Ernst, J.; Ziv, B.J. STEM: A tool for the analysis of short time series gene expression data. *BMC Bioinform.* **2006**, *7*, 191. [CrossRef]
53. Zhuo, C.L.; Wang, T.; Guo, Z.F.; Lu, S.Y. Overexpression of MfPIP2-7 from Medicago falcata promotes cold tolerance and growth under NO 3 deficiency in transgenic tobacco plants. *BMC Plant Biol.* **2016**, *16*, 138. [CrossRef] [PubMed]
54. Zhuo, C.; Wang, T.; Lu, S.; Zhao, Y.; Li, X.; Guo, Z. A cold responsive galactinol synthase gene from *Medicago falcata* (MfGolS1) is induced by myo-inositol and confers multiple tolerances to abiotic stresses. *Physiol. Plant.* **2013**, *149*, 67–78. [CrossRef] [PubMed]
55. Livak, K.J.; Schmittgen, T.D. Analysis of relative gene expression data using real-time quantitative PCR and the $2^{-\Delta\Delta Ct}$ Method. *Methods* **2001**, *25*, 402–408. [CrossRef] [PubMed]

plants

MDPI

Article

Identification of Morphogenesis-Related NDR Kinase Signaling Network and Its Regulation on Cold Tolerance in Maize

Ran Tian [1,†], Sidi Xie [1,†], Junjie Zhang [2], Hanmei Liu [2], Yangping Li [1], Yufeng Hu [1], Yubi Huang [1,*] and Yinghong Liu [3,*]

1 State Key Laboratory of Crop Gene Exploration and Utilization in Southwest China, Sichuan Agricultural University, Chengdu 611130, China; tianranchn@163.com (R.T.); xiesidichn@163.com (S.X.); liyangping163@163.com (Y.L.); huyufeng@sohu.com (Y.H.)

2 College of Life Science, Sichuan Agricultural University, Ya'an 625014, China; junjiezh@163.com (J.Z.); hanmeil@163.com (H.L.)

3 Maize Research Institute, Sichuan Agricultural University, Chengdu 611130, China

* Correspondence: 10024@sicau.edu.cn (Y.H.); 13964@sicau.edu.cn (Y.L.)

† These authors contributed equally to this work.

Abstract: The MOR (Morphogenesis-related NDR kinase) signaling network, initially identified in yeast, exhibits evolutionary conservation across eukaryotes and plays indispensable roles in the normal growth and development of these organisms. However, the functional role of this network and its associated genes in maize (*Zea mays*) has remained elusive until now. In this study, we identified a total of 19 maize MOR signaling network genes, and subsequent co-expression analysis revealed that 12 of these genes exhibited stronger associations with each other, suggesting their potential collective regulation of maize growth and development. Further analysis revealed significant co-expression between genes involved in the MOR signaling network and several genes related to cold tolerance. All MOR signaling network genes exhibited significant co-expression with *COLD1* (*Chilling tolerance divergence1*), a pivotal gene involved in the perception of cold stimuli, suggesting that COLD1 may directly transmit cold stress signals to MOR signaling network genes subsequent to the detection of a cold stimulus. The findings indicated that the MOR signaling network may play a crucial role in modulating cold tolerance in maize by establishing an intricate relationship with key cold tolerance genes, such as *COLD1*. Under low-temperature stress, the expression levels of certain MOR signaling network genes were influenced, with a significant up-regulation observed in *Zm00001d010720* and a notable down-regulation observed in *Zm00001d049496*, indicating that cold stress regulated the MOR signaling network. We identified and analyzed a mutant of *Zm00001d010720*, which showed a higher sensitivity to cold stress, thereby implicating its involvement in the regulation of cold stress in maize. These findings suggested that the relevant components of the MOR signaling network are also conserved in maize and this signaling network plays a vital role in modulating the cold tolerance of maize. This study offered valuable genetic resources for enhancing the cold tolerance of maize.

Keywords: maize; MOR signaling network; cold stress; cold tolerance

1. Introduction

The MOR signaling network is indispensable for the proper growth and development of eukaryotes [1,2]. The signaling pathway, initially discovered in yeast, plays a crucial role in governing cell morphogenesis, polarity, and cytokinesis in *Saccharomyces cerevisiae* [3–9]. Mutations in the MOR components of yeast result in cellular demise or impairments in cellular polarity [6,10]. The MOR signaling network is highly conserved across eukaryotes and likely plays a pivotal role in the regulation of plant stem cell maintenance as well as cell polarization [1,2].

The MOR signaling network comprises two protein kinases, KIC1 (Kinase that interacts with cell division cycle1) and CBK1 (Cell wall biosynthesis kinase1), along with three associated proteins, MOB (Mps one binder), MO25 (Mouse embryo scaffolding protein25) and TAO3/FRY (Transcriptional activator of OCH1/Furry) [1,6,11,12]. Among them, CBK1 serves as the pivotal component of this regulatory pathway. This protein is a serine-threonine protein kinase belonging to the NDR (Nuclear Dbf2 Related) kinase family and exerts its influence on cellular growth and development through phosphorylation of downstream effector proteins [1,2,13]. The activation of CBK1 is facilitated by MOB, and its interaction with CBK1 is essential for the regulation of kinase activity [6,8]. KIC1 is a GCK (germinal center kinase) in the Ste20 (sterile 20 protein) kinase family and is a MAP4K (mitogen-activated protein kinase kinase kinase kinase) which phosphorylates CBK1, thereby activating its function [12,14]. The protein MO25 functions as an activator for KIC1, while TAO3/FRY serves as a mediator linking the interaction between KIC1-MO25 and CBK1-MOB, thereby facilitating the phosphorylation of CBK1 by KIC1 [9,12,15].

Cold stress significantly impedes the growth, development, and distribution of plants, thus, posing a substantial threat to agricultural production [16]. Plants have developed diverse molecular mechanisms to effectively respond to cold stress, and several key regulatory factors in response to this environmental condition have been identified [17–20]. The perception of cold stimuli involves the cell membranes, calcium channels, and COLD1, a gene encoding G-protein regulator [18,19,21]. After perceiving the cold stimulus, plants initiate a diverse array of regulatory networks to induce the expression of *COR* (*Cold regulated*) genes [19]. The CBF/DREB1 (C-repeat binding factor/Dehydration-responsive element binding factor1)-dependent transcriptional regulatory pathway represents the central mechanism underlying plant responses to cold stress [22]. The up-regulation of the CBF/DREB1 gene in response to low temperature triggers the activation of the COR gene's expression and subsequent accumulation of protective substances, including osmolytes and cryoprotective proteins, thus enhancing cold tolerance [22]. ICE1 (Inducer of CBF expression1) and ICE2 facilitate the expression of CBF/DREB1 genes, thereby positively regulating cold tolerance [23–25]. The phosphorylation of ICE1 is a critical factor in the regulation of plant cold tolerance, and under cold stress conditions, the *Arabidopsis* protein kinase OST1/SnRK2.6 (Open stomata1/SNF1-related protein kinase2.6) phosphorylates ICE1 to enhance its transcriptional activity and stability [26]. Moreover, *Arabidopsis* MPK3 (mitogen-activated protein kinase3) and MPK6 phosphorylate ICE1, thereby attenuating its protein stability and impairing its target binding activity [27,28]. However, in rice, OsMPK3 exerts a positive regulatory role in enhancing cold resistance by impeding the degradation of OsICE1 during periods of cold stress [29].

Maize (*Zea mays*) is globally recognized as one of the most pivotal crops [30–32], with seed germination and seedling growth being particularly susceptible to cold stress, especially during the early spring season [33–36]. The growth of maize leaves and seedlings was adversely affected by the reduced rate of cell division caused by low temperatures [37]. When the temperature drops below −2.2 °C for a few minutes and remains below 0 °C for more than 4 h, the stems, leaves, and ears suffer irreparable damage [34]. The formation rate of leaves is decelerated by low temperatures, leading to a reduction in the overall leaf count [38]. Moreover, the growth of maize roots is also influenced by lower temperatures [39]. Consequently, exposure to lower temperatures can result in impaired germination, inhibited seedling growth, and even tissue or whole plant mortality, thereby potentially leading to crop failure [33,40,41]. Currently, several key genes involved in the response of maize to cold stress have been identified, such as *RR1* (*Response regulator 1*), *CesA* (*Cellulose synthase*), *MPK8*, *bZIP68* (*basic leucine zipper 68*), *bZIP113*, *TSAH1* (*Tryptophan synthase A homolog1*), *DREB1s*, and *ICE1* [42–46]. The identification of these genes holds significant implications for enhancing the cold tolerance of maize. However, the current comprehension regarding cold tolerance-related genes in maize remains incomplete. The exploration of more cold tolerance-related genes and regulatory mechanisms is of great significance for the development of cold-tolerant maize varieties. The MOR signaling network is necessary for normal growth and development of eukaryotes, potentially serving as a pivotal regulator in plant stem cell maintenance and cell polarization [1]. Furthermore, the MOR signaling network genes of *Arabidopsis* exhibit predominant expression in the shoot apical meristem and inflorescence meristem [1], and these developmental stages of maize are susceptible to cold stress [33–35]. The MOR signaling network genes potentially participate in the regulation of cold stress response in maize.

In this study, the genes comprising the MOR signaling network in maize were identified through sequence similarity searches using known components of *Arabidopsis*. The co-expression analysis was employed to ascertain the potential involvement of MOR signaling network genes in the regulation of cold stress in maize. Moreover, utilizing qRT-PCR (quantitative real-time polymerase chain reaction) analysis, it was ascertained that cold stress exerts regulatory effects on the expression of genes within the MOR signaling network. Subsequently, a mutant of one of these genes was identified and analyzed, thus elucidating its pivotal role in governing cold tolerance in maize. The results of this study will contribute to the elucidation of the regulatory mechanism underlying cold tolerance in maize and provide valuable gene resources for the breeding of cold-tolerant maize varieties.

2. Results

2.1. Identification of Pivotal Components of Maize MOR Signaling Network

The *Arabidopsis* MOR signaling network genes were employed to query the maize genome annotation database using BLASTp, resulting in the identification of a total of 19 maize MOR signaling network genes. Among these genes, only one exhibits homology to *Arabidopsis* KIC1 kinase, another one shows homology to TAO3/FRY protein, four share homology with MO25 protein, six exhibit homology with MOB protein, and seven show homology with CBK1 protein (Figure 1A; Table S1). These 19 genes together constitute the MOR signaling network of maize, which may play a crucial role in growth and development. To further ascertain whether the identified MOR pathway genes belong to the same signaling network, we conducted a co-expression analysis among these genes. The KIC1 coding gene *Zm00001d034055*; TAO3/FRY coding gene *Zm00001d051839*; MO25 coding genes *Zm00001d006710*, *Zm00001d021954*, and *Zm00001d007058*; MOB coding genes *Zm00001d049496* and *Zm00001d024510*; as well as CBK1 coding genes *Zm00001d039526*, *Zm00001d008791*, *Zm00001d009940*, *Zm00001d031040*, and *Zm00001d010720* exhibit significant co-expression with each other (Figure 1B; Table S2). The results of co-expression analysis unveiled a robust interrelationship among these 12 genes out of the 19 MOR signaling network genes identified in this study, implying their potential collective regulation on the growth and development of maize. Furthermore, the interactions among these 12 genes were analyzed utilizing the Pathway Mapping function provided by the MaizeNetome website (http://minteractome.ncpgr.cn/, accessed on 15 August 2023) [47], unveiling potential interconnections and implying their participation in a common regulatory pathway (Figure 1C). Therefore, we have chosen these 12 genes as essential components of the maize MOR signaling network for further examination.

2.2. Co-Expression Analysis of MOR Signaling Network Genes and Cold Tolerance-Related Genes Revealed the Potential Regulatory Mechanism of Cold Tolerance in Maize

To elucidate the contribution of MOR signaling network genes to cold stress tolerance in maize, we conducted a co-expression analysis between MOR signaling network genes and cold tolerance-related genes. The findings revealed a significant co-expression between genes involved in the MOR signaling network and multiple cold tolerance-related genes (Figure 2A; Table S3). All MOR signaling network genes showed significant co-expression with *COLD1*, a pivotal gene involved in the perception of cold stimuli [18]. These findings suggested that upon detection of a cold stimulus, COLD1 may directly transmit a signal of cold stress to genes involved in the MOR signaling network. Furthermore, eight, eight, seven, and six MOR signaling network genes were significantly co-expressed with *bZIP68*, *MPK6*, *CESA4*, and *ICE2*, respectively (Figure 2A; Table S3). These genes have previously been identified as pivotal regulators of cold tolerance in maize and *Arabidopsis* [24,44–46]. These results indicated that the MOR signaling network may be involved in the regulation of cold tolerance of maize. Moreover, we used the Pathway Mapping function of the MaizeNetome website (http://minteractome.ncpgr.cn/, accessed on 15 August 2023) [47] to analyze the interaction between cold tolerance-related genes and *CBK1* genes, as CBK1 potentially belongs to the MOR signaling network member that directly governs cold tolerance-related genes. The potential interaction between CBK1 and these cold tolerance-related genes further suggests the involvement of the MOR pathway in regulating cold tolerance (Figure 2B).

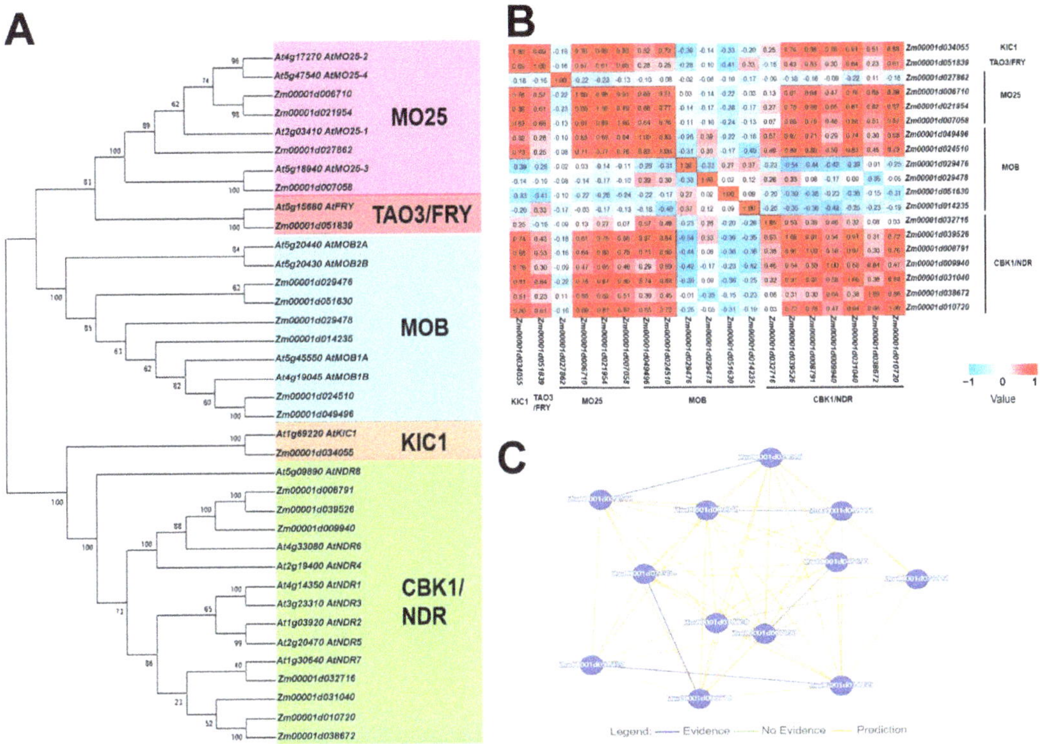

Figure 1. Identification of MOR signaling network genes in maize. (**A**) Phylogenetic analysis of MOR signaling network genes in maize and *Arabidopsis*. A total of 19 maize MOR signaling network genes were identified. The neighbor-joining algorithm is employed for the assessment of evolutionary distances. The numbers at the nodes represent the percentage of 1000 bootstraps. (**B**) Co-expression analysis of 19 maize MOR signaling network genes. In total. 12 of the genes had stronger correlations. (**C**) These 12 genes potentially interact with each other. The network map was generated using the Pathway Mapping function available on the MaizeNetome website (http://minteractome.ncpgr.cn./, accessed on 15 August 2023).

2.3. The Expression of Certain MOR Signaling Network Genes in Maize Was Regulated by Cold Stress

We subjected maize B73 seedlings to cold treatment and assessed the expression of *Zm00001d034055*, *Zm00001d051839*, *Zm00001d006710*, *Zm00001d049496*, and *Zm00001d010720*, following cold stress using qRT-PCR. These five genes encode KIC1, TAO3/FRY, MO25, MOB, and CBK1, respectively, and represent key components of the MOR pathway. The results of gene expression analysis revealed a significant up-regulation in the expression of *Zm00001d010720* following cold treatment, as compared to the control, while there was a significant down-regulation in the expression level of Zm00001d049496 (Figure 3). Furthermore, the expression levels of *Zm00001d034055*, *Zm00001d051839*, and *Zm00001d006710* did not show any significant variation following cold treatment compared to the control (Figure 3). The findings suggested that the expression of some MOR signaling network genes was regulated by cold stress, potentially contributing to the modulation of cold tolerance in maize.

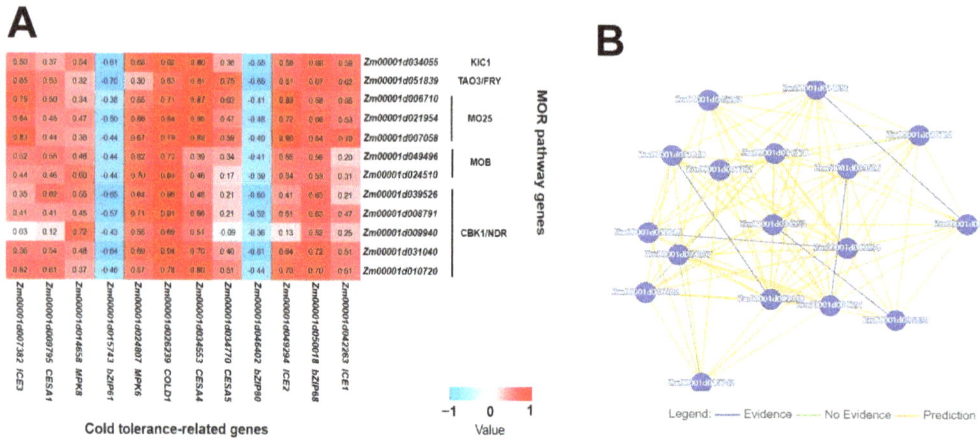

Figure 2. The MOR signaling network potentially participates in the regulation of cold tolerance in maize. (**A**) MOR signaling network genes and cold tolerance-related genes were significantly co-expressed. (**B**) There are potential interactions between CBK1 and genes related to cold tolerance. The network map was generated using the Pathway Mapping function available on the MaizeNetome website (http://minteractome.ncpgr.cn/, accessed on 15 August 2023).

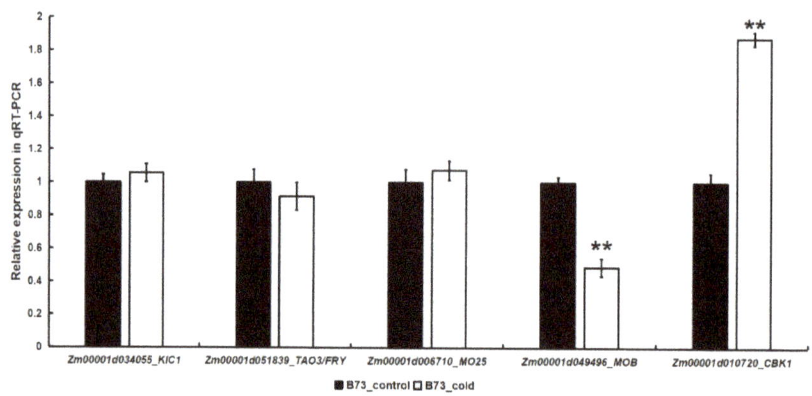

Figure 3. Cold stimulation regulated the expression of certain genes in maize in the MOR signaling network. ** significant at $p < 0.01$ by the Student's t test.

2.4. The Mutant of zm00001d010720 Exhibited Heightened Susceptibility to Cold Stress

We further identified and analyzed a mutant *zm00001d010720*, which encodes the CBK1 kinase. The mutation of *Zm00001d010720* is a C-to-T substitution in exon 5, which results in the transformation of a glutamine residue to a premature stop codon, resulting in the partial deletion of the catalytic domain of kinase in Zm00001d010720, potentially impacting its protein functionality (Figure 4A). Fourteen-day-old WT (wild type) and *zm00001d010720* seedlings were subjected to a 24-h cold treatment at 4 °C and then we observed the growth of the plants after two days of recovery at 25 °C. After cold treatment, the *zm00001d010720* seedling exhibited wilting and subsequent mortality, whereas the WT seedling displayed normal growth (Figure 4C,E,G). However, both the WT and *zm00001d010720* seedlings without cold treatment could grow normally in the same period (Figure 4B,D,F). The results demonstrated that the mutation in the *Zm00001d010720* gene, which encodes CBK1 kinase, significantly impaired the cold tolerance of maize. This finding suggested that this gene plays a regulatory role in modulating the response of maize to low-temperature stress.

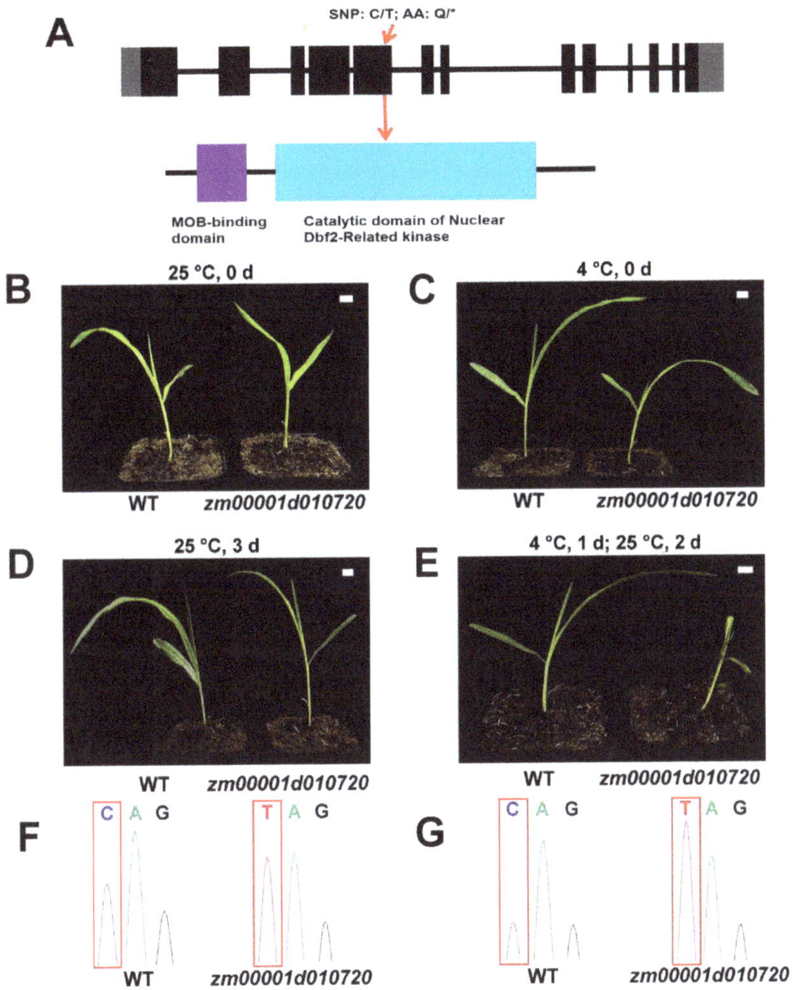

Figure 4. The mutant *zm00001d010720* showed increased susceptibility to cold stress. (**A**) Schematic diagram of *Zm00001d010720* gene with indicated mutation sites (**top**) and protein conserved domains (**bottom**). Black boxes represent coding regions, gray boxes represent the 5′ and 3′ untranslated regions, and lines represent introns. SNP, single-nucleotide polymorphism; AA, amino acids; *, stop gained. (**B**) Before cold treatment, the phenotypes of wild type and mutant seedlings in control group. Scale bar, 1 cm. (**C**) Before cold treatment, the phenotypes of wild type and mutant seedlings in cold stress group. Scale bar, 1 cm. (**D**) The phenotype of wild type and mutant seedlings in control group after 3 days. Scale bar, 1 cm. (**E**) The phenotypes of wild type and mutant seedlings after treatment at 4 °C for one day and recovery at 25 °C for two days. The mutant seedling displayed wilting and subsequent mortality. Scale bar, 1 cm. (**F**) Mutation site analysis of wild type and mutant seedlings in control group. (**G**) Mutation site analysis of wild type and mutant seedlings in cold stress group.

3. Discussion

3.1. Relevant Components of the MOR Signaling Network Exhibit Conservation in Maize

In this study, we identified a total of 19 MOR signaling network genes in maize, 12 of which were more closely related to each other and they may collaborate to govern the growth and development of maize, suggesting the conservation of MOR signaling network components in maize (Figure 1; Tables S1 and S2). The MOR signaling pathway was first discovered in yeast and is necessary for normal growth and development [3–9]. This signaling pathway has received limited attention in plants and has only been reported in *Arabidopsis* [1,2]. The MOR signaling network genes of *Arabidopsis* are mainly expressed as the shoot apical meristem and inflorescence meristem, which may play a crucial regulatory role in plant stem cell maintenance and cell polarization [1]. Currently, the elucidation of this signaling pathway in maize and its regulatory mechanisms governing plant cold tolerance remains elusive. However, it is expected that the expression pattern of these genes in maize will be similar to that observed in *Arabidopsis*. Additionally, at these developmental stages, maize is susceptible to cold stress hazards [33–35] and MOR signaling network genes possess the potential to participate in the regulation of the maize cold stress response. Furthermore, the functional characterization of these 12 MOR signaling network genes in maize remains unreported. However, the function of its homologous genes in *Arabidopsis* has been partially documented and these genes in maize may also exhibit similar functions to their counterparts in *Arabidopsis*. The *Arabidopsis* KIC1 protein is implicated in the regulation of cellular polarity, cell proliferation, cell expansion, and antibacterial immune responses [48–50]. The *Arabidopsis* TAO3/FRY protein is potentially implicated in the G-protein signaling pathway and exerts an influence on morphogenesis [51]. Furthermore, the *Arabidopsis* MOB protein interacts with KIC1 and CBK1 and is involved in the regulation of cell proliferation, cell expansion, pollen development and germination, and plant growth and development [48,52,53]. And, the *Arabidopsis* CBK1 protein plays a pivotal role in various aspects of growth and development, including embryogenesis, pollen development, and germination [53–55]. The findings of these studies suggested that genes involved in the MOR signaling network may also play a vital role in regulating maize growth and development.

3.2. The MOR Signaling Network Genes Play a Crucial Role in the Regulation of Cold Tolerance in Maize

Our study revealed significant co-expression between genes related to the MOR signaling network and multiple genes associated with cold tolerance in maize, with all MOR signaling network genes exhibiting a close relationship to *COLD1*, a pivotal gene involved in perceiving cold stimuli (Figure 2A; Table S3). After perceiving the cold stimulus, maize COLD1 may transmit a signal to the MOR signaling network genes, thereby triggering a cold tolerance response that is dependent on this specific signaling cascade. *COLD1* encodes the G-protein regulator [18,19,21], and the G-protein signaling pathway represents a crucial mechanism for transducing extracellular signals into intracellular responses [51]. The *Arabidopsis* TAO3/FRY protein exhibits a close association with the G-proteins [51], suggesting that COLD1 may potentially mediate the transmission of cold signals via TAO3/FRY. Through its interaction with KIC1 and CBK1, TAO3/FRY facilitates the phosphorylation of CBK1 by KIC1, thereby activating the MOR pathway to regulate growth, development, and stress response [9,12,15]. KIC1 can induce extracellular ROS (reactive oxygen species) burst, thereby positively modulating immune responses in *Arabidopsis* [49]. Furthermore, ROS also plays a crucial role in the regulation of cold stress response [20]. These results suggested that KIC1 may play a role in the regulation of cold stress through the modulation of ROS. In this study, the expression levels of some MOR pathway genes were modulated in response to cold stress, exhibiting a significant up-regulation in *Zm00001d010720* and a notable down-regulation in *Zm00001d049496* (Figure 3), indicating that cold stress regulated the MOR signaling network. These two genes encode the CBK1 kinase and MOB protein, respectively, which play a crucial role in regulating plant growth and development [48,52–55] and may modulate the cold tolerance of maize during the seedling. Our further study found that the mutation of *Zm00001d010720* rendered maize more susceptible to cold stress (Figure 4). This observation indicated that the MOR signaling network genes exert a positive regulatory influence on cold stress in maize. The mutation site of Zm00001d010720 is situated at amino acid position 302 within the protein-coding sequence, resulting in premature termination of the protein sequence at this site and subsequent deletion of several crucial protein binding sites, such as active sites and polypeptide substrate binding sites (Figure S1). In the future, we can identify proteins that interact with this protein of the WT and *zm00001d010720* and subsequently compare their differences to identify key downstream proteins associated with cold tolerance. Furthermore, we also analyzed upstream transcription factors that may regulate the expression of *Zm00001d010720*. Through co-expression analysis, we identified 252 transcription factors that exhibited co-expressions with *Zm00001d010720* (Table S4), further using the maize gene expression data from qTeller (https://qteller.maizegdb.org/, accessed on 13 August 2023). In total, 39 of these transcription factors were identified as having at least twofold changes in expression after cold stress, and these 39 transcription factors were more likely to act as regulatory genes upstream of *Zm00001d010720* (Table S4). We also used the plant promoter analysis website Plantpan4.0 (http://plantpan.itps.ncku.edu.tw/plantpan4/index.html, accessed on 13 August 2023) to analyze the promoter region of *Zm00001d010720* and identified that 8 of these 39 transcription factors may have binding sites in the promoter region of *Zm00001d010720* (Figure S2; Table S4). The presence of *bZIP68* (*Zm00001d050018*), a well-documented key gene associated with maize cold tolerance [46], among these eight identified transcription factors suggests that *Zm00001d010720* plays a pivotal role in the regulatory mechanisms governing cold tolerance in maize. In conjunction with the above studies, we postulate that the cold signal is transmitted to the MOR signaling pathway after being sensed by COLD1, potentially facilitating both KIC1-mediated ROS generation and phosphorylation of ICE2, ICE3, and other proteins via CBK1 to enhance the cold tolerance of maize (Figure 5).

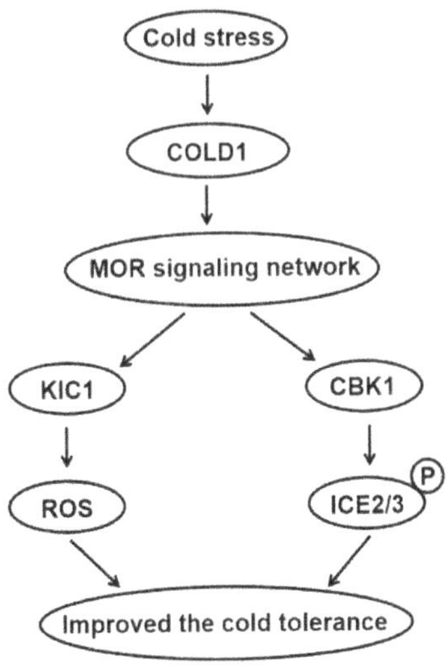

Figure 5. A putative model of the MOR signaling network regulating cold tolerance of maize. After being sensed by COLD1, the cold signal is transmitted to the MOR signaling network, potentially facilitating KIC1-mediated ROS generation, and phosphorylation of ICE2 and ICE3 via CBK1, thus enhancing the cold tolerance of maize. CBK1, Cell wall biosynthesis kinase; COLD1, Chilling tolerance divergence1; ICE2/3, Inducer of C-repeat binding factor expression2/3; KIC1, Kinase that interacts with cell division cycle1; MOR, Morphogenesis-related NDR kinase; ROS, reactive oxygen species.

4. Materials and Methods

4.1. Plant Materials

The materials used in this study included maize inbred line B73 (stored within our laboratory) and an ethylmethane sulfonate (EMS)-mutagenized stop-gained mutant (EMS4-0ab1d7, *zm00001d010720*) which were obtained from the Maize EMS-induced Mutant Database (MEMD; https://elabcaas.cn/memd/public/index.html#/, accessed on 20 August 2023) [56].

4.2. Identification of MOR Signaling Network Genes in Maize

The maize genome dataset downloaded from maizeGDB (Maize Genetics and Genomics Database, https://www.maizegdb.org/, accessed on 13 August 2023) was searched using BLASTp to identify the MOR signaling network genes, with the query sequences derived from the MOR signaling network genes in *Arabidopsis* [1].

4.3. Phylogenetic Analysis

The protein sequences of maize MOR signaling network genes were downloaded from maizeGDB, and the protein sequence of *Arabidopsis* MOR signaling network genes was downloaded from NCBI (National Center for Biotechnology Information, https://www.ncbi.nlm.nih.gov/, accessed on 13 August 2023). Protein sequences were aligned using MUSCLE in the MEGA5.1 software (v.5.1.1). The neighbor-joining algorithm was employed to assess evolutionary distances, while phylogeny testing was performed using the bootstrap method with 1000 replicates.

4.4. Co-Expression Analysis

The maize gene expression data from qTeller (https://qteller.maizegdb.org/, accessed on 13 August 2023) was utilized to perform a co-expression analysis of MOR signaling network genes with each other, as well as the co-expression analysis between MOR signaling network genes and cold tolerance-related genes (*COLD1, ICE1, ICE2, ICE3, MPK6, MPK8, bZIP61, bZIP68, bZIP80, CesA1, CesA4,* and *CesA5*) using SPSS software (v.24.0). Given the direct manifestation of leaf damage in maize following cold stress [44–46], we extracted 24 leaf-related expression data from the qTeller for co-expression analysis (Table S5). Significant co-expression was observed when the Pearson's correlation coefficient was $\geq$0.60 [57].

4.5. RNA Extraction and qRT-PCR Analysis

The seeds of B73 were sown in pots and grown under controlled conditions at a temperature of 25 °C with a photoperiod of 16 h light and 8 h darkness. When the seedlings were fourteen days old, they were subjected to a cold stress treatment at 4 °C for a duration of 24 h. B73 cold-treated leaves and control leaves were taken, respectively, and 3 biological replicates were collected. All samples were immediately frozen in liquid nitrogen and stored at −80 °C. The total RNA from each sample was extracted using Trizol reagent (Invitrogen, Carlsbad, CA, USA). The expression of *Zm00001d034055, Zm00001d051839, Zm00001d006710, Zm00001d049496*, and *Zm00001d010720* in cold stressed B73 and control was validated using qRT-PCR. The Fast Quant RT Kit (TianGen, Beijing, China) was used to synthesize the first strand cDNAs. The qRT-PCR was then conducted using the Bio-Rad iQ5 (Bio-Rad, Hercules, CA, USA) according to the SuperReal PreMix Plus (SYBR Green) instructions (TianGen, Beijing, China). All reactions were performed with three technical replicates, and the expression levels were normalized using *GAPDH* (*Glyceraldehyde-3-phosphate dehydrogenase*) as an internal reference. The qRT-PCR primers are listed in Table S6.

4.6. Identification and Analysis of Mutant zm00001d010720

An EMS-mutagenized stop-gained mutant (EMS4-0ab1d7) of *Zm00001d010720* was obtained from MEMD (https://elabcaas.cn/memd/public/index.html#/, accessed on

20 August 2023) [56]. The material was grown at the Chongzhou Modern Agricultural Research and Development Base, Sichuan Agricultural University. The row length was 3 m, and the row width was 0.6 m with a plant spacing of 0.3 m within rows. The genotypes of each plant were determined using PCR, and the homozygous mutant seeds were obtained through self-pollination. The primers are listed in Table S6. The homozygous seeds of *zm00001d010720* and B73 were sown in pots and cultivated under controlled conditions at a temperature of 25 °C, with a photoperiod consisting of 16 h of light followed by 8 h of darkness. The fourteen-day-old seedlings were treated with cold stress at 4 °C for 24 h. After cold stress, the plants recovered at 25 °C for 2 days, and then the growth was observed.

5. Conclusions

In summary, we have successfully identified 19 maize MOR signaling network genes for the first time and observed that 12 of them exhibit a higher degree of co-expression, suggesting the conservation of these signaling network components in maize and their potential significance in maize growth and development. The subsequent analysis revealed the involvement of MOR signaling network genes in the regulation of cold tolerance. The expression levels of some MOR pathway genes were regulated under cold stress, and the mutation in *Zm00001d010720*, which codes CBK1, confers increased cold stress sensitivity upon maize. These findings suggested that genes associated with the MOR signaling network play a pivotal role in modulating cold tolerance in maize. This study has contributed novel genetic resources to facilitate the development of cold-tolerant maize varieties.

Supplementary Materials: The following supporting information can be downloaded at: https://www.mdpi.com/article/10.3390/plants12203639/s1, Table S1: List of 19 MOR signaling network genes in maize; Table S2: Co-expression analysis of 19 maize MOR signaling network genes; Table S3: Co-expression analysis of MOR signaling network genes and cold tolerance-related genes; Table S4: Upstream transcription factors analysis of Zm00001d010720; Table S5: Expression data for co-expression analysis in this study; Table S6: Primers used in this study; Figure S1: Analysis of protein conserved domain of Zm00001d010720; Figure S2: Promoter analysis of *Zm00001d010720*.

Author Contributions: Data curation, R.T. and S.X.; Formal analysis, R.T. and J.Z.; Funding acquisition, Y.H. (Yubi Huang); Investigation, R.T., S.X., Y.L. (Yangping Li) and Y.H. (Yufeng Hu); Methodology, R.T., S.X. and Y.L. (Yinghong Liu); Project administration, Y.H. (Yubi Huang); Software, J.Z. and H.L.; Validation, R.T., S.X. and Y.L. (Yinghong Liu); Visualization, H.L.; Writing—original draft, R.T.; Writing—review and editing, S.X. and J.Z. All authors have read and agreed to the published version of the manuscript.

Funding: This research was funded by the Sichuan innovation team of national modern agricultural industry technology system (SCCXTD-2023-02) and the National Natural Science Foundation of China (32072071).

Data Availability Statement: Data is contained within the article and Supplementary Materials.

Conflicts of Interest: The authors declare no conflict of interest.

References

1. Zermiani, M.; Begheldo, M.; Nonis, A.; Palme, K.; Mizzi, L.; Morandini, P.; Nonis, A.; Ruperti, B. Identification of the arabidopsis RAM/MOR signalling network: Adding new regulatory players in plant stem cell maintenance and cell polarization. *Ann. Bot.* **2015**, *116*, 69–89. [CrossRef] [PubMed]
2. Bizotto, F.M.; Ceratti, R.S.; Braz, A.S.K.; Masuda, H.P. Evolutionary history of Mo25 gene in plants, a component of RAM/MOR signaling network. *Mech. Develop.* **2018**, *153*, 64–73. [CrossRef] [PubMed]
3. Racki, W.J.; Becam, A.M.; Nasr, F.; Herbert, C.J. CBK1p, a protein similar to the human myotonic dystrophy kinase, is essential for normal morphogenesis in *Saccharomyces cerevisiae*. *Embo. J.* **2000**, *19*, 4524–4532. [CrossRef] [PubMed]
4. Bidlingmaier, S.; Weiss, E.L.; Seidel, C.; Drubin, D.G.; Snyder, M. The CBK1p pathway is important for polarized cell growth and cell separation in *Saccharomyces cerevisiae*. *Mol. Cell. Biol.* **2001**, *21*, 2449–2462. [CrossRef] [PubMed]

5. Weiss, E.L.; Kurischko, C.; Zhang, C.; Shokat, K.; Drubin, D.G.; Luca, F.C. The *Saccharomyces cerevisiae* MOB2p-CBK1p kinase complex promotes polarized growth and acts with the mitotic exit network to facilitate daughter cell-specific localization of ACE2p transcription factor. *J. Cell Biol.* **2002**, *158*, 885–900. [CrossRef] [PubMed]
6. Nelson, B.; Kurischko, C.; Horecka, J.; Mody, M.; Nair, P.; Pratt, L.; Zougman, A.; McBroom, L.D.B.; Hughes, T.R.; Boone, C.; et al. RAM: A conserved signaling network that regulates Ace2p transcriptional activity and polarized morphogenesis. *Mol. Biol. Cell* **2003**, *14*, 3782–3803. [CrossRef]
7. Jansen, J.M.; Barry, M.F.; Yoo, C.K.; Weiss, E.L. Phosphoregulation of CBK1 is critical for RAM network control of transcription and morphogenesis. *J. Cell Biol.* **2006**, *175*, 755–766. [CrossRef]
8. Maerz, S.; Seiler, S. Tales of RAM and MOR: NDR kinase signaling in fungal morphogenesis. *Curr. Opin. Microbiol.* **2010**, *13*, 663–671. [CrossRef]
9. Saputo, S.; Chabrier-Rosello, Y.; Luca, F.C.; Kumar, A.; Krysan, D.J. The RAM network in pathogenic fungi. *Eukaryot. Cell* **2012**, *11*, 708–717. [CrossRef]
10. Jorgensen, P.; Nelson, B.; Robinson, M.D.; Chen, Y.; Andrews, B.; Tyers, M.; Boone, C. High-Resolution Genetic Mapping with Ordered Arrays of *Saccharomyces cerevisiae* Deletion Mutants. *Genetics* **2002**, *162*, 1091–1099. [CrossRef]
11. Colman-Lerner, A.; Chin, T.E.; Brent, R. Yeast Cbk1 and Mob2 activate daughter-specific genetic programs to induce asymmetric cell fates. *Cell* **2001**, *107*, 739–750. [CrossRef] [PubMed]
12. Kurischko, C.; Weiss, G.; Ottey, M.; Luca, F.C. A role for the *Saccharomyces cerevisiae* regulation of ACE2 and polarized morphogenesis signalingnetwork in cell integrity. *Genetics* **2005**, *171*, 443–455. [CrossRef] [PubMed]
13. Hergovich, A.; Stegert, M.R.; Schmitz, D.; Hemmings, B.A. NDR kinases regulate essential cell processes from yeast to humans. *Nat. Rev. Mol. Cell Bio.* **2006**, *7*, 253–264. [CrossRef] [PubMed]
14. Dan, I.; Watanabe, N.M.; Kusumi, A. The Ste20 group kinases as regulators of MAP kinase cascades. *Trends Cell Biol.* **2001**, *11*, 220–230. [CrossRef] [PubMed]
15. Weiss, E.L. Mitotic exit and separation of mother and daughter cells. *Genetics* **2012**, *192*, 1165–1202. [CrossRef]
16. Wang, D.; Xiao, Y.; Chen, H.; Huang, C.; Chen, P.; Chen, D.; Deng, W.; Wang, J. Combination of genomics, transcriptomics identifies candidate loci related to cold tolerance in Dongxiang wild rice. *Plants* **2022**, *11*, 2329. [CrossRef]
17. Chinnusamy, V.; Zhu, J.; Zhu, J.K. Cold stress regulation of gene expression in plants. *Trends Plant Sci.* **2007**, *12*, 444–451. [CrossRef]
18. Ma, Y.; Dai, X.; Xu, Y.; Luo, W.; Zheng, X.; Zeng, D.; Pan, Y.; Lin, X.; Liu, H.; Zhang, D.; et al. COLD1 confers chilling tolerance in rice. *Cell* **2015**, *160*, 1209–1221. [CrossRef]
19. Ding, Y.; Shi, Y.; Yang, S. Molecular regulation of plant responses to environmental temperatures. *Mol. Plant* **2020**, *13*, 544–564. [CrossRef]
20. Ding, Y.; Yang, S. Surviving and thriving: How plants perceive and respond to temperature stress. *Dev. Cell* **2022**, *57*, 947–958. [CrossRef]
21. Guo, X.; Liu, D.; Chong, K. Cold signaling in plants: Insights into mechanisms and regulation. *J. Integr. Plant Biol.* **2018**, *60*, 745–756. [CrossRef]
22. Shi, Y.; Ding, Y.; Yang, S. Molecular regulation of CBF signaling in cold acclimation. *Trends Plant Sci.* **2018**, *23*, 623–637. [CrossRef]
23. Chinnusamy, V.; Ohta, M.; Kanrar, S.; Lee, B.H.; Hong, X.; Agarwal, M.; Zhu, J.K. ICE1: A regulator of cold-induced transcriptome and freezing tolerance in *Arabidopsis*. *Gene. Dev.* **2003**, *17*, 1043–1054. [CrossRef]
24. Fursova, O.V.; Pogorelko, G.V.; Tarasov, V.A. Identification of *ICE2*, a gene involved in cold acclimation which determines freezing tolerance in *Arabidopsis thaliana*. *Gene* **2009**, *429*, 98–103. [CrossRef] [PubMed]
25. Tang, K.; Zhao, L.; Ren, Y.; Yang, S.; Zhu, J.K.; Zhao, C. The transcription factor ICE1 functions in cold stress response by binding to the promoters of *CBF* and *COR* genes. *J. Integr. Plant Biol.* **2020**, *62*, 258–263. [CrossRef] [PubMed]
26. Ding, Y.; Li, H.; Zhang, X.; Xie, Q.; Gong, Z.; Yang, S. OST1 kinase modulates freezing tolerance by enhancing ICE1 stability in *Arabidopsis*. *Dev. Cell* **2015**, *32*, 278–289. [CrossRef] [PubMed]
27. Li, H.; Ding, Y.; Shi, Y.; Zhang, X.; Zhang, S.; Gong, Z.; Yang, S. MPK3- and MPK6-mediated ICE1 phosphorylation negatively regulates ICE1 stability and freezing tolerance in *Arabidopsis*. *Dev. Cell* **2017**, *43*, 630–642. [CrossRef]
28. Zhao, C.; Wang, P.; Si, T.; Hsu, C.C.; Wang, L.; Zayed, O.; Yu, Z.; Zhu, Y.; Dong, J.; Tao, W.A.; et al. MAP kinase cascades regulate the cold response by modulating ICE1 protein stability. *Dev. Cell* **2017**, *43*, 618–629. [CrossRef] [PubMed]
29. Zhang, C.Y.; Zhang, Z.Y.; Li, J.H.; Li, F.; Liu, H.H.; Yang, W.S.; Chong, K.; Xu, Y.Y. OsMAPK3 phosphorylates OsbHLH002/OsICE1 and inhibits its ubiquitination to activate *OsTPP1* and enhances rice chilling tolerance. *Dev. Cell* **2017**, *43*, 731–743. [CrossRef]
30. Gao, R.X.; Hu, M.J.; Zhao, H.M.; Lai, J.S.; Song, W.B. Genetic dissection of ear-related traits using immortalized F2 population in maize. *J. Integr. Agr.* **2022**, *21*, 2492–2507. [CrossRef]
31. Zhao, Y.P.; Zhao, B.B.; Wu, G.X.; Ma, X.J.; Wang, B.B.; Kong, D.X.; Wei, H.B.; Wang, H.Y. Creation of two hyperactive variants of phytochrome B1 for attenuating shade avoidance syndrome in maize. *J. Integr. Agr.* **2022**, *21*, 1253–1265. [CrossRef]
32. Xie, S.D.; Tian, R.; Zhang, J.J.; Liu, H.M.; Li, Y.P.; Hu, Y.F.; Yu, G.W.; Huang, Y.B.; Liu, Y.H. *Dek219* encodes the DICER-LIKE1 protein that affects chromatin accessibility and kernel development in maize. *J. Integr. Agr.* **2023**, *22*, 2961–2980. [CrossRef]
33. Allen, D.J.; Ort, D.R. Impacts of chilling temperatures on photosynthesis in warm-climate plants. *Trends Plant Sci.* **2001**, *6*, 36–42. [CrossRef] [PubMed]

34. Sánchez, B.; Rasmussen, A.; Porter, J.R. Temperatures and the growth and development of maize and rice: A review. *Global Change Biol.* **2014**, *20*, 408–417. [CrossRef]
35. Riva-Roveda, L.; Escale, B.; Giauffret, C.; Périlleux, C. Maize plants can enter a standby mode to cope with chilling stress. *BMC Plant Biol.* **2016**, *16*, 212. [CrossRef] [PubMed]
36. Zhao, X.; Zhao, C.; Niu, Y.; Chao, W.; He, W.; Wang, Y.; Mao, T.; Bai, X. Understanding and comprehensive evaluation of cold resistance in the seedlings of multiple maize genotypes. *Plants* **2022**, *11*, 1881. [CrossRef]
37. Ben-Haj-Salah, H.; Tardieu, F. Temperature affects expansion rate of maize leaves without change in spatial distribution of cell length. *Plant Physiol.* **1995**, *109*, 861–870. [CrossRef]
38. Warrington, I.J.; Kanemasu, E.T. Corn growth response to temperature and photoperiod II. Leaf-initiation and leaf-appearance rates. *Agron. J.* **1983**, *75*, 755–761. [CrossRef]
39. Hund, A.; Fracheboud, Y.; Soldati, A.; Stamp, P. Cold tolerance of maize seedlings as determined by root morphology and photosynthetic traits. *Eur. J. Agron.* **2008**, *28*, 178–185. [CrossRef]
40. Nguyen, H.T.; Leipner, J.; Stamp, P.; Guerra-Peraza, O. Low temperature stress in maize (*Zea mays* L.) induces genes involved in photosynthesis and signal transduction as studied by suppression subtractive hybridization. *Plant Physiol. Bioch.* **2009**, *47*, 116–122. [CrossRef]
41. Hussain, H.A.; Men, S.; Hussain, S.; Zhang, Q.; Ashraf, U.; Anjum, S.A.; Ali, I.; Wang, L. Maize tolerance against drought and chilling stresses varied with root morphology and antioxidative defense system. *Plants* **2020**, *9*, 720. [CrossRef] [PubMed]
42. He, R.Y.; Zheng, J.J.; Chen, Y.; Pan, Z.Y.; Yang, T.; Zhou, Y.; Li, X.F.; Nan, X.; Li, Y.Z.; Cheng, M.J.; et al. QTL-seq and transcriptomic integrative analyses reveal two positively regulated genes that control the low-temperature germination ability of MTP-maize introgression lines. *Theor. Appl. Genet.* **2023**, *136*, 116. [CrossRef] [PubMed]
43. Qin, F.; Sakuma, Y.; Li, J.; Liu, Q.; Li, Y.Q.; Shinozaki, K.; Yamaguchi-Shinozaki, K. Cloning and functional analysis of a novel DREB1/CBF transcription factor involved in cold-responsive gene expression in *Zea mays* L. *Plant Cell Physiol.* **2004**, *45*, 1042–1052. [CrossRef] [PubMed]
44. Zeng, R.; Li, Z.; Shi, Y.; Fu, D.; Yin, P.; Cheng, J.; Jiang, C.; Yang, S. Natural variation in a type-A response regulator confers maize chilling tolerance. *Nat. Commun.* **2021**, *12*, 4713. [CrossRef] [PubMed]
45. Jiang, H.; Shi, Y.; Liu, J.; Li, Z.; Fu, D.; Wu, S.; Li, M.; Yang, Z.; Shi, Y.; Lai, J.; et al. Natural polymorphism of ZmICE1 contributes to amino acid metabolism that impacts cold tolerance in maize. *Nat. Plants* **2022**, *8*, 1176–1190. [CrossRef]
46. Li, Z.; Fu, D.; Wang, X.; Zeng, R.; Zhang, X.; Tian, J.; Zhang, S.; Yang, X.; Tian, F.; Lai, J.; et al. The transcription factor *bZIP68* negatively regulates cold tolerance in maize. *Plant Cell* **2022**, *34*, 2833–2851. [CrossRef]
47. Han, L.; Zhong, W.; Qian, J.; Jin, M.; Tian, P.; Zhu, W.; Zhang, H.; Sun, Y.; Feng, J.W.; Liu, X.; et al. A multi-omics integrative network map of maize. *Nat. Genet.* **2023**, *55*, 144–153. [CrossRef]
48. Xiong, J.; Cui, X.; Yuan, X.; Yu, X.; Sun, J.; Gong, Q. The Hippo/STE20 homolog SIK1 interacts with MOB1 to regulate cell proliferation and cell expansion in *Arabidopsis*. *J. Exp. Bot.* **2016**, *67*, 1461–1475. [CrossRef] [PubMed]
49. Zhang, M.; Chiang, Y.H.; Toruño, T.Y.; Lee, D.; Ma, M.; Liang, X.; Lal, N.K.; Lemos, M.; Lu, Y.J.; Ma, S.; et al. The MAP4 kinase SIK1 ensures robust extracellular ROS burst and antibacterial immunity in plants. *Cell Host Microbe* **2018**, *24*, 379–391. [CrossRef]
50. Zhang, P.; Yu, X.; Bai, J.; Gong, Q. The *Arabidopsis* STE20/Hippo kinase SIK1 regulates polarity independently of PIN proteins. *Biochem. Bioph. Res. Co.* **2021**, *549*, 21–26. [CrossRef]
51. Klopffleisch, K.; Phan, N.; Augustin, K.; Bayne, R.S.; Booker, K.S.; Booker, J.R.; Carpita, N.C.; Carr, T.; Chen, J.G.; Cooke, T.R.; et al. *Arabidopsis* G-protein interactome reveals connections to cell wall carbohydrates and morphogenesis. *Mol. Syst. Biol.* **2011**, *7*, 532. [CrossRef] [PubMed]
52. Guo, Z.; Yue, X.; Cui, X.; Song, L.; Cheng, Y. *AtMOB1* genes regulate jasmonate accumulation and plant development. *Plant Physiol.* **2020**, *182*, 1481–1493. [CrossRef] [PubMed]
53. Zhou, P.M.; Liang, Y.; Mei, J.; Liao, H.Z.; Wang, P.; Hu, K.; Chen, L.Q.; Zhang, X.Q.; Ye, D. The *Arabidopsis* AGC kinases NDR2/4/5 interact with MOB1A/1B and play important roles in pollen development and germination. *Plant J.* **2021**, *105*, 1035–1052. [CrossRef]
54. Bögre, L.; Okrész, L.; Henriques, R.; Anthony, R.G. Growth signalling pathways in *Arabidopsis* and the AGC protein kinases. *Trends Plant Sci.* **2003**, *8*, 424–431. [CrossRef] [PubMed]
55. Yoon, H.S.; Fujino, K.; Liu, S.; Takano, T.; Tsugama, D. NDR/LATS-family protein kinase genes are indispensable for embryogenesis in *Arabidopsis*. *FEBS Open Bio.* **2021**, *11*, 2600–2606. [CrossRef]
56. Lu, X.; Liu, J.; Ren, W.; Yang, Q.; Chai, Z.; Chen, R.; Wang, L.; Zhao, J.; Lang, Z.; Wang, H.; et al. Gene-indexed mutations in maize. *Mol. Plant* **2018**, *11*, 496–504. [CrossRef] [PubMed]
57. Xie, S.; Zhang, X.; Zhou, Z.; Li, X.; Huang, Y.; Zhang, J.; Weng, J. Identification of genes alternatively spliced in developing maize endosperm. *Plant Biol.* **2018**, *20*, 59–66. [CrossRef]

plants

MDPI

Article

Overexpression of *OsPIN9* Impairs Chilling Tolerance via Disturbing ROS Homeostasis in Rice

Qiqi Ouyang, Yanwen Zhang, Xiaoyi Yang, Chong Yang, Dianyun Hou, Hao Liu and Huawei Xu *

College of Agriculture, Henan University of Science and Technology, Luoyang 471000, China; 17538391647@163.com (Q.O.); 18438616118@163.com (Y.Z.); yangxy@163.com (X.Y.); yc051722@163.com (C.Y.); dianyun518@163.com (D.H.); liuhao_1990@126.com (H.L.)

* Correspondence: xhw@haust.edu.cn

Abstract: The auxin efflux transporter PIN-FORMED (PIN) family is one of the major protein families that facilitates polar auxin transport in plants. Here, we report that overexpression of *OsPIN9* leads to altered plant architecture and chilling tolerance in rice. The expression profile analysis indicated that *OsPIN9* was gradually suppressed by chilling stress. The shoot height and adventitious root number of *OsPIN9*-overexpressing (OE) plants were significantly reduced at the seedling stage. The roots of OE plants were more tolerant to *N*-1-naphthylphthalamic acid (NPA) treatment than WT plants, indicating the disturbance of auxin homeostasis in OE lines. The chilling tolerance assay showed that the survival rate of OE plants was markedly lower than that of wild-type (WT) plants. Consistently, more dead cells, increased electrolyte leakage, and increased malondialdehyde (MDA) content were observed in OE plants compared to those in WT plants under chilling conditions. Notably, OE plants accumulated more hydrogen peroxide (H_2O_2) and less superoxide anion radicals (O_2^-) than WT plants under chilling conditions. In contrast, catalase (CAT) and superoxide dismutase (SOD) activities in OE lines decreased significantly compared to those in WT plants at the early chilling stage, implying that the impaired chilling tolerance of transgenic plants is probably attributed to the sharp induction of H_2O_2 and the delayed induction of antioxidant enzyme activities at this stage. In addition, several *OsRboh* genes, which play a crucial role in ROS production under abiotic stress, showed an obvious increase after chilling stress in OE plants compared to that in WT plants, which probably at least in part contributes to the production of ROS under chilling stress in OE plants. Together, our results reveal that *OsPIN9* plays a vital role in regulating plant architecture and, more importantly, is involved in regulating rice chilling tolerance by influencing auxin and ROS homeostasis.

Keywords: *OsPIN9*; polar auxin transport; chilling tolerance; ROS homeostasis; antioxidant enzymes; rice (*Oryza sativa* L.)

1. Introduction

Due to extreme weather events, plants are progressively subjected to various abiotic stresses, such as chilling stress. Chilling stress restricts the geographical distribution and affects plant productivity [1,2]. Rice (*Oryza sativa* L.) originates from tropical and subtropical regions and is more sensitive to chilling stress than other cereal crops [3–5]. Previous data showed that rice yield decreased by 30–40% due to cold stress in temperate areas [6]. Therefore, improving rice chilling tolerance for maintaining food security is an urgent target [7]. In recent years, excellent progress has been made in understanding the physiological and molecular mechanisms underlying chilling stress [1,8–10]. Chilling stress triggers Ca^{2+} influx, which plays a key role in chilling tolerance [1,11–13]. For example, COLD1 (CHILLING-TOLERANCE DIVERGENCE 1) and OsCNGC9 (CYCLIC NUCLEOTIDE-GATED CHANNEL 9) have been demonstrated to regulate rice chilling tolerance by mediating Ca^{2+} influx under chilling stress [7,14]. Growing evidence shows that low temperatures dramatically induce the expression of the well-known C-repeat binding factor (*CBF*)/dehydration-responsive element binding factor (*DREB*) genes, which play a crucial role in plant chilling tolerance [15–18]. Conversely, evidence also showed that the transcriptional abundance of these transcription factors plays a minor role in cold tolerance [19–21]. Additionally, reactive oxygen species (ROS) are pivotal in various abiotic stress adaptations [22,23]. Low-level ROS at the early stress stage act as signals to trigger diverse stress responses, whereas excessive ROS cause cell damage; therefore, ROS homeostasis should be well controlled during abiotic stresses [10], including chilling tolerance [21,24–26]. In line with this, it is evidenced that more ROS are accumulated at the relative earlier chilling stage, while less ROS accumulation is detected at the late chilling stage in high chilling tolerance rice than in low chilling tolerance rice, which suggests that ROS homeostasis plays a vital role in regulating rice chilling tolerance [21].

The phytohormone auxin (indole-3-acetic acid, IAA) functions in almost all plant developmental processes and is involved in responses to many environmental clues, such as light, gravity, and temperature fluctuations [27–30]. Most auxin is synthesized in primordial and young leaves [31–33] and is transported basipetally to facilitate plant growth and development. The plant-specific PIN-FORMED (PIN) efflux carriers are the most important players in this process [34], which cooperated with the AUXIN1 (AUX1)/LIKE AUX1 (LAX) influx carriers, the phosphoglycoprotein (PGP/MDR/ABCB) efflux/influx transporters [35,36] and PIN-LIKES (PILS) family proteins [37] to drive polar auxin transport.

Significant progress has been made in the role of *PIN* genes, which are mainly derived from the dissection of *PIN* genes in the model plants *Arabidopsis thaliana* and rice [38–42]. The *Arabidopsis* genome possesses eight *PIN* genes, and the function of these *AtPIN* genes has been deeply investigated [38,43,44]. In contrast, the role of rice *OsPIN* genes is largely unknown. Twelve *OsPIN* genes have been identified in the rice genome [45,46], and several *OsPIN* genes have been functionally identified to play key roles in regulating rice growth and development by influencing polar auxin transport. *OsPIN1b*, which is also named *OsPIN1*, is mainly expressed in the vascular tissues and root primordial and functions in regulating the emergence of adventitious roots [47]. Our previous study showed that *OsPIN1b*, also designated as *OsPIN1a* [46], is associated with root negative phototropism [48]. Further investigation indicated that *OsPIN1a* and *OsPIN1b* redundantly regulate root and inflorescence development, while *OsPIN1c* and *OsPIN1d* are redundantly involved in modulating panicle formation [49]. *OsPIN2* regulates rice root system architecture and gravitropism by influencing auxin transport and distribution in roots, especially in columella cells [50–52]. *OsPIN5b* is constitutively expressed in vegetative and reproductive tissues and functions in modulating rice architecture and yield [53]. *OsPIN10a* (also designated as *OsPIN3t*) is mainly expressed in vascular tissue and functions in drought stress response and drought tolerance [54]. Recently, the monocot-specific *OsPIN9* gene has been indicated to function in regulating tiller bud outgrowth and affecting tiller number [55]. In addition, the expression profile of *PIN* genes, such as tissue-specific analysis and response to abiotic stress, in the other two monocot plants, sorghum (*Sorghum bicolor*) and maize (*Zea mays*), was deeply investigated [56–58]. Taken together, although excellent progress has been made in the functional dissection of *OsPIN* genes, which mainly focuses on rice growth and development, whether and how *OsPIN* genes are involved in regulating abiotic stresses, especially chilling stress, is still elusive.

The underlying mechanism of auxin in regulating cold tolerance is largely unknown, although auxin plays key roles in regulating plant growth and development. Increasing evidence has shown that cold stress is tightly related to auxin homeostasis. For example, an earlier study indicated that temperature could influence exogenous auxin transport velocity in some plant species [59]. Although the underlying mechanism of auxin in regulating cold tolerance is almost completely unknown, cold stress not only differentially regulates the expression of auxin-responsive genes but also markedly increases auxin content in rice [30,60]. Additionally, auxin transport could be suppressed and restored by low temperature and room temperature, respectively [61]. Inhibition of intracellular trafficking of auxin efflux carriers substantially influences auxin homeostasis under cold conditions [29]. Further research demonstrated that GNOM, a SEC7 containing ARF-GEF, which is closely associated with endosomal trafficking of auxin efflux carriers, can positively regulate *Arabidopsis* cold tolerance [62]. Collectively, these results strongly suggest that auxin homeostasis and transport are implicated in regulating plant chilling tolerance. However, little is known about the underlying molecular mechanisms involved in the role of auxin homeostasis and transport in chilling stress.

Previous studies demonstrated that *OsPIN9* functions in polar auxin transport and is involved in regulating adventitious root number and tiller number [55]. Recently, Manna et al. (2022) [63] reported that salt and drought treatment significantly induced *OsPIN9* expression, indicating the potential role of *OsPIN9* in abiotic stress adaptation. In line with this, our previous study showed that mutation of *OsPIN9* by CRISPR/Cas9 technology confers enhanced chilling tolerance in rice [64]. However, whether and how *OsPIN9* overexpression regulates rice chilling tolerance is unknown. In addition, since off-target effects or newly formed mutation proteins derived from CRISPR/Cas9 technology have the potential to cover the true function of target genes, it is necessary to further verify gene function by upregulating its expression. Here, we functionally examined the role of *OsPIN9* in modulating chilling tolerance by overexpressing technology. Our results showed that overexpression of *OsPIN9* not only leads to the alteration of rice architecture but also impairs rice chilling tolerance. We provided strong evidence that *OsPIN9* plays a crucial role in modulating rice chilling tolerance and shed light on the relationship between polar auxin transport and chilling tolerance in rice.

2. Results

2.1. Expression Patterns of *OsPIN* Genes under Chilling Stress

Increasing evidence shows that plant chilling tolerance is tightly associated with polar auxin transport [29,59,61,62]. However, the mechanism by which *OsPIN* genes respond to chilling stress is still largely unknown. To this end, quantitative real-time polymerase chain reaction (qRT-PCR) was performed to evaluate the responses of *OsPIN* genes in rice roots upon chilling stress. Nine *OsPIN* genes (*OsPIN1b*, *OsPIN1c*, *OsPIN1d*, *OsPIN2*, *OsPIN5a*, *OsPIN5b*, *OsPIN9*, *OsPIN10a*, and *OsPIN10b*), which can be detected in rice roots [39,45,46], were selected to analyze the expression profiles under chilling stress. *OsPIN1b* displayed a different expression pattern under chilling treatment compared to *OsPIN1c* and *OsPIN1d*, albeit belonging to the same subfamily. In detail, *OsPIN1b* was induced to the highest expression after chilling for 6 h and then decreased sharply, while *OsPIN1c* and *OsPIN1d* were suppressed after chilling for 3 h and kept to a relatively constant level thereafter. *OsPIN2* remained at a constant level under chilling treatment. Similar to *OsPIN1b*, *OsPIN5a* and *OsPIN5c* also showed increased expression first and then decreased. Three monocot-specific *PIN* genes, *OsPIN9*, *OsPIN10a*, and *OsPIN10b*, were greatly inhibited after chilling treatment and gradually decreased with treatment duration (Figure 1). Intriguingly, the paralogous *OsPIN* genes, *OsPIN1c/1d*, *OsPIN5a/5c*, and *OsPIN10a/10b*, showed a similar expression profile under chilling stress, implying that they might play a synergistic role in chilling tolerance.

2.2. Generating *OsPIN9*-Overexpressing Transgenic Rice and Phenotypes of Transformants

The monocot-specific *OsPIN9*, which was dramatically suppressed under chilling treatment, was chosen for further study. To obtain insight into the function of *OsPIN9* in chilling tolerance, a plasmid carrying the *OsPIN9* ORF, driven by a strong constitutive *Ubiquitin* promoter (*pUbi*) (Figure 2A), was introduced into rice cultivar Nipponbare by *Agrobacterium*. The transgenic plants were confirmed by PCR using genomic DNA as a template with specific primers. qRT-PCR analysis showed that the *OsPIN9* transcript significantly increased by 3.6–1237 times in the five tested transgenic line roots compared with that in WT plants (Figure 2B). Among which, A4 and B14 represented the high expression levels, A1 represented the moderate expression level, and A2 and B15 represented the low expression levels. Three lines, A4, A1, and B15, which represented high, moderate, and low expression levels of *OsPIN9*, were employed to further analyze the expression of *OsPIN9* in leaves, roots, and stem bases. The highest increase in expression was detected in leaves, followed by roots and stem bases (Figure 2C). Two T_2 homozygous transgenic lines, A4 and B14, with high expression levels of *OsPIN9*, designated as OE1 and OE2, respectively, were used for further investigation.

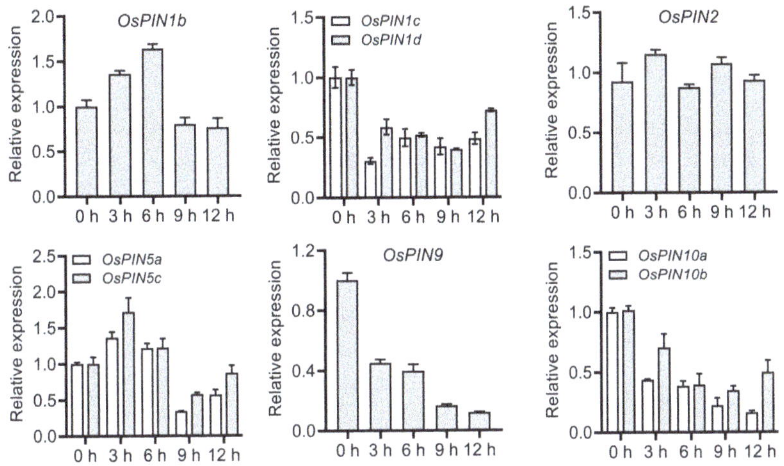

Figure 1. Expression profiles of *OsPIN* genes under chilling treatment. Values are means ± standard deviation (SD) (n = 3).

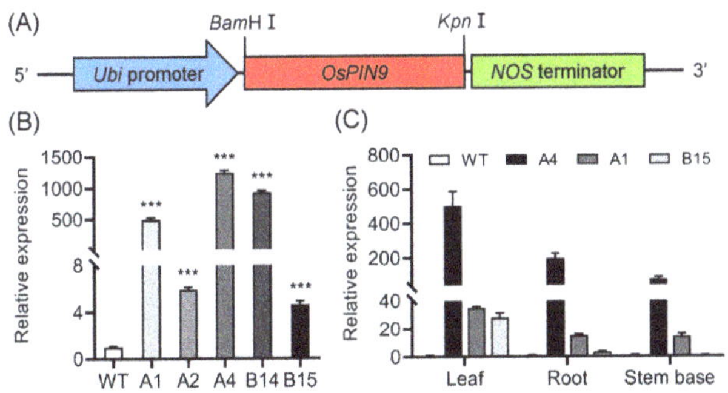

Figure 2. Generation of *OsPIN9*-overexpressing transgenic lines. (**A**) Schematic diagram of *pUbi:OsPIN9*. (**B**) Expression of *OsPIN9* in different transgenic lines. The expression level of *OsPIN9* in WT was set to one. The data were analyzed by ANOVA and Tukey's tests at a $p < 0.05$ significance level. ***: $p < 0.001$. (**C**) Expression analysis of *OsPIN9* in diverse tissues of *OsPIN9* transgenic lines. Values are means ± standard deviation (SD) (n = 3).

Overexpression of *OsPIN9* significantly decreased shoot height, root length, and adventitious root number after germination for 7 d in comparison with those in WT plants (Figure 3A). In detail, compared with the WT plants, the shoot height, root length, and adventitious root number of the OE plants were significantly decreased by 23–26%, 11–15%, and 17–18%, respectively. After germination for 14 d, shoot height and adventitious root number of OE lines declined by 21–25% and 13–14%, respectively, compared to those in WT plants. However, the root length was similar to WT plants at this stage (Figure 3B), indicating that rice plants can dynamically adapt to the change in auxin transport to some extent.

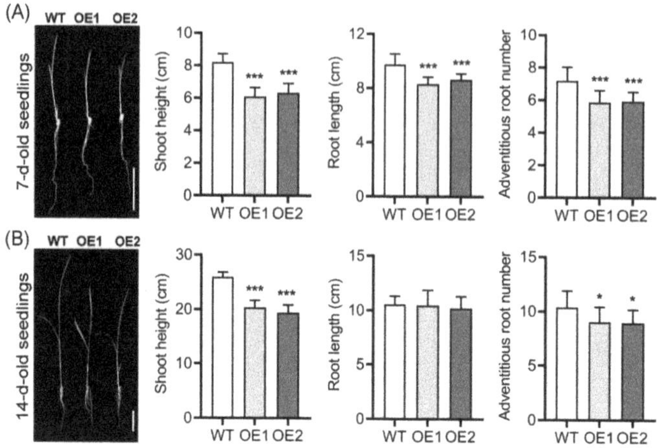

Figure 3. Phenotypes of wild-type (WT) and *OsPIN9*-overexpressing lines (OE) in 7 day old seedlings (**A**) and 14 day old seedlings (**B**). Bar = 4 cm. Values are means ± standard deviation (SD) (n = 24). The data were analyzed by ANOVA and Tukey's tests at a $p < 0.05$ significance level. *: $p < 0.05$; ***: $p < 0.001$.

2.3. Overexpression of OsPIN9 Disturbs Auxin Homeostasis in Rice

It was reported that overexpression of *OsPIN9* accelerated auxin transport from shoots to roots and resulted in an increase and decrease of IAA in roots and shoots, respectively [55]. *N*-1-naphthylphthalamic acid (NPA) is an inhibitor of polar auxin transport and can interact with PIN carriers directly [65]. To further assess the role of *OsPIN9* in auxin transport, germinated wild-type and transgenic seeds were cultured in Kimura B complete nutrient solution [39] supplemented with 0.5 μM NPA. After 7 d treatment, NPA obviously inhibited plant growth; the shoot height of OE lines was significantly decreased compared to WT plants both under normal conditions and NPA treatment. Contrastingly, the root length of OE lines was shorter than that of WT under normal conditions, while it increased significantly more than that of WT plants under NPA treatment. In detail, the root length of OE lines was significantly decreased by 10–12% when compared with WT under normal conditions, whereas it was about 39% and 49% longer compared with control plants after NPA treatment, indicating the roots of OE lines were more resistant to NPA than those of WT plants (Figure 4). Overexpression of *OsPIN9* significantly lowered the adventitious root number under control conditions, while NPA treatment severely decreased the adventitious root number and completely abolished the difference in adventitious root number between WT and OE lines. In addition, the ability to grow upright in some OE plant shoots was obviously inhibited and showed a partially impaired negative gravitropism compared with that in WT plants, indicating that the shoots of OE lines are probably more sensitive to NPA than those of WT plants (Figure 4). Collectively, these results demonstrate that auxin homeostasis is disturbed in OE lines.

2.4. Overexpression of OsPIN9 Impairs Chilling Tolerance in Rice

Due to the fact that polar auxin transport is closely related to chilling stress [66], and PIN carriers perform a rate-limiting role in regulating auxin efflux [34], as well as the quick suppression of *OsPIN9* under chilling stress (Figure 1), the role of *OsPIN9* in chilling tolerance was evaluated in WT and OE plants. Fourteen day old rice seedlings were moved from 28 °C to 4 °C. After chilling for 36 h, most WT leaves remained normal, but almost all leaves of OE plants displayed rolling and wilting (Figure 5A). The survival rate of OE plants was significantly lower than that of WT plants after 3 d of recovery at 28 °C (Figure 5A). Consistently, after 57 h of chilling treatment followed by 3 d of recovery, the survival rate of WT was dramatically higher than that of OE plants (Figure 5B). In detail,

WT plants showed a 98% survival rate, while the survival rates of OE1 and OE2 were only about 17% and 26%, respectively. These results suggest that the expression of *OsPIN9* is involved in negatively modulating rice chilling tolerance.

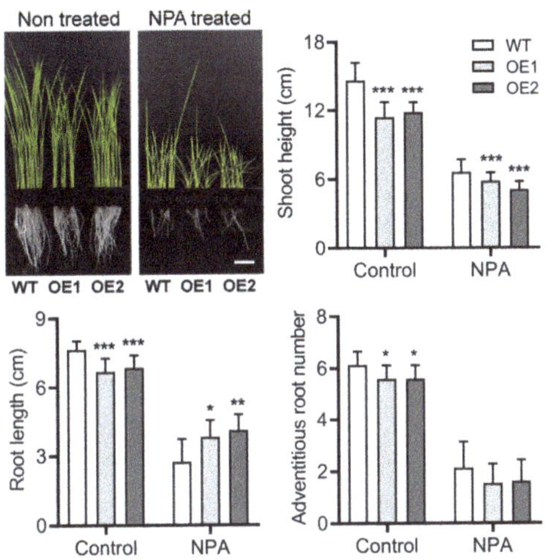

Figure 4. Sensitivity analysis of wild-type (WT) and *OsPIN9*-overexpressing (OE) plants under NPA treatment. Bar = 2 cm. Values are means ± standard deviation (SD) ($n = 24$). The data were analyzed by ANOVA and Tukey's tests at a $p < 0.05$ significance level. *: $p < 0.05$; **: $p < 0.01$; ***: $p < 0.001$.

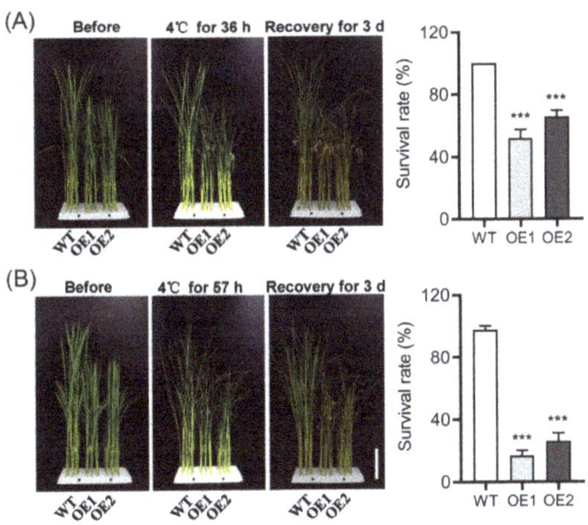

Figure 5. Chilling tolerance analysis of WT and *OsPIN9*-overexpressing (OE) rice. Wild-type (WT) and OE plants were treated with 4 °C for 36 h (**A**) and 57 h (**B**), followed by another 3 days of recovery under normal conditions, and then the survival rate was analyzed statistically. Bar = 5 cm. Values are means ± standard deviation (SD) ($n = 24$). The data were analyzed by ANOVA and Tukey's tests at a $p < 0.05$ significance level. ***: $p < 0.001$.

We then detected several physiological indicators, including cell death, electrolyte leakage, and malondialdehyde (MDA) contents, to evaluate the influence of overexpression of *OsPIN9* on rice chilling tolerance. Trypan blue staining, which is an indicator of cell death, was performed to monitor the cell damage. There was no apparent difference in trypan blue staining in leaves between WT and OE plants under normal growth conditions, while staining was more intensive in the leaves of OE plants compared to that of WT plants after chilling for 36 h and 57 h (Figure 6A), indicating that the cell death in OE plants is greater than that in WT plants. In line with this, another two indicators, electrolyte leakage and MDA contents, were both significantly increased in OE plants compared to those in WT plants after chilling for 36 h and 57 h (Figure 6B,C). Taken together, these results strongly suggest that overexpression of *OsPIN9* substantially impairs rice chilling tolerance.

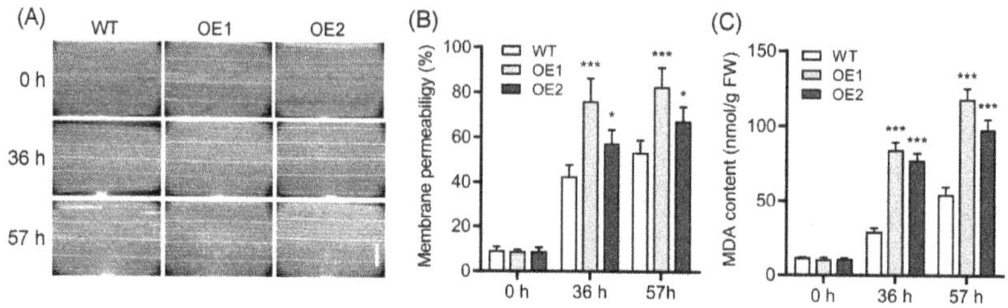

Figure 6. Overexpression of *OsPIN9* causes cell damage in rice seedling leaves. (**A**) Trypan blue staining. Bar = 1 mm. (**B**) Electrolyte leakage. (**C**) Malondialdehyde contents. Values are means ± standard deviation (SD) ($n = 6$). The data were analyzed by ANOVA and Tukey's tests at a $p < 0.05$ significance level. *: $p < 0.05$; ***: $p < 0.001$.

2.5. The Role of OsDREB1 Regulon and Ca^{2+} Signaling-Related Genes in Chilling Stress

The expression of *OsDREB1* genes could be dramatically induced by low temperatures [19,67], and more and more evidence has demonstrated that *OsDREB1* genes play a vital role in chilling stress [14,19,67]. We then detected the expression levels of *OsDREB1A*, *OsDREB1B*, and *OsDREB1C* in WT and OE plants upon chilling stress. Under normal conditions, these *OsDREB1* genes were significantly more induced in OE plants than in WT plants (Figure 7A). However, after chilling for 8 h, the expression levels of these genes in OE plants were similar to or even lower than those in WT plants (Figure 7A). Similarly, trehalose-6-phosphate phosphatase 1 (*OsTPP1*), which has been demonstrated to play a vital role in rice cold tolerance [9,68], displayed a similar expression profile with *OsDREB1* genes in WT and OE plants under both normal and chilling conditions (Figure 7B). Additionally, increasing evidence shows that Ca^{2+} signaling is involved in chilling tolerance regulation [9,69], and *COLD1* and *OsCNGC9* have been demonstrated to facilitate cytoplasmic Ca^{2+} elevation and positively regulate rice cold tolerance [7,14]. To evaluate the role of these Ca^{2+} signaling-related genes in OE lines under chilling stress, we further examined the expression of these two genes under normal and chilling conditions. The results showed that both *COLD1* and *OsCNGC9* in OE lines were slightly induced compared to those in WT plants under normal conditions, while strikingly suppressed under chilling stress (Figure 7C). Collectively, these data suggest that the defense response might be triggered in advance under normal conditions, and the *DREB1* regulon and Ca^{2+} signaling are likely to be implicated in regulating the chilling tolerance of OE lines.

2.6. OsPIN9-Overexpressing Plants Accumulated More H_2O_2 Rather Than O_2^- under Chilling Stress

In addition to the well-known CBF/DREB regulon, reactive oxygen species (ROS) also play a crucial role in chilling stress [21]. To further investigate the underlying mechanism of *OsPIN9* under chilling stress, ROS content assays were performed by diaminobenzidine tetrahydrochloride (DAB 4HCl) staining for H_2O_2 and NBT staining for O_2^- under normal and chilling conditions. The result showed that OE plants accumulated a similar level of H_2O_2 and O_2^- compared to WT plants before chilling treatment (Figure 8A). After chilling for 36 h, the leaves of OE plants accumulated more H_2O_2 and less O_2^- compared to WT plants (Figure 8B). Consistently, more H_2O_2 and less O_2^- were still detected in OE leaves compared to WT plants after chilling for 57 h (Figure 8C). These results indicate that OE lines accumulate more H_2O_2 than O_2^- when compared with WT plants, which probably causes cell damage and impairs rice chilling tolerance.

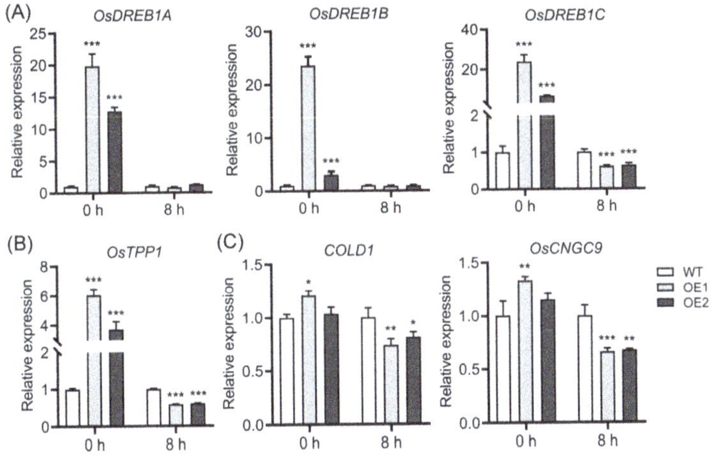

Figure 7. Expression analysis of *OsDREB1* genes, *OsTPP1*, and Ca^{2+} signaling genes under normal and chilling conditions in wild-type (WT) and *OsPIN9*-overexpressing (OE) plants. (**A**) *OsDREB1s* gene expression analysis. (**B**) *OsTPP1* expression analysis. (**C**) Ca^{2+} signaling gene expression analysis. Values are means ± standard deviation (SD) (n = 3). The data were analyzed by ANOVA and Tukey's tests at a $p < 0.05$ significance level. *: $p < 0.05$; **: $p < 0.01$; ***: $p < 0.001$.

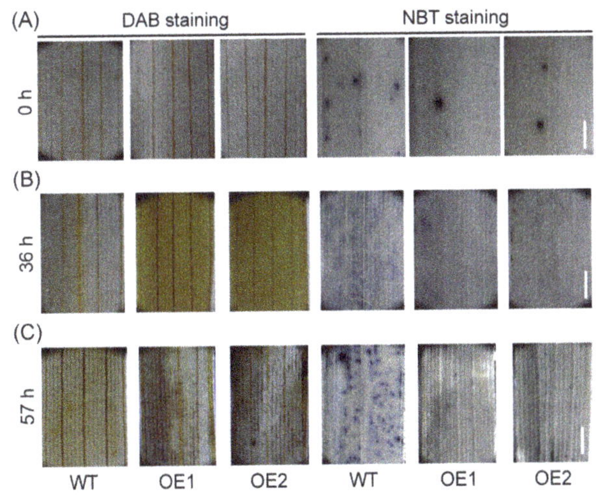

Figure 8. ROS detection in wild-type (WT) and *OsPIN9*-overexpressing (OE) leaves before and after chilling treatments. H_2O_2 and O_2^- in WT and OE leaves were monitored by DAB and NBT staining, respectively, before chilling stress (**A**) and after chilling for 36 h (**B**) and 57 h (**C**). Bar = 1 mm.

2.7. The Delayed Induction of Antioxidant Enzymes Probably Leads to the Accumulation of H_2O_2 in OE Lines

ROS are vital signaling molecules that play a crucial role in biotic and abiotic stress responses, while their content should be carefully controlled in plant cells via an array of enzymatic and non-enzymatic antioxidants [23]. Given that ROS homeostasis is disturbed in OE lines, we then analyzed the activities of three antioxidant enzymes, including superoxide dismutase (SOD), catalase (CAT), and peroxidase (POD), which play a key role in ROS scavenging [10,23], before chilling treatment and after chilling for 36 h and 72 h. There were no significant differences in SOD and CAT activities in WT and OE plants, while POD activity was significantly increased in OE lines compared to WT plants under normal conditions (Figure 9A). After chilling for 36 h, the activities of CAT and SOD decreased significantly in OE plants compared to those in WT plants. To our surprise, the POD activity was still strikingly higher in OE lines than in WT plants (Figure 9B). After chilling for 57 h, the activities of the three enzymes in OE lines all increased greatly compared to those in WT plants (Figure 9C). These results imply that it is likely that the decreased activities of SOD and CAT rather than increased POD activity led to the accumulation of H_2O_2 in OE lines at the early chilling stage, and the delayed induction of SOD and CAT is ineffective to quench excess ROS at the late chilling stage and causes cell damage in OE lines. However, considering that SOD usually functions in the conversion of O_2^- to H_2O_2, the source of the accumulated H_2O_2 during chilling stress needs further investigation.

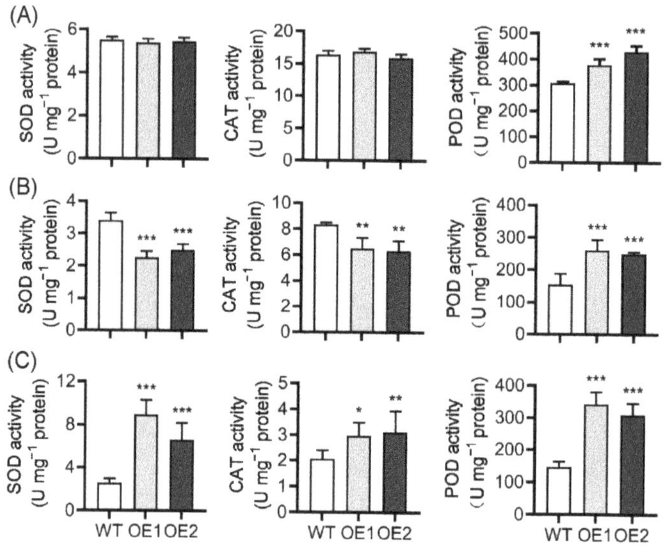

Figure 9. Analysis of antioxidant enzyme activities in wild-type (WT) and *OsPIN9*-overexpressing (OE) plants before and after chilling treatments. Enzyme activities were detected before chilling stress (**A**) and after chilling for 36 h (**B**) and 57 h (**C**) in WT and OE lines. Values are means ± standard deviation (SD) ($n = 6$). The data were analyzed by ANOVA and Tukey's tests at a $p < 0.05$ significance level. *: $p < 0.05$; **: $p < 0.01$; ***: $p < 0.001$.

2.8. OsRobh Genes Show Differential Expression in OsPIN9-Overexpressing Lines

Previous reports showed that respiratory burst oxidase homologs (Rboh), also known as NADPH oxidase, play critical roles in various cellular activities and responses to abiotic stresses by regulating the production of ROS [70–72]. For the higher production of H_2O_2 detected in OE lines, we assayed the expression of *OsRboh* genes before and after chilling stress treatment to evaluate the source of H_2O_2 in OE lines. As shown in Figure 10, under normal conditions, most *OsRboh* genes in transgenic lines kept similar expression levels compared to those in WT plants, except for *OsRboh1*, *OsRboh2*, and *OsRboh9*, which showed a slight increase in OE lines. After chilling for 24 h, more *OsRboh* genes were induced, including *OsRboh2, OsRboh4, OsRboh6, and OsRboh8*, and showed a relative higher expression level in OE lines, which may contribute to the production of H_2O_2 in transgenic lines under chilling conditions.

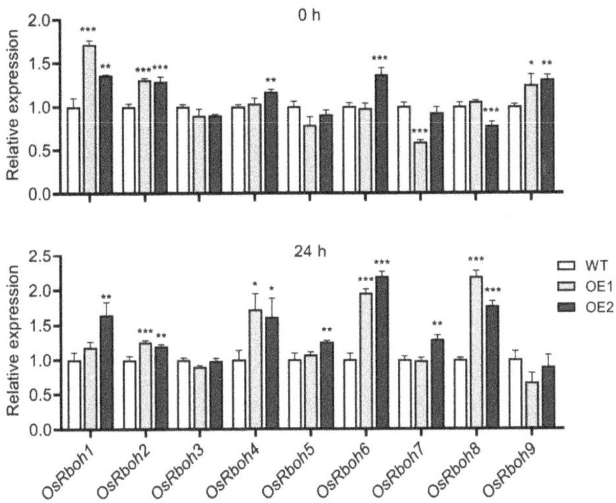

Figure 10. *OsRboh* expression analysis in wild-type (WT) and *OsPIN9*-overexpressing (OE) lines before and after chilling treatment. The expression level of *OsRboh* genes in WT was set to one. The data were analyzed by ANOVA and Tukey's tests at a $p < 0.05$ significance level. *: $p < 0.05$; **: $p < 0.01$; ***: $p < 0.001$.

3. Discussion

Although excellent progress has been made in the molecular mechanisms of plant cold adaptation [8–10], auxin, the first phytohormone discovered and one of the most important phytohormones in plant growth and developmental regulation, its role in modulating chilling adaptation is still elusive. In this present study, the role of a monocot-specific *OsPIN9* gene, which mainly functions in regulating auxin transport basipetally, in chilling stress was investigated by overexpression technology.

Auxin plays a vital role in regulating almost all cellular processes, and in many cases, an appropriate auxin level and distribution within plant tissues are direct determinators affecting plant architecture [73]. *PIN* genes have been demonstrated to play a central role in auxin efflux and, consequently, influence auxin content and distribution in various tissues. Expression levels of *PIN* genes are usually closely associated with plant architecture [74,75]. The most representative is *AtPIN1*, and disruption of *AtPIN1* seriously influences inflorescence development and leads to the formation of needle-like inflorescence in *Arabidopsis* [76]. The rice genome possesses 12 *OsPIN* genes [45,46], and although the possibility of redundant functions of *OsPIN* genes exists [49], mutation of most *OsPIN* genes substantially influences rice architecture [47,49,53–55,77]. The previous report demonstrated that the OsPIN9 carrier functions in auxin transport basipetally, overexpression of *OsPIN9* substantially increases auxin content in rice roots [55], so we speculate that the auxin content in OE plant roots should be increased compared to that in WT plants. In line with this, the OE plant roots displayed a more resistant phenotype under NPA treatment (Figure 4), implying an increased auxin content in the OE roots. It was reported that overexpression of *OsPIN9* increased the adventitious root number [55]. Unexpectedly, we observed a lower number of adventitious roots in these two OE lines, which is different from the previous report [55]. Considering that adventitious root number is tightly associated with auxin content [55], and auxin regulates plant growth and development usually in a concentration-dependent manner, relative high auxin content accelerates plant growth, while excess auxin content strongly inhibits plant growth. Therefore, we speculate that the dramatic expression of *OsPIN9* probably leads to excess accumulation of auxin in OE roots, which suppresses adventitious root number (Figure 3). In agreement with this, the relative lower expression of *OsPIN9* (A2 line) significantly increased adventitious root number when compared to WT plants (Figure S1).

Rice is sensitive to chilling stress [7,78–80], and severe chilling disasters hamper rice growth and strikingly impair rice productivity [10]. Previous reports have suggested that chilling stress is implicated in auxin transport in plants [29,30,61,62]. However, there is no direct genetic evidence regarding the role of auxin transport in the regulation of chilling stress. In this study, we provided direct evidence that OsPIN9, which functions in auxin transport basipetally and could be quickly suppressed by chilling stress (Figure 1), is involved in rice chilling tolerance. The chilling treatment assay showed that OE lines are more sensitive to low temperatures than WT plants (Figure 5); further experiments confirmed that OE lines possess more dead cells (Figure 6A), higher membrane permeability (Figure 6B), and increased MDA content (Figure 6C) compared to WT plants after chilling treatment. Increased MDA content usually acts as a marker of lipid peroxidation [81], especially under various abiotic stresses [82,83], including chilling stress. Apart from rice, MDA levels were also employed to evaluate chilling tolerance in other plants, such as bermudagrass (*Cynodon dactylon*) [84] and *Pyrus betulaefolia* [85]. Collectively, these results strongly demonstrate that *OsPIN9* is involved in regulating rice chilling tolerance, which is probably mediated by influencing auxin transport. Over the last decade, significant progress has been made in the dissection of the underlying molecular mechanisms of plant chilling adaptation. At least the CBF/DREB-dependent cold-signaling pathway, the Ca^{2+} signaling pathway, the ROS-dominated adaptation mechanism, and the phytohormone regulation mechanism [1,9,86] have been demonstrated to regulate plant cold adaptation. Several lines of evidence have strongly suggested that overexpression of *OsPIN9* notably impairs rice chilling tolerance (Figures 5 and 6), while how *OsPIN9* influences plant chilling tolerance is completely unknown. To this end, we first examined whether the impaired chilling tolerance of OE lines is associated with the CBF/DREB-dependent cold signaling pathway and the Ca^{2+} signaling pathway by employing qRT-PCR technology. The expression of *OsDREB1A*, *OsDREB1B*, and *OsDREB1C* was detected before and after chilling stress due to the sharp induction of these genes under chilling conditions [19,67]. Surprisingly, the *OsDREB1* genes, *OsTPP1*, *COLD1*, and *OsCNGC9* were all induced before treatment and kept at similar or even lower levels in OE lines compared to those in WT plants after chilling treatment (Figure 7). Considering that *OsDREB1* genes are mainly induced by low temperatures [19], and higher expression of *OsTPP1*, *COLD1*, and *OsCNGC9* also facilitates plant cold adaptation [7,14,68], we speculate that a stress response probably occurred in OE lines before chilling treatment. In contrast, all genes show a similar or lower expression level in OE lines compared to WT plants under chilling conditions, suggesting that the DREB regulon and Ca^{2+} signaling might be implicated in the impaired chilling tolerance of OE lines. Further investigations are needed to address this speculation.

Next, we assayed ROS levels to clarify the role of ROS homeostasis in WT and OE lines under chilling stress. Chilling stress triggers ROS accumulation [21,87]. Low-level H_2O_2 can act as a signal to trigger the stress response at the early stress stage, whereas at the later stress stage, high-level H_2O_2 can damage plant cells [10,87]. Surprisingly, we only observed the accumulation of H_2O_2 in OE leaves at both early and late chilling stages, while O_2^- was not accumulated during the chilling stages (Figure 8), indicating that ROS homeostasis is disturbed in OE plant cells and that H_2O_2 rather than O_2^- causes cell damage in OE lines. It has been demonstrated that ROS homeostasis plays a vital role in plant abiotic stress adaptations [88,89], and ROS content must be carefully controlled by many ROS scavenging and detoxification systems during stress conditions [10]. Antioxidant enzyme activity measurement showed that SOD and CAT activities were significantly decreased and increased in OE lines compared to those in WT plants at the early and late chilling stages, respectively, while POD activity was always strikingly higher than that in WT plants (Figure 9B,C), even before chilling treatment (Figure 9A), indicating that the prompt trigger of antioxidant enzymes, mainly SOD and CAT in OE lines, plays a vital role in quenching accumulated ROS at the early chilling stage. Additionally, although antioxidant enzymes showed significantly higher activities in OE lines than in WT plants at the late chilling stage (Figure 9C), which could not detoxify excess H_2O_2 quickly (Figure 8C), indicating that the delayed induction of the antioxidant enzymes is not enough to scavenge ROS at the late chilling stage, and the constantly high levels of H_2O_2 then damage cells.

To further evaluate the source of H_2O_2 in OE lines under chilling conditions, we assayed the expression of *OsRboh* genes before and after chilling stress. Since only a slight increase of *OsRboh1*, *OsRboh2*, and *OsRboh9* was detected in OE lines before chilling stress, it is reasonable that less ROS accumulation was observed under normal conditions. In contrast, more *OsRboh* genes were induced after chilling treatment, indicating that *OsRboh* is at least in part involved in ROS production after chilling stress. We also noticed that these induced *OsRboh* genes were still kept at low expression levels (up to about twofold) even after chilling treatment (Figure 10). Although evidence has shown that *Rboh* plays a vital role in ROS production under various abiotic stress conditions [70,71], we cannot rule out the possibility that other ROS-production sources probably exist that contribute to the ROS burst after chilling stress in OE lines.

4. Materials and Methods

4.1. Plant Materials, Growth Conditions, and Chilling Treatment

Rice *japonica* variety Nipponbare was employed for the physiological experiments and rice transformation. Hydroponic experiments were performed according to our previous report [90]. Briefly, rice seeds were sterilized first and then cultured in darkness for 3–4 days at 30 °C. The germinated seeds were transferred to Kimura B complete nutrient solution in plant growth chambers with a 12 h of light (30 °C)/12 h of darkness (25 °C) photoperiod and 60–70% relative humidity.

To assay the responses of *OsPIN* genes to chilling treatment, 14 day old seedlings were transferred to low temperatures (4 °C), and root samples were collected at the indicated time points. The samples were then frozen quickly using liquid nitrogen and preserved at −80 °C for gene expression analysis.

To evaluate rice chilling tolerance, 14 day old WT and OE seedlings were transferred to 4 °C for 36 h and 57 h, followed by a 3 d recovery, and then the survival rate and physiological indicators were analyzed.

4.2. Vector Construction and Generation of the Transgenic Plants

The full-length *OsPIN9* gene was amplified by PrimeSTAR HS DNA Polymerase (TaKaRa Biotechnology Co., Ltd., Dalian, China) from rice cDNA using the specific primers listed in Table S1. A plant expression vector, pCAMBIA1301-pUbi, which was kindly provided by Dr. Yao-Guang Liu (College of Life Sciences, South China Agricultural University, Guangzhou, China), was used for overexpressing *OsPIN9*. To insert into the expression vector pCAMBIA1301-pUbi, the restriction sites *Bam*H I and *Kpn* I were added at the 5′ end of forward and reverse primers, respectively. The recombinant construct was named *pUbi:OsPIN9* and confirmed by restriction enzyme digestion and sequencing. The constructed plasmid was then transformed into *Agrobacterium tumefaciens* (EHA105) and employed for rice transformation according to the previous report [91]. Homozygous T_3-generation transgenic plants were used for further studies.

4.3. Quantitative RT-PCR

qRT-PCR was performed according to a previous report [39]. In short, total RNA was extracted from the collected samples using RNAiso Plus (TAKARA Bio Inc., Dalian, China), and reverse transcription was conducted using HiScript III RT SuperMix for qPCR (Nanjing Vazyme Biotech Company, Ltd., Nanjing, China). The online website INTEGRATED DNA TECHNOLOGIES (https://sg.idtdna.com, accessed on 1 May 2019) was used to design the gene-specific primers. qRT-PCR was conducted using AceQ Universal SYBR qPCR Master Mix (Nanjing Vazyme Biotech Company, Ltd.) and the Lightcycle® 96 system. At least three biological replicates and three technical repetitions were performed to assay the gene expression. The *OsACTIN1* gene (Os03g0718100) was employed as an internal control. All primers used in this study are listed in Table S1, and gene names and ID numbers used for qRT-PCR in this study are listed in Table S2.

4.4. Exogenous NPA Treatment

NPA is usually employed as a classical inhibitor of polar auxin transport to elucidate the underlying mechanisms of plant growth and development associated with polar auxin transport [65]. For analyzing the responses of WT and OE lines to NPA treatment, germinated seeds were transferred into Kimura B complete nutrient solution containing 0.5 μM NPA and cultured for 7 d, and then the shoot height, root length, and adventitious root number were assessed. WT and OE lines cultured under normal conditions were used as controls in this experiment.

4.5. Physiological Analysis

Trypan blue staining was conducted as described previously [92] with minor modifications. Rice leaves were sampled before and after chilling treatment and stained with a lactophenol-trypan blue solution (10 mL lactic acid, 10 mL glycerol, 10 g phenol, and 10 mg trypan blue, mixed in 10 mL distilled water). The leaves were boiled in a water bath for 10 min in trypan blue solution and then kept at room temperature for 1 h. Decolorization was performed using chloral hydrate solution (25 g chloral hydrate dissolved in 10 mL distilled water). Electrolyte conductivity and MDA content were detected according to the previous report [93]. 3′3-diaminobenzidine (DAB) and nitro blue tetrazolium (NBT) staining were performed as previously reported [21]. Briefly, rice leaves were first cut into sections (about 2 cm in length) and then soaked in DAB and NBT solutions, respectively, overnight at 37 °C. Alcohol (95%) was used for decolorization. CAT, POD, and SOD activities were assayed according to the previous methods [94] with minor modifications. In short, 0.1 g of rice leaves were broken with liquid nitrogen and then homogenized in 1 mL of extraction buffer (50 mM phosphate, pH 7.8). After centrifugation, the supernatant was used for enzyme activity assays. CAT activity was measured in a reaction solution consisting of 50 mM phosphate buffer (pH 7.0), 0.2% H_2O_2, and enzyme extract. POD activity was measured in a reaction mixture (0.2% H_2O_2, 0.2% guaiacol, and enzyme extract) at 470 nm. SOD activity was assayed in a reaction mixture containing 50 mM phosphate, pH 7.8, 130 mM L-methionine, 750 μM NBT, 100 μM EDTA, 20 μM riboflavin, and enzyme extract at 560 nm. The Coomassie Brilliant Blue G-250 staining method was employed to measure protein content [95].

4.6. Data Analysis

Each experiment was biologically replicated at least three times. The one-way analysis of variance (ANOVA) method was used to statistically analyze experimental data by GraphPad PRISM 8 version 8.0.2 (GraphPad Software Inc., San Diego, CA, USA) at the significance levels of $p < 0.05$ (*), $p < 0.01$ (**), and $p < 0.001$ (***), and all data are given as means ± SD.

5. Conclusions

In conclusion, in this report, we systematically monitored the responses of *OsPIN* genes to chilling treatment and generated transgenic rice plants with high expression levels of *OsPIN9*. The transgenic plants showed impaired chilling tolerance compared to WT plants, which is probably mainly caused by the sharp accumulation of H_2O_2 and delayed induction of SOD and CAT. *OsRboh* genes, at least in part, contribute to the production of ROS in OE lines under chilling stress conditions. The DREB regulon and Ca^{2+} signaling pathway are implicated in the chilling tolerance of OE lines, while the detailed functional mechanisms need further investigation. Several issues need further discussion, for example, how the relationship between auxin homeostasis and ROS homeostasis is derived and where the accumulated H_2O_2 is derived from? What is the role of O_2^- and POD in OE lines under chilling stress? Further investigation focused on these detailed issues may lead to insight into the underlying molecular mechanisms of rice chilling tolerance and may provide potential targets for breeding chilling-resistant crops.

Supplementary Materials: The following supporting information can be downloaded at: https://www.mdpi.com/article/10.3390/plants12152809/s1, Figure S1: Adventitious root number analysis in the transgenic A2 line; Table S1: Primers used in this study; Table S2: Gene names and ID numbers used for qRT-PCR in this study.

Author Contributions: Conceptualization and research design, H.X.; experimental, Q.O., Y.Z., X.Y. and C.Y.; analyzing the data, D.H., H.L. and H.X.; manuscript—writing, H.X.; writing—review (completion), D.H., H.L. and H.X. All authors have read and agreed to the published version of the manuscript.

Funding: This work was funded by the National Natural Science Foundation of China (grant number 3110019) and the Natural Science Foundation of Henan Province (grant number 182300410012).

Data Availability Statement: Not applicable.

Acknowledgments: We sincerely thank Chunzhao Zhao (Shanghai Center for Plant Stress Biology, CAS Center for Excellence in Molecular Plant Sciences, Chinese Academy of Sciences) for constructive discussion. The authors are grateful to all lab members for their useful suggestions and encouragement.

Conflicts of Interest: The authors declare no conflict of interest.

References

1. Ding, Y.; Shi, Y.; Yang, S. Molecular regulation of plant responses to environmental temperatures. *Mol. Plant* **2020**, *13*, 544–564. [CrossRef]
2. Shi, Y.; Ding, Y.; Yang, S. Molecular regulation of CBF signaling in cold acclimation. *Trends Plant Sci.* **2018**, *23*, 623–637. [CrossRef]
3. Wang, W.; Vinocur, B.; Altman, A. Plant responses to drought, salinity and extreme temperatures: Towards genetic engineering for stress tolerance. *Planta* **2003**, *218*, 1–14. [CrossRef] [PubMed]
4. Wang, D.; Liu, J.L.; Li, C.G.; Kang, H.X.; Wang, Y.; Tan, X.Q.; Liu, M.H.; Deng, Y.F.; Wang, Z.L.; Liu, Y.; et al. Genome-wide association mapping of cold tolerance genes at the seedling stage in rice. *Rice* **2016**, *9*, 61. [CrossRef] [PubMed]
5. Zhang, Q.; Chen, Q.; Wang, S.; Hong, Y.; Wang, Z. Rice and cold stress: Methods for its evaluation and summary of cold tolerance-related quantitative trait loci. *Rice* **2014**, *7*, 24. [CrossRef] [PubMed]
6. Andaya, V.C.; Mackill, D.J. Mapping of QTLs associated with cold tolerance during the vegetative stage in rice. *J. Exp. Bot.* **2003**, *54*, 2579–2585. [CrossRef] [PubMed]
7. Ma, Y.; Dai, X.; Xu, Y.; Luo, W.; Zheng, X.; Zeng, D.; Pan, Y.; Lin, X.; Liu, H.; Zhang, D.; et al. COLD1 confers chilling tolerance in rice. *Cell* **2015**, *160*, 1209–1221. [CrossRef]
8. Ding, Y.; Yang, S. Surviving and thriving: How plants perceive and respond to temperature stress. *Dev. Cell* **2022**, *57*, 947–958. [CrossRef]
9. Guo, X.; Liu, D.; Chong, K. Cold signaling in plants: Insights into mechanisms and regulation. *J. Integr. Plant Biol.* **2018**, *60*, 745–756. [CrossRef]
10. Zhang, J.; Li, X.M.; Lin, H.X.; Chong, K. Crop improvement through temperature resilience. *Annu. Rev. Plant Biol.* **2019**, *70*, 753–780. [CrossRef]
11. Knight, H.; Trewavas, A.J.; Knight, M.R. Cold calcium signaling in Arabidopsis involves two cellular pools and a change in calcium signature after acclimation. *Plant Cell* **1996**, *8*, 489–503. [CrossRef]
12. Gong, M.A.V.D.; Knight, M.R.; Trewavas, A.J.; Ah, V.D.L. Heat-shock-induced changes in intracellular Ca^{2+} level in tobacco seedlings in relation to thermotolerance. *Plant Physiol.* **1998**, *116*, 429–437. [CrossRef]
13. Zhu, J.K. Abiotic stress signaling and responses in plants. *Cell* **2016**, *167*, 313–324. [CrossRef]
14. Wang, J.; Ren, Y.; Liu, X.; Luo, S.; Zhang, X.; Liu, X.; Lin, Q.; Zhu, S.; Wan, H.; Yang, Y.; et al. Transcriptional activation and phosphorylation of OsCNGC9 confer enhanced chilling tolerance in rice. *Mol. Plant* **2021**, *14*, 315–329. [CrossRef]
15. Sun, Q.; Huang, R.; Zhu, H.; Sun, Y.; Guo, Z. A novel *Medicago* truncatula calmodulin-like protein (MtCML42) regulates cold tolerance and flowering time. *Plant J.* **2021**, *108*, 1069–1082. [CrossRef]
16. Ding, Y.; Shi, Y.; Yang, S. Advances and challenges in uncovering cold tolerance regulatory mechanisms in plants. *New Phytol.* **2019**, *222*, 1690–1704. [CrossRef]
17. Stockinger, E.J.; Gilmour, S.J.; Thomashow, M.F. *Arabidopsis thaliana CBF1* encodes an AP2 domain-containing transcriptional activator that binds to the C-repeat/DRE, a cis-acting DNA regulatory element that stimulates transcription in response to low temperature and water deficit. *Proc. Natl. Acad. Sci. USA* **1997**, *94*, 1035–1040. [CrossRef]

18. Liu, Q.; Kasuga, M.; Sakuma, Y.; Abe, H.; Miura, S.; Yamaguchi-Shinozaki, K.; Shinozaki, K. Two transcription factors, DREB1 and DREB2, with an EREBP/AP2 DNA binding domain separate two cellular signal transduction pathways in drought- and low-temperature-responsive gene expression, respectively, in Arabidopsis. *Plant Cell* **1998**, *10*, 1391–1406. [CrossRef]
19. Mao, D.; Chen, C. Colinearity and similar expression pattern of rice *DREB1s* reveal their functional conservation in the cold-responsive pathway. *PLoS ONE* **2012**, *7*, e47275. [CrossRef]
20. Zhao, C.; Lang, Z.; Zhu, J.K. Cold responsive gene transcription becomes more complex. *Trends Plant Sci.* **2015**, *20*, 466–468. [CrossRef]
21. Zhang, J.; Luo, W.; Zhao, Y.; Xu, Y.; Song, S.; Chong, K. Comparative metabolomic analysis reveals a reactive oxygen species-dominated dynamic model underlying chilling environment adaptation and tolerance in rice. *New Phytol.* **2016**, *211*, 1295–1310. [CrossRef]
22. Baxter, A.; Mittler, R.; Suzuki, N. ROS as key players in plant stress signalling. *J. Exp. Bot.* **2014**, *65*, 1229–1240. [CrossRef]
23. Mittler, R.; Zandalinas, S.I.; Fichman, Y.; Van Breusegem, F. Reactive oxygen species signalling in plant stress responses. *Nat. Rev. Mol. Cell Biol.* **2022**, *23*, 663–679. [CrossRef]
24. Du, H.; Wu, N.; Chang, Y.; Li, X.; Xiao, J.; Xiong, L. Carotenoid deficiency impairs ABA and IAA biosynthesis and differentially affects drought and cold tolerance in rice. *Plant Mol. Biol.* **2013**, *83*, 475–488. [CrossRef]
25. Lee, B.H.; Lee, H.; Xiong, L.; Zhu, J.K. A mitochondrial complex I defect impairs cold-regulated nuclear gene expression. *Plant Cell* **2002**, *14*, 1235–1251. [CrossRef]
26. Dong, C.H.; Zolman, B.K.; Bartel, B.; Lee, B.H.; Stevenson, B.; Agarwal, M.; Zhu, J.K. Disruption of *Arabidopsis CHY1* reveals an important role of metabolic status in plant cold stress signaling. *Mol. Plant* **2009**, *2*, 59–72. [CrossRef]
27. Kimura, M.; Kagawa, T. Phototropin and light-signaling in phototropism. *Curr. Opin. Plant Biol.* **2006**, *9*, 503–508. [CrossRef]
28. Palme, K.; Dovzhenko, A.; Ditengou, F.A. Auxin transport and gravitational research: Perspectives. *Protoplasma* **2006**, *229*, 175–181. [CrossRef]
29. Shibasaki, K.; Uemura, M.; Tsurumi, S.; Rahman, A. Auxin response in *Arabidopsis* under cold stress: Underlying molecular mechanisms. *Plant Cell* **2009**, *21*, 3823–3838. [CrossRef]
30. Du, H.; Liu, H.; Xiong, L. Endogenous auxin and jasmonic acid levels are differentially modulated by abiotic stresses in rice. *Front. Plant Sci.* **2013**, *4*, 397. [CrossRef]
31. Woodward, A.W.; Bartel, B. Auxin: Regulation, action, and interaction. *Ann. Bot.* **2005**, *95*, 707–735. [CrossRef]
32. Zhao, Y. Auxin biosynthesis and its role in plant development. *Annu. Rev. Plant Biol.* **2010**, *61*, 49–64. [CrossRef]
33. Zhao, Y. The role of local biosynthesis of auxin and cytokinin in plant development. *Curr. Opin. Plant Biol.* **2008**, *11*, 16–22. [CrossRef]
34. Petrasek, J.; Mravec, J.; Bouchard, R.; Blakeslee, J.J.; Abas, M.; Seifertova, D.; Wisniewska, J.; Tadele, Z.; Kubes, M.; Covanova, M.; et al. PIN proteins perform a rate-limiting function in cellular auxin efflux. *Science* **2006**, *312*, 914–918. [CrossRef]
35. Zazimalova, E.; Murphy, A.S.; Yang, H.; Hoyerova, K.; Hosek, P. Auxin transporters-why so many? *Cold Spring Harb. Perspect. Biol.* **2010**, *2*, a001552. [CrossRef]
36. Benkova, E.; Michniewicz, M.; Sauer, M.; Teichmann, T.; Seifertova, D.; Jurgens, G.; Friml, J. Local, efflux-dependent auxin gradients as a common module for plant organ formation. *Cell* **2003**, *115*, 591–602. [CrossRef]
37. Sauer, M.; Kleine-Vehn, J. PIN-FORMED and PIN-LIKES auxin transport facilitators. *Development* **2019**, *146*, dev168088. [CrossRef]
38. Krecek, P.; Skupa, P.; Libus, J.; Naramoto, S.; Tejos, R.; Friml, J.; Zazimalova, E. The PIN-FORMED (PIN) protein family of auxin transporters. *Genome Biol.* **2009**, *10*, 249. [CrossRef]
39. Xu, H.W.; Zhang, Y.W.; Yang, X.Y.; Wang, H.H.; Hou, D.Y. Tissue specificity and responses to abiotic stresses and hormones of *PIN* genes in rice. *Biologia* **2022**, *77*, 1459–1470. [CrossRef]
40. Abdollahi Sisi, N.; Ruzicka, K. ER-localized PIN carriers: Regulators of intracellular auxin homeostasis. *Plants* **2020**, *9*, 1527. [CrossRef]
41. Zwiewka, M.; Bilanovicova, V.; Seifu, Y.W.; Nodzynski, T. The nuts and bolts of PIN auxin efflux carriers. *Front. Plant Sci.* **2019**, *10*, 985. [CrossRef]
42. Lin, Y.Q.; Qi, Y.H. Advances in auxin efflux carrier PIN proteins. *Chin. Bull. Bot.* **2021**, *56*, 151–165.
43. Petrasek, J.; Friml, J. Auxin transport routes in plant development. *Development* **2009**, *136*, 2675–2688. [CrossRef]
44. Robert, H.S.; Friml, J. Auxin and other signals on the move in plants. *Nat. Chem. Biol.* **2009**, *5*, 325–332. [CrossRef]
45. Wang, J.R.; Hu, H.; Wang, G.H.; Li, J.; Chen, J.Y.; Wu, P. Expression of *PIN* genes in rice (*Oryza sativa* L.): Tissue specificity and regulation by hormones. *Mol. Plant* **2009**, *2*, 823–831. [CrossRef]
46. Miyashita, Y.; Takasugi, T.; Ito, Y. Identification and expression analysis of *PIN* genes in rice. *Plant Sci.* **2010**, *178*, 424–428. [CrossRef]
47. Xu, M.; Zhu, L.; Shou, H.; Wu, P. A *PIN1* family gene, *OsPIN1*, involved in auxin-dependent adventitious root emergence and tillering in rice. *Plant Cell Physiol.* **2005**, *46*, 1674–1681. [CrossRef]
48. Xu, H.W.; Mo, Y.W.; Wang, W.; Wang, H.; Wang, Z. *OsPIN1a* gene participates in regulating negative phototropism of rice root. *Rice Sci.* **2014**, *21*, 83–89. [CrossRef]
49. Li, Y.; Zhu, J.; Wu, L.; Shao, Y.; Wu, Y.; Mao, C. Functional divergence of *PIN1* paralogous genes in rice. *Plant Cell Physiol.* **2019**, *60*, 2720–2732. [CrossRef]

50. Inahashi, H.; Shelley, I.J.; Yamauchi, T.; Nishiuchi, S.; Takahashi-Nosaka, M.; Matsunami, M.; Ogawa, A.; Noda, Y.; Inukai, Y. *OsPIN2*, which encodes a member of the auxin efflux carrier proteins, is involved in root elongation growth and lateral root formation patterns via the regulation of auxin distribution in rice. *Physiol. Plant* **2018**, *164*, 216–225. [CrossRef]
51. Wang, L.; Guo, M.; Li, Y.; Ruan, W.; Mo, X.; Wu, Z.; Sturrock, C.J.; Yu, H.; Lu, C.; Peng, J.; et al. *LARGE ROOT ANGLE1*, encoding *OsPIN2*, is involved in root system architecture in rice. *J. Exp. Bot.* **2018**, *69*, 385–397. [CrossRef]
52. Li, W.Q.; Zhang, M.J.; Qiao, L.; Chen, Y.B.; Zhang, D.P.; Jing, X.Q.; Gan, P.F.; Huang, Y.B.; Gao, J.R.; Liu, W.T.; et al. Characterization of *wavy root* 1, an agravitropism allele, reveals the functions of OsPIN2 in fine regulation of auxin transport and distribution and in ABA biosynthesis and response in rice (*Oryza sativa* L.). *Crop. J.* **2022**, *10*, 980–992. [CrossRef]
53. Lu, G.; Coneva, V.; Casaretto, J.A.; Ying, S.; Mahmood, K.; Liu, F.; Nambara, E.; Bi, Y.M.; Rothstein, S.J. *OsPIN5b* modulates rice (*Oryza sativa*) plant architecture and yield by changing auxin homeostasis, transport and distribution. *Plant J.* **2015**, *83*, 913–925. [CrossRef]
54. Zhang, Q.; Li, J.; Zhang, W.; Yan, S.; Wang, R.; Zhao, J.; Li, Y.; Qi, Z.; Sun, Z.; Zhu, Z. The putative auxin efflux carrier *OsPIN3t* is involved in the drought stress response and drought tolerance. *Plant J.* **2012**, *72*, 805–816. [CrossRef]
55. Hou, M.; Luo, F.; Wu, D.; Zhang, X.; Lou, M.; Shen, D.; Yan, M.; Mao, C.; Fan, X.; Xu, G.; et al. OsPIN9, an auxin efflux carrier, is required for the regulation of rice tiller bud outgrowth by ammonium. *New Phytol.* **2021**, *229*, 935–949. [CrossRef]
56. Shen, C.; Bai, Y.; Wang, S.; Zhang, S.; Wu, Y.; Chen, M.; Jiang, D.; Qi, Y. Expression profile of PIN, AUX/LAX and PGP auxin transporter gene families in Sorghum bicolor under phytohormone and abiotic stress. *FEBS J.* **2010**, *277*, 2954–2969. [CrossRef]
57. Yue, R.; Tie, S.; Sun, T.; Zhang, L.; Yang, Y.; Qi, J.; Yan, S.; Han, X.; Wang, H.; Shen, C. Genome-wide identification and expression profiling analysis of *ZmPIN, ZmPILS, ZmLAX* and *ZmABCB* auxin transporter gene families in maize (*Zea mays* L.) under various abiotic stresses. *PLoS ONE* **2015**, *10*, e0118751. [CrossRef]
58. Forestan, C.; Farinati, S.; Varotto, S. The maize *PIN* gene family of auxin transporters. *Front. Plant Sci.* **2012**, *3*, 16. [CrossRef]
59. Morris, D.A. The effect of temperature on the velocity of exogenous auxin transport in intact chilling-sensitive and chilling-resistant plants. *Planta* **1979**, *146*, 603–605. [CrossRef]
60. Jain, M.; Khurana, J.P. Transcript profiling reveals diverse roles of auxin-responsive genes during reproductive development and abiotic stress in rice. *FEBS J.* **2009**, *276*, 3148–3162. [CrossRef]
61. Nadella, V.; Shipp, M.J.; Muday, G.K.; Wyatt, S.E. Evidence for altered polar and lateral auxin transport in the *gravity persistent signal* (*gps*) mutants of *Arabidopsis*. *Plant Cell Environ.* **2006**, *29*, 682–690. [CrossRef]
62. Ashraf, M.A.; Rahman, A. Cold stress response in *Arabidopsis thaliana* is mediated by GNOM ARF-GEF. *Plant J.* **2019**, *97*, 500–516. [CrossRef]
63. Manna, M.; Rengasamy, B.; Ambasht, N.K.; Sinha, A.K. Characterization and expression profiling of PIN auxin efflux transporters reveal their role in developmental and abiotic stress conditions in rice. *Front. Plant Sci.* **2022**, *13*, 1059559. [CrossRef]
64. Xu, H.; Yang, X.; Zhang, Y.; Wang, H.; Wu, S.; Zhang, Z.; Ahammed, G.J.; Zhao, C.; Liu, H. CRISPR/Cas9-mediated mutation in auxin efflux carrier *OsPIN9* confers chilling tolerance by modulating reactive oxygen species homeostasis in rice. *Front. Plant Sci.* **2022**, *13*, 967031. [CrossRef]
65. Abas, L.; Kolb, M.; Stadlmann, J.; Janacek, D.P.; Lukic, K.; Schwechheimer, C.; Sazanov, L.A.; Mach, L.; Friml, J.; Hammes, U.Z. Naphthylphthalamic acid associates with and inhibits PIN auxin transporters. *Proc. Natl. Acad. Sci. USA* **2021**, *118*, e2020857118. [CrossRef]
66. Rahman, A. Auxin: A regulator of cold stress response. *Physiol. Plant* **2013**, *147*, 28–35. [CrossRef]
67. Dubouzet, J.G.; Sakuma, Y.; Ito, Y.; Kasuga, M.; Dubouzet, E.G.; Miura, S.; Seki, M.; Shinozaki, K.; Yamaguchi-Shinozaki, K. *OsDREB* genes in rice, *Oryza sativa* L., encode transcription activators that function in drought-, high-salt- and cold-responsive gene expression. *Plant J.* **2003**, *33*, 751–763. [CrossRef]
68. Zhang, Z.; Li, J.; Li, F.; Liu, H.; Yang, W.; Chong, K.; Xu, Y. OsMAPK3 phosphorylates OsbHLH002/OsICE1 and inhibits its ubiquitination to activate *OsTPP1* and enhances rice chilling tolerance. *Dev. Cell* **2017**, *43*, 731–743.e5. [CrossRef]
69. Dong, Q.; Wallrad, L.; Almutairi, B.O.; Kudla, J. Ca^{2+} signaling in plant responses to abiotic stresses. *J. Integr. Plant Biol.* **2022**, *64*, 287–300. [CrossRef]
70. Xu, Y.; Wang, R.; Wang, Y.; Zhang, L.; Yao, S. A point mutation in *LTT1* enhances cold tolerance at the booting stage in rice. *Plant Cell Environ.* **2020**, *43*, 992–1007. [CrossRef]
71. Hu, C.H.; Wang, P.Q.; Zhang, P.P.; Nie, X.M.; Li, B.B.; Tai, L.; Liu, W.T.; Li, W.Q.; Chen, K.M. NADPH oxidases: The vital performers and center hubs during plant growth and signaling. *Cells* **2020**, *9*, 437. [CrossRef]
72. Marino, D.; Dunand, C.; Puppo, A.; Pauly, N. A burst of plant NADPH oxidases. *Trends Plant Sci.* **2012**, *17*, 9–15. [CrossRef]
73. Olatunji, D.; Geelen, D.; Verstraeten, I. Control of endogenous auxin levels in plant root development. *Int. J. Mol. Sci.* **2017**, *18*, 2587. [CrossRef]
74. Michniewicz, M.; Brewer, P.B.; Friml, J.I. Polar auxin transport and asymmetric auxin distribution. *Arab. Book* **2007**, *5*, e0108. [CrossRef]
75. Friml, J. Auxin transport-shaping the plant. *Curr. Opin. Plant Biol.* **2003**, *6*, 7–12. [CrossRef]
76. Okada, K.; Ueda, J.; Komaki, M.K.; Bell, C.J.; Shimura, Y. Requirement of the auxin polar transport system in early stages of *Arabidopsis* floral bud formation. *Plant Cell* **1991**, *3*, 677–684. [CrossRef]
77. Chen, Y.; Fan, X.; Song, W.; Zhang, Y.; Xu, G. Over-expression of *OsPIN2* leads to increased tiller numbers, angle and shorter plant height through suppression of *OsLAZY1*. *Plant Biotechnol. J.* **2012**, *10*, 139–149. [CrossRef]

78. Kovach, M.J.; Sweeney, M.T.; McCouch, S.R. New insights into the history of rice domestication. *Trends Genet.* **2007**, *23*, 578–587. [CrossRef]
79. Saito, K.; Hayano-Saito, Y.; Kuroki, M.; Sato, Y. Map-based cloning of the rice cold tolerance gene *Ctb1*. *Plant Sci.* **2010**, *179*, 97–102. [CrossRef]
80. Sang, T.; Ge, S. Genetics and phylogenetics of rice domestication. *Curr. Opin. Genet. Dev.* **2007**, *17*, 533–538. [CrossRef]
81. Demidchik, V. Mechanisms of oxidative stress in plants: From classical chemistry to cell biology. *Environ. Exp. Bot.* **2015**, *109*, 212–228. [CrossRef]
82. Hasanuzzaman, M.; Bhuyan, M.; Parvin, K.; Bhuiyan, T.F.; Anee, T.I.; Nahar, K.; Hossen, M.S.; Zulfiqar, F.; Alam, M.M.; Fujita, M. Regulation of ROS metabolism in plants under environmental stress: A review of recent experimental evidence. *Int. J. Mol. Sci.* **2020**, *21*, 8695. [CrossRef]
83. Ahmad, P.; Alyemeni, M.N.; Al-Huqail, A.A.; Alqahtani, M.A.; Wijaya, L.; Ashraf, M.; Kaya, C.; Bajguz, A. Zinc oxide nanoparticles application alleviates arsenic (As) toxicity in soybean plants by restricting the uptake of as and modulating key biochemical attributes, antioxidant enzymes, ascorbate-glutathione cycle and glyoxalase system. *Plants* **2020**, *9*, 825. [CrossRef]
84. Huang, X.; Cao, L.; Fan, J.; Ma, G.; Chen, L. CdWRKY2-mediated sucrose biosynthesis and CBF-signalling pathways coordinately contribute to cold tolerance in bermudagrass. *Plant Biotechnol. J.* **2021**, *20*, 660–675. [CrossRef]
85. Xing, C.; Liu, Y.; Zhao, L.; Zhang, S.; Huang, X. A novel MYB transcription factor regulates ascorbic acid synthesis and affects cold tolerance. *Plant Cell Environ.* **2019**, *42*, 832–845. [CrossRef]
86. Liu, J.; Shi, Y.; Yang, S. Insights into the regulation of C-repeat binding factors in plant cold signaling. *J. Integr. Plant Biol.* **2018**, *60*, 780–795. [CrossRef]
87. Prasad, T.K.; Anderson, M.D.; Stewart, C.R. Evidence for chilling-induced oxidative stress in maize seedling and a regulatory role for hydrogen peroxide. *Plant Cell* **1994**, *105*, 619–627.
88. Considine, M.J.; Foyer, C.H. Stress effects on the reactive oxygen species-dependent regulation of plant growth and development. *J. Exp. Bot.* **2021**, *72*, 5795–5806. [CrossRef]
89. Xia, X.J.; Zhou, Y.H.; Shi, K.; Zhou, J.; Foyer, C.H.; Yu, J.Q. Interplay between reactive oxygen species and hormones in the control of plant development and stress tolerance. *J. Exp. Bot.* **2015**, *66*, 2839–2856. [CrossRef]
90. Xu, H.W.; Ji, X.M.; He, Z.H.; Shi, W.P.; Zhu, G.H.; Niu, J.K.; Li, B.S.; Peng, X.X. Oxalate accumulation and regulation is independent of glycolate oxidase in rice leaves. *J. Exp. Bot.* **2006**, *57*, 1899–1908. [CrossRef]
91. Hiei, Y.; Ohta, S.; Komari, T.; Kumashiro, T. Efficient transformation of rice (*Oryza sativa* L.) mediated by *Agrobacterium* and sequence analysis of the boundaries of the T-DNA. *Plant J.* **1994**, *6*, 271–282. [CrossRef]
92. Choi, H.W.; Kim, Y.J.; Lee, S.C.; Hong, J.K.; Hwang, B.K. Hydrogen peroxide generation by the pepper extracellular peroxidase CaPO2 activates local and systemic cell death and defense response to bacterial pathogens. *Plant Physiol.* **2007**, *145*, 890–904. [CrossRef]
93. Hu, Z.; Huang, X.; Amombo, E.; Liu, A.; Fan, J.; Bi, A.; Ji, K.; Xin, H.; Chen, L.; Fu, J. The ethylene responsive factor CdERF1 from bermudagrass (*Cynodon dactylon*) positively regulates cold tolerance. *Plant Sci.* **2020**, *294*, 110432. [CrossRef]
94. Wu, F.B.; Zhang, G.P.; Dominy, P. Four barley genotypes respond differently to cadmium: Lipid peroxidation and activities of antioxidant capacity. *Environ. Exp. Bot.* **2003**, *50*, 67–78. [CrossRef]
95. Bradford, M.M. A rapid and sensitive method for the quantitation of microgram quantities of protein utilizing the principle of protein-dye binding. *Anal. Biochem.* **1976**, *72*, 248–254. [CrossRef]

plants

MDPI

Review

The Potential of CRISPR/Cas Technology to Enhance Crop Performance on Adverse Soil Conditions

Humberto A. Gajardo [1,†], Olman Gómez-Espinoza [1,2,†], Pedro Boscariol Ferreira [3], Helaine Carrer [3] and León A. Bravo [1,*]

1 Laboratorio de Fisiología y Biología Molecular Vegetal, Instituto de Agroindustria, Departamento de Ciencias Agronómicas y Recursos Naturales, Facultad de Ciencias Agropecuarias y Medioambiente & Center of Plant, Soil Interaction and Natural Resources Biotechnology, Scientific and Technological Bioresource Nucleus, Universidad de La Frontera, Temuco 1145, Chile; h.gajardo01@ufromail.cl (H.A.G.); olmang03@gmail.com (O.G.-E.)

2 Centro de Investigación en Biotecnología, Escuela de Biología, Instituto Tecnológico de Costa Rica, Cartago 30101, Costa Rica

3 Department of Biological Sciences, Luiz de Queiroz College of Agriculture (ESALQ), University of São Paulo, Piracicaba 13418-900, Brazil; boscariolfp@gmail.com (P.B.F.); hecarrer@usp.br (H.C.)

* Correspondence: leon.bravo@ufrontera.cl

† These authors contributed equally to this work.

Abstract: Worldwide food security is under threat in the actual scenery of global climate change because the major staple food crops are not adapted to hostile climatic and soil conditions. Significant efforts have been performed to maintain the actual yield of crops, using traditional breeding and innovative molecular techniques to assist them. However, additional strategies are necessary to achieve the future food demand. Clustered regularly interspaced short palindromic repeat/CRISPR-associated protein (CRISPR/Cas) technology, as well as its variants, have emerged as alternatives to transgenic plant breeding. This novelty has helped to accelerate the necessary modifications in major crops to confront the impact of abiotic stress on agriculture systems. This review summarizes the current advances in CRISPR/Cas applications in crops to deal with the main hostile soil conditions, such as drought, flooding and waterlogging, salinity, heavy metals, and nutrient deficiencies. In addition, the potential of extremophytes as a reservoir of new molecular mechanisms for abiotic stress tolerance, as well as their orthologue identification and edition in crops, is shown. Moreover, the future challenges and prospects related to CRISPR/Cas technology issues, legal regulations, and customer acceptance will be discussed.

Keywords: genetic engineering; extremophytes; food security; abiotic stress

1. Introduction

Humans have depended on plants throughout their existence. Since the beginning of agriculture and the domestication of plants, agronomic management and traditional breeding have provided humanity with the modern varieties that feed the world today [1]. In contrast to the primary evolution of land plants, which occurred under unfavorable conditions (e.g., drought, fluctuating light, and temperature) [2], domestication occurred under relatively stress-free, managed conditions [3,4]. Later, during the green revolution of the mid-twentieth century, agricultural breeding radically modified plant architecture to achieve high yields [5]. As a result, the current world situation is that crop plants are much more used as food and feed than wild species [6].

Unfortunately, improvement of yield-related traits can compromise resource allocation to other traits, impairing biotic and abiotic stress tolerance [4]. This trade-off between traits hinders the capacity of crop species to mitigate the effect of changing environmental conditions [7]. Therefore, crop domestication increased the likelihood of these species being more sensitive to stresses than their wild relatives [8,9]. Within this framework, some crops can only achieve high yields with management-intense modern agricultural practices [10].

Indeed, crop production is already being affected across several regions worldwide due to climate change. The rising frequency of extreme climate events threatens further damage to food and feed production [11]. Many species are and will be affected by combinations of elevated atmospheric CO_2 concentration, increased temperatures, and changing seasonal rainfall patterns [12]. Between 2013 and 2016, for example, all Caribbean islands experienced an extensive drought that pushed more than two million people into food insecurity [13], and over 50% of the crops were lost in some of these regions [14]. Additionally, drought cost the United States of America (USA) USD $250 billion in damages, one of the costliest natural disasters [15]. Aside from natural causes, agricultural practices, such as artificial fertilization, burning agricultural residues, trading, long-distance transportation, and pesticides, are responsible for significant carbon and methane emissions and environmental pollution [16]. These factors, combined, are believed to have accelerated climate change, and there is an urgent need to adopt practices to reduce the future impacts of extreme climate events [17].

Currently, abiotic stresses, such as drought, salinity, and flooding, already limit food production severely, resulting in yearly global losses of over USD $100 billion to the agricultural sector [18,19]. Coupled with the abovementioned stresses, heavy metal accumulation and nutrient deficiencies promote hostile soils for food, feed, and fuel production [18,20]. While healthy soils are pivotal to sustainable crop yield, one-third of global soils face progressive degradation [21,22]. Therefore, effective adaptation strategies are needed to mitigate the negative impacts of these soils on crop production. As such, technology-based approaches are a faster alternative to traditional techniques and management strategies [23].

Current and developing technologies that might aid in creating or enhancing stress resilience in crops include molecular-assisted plant breeding [24], genetic manipulation of traits by transgenesis or gene editing [25,26], plant-microbe engineering [27,28], de novo domestication of wild species [29], and artificial apomixis [30]. In most of these approaches, natural genetic variability and gene orthology are sources of targets for genetic manipulation to enhance crops. Natural variation in the gene, in its *cis*-regulatory elements, protein-coding sequence, transcription start and termination sites, and splice sites, can explain intra-species variability regarding stress tolerance, for example [31]. In addition, since the genome-wide functional characterization is unavailable for any single species, researchers use the orthology-function conjecture, wherein orthologous genes might perform similar functions in different species [32]. In both cases, genetic information from intra- or interspecies variation guides the strategies to engineer desirable traits. In this context, an under-utilized genetic resource resides in crop wild relatives and extremophytes that can be naturally tolerant to extreme conditions [33]. A broader knowledge of extremophytes' genetics could provide even more information to the plant biotechnology toolbox.

Given this background, this review will focus on recent advances in the applications of gene editing by Clustered Regularly Interspaced Short Palindromic Repeat (CRISPR)/CRISPR-associated protein (Cas) systems in crops to cope with hostile soil conditions, such as drought, flooding, salinity, accumulation of heavy metals or toxic elements, and nutrient deficiency. Since CRISPR/Cas technologies enable the targeted and accurate genetic modification of crops without the incorporation of foreign DNA, they increase the speed of crop improvement [34] and are gaining popularity instead of classic transgenesis [35]. The use of extremophytes as reservoirs of natural variants and orthologue targets for CRISPR/Cas applications, as well as future challenges and prospects of this technology, are also discussed.

2. A Broad Overview of CRISPR/Cas Technologies in Plants

Precise genome editing techniques and applications have radically changed after the development of CRISPR/Cas technologies. The CRISPR defense systems were first noticed in bacterial genomes in 1987 [36], as part of the natural adaptive immunity in bacteria and

Archaea [37,38]. In general, CRISPR/Cas-mediated immunity occurs in three steps: (1) the CRISPR-containing organism acquires deoxyribonucleic acid (DNA) or ribonucleic acid (RNA) fragments from invading bacteriophages or plasmids, then (2) it uses the stored nucleic acid to generate CRISPR RNAs (crRNAs) to (3) guide the RNA (gRNA) toward the Cas-mediated inactivation-by-cleavage of future invading viruses (Figure 1) [39].

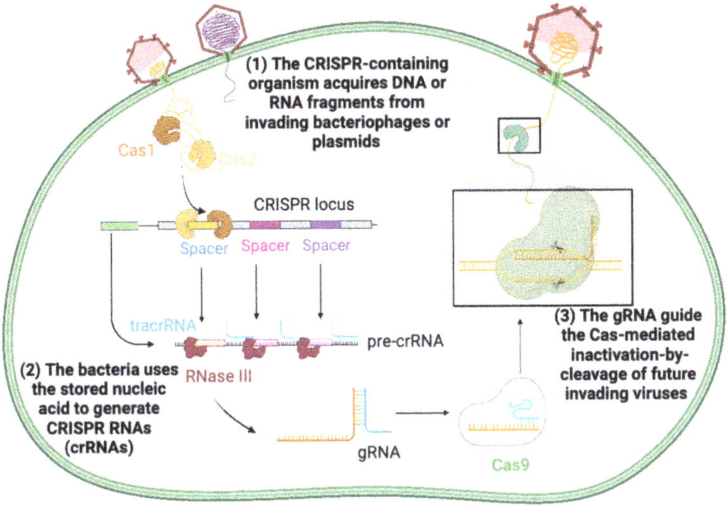

Figure 1. CRISPR/Cas-mediated immunity in bacteria: Three main phases. Image adapted from CRISPR-Cas9 adaptive immune system of *Streptococcus pyogenes* against bacteriophages template by BioRender.com (2023). Retrieved from https://app.biorender.com/biorender-templates, accessed on 31 March 2023.

In 2012, Jinek et al. [37] showed that the Type II CRISPR/Cas9 system of *Streptococcus pyogenes* could be mimicked using a single chimeric gRNA instead of the natural trans-activating crRNA (tracrRNA):crRNA duplex. Similar work was concurrently published by Gasiunas et al. (2012) [40], utilizing the CRISPR/Cas9 system from the bacterium *Streptococcus thermophilus*. Both publications proposed using the artificial CRISPR/Cas9 system to induce double strand breaks (DSB) at target locations in a genome of interest and to take advantage of the error-prone DSB repair pathways to generate genetic variability. Since then, CRISPR/Cas technologies have become synonymous with a relatively cheap, global gene-editing tool, and several modifications and enhancements have been made in this first decade of use [39]. In addition, since the CRISPR/Cas system is diverse, and different combinations of Cas proteins participate in the immune process depending on the host species [41], more systems and variations can be discovered.

In plants, the first five reports of CRISPR/Cas9-based genome editing were published in August 2013 and focused on demonstrating the vast versatility of the technology in the area of plant biology (proof of concept) in the model species *Arabidopsis thaliana*, *Nicotiana benthamiana*, and *Oryza sativa* [42–46]. Shortly after, the CRISPR/Cas9 system became a helpful tool for the functional annotation of plant genes [47], and the first reports of its use in model crops showed successful results in sorghum [48], wheat [49], maize [50], and soybean [51,52]. Since 2012, the number of publications related to genome editing in plants using this technology has grown exponentially (Figure 2), and 925 Research Articles were published in 2022 alone.

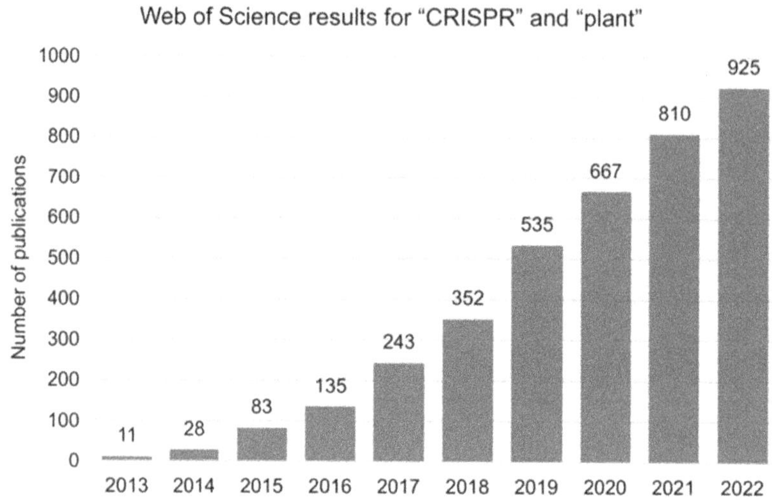

Figure 2. New original research papers per year in the Web of Science database (webofscience.com) from 2013 to 2021 containing the keywords "plant" and "CRISPR".

In the majority of these publications, the general protocol needed to achieve a CRISPR gene-edited line in plants is comprised of four main steps: (1) guide RNA design and optional design validation, (2) gene editing by expression of *Cas* and *gRNA* plant cells, (3) tissue culture for plant regeneration, and (4) evaluation of mutations by sequencing and selection of transgene-free mutated lines (Figure 3). For successful gene editing, 20 nucleotides specific to the target DNA sequence must be provided in the gRNA, applying standard RNA–DNA complementary base-pairing rules [53]. The target sites must be contiguous to a Protospacer Adjacent Motif (PAM) sequence, which varies depending on the Cas nuclease chosen [54]. In the case of the Cas9 nuclease, activity is directed to any DNA region preceding a 5′-NGG-3′ PAM sequence [55]. Several bioinformatics tools have been developed to help design gRNAs and predict their off-target potential [56]. An optional validation of the gRNA can be performed by transiently expressing the *Cas/gRNA* in plant protoplasts. In this stage, genomic DNA from the protoplast culture is submitted to sequencing to evaluate the presence of mutations in the desired region [48]. After proper gRNA design, a *Cas/gRNA* transgene is used for the genetic transformation of a target plant, using the appropriate explants and tissue culture protocols (Figure 3b). Then, the last step is to assess whether independent lines carry mutations in the target site by conventional PCR, followed by Sanger sequencing or Illumina deep sequencing. An additional step is usually added to eliminate the *Cas/gRNA* cassette to prevent off-target mutations induced by the nuclease's constant expression and to reduce concerns about 'genome-editing' plants. Thus, one of the most used strategies to obtain mutant lines without the *Cas/gRNA*-expressing transgene is selection by Mendelian segregation, facilitated by visual markers, such as fluorescent proteins (Figure 3c) [57]. However, other transgene-free methods are desired, since traditional methods are laborious and time-consuming. These methods include technologies for self-elimination of transgenes, direct delivery of the *Cas/gRNA* ribonucleoprotein (RNP), or expression of *Cas/gRNA* by viral vectors [58].

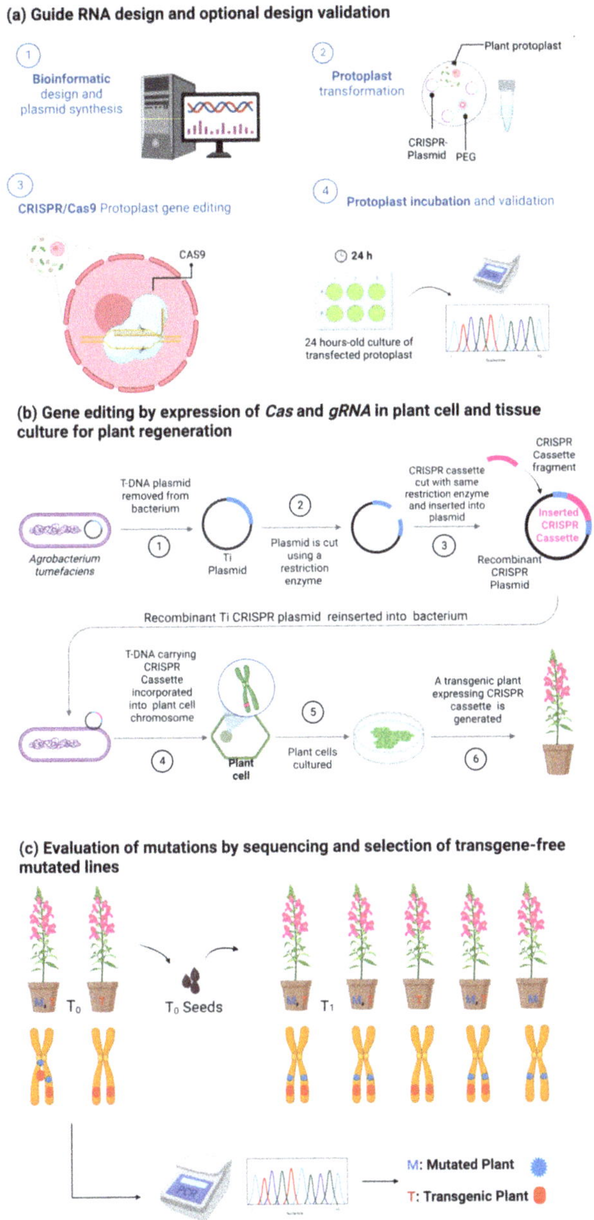

Figure 3. Standard protocol for generating transgene-free gene-edited plants using CRISPR/Cas9. (**a**) The use of protoplasts is the most common technique for validating CRISPR/Cas9 construct designs and generating transient gene expression. (**b**) Then, Agrobacterium-mediated transformation is the common technique to generate CRISPR/Cas9 mutated plants with a stable gene expression. (**c**) Finally, elimination of transgenic sequences is performed to generate "null segregants" via Mendelian segregation. Images (**a**,**b**) are adapted from Agrobacterium-Mediated Transformation and CRISPR-Cas9 Gene Editing in *Trypanosoma cruzi* templates by BioRender.com (2022). Retrieved from https://app.biorender.com/biorender-templates, accessed on 31 March 2023.

Following the success of CRISPR/Cas technologies in single gene editing, several modifications were developed to enhance and diversify their use [39,59]. The multiplex CRISPR approach is a strategy that enables the editing of multiple genes in a single transformation event and uses multiple gRNAs delivered at once to achieve this goal [60–63]. Abdallah et al. [60] used three distinct gRNAs to knockout five TaSal1 genes in wheat, resulting in drought-tolerant seedlings. In another example, Lorenzo et al. [64] used 12 gRNAs, in combination, to target the knockout of 12 different growth-related genes, producing lines with enhanced yield in maize. Another approach is to produce point mutations using CRISPR base editing, which does not create breaks in the DNA. This technique uses a catalytically defective Cas enzyme, with nickase activity in a single DNA strand (nCas). Then, nCas is fused with enzymes with deaminase activity that modify the bases in the window targeted by the gRNA [54]. A different technology, Prime Editing CRISPR (CRISPR-PE), allows the insertion of sequences at the target location by providing a RNA template in the gRNA (then called pegRNA) when nCas produces a single strand break. In this technique, nCas is fused with a Reverse Transcriptase that uses the template pegRNA to insert the modification on the generated single strand [54]. Other systems, such as CRISPR-Combo [65], combine gene editing capabilities with CRISPR/Cas-based gene expression activation to boost plant genome engineering. In this case, gRNAs with different protospacer lengths determine whether the target is cleaved by the Cas9 (20 nt protospacer) or activated by the MS2-SunTag activator (15 nucleotides protospacer). The authors tested several applications of the system, and one example is the concurrent activation of *AtFT*, accelerating flowering time and inactivation of herbicide target genes *AtALS* and *AtACC2*. By selecting only early flowering plants, transgene-free herbicide-resistant mutants were easily detected [65]. The different CRISPR/Cas technologies can be used in genome-wide screens, which provide a targeted approach in generating a large number of mutants that can be selected by their phenotype in a desired condition [66]. The identification of causal genes is then easily performed by using the gRNAs as barcodes in deep sequencing [67]. Gaillochet et al. [66] and Pan et al. [68] review CRISPR screening in plants in more detail, demonstrating its potential to identify genes and generate varieties tolerant to multiple stresses, since mutants can be selected by phenotype after they are established. The toolbox of CRISPR/Cas technologies is still expanding, and this plethora of strategies to manipulate plant genomes shows promise to generate transgene-free stress-resilient crops.

From this perspective, genome manipulation techniques have always supported basic and applied crop research and are crucial for modern agricultural production [69]. Although these techniques require prior physiology and molecular genetics knowledge of the plant species under study, at least 42 plant species have been successfully edited by CRISPR/Cas technologies [70]. An attractive characteristic of CRISPR/Cas systems is their success in genome editing polyploid species, which is the case for most crop and biofuel species [71]. With the theoretical knowledge and the know-how of gene-editing techniques, custom modifications can be targeted to specific genes to improve desired traits in a highly predictive manner. The following sections summarize stress-related genetic discoveries in crops achieved by CRISPR/Cas and discuss the potential of these technologies for engineering crops with higher tolerance to extreme conditions affecting the soil–plant interface.

3. Advances in Engineering Commercial Crop Genomes to Cope with Different Hostile Soil Conditions

The decline in soil quality poses a significant challenge to agriculture [72], and CRISPR/Cas systems can be valuable tools to address abiotic stress-related traits in plants. These traits are often controlled by regulatory genes, which can be knocked out or down to improve tolerance [73]. However, natural environments typically present a combination of different stresses simultaneously, and while most progress has been made in studying individual stresses, genome-wide association studies and transcriptomic information have identified numerous candidate genes involved in stress tolerance regulation that are potential targets for CRISPR/Cas applications [74]. Cross-species analysis of stress responses shows a conserved core genetic response to stresses [75–77], which may help develop genotype-independent strategies to cope with a changing climate. CRISPR-edited single genes that confer tolerance to individual stresses can be used as a starting point for a multiplexed approach, where combinations of mutations can confer combined stress tolerance.

3.1. Drought Stress Tolerance

Water deficiency is a chronic abiotic crop stress that impacts plant growth and development, constituting about 70% of potential crop yield and productivity losses globally [78]. Drought exists either due to significantly less rainfall or a significant decrease in the quantity of moisture, and it is considered a substantial abiotic stress, hindering agriculture and forestry [78]. Indeed, modeled climate change projections show an even worse scenario for drought, independent of the decrease or increase in greenhouse emissions in most of the world [79,80]. Soil drought can have significant negative impacts on crops by reducing plant growth, altering plant architecture, delaying or inhibiting plant development, reducing reproductive success, and increasing susceptibility to diseases and pests [78].

Plant responses to drought might be classified in five critical processes: sensing, avoidance, tolerance, scape, and recovery. After drought sensing, two pathways can be activated: an abscisic acid (ABA)-dependent pathway or an ABA-independent pathway, triggering the activation of transcription factors and specific drought responsive genes [81,82]. Drought avoidance involves morpho-physiological changes, such as stomatal closure, leaf area or leaf number reduction, wax synthesis, and increased root systems [82]. On the other hand, drought tolerance involves mechanisms to cope with severe drought at different phenological stages, such as changes in stomatal density, gene expression of drought responsive genes, and synthesis of osmoprotectans. In addition, the reduction of photosynthesis rate under drought leads to an imbalance in energy, inducing the production of Reactive Oxygen Species (ROS), which are signaling molecules of stress that can damage the plant cellular machinery [83]. An additional tolerance mechanism is the biosynthesis of antioxidant molecules and expression of the enzymatic antioxidant system to ameliorate the oxidative cellular stress triggered by drought conditions [83]. Short life cycles or early flowering, on the other hand, are examples of escape mechanism to drought, and they could be interesting targets for gene editing. Finally, recovery, the capacity of the plant to survive a severe drought event, involves processes of cellular protection, repair, and stress priming to promote photosynthesis recovery, and it has been extensively studied in resurrection plants [84]. These cellular processes are all targets for genetic manipulation, and several elements have already been studied by CRISPR/Cas technologies. A comprehensive summary of research that employed CRISPR/Cas-directed mutagenesis strategies to study drought stress tolerance in crops is presented in Table 1. All of the studies have proven to either enhance or reduce performance of the mutant plants in comparison to the wild type by experiments in growth chambers, greenhouses, or in the field. Proof-of-concept studies and cross-species gene validation studies were excluded from Table 1 for brevity.

Table 1. Studies employing CRISPR/Cas on genes related to drought stress. TF: Transcription Factor; GA: Gibberellic acid; ABA: Abscisic acid; BR: Brassinosteroid; SA: Salicylic acid; KO = Knockout; KD = Knockdown; I.N.F. = Information Not Found, KU: Knokup.

Species	Target Locus	Pathway/Function	Effect on Tolerance	Result	Reference
Brassica napus	*BnaA6.RGA*	Growth regulation/DELLA transcription regulator	Enhanced	Gain-of-function	[85]
	BnaA6.RGA *BnaC7.RGA* *BnaA9.RGA* *BnaC9.RGA*	Growth regulation/DELLA transcription regulator	Reduced	KO	[85]
Cucumis sativus	*CsAKT1*	Osmoregulation/K+ transporter	Reduced	KO	[86]
Fragaria vesca	*FvICE1*	Cold stress response/TF	Reduced	KO	[87]
Glycine max	*GmHdz4*	Drought stress response/HD-ZIP I TF	Enhanced	KO	[88]
	GmLHY1a *GmLHY1b* *GmLHY2a* *GmLHY2b*	Regulation of circadian rhythm/TF	Enhanced	KO	[89]
	GmCOL1a	Flowering time/CONSTANS-like TF	Reduced	KO	[90]
	GmMYB118	Flavonoid biosynthesis/MYB TF	Reduced	Amino acid change	[91]
	GmNAC12	Abiotic stress response/NAC TF	Reduced	KO	[92]
	GmNAC8	Nodulation, abiotic stress response/NAC TF	Reduced	KO	[93]
Medicago sativa	*MsSPL8*	Nodulation, growth, GA pathway/SPL TF	Enhanced	KD	[94]
Nicotiana tabacum	*NtAITR1* *NtAITR2* *NtAITR3* *NtAITR5* *NtAITR6*	ROS homeostasis/ABA-induced transcription repressors	Enhanced	I.N.F.	[95]
	NtPOD63L	Cell wall integrity/class III peroxidase	Enhanced	KO	[96]
	NtRAV4	Growth, development, stress response/RAV TF	Enhanced	KO	[97]

Table 1. *Cont.*

Species	Target Locus	Pathway/Function	Effect on Tolerance	Result	Reference
Oryza sativa	*Ghd2*	Grain development, flowering/CCT TF	Enhanced	KO	[98]
	JMJ710	Flowering time/Histone demethylase	Enhanced	KO	[99]
	osa-MIR535	Phosphate homeostasis, root development/Drought-induced miRNAs	Enhanced	KO	[100]
	OsABA8ox2	Biosynthesis of ABA/ABA hydroxylase	Enhanced	KO	[101]
	OsDST	ABA-dependent stress signaling/Zinc finger TF	Enhanced	Domain deletion	[102]
	OsERA1	BR signaling/GASA growth regulator	Enhanced	I.N.F.	[103]
	OsFTL4	Flowering/PEBP, florigen	Enhanced	KO	[104]
	OsIPK1	Growth, development, ion homeostasis/Kinase	Enhanced	11-aminoacid deletion	[105]
	OsNAC016	Growth, development, hormone signaling, abiotic stress response/NAC TF	Enhanced	KO	[106]
	OsNAC092	Biotic and abiotic stress response/NAC TF	Enhanced	KO	[107]
	OsNR1.2	Nitrogen metabolism/Nitrate reductase	Enhanced	KO	[108]
	OsPPR035	Energy metabolism, stress response/Mitochondrial RNA editing	Enhanced	KO	[109]
	OsPPR406	Energy metabolism, stress response/Mitochondrial RNA editing	Enhanced	KO	[109]
	OsPYL9	Stress responses/ABA receptor	Enhanced	KO	[110]
	OsWRKY5	ABA signaling/WRKY TF	Enhanced	KO	[111]
	SRL1,2	Root development, stress response/LRR-RLK protein	Enhanced	KD	[112]
	osa-MIR171	Flavonoid biosynthesis/microRNA	Reduced	KO	[113]
	osa-MIR818b	Stress response/Drought-induced miRNAs	Reduced	KD	[114]
	OsADR3	Spikelet development/MADS-box TF	Reduced	KO	[115]
	OsASLRK	Root development/Armadillo-like Repeat Kinesin	Reduced	KD	[116]
	OsbZIP86	Stress response/bZIP TF	Reduced	KO	[117]
	OsCCR10	Biosynthesis of lignin/cinnamoyl-CoA reductase	Reduced	KO	[118]
	OsDIP1	Root water uptake/Aquaporin	Reduced	KO	[119]
	OsFTIP6	Flowering, leaf senescence, plant architecture/Florigen transporter	Reduced	KO	[120]
	OsGRP3	RNA processing/Glycine-rich RNA-binding protein	Reduced	KO	[121]
	OsHB22	Growth, development, abiotic stress response/HD-ZIP TF	Reduced	KO	[120]
	OsMYB60	Osmoprotectants and antioxidants biosynthesis/MYB TF	Reduced	KO	[122]
	OsMYBR57	Drought stress response/MYB-Related TF	Reduced	KO	[120]
	OsNAC006	Abiotic stress response/NAC TF	Reduced	KO	[123]
	OsNAC17	Development, stress response/NAC TF	Reduced	KO	[124]
	OsNPF8.1	Nutrient acquisition/Phosphate transporter	Reduced	KO	[125]
	OsPM1	Ion homeostasis/Plasma membrane protein	Reduced	KO	[126]
	OsPUB67	Protein degradation, root development/U-box E3 ubiquitin ligase	Reduced	KO	[127]
	OsRINGzf1	Protein degradation/RING zinc finger E3 ligase	Reduced	KO	[128]
	OsRNS4	Biotic and abiotic stress response/S-like RNAse	Reduced	KD	[116]
	OsSAPK2	Stress/ABA–activated protein kinase	Reduced	KO	[129]
	OsSAPK3	Stress/ABA–activated protein kinase	Reduced	KO	[130]
	IPA1/OsSPL14	Growth, development, environmental stimuli response/SPL TF	Reduced	KO	[131]
	OsAO3	ABA biosynthesis/Aldehyde oxidase	Reduced	KO	[132]

Table 1. *Cont.*

Species	Target Locus	Pathway/Function	Effect on Tolerance	Result	Reference
Populus clone 717-1B4 (*Populus tremula* × *Populus alba*)	*PdGNC*	Carbon and nitrogen metabolism/TF	Reduced	KO	[133]
Populus clone NE-19 (*Populus nigra* × (*Populus deltoides* × *P. nigra*))	*PdNF-YB21*	Flowering, growth, abiotic stress response/NF-Y TF	Reduced	KO	[134]
Populus trichocarpa	*PtrADA2b-3*	Chromatin modification/Histone acetyltransferase adaptor	Reduced	KO	[135]
Solanum lycopersicum	*SlALD1*	Stress responses/Pipecolic acid	Enhanced	KO	[136]
	SlARF4	Auxin signaling/Auxin response factor	Enhanced	KO	[137]
	SlRR26	Cytokinin pathway/Type-B Response Regulator	Enhanced	KO	[138]
	SlSNAT2	Negative regulation of rbcL/RUBISCO lysine acetylase	Enhanced	KO	[139]
	SlLBD40	Lateral root development/LBD TF	Reduced	KO	[140]
	SlMAPK3	Biotic and abiotic stress response/Mitogen-Activated Protein Kinase	Reduced	KO	[141]
	SlNPR1	Plant immunity/SA receptor	Reduced	KO	[142]
	SP3C	Anti-florigen/PEBP	Reduced	KO	[143]
Solanum tuberosum	*StFLORE*	Flowering/long non-coding RNA	Reduced	KD	[144]
Triticum aestivum	*TaSal1* (*6 homeologs*)	Monophosphate 3′-phosphoadenosine 5′phosphate (PAP) signaling	Enhanced	KO	[145]
	TaCER1-6A	Cuticle biosynthesis	Reduced	KO	[146]
	TaIPT8	Cytokinin biosynthesis/isopentenyltransferase	Reduced	KO	[147]
	TaPYL1-1B	Abscisic acid receptor	Reduced	KD	[148]
Vitis vinifera	*VvEPFL9-1*	Stomata formation	Enhanced	KO	[149]
Zea mays	*ARGOS8*	Negative regulator of ethylene responses	Enhanced	KU	[150]
	ZmLBD5	LBD Transcription factor	Enhanced	KO	[151]
	ZmLRT	lateral root Development/miR166a-encoding gene	Enhanced	KO	[152]
	ZmPP84	PP2C Phosphatase	Enhanced	KO	[153]
	ZmSAG39	Papain-like cysteine proteases	Enhanced	KO	[154]
	ZmTCP14	TCP Transcription factor	Enhanced	KO	[155]
	ZmATHB-6	Homeobox Transcription Factor	Reduced	KO	[156]
	ZmEREB46	Ethylene-responsive Transcription factor	Reduced	KO	[157]
	ZmRBOHC	NADPH oxidase	Reduced	KO	[158]
	ZmRtn16	Reticulon-like protein	Reduced	KO	[159]
	ZmSRL5	Cuticle biosynthesis	Reduced	KO	[160]
	ZmSRO1d-S	Oxidative and abiotic stress response	Reduced	KO	[158]

Most studies employing CRISPR/Cas genome editing to study drought resistance so far have occurred in rice varieties (Table 1). An in-frame deletion of a *DROUGHT AND SALT TOLERANCE* (*DST*) gene using CRISPR/Cas9 in *O. sativa* subsp. *indica* caused deletion of the C-terminal EAR motif in the protein product, which produced plants with broader leaves and reduced stomatal density in comparison with the wild type, resulting in enhanced water retention under dehydration stress [102]. The rice CONSTANS-like transcription factor, *Ghd2*, regulates drought-induced leaf senescence, and its knockout (KO) by CRISPR/Cas9 enhances drought tolerance by delaying the senescence process [98]. While these examples are of mutations that increased tolerance to drought, several CRISPR studies revealed genes whose KO impairs tolerance. The KO of gene *OsNPF8.1*, a nitrate transporter, reduced tolerance to drought and salt stress, as well as lower grain yield with less N accumulation in comparison with the control genotype [125].

Another important characteristic of CRISPR/Cas technologies is their ability to KO microRNAs, previously difficult to achieve by classic transgenesis due to their short sequence [161]. Um et al. (2022) [113] used CRISPR/Cas9 to generate KO rice lines for *osa-MIR171*, which showed sensitivity to drought in comparison to the wild type. With additional experiments, the authors show that *osa-MIR171* regulates the expression of flavonoid biosynthesis genes, which are known participants of stress response pathways [162]. Contrastingly, the CRISPR/Cas9 KO of *osa-miR535* in rice enhances the tolerance of plants to dehydration and PEG stresses in comparison to unedited plants [100]. Interestingly, *osa-mir535* is a highly conserved miRNA present in more than 50 plant species [163], which makes it an interesting target to engineer drought tolerance in crops.

In *Solanum lycopersicum* (tomato), pipecolic acid (Pip) biosynthetic gene, *SlALD1*, CRISPR-generated mutants show elevated drought resistance compared with the wild-type, a phenotype associated with CO_2 assimilation, photosystems activities, and antioxidant enzyme activities [136]. Still, in tomatoes, the KO of the jasmonic acid-responsive transcription factor *SlLBD40* by CRISPR/Cas9 enhanced drought tolerance in comparison to unedited plants [140]. Similarly, in maize, CRISPR/Cas9 KO mutants of another LBD transcription factor, *ZmLDB5*, have higher grain yield under drought stress compared to the wild type and do not exhibit differences in well-watered conditions [151]. In soybean, a quadruple KO of circadian rhythm transcription factors *GmLHY1a*, *GmLHY1b*, *GmLHY2a*, and *GmLHY2b* produced plants with enhanced drought tolerance and delayed maturity in comparison to the unedited genotype [89]. It is important to note that, although most CRISPR/Cas research so far has focused on model crops, such as rice and maize, other species are being explored, including oilseed rape [85], cucumber [86], strawberry [87], alfalfa [94], tobacco [95–97], poplar [133–135], potato [144], wheat [145–148], and grape [149] (Table 1). Collectively, these studies provide a growing database of mutant alleles that modulate responses to drought in crops, which could be used to breed stress-resilient cultivars. Furthermore, based on orthology principles, similar mutations could be effective across different species.

3.2. Flooding and Waterlogging Tolerance

Although drought and flood are viewed as opposing stresses and are usually studied separately, they share molecular pathways of tolerance, and both stresses reduce energy-consuming processes, facilitating the allocation of energetic resources to stress adaptation [164,165]. Similar to plants in drought stress, shoots need to adapt to dehydration caused by impaired root hydraulics and leaf water loss after a flood event [165]. Due to climate change, there is evidence suggesting that compound extremes, such as the co-occurrence of droughts and floods, are increasing in many parts of the world, including the United States, Europe, and Asia [166]. When the El-Niño South Oscillation perseveres, for example, it leads to prolonged flooding in some areas [80], and drought–flood abrupt alternation is becoming more unpredictable [167]. The co-occurrence of droughts and floods can amplify their impacts and create complex challenges for ecosystems, agriculture, water resources, and human settlements.

In addition to flooding, characterized by the presence of standing water above the soil surface, another phenomenon of excess water in the soil is waterlogging, which is the lack of drainage [168]. Both flooding and waterlogging can cause significant damage to crops and soil, but they have different impacts on plant growth and development [169]. While some plants may be adapted to tolerate occasional flooding, most plants are highly sensitive to waterlogging and can suffer from reduced growth, root damage, and even death [170]. Unfortunately, until the publication of this review, no studies involving CRISPR/Cas have tackled waterlogging, and only two examples of specific flood-related CRISPR studies were found (Table 2). In both cases, the evaluated gene KOs present reduced tolerance to floods. The *OsGF14h* gene encodes a 14-3-3 protein in weedy *O. sativa* subsp. *japonica* cultivar WR04-6, and its KO mutant in this background is sensitive to anaerobic conditions imposed by flooding stress. Interestingly, already sensitive modern cultivars SN9816 and Nipponbare show six polymorphic sites in the coding sequence of *OsGF14h*, which produce an incomplete isoform of the 14-3-3 protein [171]. The second study knocked out the ethylene-response factor-like gene *SUB1A* in a flooding-tolerant cultivar of *O. sativa* subsp. *indica*, Chiherang-Sub1, resulting in sensitivity to the flooding experiment, similar to the wild-type cultivar Chiherang [172]. Another interesting study by Ye et al. [173] verified that CRISPR-mediated KO of gene *OsCBL10* is embryo-lethal, but natural variations in the gene's promoter were associated with flooding tolerance.

Table 2. Studies employing CRISPR/Cas on genes related to flooding stress. GA: Gibberellic Acid; ABA: Abscisic Acid; KO = Knockout.

Species	Target Locus	Pathway/Function	Effect on Tolerance	Result	Reference
Oryza sativa	*OsGF14h*	ABA and GA signaling/14-3-3 protein	Reduced	KO	[171]
	SUB1A	Ethylene-responsive transcription factor	Reduced	KO	[172]

In summary, since drought and flood may coexist and possibly share regulatory mechanisms in plants, it is urgent to revisit already characterized mutants tolerant to drought concerning flood tolerance. This would be an effective strategy to accelerate the discovery of genes conferring this trait. Furthermore, genetic resources for flood tolerance might be found in crop wild relatives. All major crop families possess members that show adaptation to seasonal wetlands, including members of genera *Oryza* and *Zea* (*Poaceae*), *Lotus* (*Fabaceae*), *Solanum* (*Solanaceae*), and *Rorippa* (*Brassicaceae*), which can provide insight into plastic survival strategies lost during crop domestication or selection for production agriculture [174].

3.3. Salinity Stress Tolerance

Another global problem in agriculture, affecting over 400 million ha worldwide, with direct implications on crop yield and food security, is soil salinity [72]. This phenomenon can be caused by irrigation with saline water or over-irrigation [175], excessive fertilization [176], conversion of natural habitats to agricultural land [177,178], geological factors [179], climate effects, sea level increase, flooding, or tsunamis [180,181]. The effects of salinity on crops have been extensively studied at the physiological and molecular levels, including osmotic and toxic consequences on different phenological stages [182–184].

Physiological consequences of salinity stress include ion toxicity, which impairs the uptake and transport of essential nutrients [185], osmotic stress, which reduces water potential in cells [186], oxidative stress [187], changes in the expression of genes involved in growth and development [188], and alteration of hormones impacting growth and stress responses [189]. At the early phase of perception, sodium/hydrogen exchangers (NHXs) and high-affinity potassium transporters (HKTs) import Na^+ ions and activate a Na^+ sensing module [190]. Then, early signaling is activated, involving K^+, Ca^{2+}, cGMP, phospholipids, ROS, and protein kinases that can activate hormones and gene responses downstream [191].

This signal cascade allows the expression of different adaptive mechanisms, such as growth and developmental response, ion exclusion and sequestration, and the synthesis of compatible solutes to cope with osmotic stress [183]. Some tolerant plant species have developed specific mechanisms to excrete salt ions through specialized structures [192]. Other species can induce the synthesis of osmoprotectant metabolites, such as proline, glycine betaine, γ-GABA, spermidine, spermine, putrescine, mannitol, sucrose, trehalose, and enzymatic and non-enzymatic antioxidant molecules [193–196].

Salinity stress is the second most common abiotic stress with available CRISPR/Cas data, with many of the same genes also implicated in drought tolerance (Table 3, Figure 4). There are five genes with CRISPR mutants that enhance both drought and salinity tolerance: *OsPPR035* and *OsPPR406* [109], *OsDST* [102], *osa-MIR535* [100], and *OsIPK1* [105], all in rice. There are also four genes with mutants having reduced stress tolerance: *OsNPF8.1* [125] and *OsDIP1* [119], in rice, as well as *GmMYB118* [91] and *GmCOL1a* [90], in soybean. This overlap is largely explained by the shared genetic networks involved in the ABA-dependent and ABA-independent pathways of the abiotic stress response [197]. The CRISPR/Cas9-generated in-frame deletion of 33 bp in gene *OsIPK1* controlled the synthesis of phytic acid and conferred salt and drought tolerance without apparent penalties in yield [105]. The expression of ABA-independent TF *OsDREB1A* is upregulated in both stresses in *osipk1_1* mutants, corroborating the overlap between shared stress regulation.

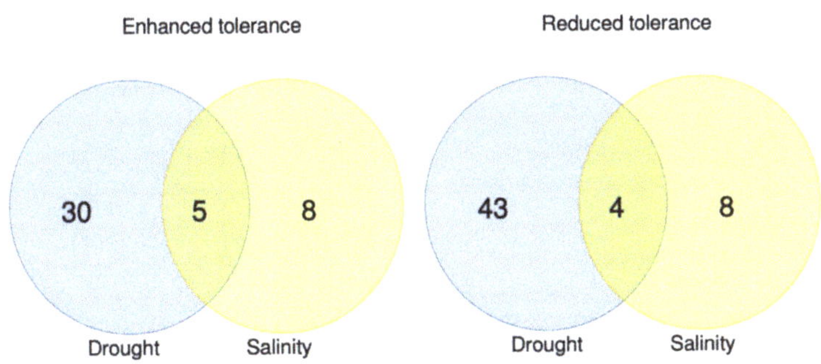

Figure 4. Overlap between genes involved in drought and salinity stresses in CRISPR studies presented in this review. There are a total of 35 studies of gene editing promoting enhanced drought tolerance, of which five overlap with enhanced salinity tolerance. Studies of reduced drought tolerance sum to 47, of which four also show reduced salinity tolerance. For salinity, a total of 13 genes shows enhanced tolerance, while 12 show reduced tolerance.

In addition to the shared genes, 16 other studies are summarized in Table 3. One of these studies in rice generated 14 CRISPR-mediated mutations in gene *OsRR22*, a B-type response regulator TF involved in cytokinin signal transduction and metabolism [198]. These mutations confer salt tolerance at the seedling and mature stages compared with wild-type plants, without effects on other agronomic traits [199]. In soybean, CRISPR/Cas9 was used to validate the participation of TF *GmNAC06* in salt stress, since KO mutants display poor performance under experimental conditions in comparison to the wild-type and overexpression lines [200]. Contrastingly, enhanced performance in laboratory and field salinity stress experiments was found for double and quadruple KO soybean mutants *gmaitr36* and *gmaitr23456*, respectively [201]. This was achieved by a multiplexed approach of CRISPR/Cas9-mediated KO of *GmAITR* genes, which are ABA-induced transcription repressors involved in regulating ABA signaling [202].

Table 3. Studies employing CRISPR/Cas on genes related to salinity stress. TF: Transcription Factor; ABA: Abscisic Acid; KO = Knockout; KD = Knockdown; I.N.F. = Information Not Found.

Species	Target Locus	Pathway/Function	Effect on Tolerance	Result	Reference
Cucurbita moschata	*CmoPIP1-4*	Plasma membrane intrinsic protein	Reduced	KO	[203]
Glycine max	*GmAITR2* *GmAITR3* *GmAITR4* *GmAITR5* *GmAITR6*	ABA-induced transcription repressor	Enhanced	KO	[201]
	E2	Photoperiodic flowering	Enhanced	KO	[204]
	GmCOL1a	CONSTANS-like TF	Reduced	KO	[90]
	GmMYB118	MYB TF	Reduced	Amino acid change	[91]
	GmNAC06	NAC TF	Reduced	I.N.F.	[200]
Hordeum vulgare	*HVP10*	Vacuolar H+-pyrophosphatase	Reduced	KO	[205]
Oryza sativa	*osa-MIR535*	Drought-induced miRNA	Enhanced	KO	[100]
	OsbHLH024	bHLH TF	Enhanced	KO	[206]
	OsDST	Zn Finger TF	Enhanced	Domain deletion	[102]
	OsIPK1	Inositol 1,3,4,5,6-pentakisphosphate 2-kinase	Enhanced	11-amino acid deletion	[105]
	OsPPR035	Chloroplast RNA editing	Enhanced	KO	[109]
	OsPPR406	Chloroplast RNA editing	Enhanced	KO	[109]
	OsRR22	B-type RR TF	Enhanced	KO	[199]
	OsVDE	Xanthophyll cycle/Violaxanthin deoxidase	Enhanced	KD	[207]
	BEAR1	bHLH TF	Reduced	KD	[208]
	OsDIP1	TF-interacting protein	Reduced	KO	[119]
	OsGLYI3	glyoxalase	Reduced	KO	[209]
	OsNPF8.1	Peptide transporter	Reduced	KO	[125]
	OsWRKY28	WRKY TF	Reduced	KO	[210]
	OsWRKY54	WRKY TF	Reduced	KO	[211]
Solanum lycopersicum	*AIT1.1*	ABA transporter	Enhanced	KO	[212]
	SlABIG1	HD-ZIP II TF	Enhanced	KO	[213]
	SlHyPRP1	Hybrid Proline-rich protein	Enhanced	Domain deletion	[214]
	Put2	Polyamine uptake transporter	Reduced	KO	[215]

3.4. Heavy Metals or Toxic Element Tolerance

Heavy metals occur naturally in the Earth's crust, and the release of these metals into the soil can occur due to natural or anthropogenic processes. Some of the natural causes of heavy metals in soils are the weathering of parent rock, releasing trace amounts of metals, transport of heavy metals from one location to another during floods, landslides or wind erosion, and atmospheric deposition from volcanic emissions, among others [216]. The biological activity of microorganisms, plants, and animals can also concentrate heavy metals in the soil through biological processes, such as uptake and bioaccumulation [217]. Anthropogenic heavy metal accumulation in soils is far more significant than natural sources, and the use of agrochemicals is the most impactful [217]. Fertilizers, pesticides, and herbicides can contain these harmful molecules, as well as cause soil acidity and erosion, which intensifies their accumulation in soils and possibly contaminates the water table [218,219].

The accumulation of heavy metals and toxic elements in plant tissues can affect their nutritional quality, making them unsuitable for consumption or even harmful to human health. This is a significant concern in the food industry and public health because of the reported diseases associated with the consumption of heavy metal-contaminated foods and exposure to contaminated environments [220,221]. In mining countries, such as Chile, the accumulation of heavy metals derived from the copper industry, for example, has generated the contamination of soils and groundwater in localities considered nowadays as "sacrifice zones", such as Puchuncaví and Quintero-Ventanas Bay [222]. Besides, other mining-associated activities or agricultural practices, such as smelting, industrial exhaust, irrigation with mining wastewater, natural presence in some agricultural soils, and applying fertilizers and pesticides with heavy metal traces, have generated a similar problem worldwide [223]. The primary heavy metals and metalloids found in contaminated soils are copper (Cu), zinc (Zn), lead (Pb), cadmium (Cd), mercury (Hg), and arsenic (As). In southern China, for instance, the analysis of rice samples from contaminated or very industrialized areas showed a high percentage (56 to 87%) of samples contaminated with Cd [224]. Additionally, since rice is the second-most produced staple food worldwide, its contamination generates concern and health risks in different countries [223].

Since heavy metals and toxic elements can have negative impacts on crop growth and development, they can accumulate in plant tissues and lead to reduced yield, quality, and even plant death [225]. These contaminants can also affect nutrient uptake and interfere with photosynthesis, respiration, and transpiration [226]. Physiological and molecular impacts on plants may lead to growth inhibition, chlorosis, necrosis, reduced photosynthesis, and decreased crop yield. These elements can also affect the uptake and transport of essential nutrients, leading to nutrient imbalances and deficiencies. At the molecular level, heavy metals and toxic elements can induce oxidative stress, disrupt cellular homeostasis, alter gene expression, and impair enzymatic activities [227]. Additionally, heavy metals and toxic elements can alter the composition and diversity of the plant-associated microbial communities, affecting plant–microbe interactions and nutrient cycling in the soil [228].

Some metal elements are essential micronutrients for the enzymatic cellular machinery to function. However, under an unbalance of heavy metal homeostasis, some plant species have developed mechanisms to deal with the rise of their concentrations in different cellular compartments [229]. Among the mechanisms involved, we can mention the expression of Heavy Metal ATPases (HMA) proteins [230], Zn and Fe-regulated Membrane Transporter (ZIP) proteins [231], Cation Diffusion Facilitator (CDF) proteins [232], Cation/hydrogen Exchangers (CAX) proteins [233], High-affinity Copper Transport (COPT) proteins [234], Natural Resistant Associated Macrophage (NRAMPS) proteins [235], the bHLH TFs [236], and low molecular weight chelators and subcellular sequesters, such as metallothioneins, phytochelatins, amino acids, nicotinamides, glutathione, and defensins [237,238].

Most of the genes encoding the expression of the aforementioned proteins are potential targets for CRISPR/Cas9 modification to modulate heavy metal tolerance. Although major efforts have been performed using the advances in omics tools, to identify molecular targets controlling heavy metal tolerance in plants [239], few studies show heavy metal tolerance modification for major crops (Table 4). The modulation of a plant's response to this stress depends on its application. For phytoremediation, the goal is to increase the uptake of heavy metals from highly contaminated lands, while avoiding accumulation in final food products requires a decrease in the uptake of these molecules. In rice, for instance, the KO of Cd/Mn transporter *OsNRAMP5* confers Cd tolerance to a wide range of external Cd concentrations, producing shoots with sufficient nutrients and grains with lower Cd accumulation [240]. The KO of *OsNRAMP5* in two *O. sativa* subsp. *japonica* varieties generated lines with decreased accumulation of Cd in aerial organs, but reduced yield in comparison to unedited plants in both hydroponic and field experiments [241]. Similarly, KO of the rice Low Cadmium (*OsLCD*) gene also diminished Cd accumulation in the shoot, but maintained yield under high Cd concentrations in comparison to the wild genotype [242]. Another Cd-related gene, *Sl1*, was knocked out in tomatoes, and edited plants displayed increased Cd accumulation in plant tissues, as well as increased ROS activity in comparison to the wild-type and overexpression lines [243].

The *R2R3 MYB* transcription factor *OsARM1* regulates arsenic(As)-associated transporter genes, and KO lines generated by CRISPR/Cas9 improve the tolerance of rice to As in comparison to the wild-type [244]. A similar proof of concept used the KO of Antioxidant Protein 1 (*OsATX1*) gene, a Cu chaperone in rice, which induced an increase in Cu concentration in roots, thereby decreasing the root-to-shoot translocation of Cu [245]. The CRISPR/Cas technology has also been used to deal with other toxic element contamination in soils, such as radioactive Cs+. The inactivation of Cs+ transporter *OsHAK1* in rice by CRISPR/Cas9 dramatically reduced the uptake of Cs+ in highly Cs+ contaminated lands from Fukushima, Japan [246].

Table 4. Studies employing CRISPR/Cas on genes related to flooding heavy metal and toxic element stresses. TF: Transcription Factor; KO = Knockout.

Species	Target Locus	Pathway/Function	Effect on Tolerance	CRISPR Result	Reference
Oryza sativa	*OsHAK1*	Cs+-permeable transporter	Cesium resistant	KO	[246]
	OsATX1	Cu chaperone	Dosage-dependent tolerant	KO	[245]
	osa-MIR535	Drought-induced miRNAs	Enhanced	KO	[247]
	OsARM1	R2R3 MYB TF regulator of As-associated transporters genes	Enhanced	KO	[244]
	OsLCD	Unknown, Cd related	Enhanced	KO	[242]
	OsLCT1	Low affinity cation transporter	Enhanced	KO	[248]
	OsNRAMP1	Cd and Mn transporter	Enhanced	KO	[249]
	OsNRAMP5	Cd and Mn transporter	Enhanced	KO	[240,248]
	OsPMEI12	Pectin Methylesterase	Enhanced	KO	[250]
Solanum lycopersicum	*Sl1*	E3 Ubiquitin ligase	Reduced	KO	[243]

3.5. Tolerance to Barrenness

Nutrient deficiencies in soils can be triggered by a variety of factors, such as soil pH and soil organic matter, which influence the types and the abundance of essential nutrients, respectively [251]. Other factors, such as soil texture, can affect nutrient availability. Soil compaction can affect the nutrient and water access by the root system, and excessive plant uptake causes nutrient depletion in soils that are heavily cropped or in which fertilization is inadequate [252]. Agricultural practices that can lead to soil barrenness or degradation include the overuse of chemical fertilizers, leading to nutrient imbalances and soil acidification [253], as well as monocultures, which can deplete soil nutrients, leading to reduced yields and increased susceptibility to pests and diseases [254]. Moreover, soil erosion results in a loss of soil organic matter, nutrients, and soil structure, leading to reduced productivity and increased vulnerability to drought and flooding [255]. In addition, pesticide use can harm beneficial microorganisms and disrupt soil food webs, leading to reduced soil fertility and productivity over time [256]. Additionally, the use of fertilizers to boost the yield of crops has allowed for maintaining the requirements for global food security during the years past the green revolution. However, this practice is under the threat of actual climate change and geopolitical sceneries [257,258]. However, the environmental pollution and ecological degradation generated by the indiscriminate use of fertilizers [259], as well as the fertilizers' price increment generated by recent events, such as the Russian-Ukraine conflict [258], will raise the cost of the farmer's production, making this practice unsustainable over time, as we know today. Finally, natural disasters, such as floods, droughts, and wildfires, can also contribute to soil barrenness by altering soil properties and reducing nutrient availability [260].

Nutrient use efficiency (NUE) is the capability of a crop to take up the nutrients from soil, transport them, assimilate them, and use them to maximize its yield. NUE is a very complex trait, involving several plant functions and metabolic pathways. The polyploidy nature of major crops makes their manipulation a big challenge for plant researchers. Nevertheless, some studies have been performed to improve NUE using transgenic [261], siRNA [262], and gene over-expression approaches [263,264], which have shown impressive advances focused on NUE. Nowadays, physiological and genomic information can be used to select targets for CRISPR/Cas NUE improvement, showing promising results. Recent thorough reviews were published on potential targets for nutrient use efficiency [265–267], and more CRISPR-based studies might benefit from this knowledge. In total, eight studies are summarized in Table 5, including one in barley, five in rice, one in *Populus*, and one in wheat.

For example, a CRISPR cytosine base editing system (CBE) was used to generate a C-T point mutation in gene *OsNRT1.1B* of Japonica rice cv. Nipponbare, causing amino acid conversion T327M [268]. This mutation corresponds to an allele difference between rice varieties Nipponbare (T327) and IR24 (M327) [269], and the base editing conversion of the Nipponbare allele results in better NUE in comparison to the wild type [268,269]. Interestingly, the mutated DST protein in the Indica rice cv. MT1010 *dst* mutant shows enhanced drought and salinity tolerance [102], while its CRISPR KO in Japonica rice cv. ZH11 impairs NUE in comparison with the wild type, showing reduced growth in nitrogen-poor substrates [108].

In another major staple food crop, wheat, lines with mutant alleles of Abnormal Cytokinin Response 1 Repressor 1 Protein (TaARE1) were generated by CRISPR/Cas9, showing increased NUE, delayed senescence, and higher grain yield than the wild-type [270]. The same orthologous gene in barley, *HvARE1*, was mutated by CRISPR/Cas9, generating improved NUE in mutant lines *1are1-E-7-6* (amino acid substitution E78G) and *2are1-K-4* (substitution N205D) [271]. Recently, the overexpression of the *PdGNC* transcription factor in poplar was found to increase nitrate uptake, remobilization, and assimilation, improving overall NUE in this species, which was validated using CRISPR/Cas9 mutants [272].

Table 5. Studies employing CRISPR/Cas on genes related to nutrient deficiency stress. TF: Transcription Factor; KO = Knockout.

Species	Target Locus	Pathway/Function	Effect on Tolerance	Result	Reference
Hordeum vulgare	*HvARE1*	Abnormal cytokinin response 1 repressor 1 protein	Enhanced	Amino acid change	[271]
Oryza sativa	*NRT1.1B*	Nitrogen transporter gene	Enhanced	Base editing	[268]
	OsDST	Zinc finger TF	Reduced	Domain deletion	[108]
	OsNPF3.1	Nitrate/Peptide transporter	Reduced	KO	[273]
	OsNPF8.1	Nitrate/Peptide transporter	Reduced	KO	[125]
	OsNR1.2	Nitrate/Peptide transporter	Reduced	KO	[108]

Table 5. *Cont.*

Species	Target Locus	Pathway/Function	Effect on Tolerance	Result	Reference
Populus clone 717-1B4 (*Populus tremula* × *Populus alba*)	*PdGNC*	Nitrate uptake	Reduced	KO	[133]
Triticum aestivum	*TaARE1-A* *TaARE1-B* *TaARE1-D*	Abnormal cytokinin response 1 repressor 1 protein	Enhanced	KO	[270]

4. Extremophytes: Genetic Reservoirs for CRISPR/Cas Applications

Although evolutionarily distant species may exhibit different transcriptional responses to stress, they share core genetic regulatory elements [75]. In this context, studying extremophytes' genetics poses a great opportunity to find potential targets for genetic manipulation, leading to enhanced stress-related traits. Extremophiles can be defined as organisms capable of dealing with extreme conditions of pH, temperature, pressure, salinity, high concentrations of gasses (such as CO_2), metals, and ionizing radiation, for example [274,275]. The first well-studied extremophiles are microorganisms, which have already been extensively used in the bioprospection of potentially valuable enzymes, mainly in the biofuel industry [276]. A classic example is the DNA polymerase isolated from the thermophilic bacterium *Thermus aquaticus* [277], an essential enzyme in molecular biology research. Another example is the use of extremophile microbiota that induce drought/salinity resistance in plants, which have been isolated from deserts [278] and Antarctica [279]. Although there is high interest in extremozymes, bioactive compounds, and cultured extremophiles for direct use in the industry [280], little has been explored in plant genetic engineering.

Extremophyte species grow in harsh conditions, which are limiting to unadapted species. For instance, propagules of sub-Antarctic species may arrive in more extreme Antarctic regions, but few can establish new individuals that survive more than one season, and none can establish populations without human intervention [281]. This unique feature of extremophytes defies the trade-off between growth and stress resilience, since they can properly balance their resources, obtained from photosynthesis, to adapt to the extreme climatic factors to complete their life cycles [282]. In this context, deserts (warm and cold), salt pans, geothermal springs, and high mountains, common niches of extremophytes, serve as excellent model conditions to study plant performance on hostile soils [283].

Unfortunately, studies on the molecular and physiological determinants of the trade-off between growth and stress tolerance are scarce, particularly in non-model species. This gap leaves a significant source of variation for photosynthetic functioning and stress tolerance unexplored. Therefore, the unique opportunity provided by extremophytes to investigate how they differentially invest their photosynthetic resources to adapt their life cycles under extreme climatic factors can be leveraged to understand the mechanistic bases of the trade-off between productivity and stress tolerance [282,284]. Moreover, extremophytes offer a promising source of valuable traits for the biotechnology industry to improve crop productivity, as well as at least to maintain it in agricultural regions affected by climate change scenarios [280,285]. As the climate changes, extremophytes can provide insights into the future. Discovering the molecular and biochemical adaptations employed by these plants can enhance our understanding of how plants, in general, will respond to climate change [286].

Interestingly, even though molecular mechanisms controlling plant physiology during abiotic stress have been amply reviewed in model plants and crops [287,288], our knowledge of the molecular mechanisms that support extremophytes success is more limited [275]. Some of the best-studied extremophytes are the resurrection plants, for their potential as ideal models to engineer crops with enhanced drought tolerance [289,290]. Similarly, many studies have been performed on halophyte plants, including highly salt-tolerant close relatives of *A. thaliana*, allowing for direct comparisons of stress tolerance mechanisms [3].

Since established protocols for greenhouse cultivation, in vitro culture, and transformation or gene editing of extremophytes are scarce, functional genetic studies have mostly focused on the heterologous expression of extremophile proteins in model plants [31]. For instance, HIGH-AFFINITY POTASSIUM TRANSPORTER (*HKT*) genes from the halophytes *Thellungiella salsuginea* [291], *Eutrema parvula* [292], and *Suaeda salsa* [293] have been expressed ectopically in *Arabidopsis* plants, and they confer salt tolerance in comparison to the wild-type protein.

Transcriptome sequencing is another strategy to study the reprogrammed metabolism observed in some extremophytes, enabling target trait selection in close relative crops. All major crop families possess members that show adaptation to hostile soils, including members of genera *Oryza* and *Zea* (*Poaceae*), *Lotus* (*Fabaceae*), *Solanum* (*Solanaceae*), and *Arabidopsis*, *Rorippa* (*Brassicaceae*) [174,294,295]. These species can provide insights into plastic survival strategies to hostile conditions, which were lost during crop domestication or selection for intensive agriculture. The identification of the genetic factors controlling stress tolerance traits in extremophytes can guide the search for orthologs in closely related crops, which would then be modified by CRISPR/Cas technologies (Figure 5).

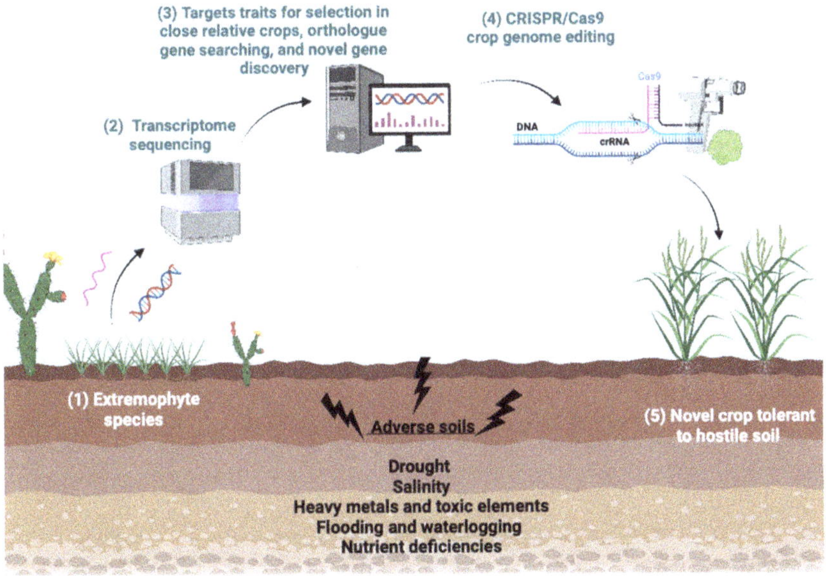

Figure 5. Schematic depicting the use of extremophytes and new sequencing technologies to identify new gene targets that can be modified in crops using CRISPR/Cas9. Image created with BioRender.com, accessed on 31 March 2023.

For instance, the transcriptomic analysis of *Populus euphratica*, a desert tree related to the commercial species poplar, showed a reprogrammed metabolism under salt stress, where genes involved in ABA regulation are differentially expressed [296]. Thereby, negative regulators of stress tolerance previously identified in extremophytes could be knocked down using a CRISPR/Cas system. Meanwhile, the sequences of promoters or positively regulatory regions of stress response genes could be modified, as was shown for the generation of HDR-based editing to produce a salt-tolerant SlHKT1;2 alleles in tomato utilizing the CRISPR/Cpf1-geminiviral replicon technique [297]. In another example, CRISPR/Cas9 KO of metallophyte *Sedum pumbizincicola* Heavy Metal ATPase 1 (*SpHMA1*) helped to characterize the function of *SpHMA1* in protecting PSII from Cd toxicity [298]. Therefore, there is still much to be explored and discovered in these extremophile species that can be used for crops to face the challenges of climate change and hostile soils.

5. Challenges and Prospects

5.1. Combined Stresses

Although great advances have been made in the study of stresses, most of these discoveries assessed plant responses to single stresses. In natural conditions, the combination of stresses is usually the norm, and climate change will also affect the intensity and frequency of these compound stresses [166]. Given these statements, it is possible to assume that combined stresses complicate the equation for stress resilience engineering. However, elegant systems, such as BREEDIT [64], aim to solve this problem by editing a combination of genes with multiplex CRISPR, resulting in additive roles in stress or yield traits. In their proof-of-concept study, a knock-out of 48 different genes involved in plant growth was conducted, combining 12 genes simultaneously, and generating over 1000 different edited lines with potential enhancements in yield [64]. If a similar strategy is used to knock out multiple genes associated with the suppression of stress tolerance, the combinations of such mutations could help establish a multi-tolerant plant line. Furthermore, sophisticated systems, such as CRISPR-Combo [65], couple gene editing with gene activation, allowing for fine-tuned metabolic engineering.

5.2. Technological Limitations and Potential Solutions

CRISPR/Cas systems have been widely acknowledged for their potential to improve crops through gene insertion, removal, point mutation, and gene replacement. However, their use in agricultural research is still in the early stages, with most reports constrained to proof-of-concept findings [299]. Even though CRISPR has been successfully applied in at least 42 plant species [70], there is still a need for a global mechanism that is genotype-independent. Several efforts are underway to improve the limitations of CRISPR/Cas technologies, such as limited PAM sites, off-target mutations, low HDR efficacy, and time consumption due to the *Agrobacterium*-mediated transformation system [300]. For instance, several mutated Cas enzymes opened the possibility of more diverse PAM sites with lower off-target potential [39], and slightly more efficient HDR could be achieved with CRISPR/Cas12a [301].

Regarding transformation limitations, advances have been achieved in both *Agrobacterium*-mediated and other methods, such as the use of viral vectors. The latter, although efficient, is limited by the size of the Cas-encoding sequences, which are very large [302]. Recently discovered Cas12f1 is considerably smaller, allowing for the use of viral vectors to produce gene editing without transgene integration [54]. These strategies hold promise for expanding the application of CRISPR/Cas9 in agriculture and addressing some of its current limitations. An important step forward in monocotyledonous and recalcitrant plant transformation, mediated by *Agrobacterium*, is the use of morphogenetic factors to induce somatic cells into initiating embryogenesis, thus partly circumventing the need for strenuous callus induction and regeneration studies [303,304]. Another important discovery is the newly described "cut-dip-budding" (CDB) system, which enables gene editing in previously recalcitrant species, which is the case for many crops and wild relatives [305]. The CBD system relies on the ability of plants to generate basal shoots from adventitious buds in roots, and it was already applied successfully in species where transformation was either difficult or impossible. An advantage of the CBD system is the absence of in vitro or sterile culture, since all steps can be performed directly in soil [305]. These technologies are promising for the application of CRISPR in wild relatives or extremophyte species to study gene function and to apply these discoveries in crop plants.

5.3. Field Evaluation of CRISPR-Modified Crops

The usefulness of the CRISPR/Cas editing techniques must be demonstrated before the large-scale distribution of any new variety possessing them [306]. However, as shown in Table S1, most studies on hostile soil tolerance in plants modified by CRISPR/Cas systems were only evaluated in the laboratory or greenhouse. Therefore, verifying whether results can be translated to crop plants grown in the field is crucial [307]. In addition, field trials provide a tremendous amount of otherwise unknown information on how plants respond to environmental changes under agricultural systems [308]. Unfortunately, the diverse landscape of legislation regarding gene-edited plants has hindered large-scale field trials, and most such tests have occurred only in China [309]. In 2018, the first field trial of a CRISPR/Cas9 gene-edited crop, *Camelina sativa*, began in Europe at the Rothamsted Research in the UK and provided a wealth of essential data and enabled the evaluation of the potential of a new trait [310]. During the experiment, the UK Department for Environment, Food & Rural Affairs reclassified gene-edited crops as GMOs, and the next field trial only occurred in 2021 [307]. Later, in 2021, field tests of low-asparagine gene-edited wheat were performed in this same research field and were essential to confirm the results observed in the laboratory [311]. Additionally, in 2021, Lee and Hutton (2021) [306] conducted field trials during three consecutive seasons using CRISPR-driven jointless pedicel, as well as fresh-market tomatoes, without detecting significant differences in fruit size yield between CRISPR-modified tomatoes and WT tomatoes [306]. Despite these studies, further field trials conducted across a broader range of regions are imperative to authenticate the scientific effectiveness of gene-edited plants and instill greater assurance and security for both producers and end consumers.

5.4. Regulation and Customer Acceptance

Important limitations on CRISPR/Cas-modified crops are the legal regulation of plant genome editing and consumer acceptance. Although CRISPR crops are being developed and grown globally, this trend is accompanied by legal, ethical, and policy debates. The technical limitations of CRISPR and whether existing GMO regulations should apply to CRISPR-edited crops are key issues [312]. In the scientific community, there is a belief that mutations generated by CRISPR/Cas9 are no different from those induced by nature or conventional breeding. Thus, plants created through this technology should not undergo the same regulatory processes as conventional GMOs. However, on a global scale, opinions differ, and some countries believe that CRISPR-generated crops should undergo the same regulations as GMOs before entering the market [313,314].

For instance, the United States and the European Union have different approaches to CRISPR-edited crops. The former is more permissive because they do not have to undergo the same regulatory process as GMOs, while in the European Union, they are considered GMOs. However, several countries have already regulated that plants generated through CRISPR with only InDels or homologous inserts can be excluded from GMO regulation [312,314–316]. Hence, the international community is considering whether certain CRISPR-edited crops can be excluded from regulatory oversight and what safety data would be required for CRISPR-edited crops to be regulated in specific countries.

The success and adoption of gene-edited foods depend ultimately on consumer acceptance, which has been a problem for GMO foods due to misinformation. Consumers worldwide display limited understanding, misconceptions, and unfamiliarity with GMO food products [317]. Consumer acceptance of gene-edited foods varies across countries. In China, 45% of respondents (n = 835) agreed that gene-edited plant products should be allowed, compared to 36% for transgenic plant products [318]. In Brazil, producers (n = 37) are prone to planting transgenic beans (84%), and consumers (n = 100) are willing to include them in their diets (79%) [319]. In the UK (n = 490) and Switzerland (n = 505), participants expressed higher acceptance levels for genome editing than for transgenic modification. Acceptance depends on perceived benefits, scientific uncertainty, and location [320]. Acceptance levels for these technologies depend mainly on whether the application is believed to be beneficial, how scientific uncertainty is perceived, and where they reside [35,316]. Surveys, such as these, and the amount of safety data required, will affect the overall cost of regulation, an essential factor to consider when bringing new CRISPR plants to market [314,321].

6. Concluding Remarks

The worldwide deterioration of soil quality has emerged as a critical challenge for agriculture, compounded by the escalating impact of climate change. This looming crisis poses a significant risk to food security, particularly as we approach the year 2050. Unfortunately, there is no single solution to address the issue of hostile soils or to ensure food production in the future. Instead, an integrated, multidisciplinary approach is necessary, leveraging specific tools and solutions to mitigate the detrimental effects of hostile soils on agriculture. By combining these solutions from diverse approaches, we can potentially safeguard agriculture and ensure global food security. These tools include CRISPR/Cas technologies, which enable the precise editing of crop genomes to develop plants that are more tolerant to the stresses of hostile soils. In the past decade, this technique has demonstrated its efficacy in accurately editing the genomes of various organisms, including plants.

As reviewed here, several scientific studies have provided concrete evidence of the effectiveness of CRISPR/Cas technologies for developing crops that are tolerant to hostile soils. These studies have demonstrated successful applications of the technology in improving plant tolerance to stressors, such as drought, heavy metals, salinity, and NUE. As a result, CRISPR/Cas systems are increasingly being considered viable solutions to these agricultural challenges. Notably, a significant amount of research on using CRISPR to develop stress-tolerant crops is being conducted in China, suggesting a potential technological advantage in this area due to its legal status on gene editing organisms.

Although CRISPR technologies for genome engineering in plants are not infallible, ongoing technical advancements are addressing its limitations. Meanwhile, the regulatory landscape is becoming more lenient, allowing for greater openness towards CRISPR-mutated crops that are transgene-free and exempt from traditional GMO regulations. Additionally, consumer acceptance of CRISPR-modified products is predicted to increase, and evidence supports the continued use of this technology for plant breeders. To achieve crops tolerant to future challenges, we suggest leveraging CRISPR technology alongside advances in sequencing and the search for new genetic targets in extremophytes. These developments, alongside novel management strategies and biotechnologies, provide promising solutions for ensuring stable food security by 2050.

Supplementary Materials: The following supporting information can be downloaded at: https://www.mdpi.com/article/10.3390/plants12091892/s1, Table S1: Studies employing CRISPR/Cas on genes related to drought, flooding and waterlogging, salinity, heavy metal and toxic elements, and barrenness tolerance, extended.

Author Contributions: H.A.G., O.G.-E. and P.B.F. conceived and designed the manuscript, conducted the literature research, compiled tables, drew figures, and drafted the manuscript. L.A.B. and H.C. acquired funding and reviewed and edited the manuscript. All authors have read and agreed to the published version of the manuscript.

Funding: This research was supported by the project: Fundação de Amparo à Pesquisa do Estado de São Paulo—Universidad de La Frontera (2020/07578-1). H.A.G. was funded by ANID National Doctorate Scholarship 21181972. O.G.-E. was funded by ANID Fondecyt Postdoctorado N°3230521.

Data Availability Statement: No new data were created or analyzed in this study. Data sharing is not applicable to this article.

Acknowledgments: The authors thank Charles L. Guy for his valuable observations and English edition of the manuscript.

Conflicts of Interest: The authors declare no conflict of interest.

References

1. Schaal, B. Plants and People: Our Shared History and Future. *Plants People Planet* **2019**, *1*, 14–19. [CrossRef]
2. De Vries, J.; Archibald, J.M. Plant Evolution: Landmarks on the Path to Terrestrial Life. *New Phytol.* **2018**, *217*, 1428–1434. [CrossRef] [PubMed]

3. Bechtold, U. Plant Life in Extreme Environments: How Do You Improve Drought Tolerance? *Front. Plant Sci.* **2018**, *9*, 543. [CrossRef] [PubMed]
4. Figueroa-Macías, J.P.; García, Y.C.; Núñez, M.; Díaz, K.; Olea, A.F.; Espinoza, L. Plant Growth-Defense Trade-Offs: Molecular Processes Leading to Physiological Changes. *Int. J. Mol. Sci.* **2021**, *22*, 693. [CrossRef] [PubMed]
5. Bailey-Serres, J.; Parker, J.E.; Ainsworth, E.A.; Oldroyd, G.E.D.; Schroeder, J.I. Genetic Strategies for Improving Crop Yields. *Nature* **2019**, *575*, 109–118. [CrossRef] [PubMed]
6. Perrino, E.V.; Perrino, P. Crop Wild Relatives: Know How Past and Present to Improve Future Research, Conservation and Utilization Strategies, Especially in Italy: A Review. *Genet. Resour. Crop Evol.* **2020**, *67*, 1067–1105. [CrossRef]
7. Dwivedi, S.L.; Reynolds, M.P.; Ortiz, R. Mitigating Tradeoffs in Plant Breeding. *iScience* **2021**, *24*, 102965. [CrossRef]
8. Koziol, L.; Rieseberg, L.H.; Kane, N.; Bever, J.D. Reduced drought tolerance during domestication and the evolution of weediness results from tolerance–growth trade-offs. *Evolution* **2012**, *66*, 3803–3814. [CrossRef]
9. Mayrose, M.; Kane, N.C.; Mayrose, I.; Dlugosch, K.M.; Rieseberg, L.H. Increased Growth in Sunflower Correlates with Reduced Defences and Altered Gene Expression in Response to Biotic and Abiotic Stress. *Mol. Ecol.* **2011**, *20*, 4683–4694. [CrossRef]
10. Meyer, R.S.; Duval, A.E.; Jensen, H.R. Patterns and Processes in Crop Domestication: An Historical Review and Quantitative Analysis of 203 Global Food Crops. *New Phytol.* **2012**, *196*, 29–48. [CrossRef]
11. Holleman, C.; Rembold, F.; Crespo, O.; Conti, V. *The Impact of Climate Variability and Extremes on Agriculture and Food Security—An Analysis of the Evidence and Case Studies. Background Paper for The State of Food Security and Nutrition in the World 2018*; Agricultural Development Economics Technical Study N°4; FAO: Rome, Italy, 2020. [CrossRef]
12. Sperry, J.S.; Venturas, M.D.; Todd, H.N.; Trugman, A.T.; Anderegg, W.R.L.; Wang, Y.; Tai, X. The Impact of Rising CO_2 and Acclimation on the Response of US Forests to Global Warming. *Proc. Natl. Acad. Sci. USA* **2019**, *116*, 25734–25744. [CrossRef]
13. Herrera, D.; Ault, T. Insights from a New High-Resolution Drought Atlas for the Caribbean Spanning 1950–2016. *J. Clim.* **2017**, *30*, 7801–7825. [CrossRef]
14. Herrera, D.A.; Ault, T.R.; Fasullo, J.T.; Coats, S.J.; Carrillo, C.M.; Cook, B.I.; Williams, A.P. Exacerbation of the 2013–2016 Pan-Caribbean Drought by Anthropogenic Warming. *Geophys. Res. Lett.* **2018**, *45*, 10–619. [CrossRef] [PubMed]
15. Ault, T.R. On the Essentials of Drought in a Changing Climate. *Science* **2020**, *368*, 256–260. [CrossRef]
16. Balogh, J.M. The Role of Agriculture in Climate Change: A Global Perspective. *Int. J. Energy Econ. Policy* **2020**, *10*, 401–408. [CrossRef]
17. Zurek, M.; Hebinck, A.; Selomane, O. Climate Change and the Urgency to Transform Food Systems. *Science* **2022**, *376*, 1416–1421. [CrossRef]
18. Shabala, S.; Bose, J.; Fuglsang, A.T.; Pottosin, I. On a Quest for Stress Tolerance Genes: Membrane Transporters in Sensing and Adapting to Hostile Soils. *J. Exp. Bot.* **2016**, *67*, 1015–1031. [CrossRef] [PubMed]
19. FAO. *The Impact of Disasters and Crises on Agriculture and Food Security*; FAO: Rome, Italy, 2021; ISBN 978-92-5-134071-4.
20. Jiménez-Mejía, R.; Medina-Estrada, R.I.; Carballar-Hernández, S.; del Carmen Orozco-Mosqueda, M.; Santoyo, G.; Loeza-Lara, P.D. Teamwork to Survive in Hostile Soils: Use of Plant Growth-Promoting Bacteria to Ameliorate Soil Salinity Stress in Crops. *Microorganisms* **2022**, *10*, 150. [CrossRef]
21. Rojas, R.V.; Achouri, M.; Maroulis, J.; Caon, L. Healthy Soils: A Prerequisite for Sustainable Food Security. *Environ. Earth Sci.* **2016**, *75*, 180. [CrossRef]
22. Perri, S.; Molini, A.; Hedin, L.O.; Porporato, A. Contrasting Effects of Aridity and Seasonality on Global Salinization. *Nat. Geosci.* **2022**, *15*, 375–381. [CrossRef]
23. Coomes, O.T.; Barham, B.L.; MacDonald, G.K.; Ramankutty, N.; Chavas, J.P. Leveraging Total Factor Productivity Growth for Sustainable and Resilient Farming. *Nat. Sustain.* **2019**, *2*, 22–28. [CrossRef]
24. Younis, A.; Ramzan, F.; Ramzan, Y.; Zulfiqar, F.; Ahsan, M.; Lim, K.B. Molecular Markers Improve Abiotic Stress Tolerance in Crops: A Review. *Plants* **2020**, *9*, 1374. [CrossRef]
25. Das, D.; Singha, D.L.; Paswan, R.R.; Chowdhury, N.; Sharma, M.; Reddy, P.S.; Chikkaputtaiah, C. Recent Advancements in CRISPR/Cas Technology for Accelerated Crop Improvement. *Planta* **2022**, *255*, 109. [CrossRef] [PubMed]
26. Gao, C. Genome Engineering for Crop Improvement and Future Agriculture. *Cell* **2021**, *184*, 1621–1635. [CrossRef] [PubMed]
27. Hakim, S.; Naqqash, T.; Nawaz, M.S.; Laraib, I.; Siddique, M.J.; Zia, R.; Mirza, M.S.; Imran, A. Rhizosphere Engineering With Plant Growth-Promoting Microorganisms for Agriculture and Ecological Sustainability. *Front. Sustain. Food Syst.* **2021**, *5*, 617157. [CrossRef]
28. Arif, I.; Batool, M.; Schenk, P.M. Plant Microbiome Engineering: Expected Benefits for Improved Crop Growth and Resilience. *Trends Biotechnol.* **2020**, *38*, 1385–1396. [CrossRef]
29. Gasparini, K.; dos Reis Moreira, J.; Peres, L.E.P.; Zsögön, A. De Novo Domestication of Wild Species to Create Crops with Increased Resilience and Nutritional Value. *Curr. Opin. Plant Biol.* **2021**, *60*, 102006. [CrossRef]
30. Xiong, J.; Hu, F.; Ren, J.; Huang, Y.; Liu, C.; Wang, K. Synthetic Apomixis: The Beginning of a New Era. *Curr. Opin. Biotechnol.* **2023**, *79*, 102877. [CrossRef]
31. Yolcu, S.; Alavilli, H.; Lee, B. Natural Genetic Resources from Diverse Plants to Improve Abiotic Stress Tolerance in Plants. *Int. J. Mol. Sci.* **2020**, *21*, 8567. [CrossRef]
32. Gabaldón, T.; Koonin, E.V. Functional and Evolutionary Implications of Gene Orthology. *Nat. Rev. Genet.* **2013**, *14*, 360–366. [CrossRef]

33. Kapazoglou, A.; Gerakari, M.; Lazaridi, E.; Kleftogianni, K.; Sarri, E.; Tani, E.; Bebeli, P.J. Crop Wild Relatives: A Valuable Source of Tolerance to Various Abiotic Stresses. *Plants* **2023**, *12*, 328. [CrossRef]
34. Rasheed, A.; Gill, R.A.; Hassan, M.U.; Mahmood, A.; Qari, S.; Zaman, Q.U.; Ilyas, M.; Aamer, M.; Batool, M.; Li, H.; et al. A Critical Review: Recent Advancements in the Use of CRISPR/Cas9 Technology to Enhance Crops and Alleviate Global Food Crises. *Curr. Issues Mol. Biol.* **2021**, *43*, 1950–1976. [CrossRef]
35. Strobbe, S.; Wesana, J.; Van Der Straeten, D.; De Steur, H. Public Acceptance and Stakeholder Views of Gene Edited Foods: A Global Overview. *Trends Biotechnol.* 2023, in press. [CrossRef]
36. Jansen, R.; van Embden, J.D.A.; Gaastra, W.; Schouls, L.M. Identification of Genes That Are Associated with DNA Repeats in Prokaryotes. *Mol. Microbiol.* **2002**, *43*, 1565–1575. [CrossRef]
37. Jinek, M.; Chylinski, K.; Fonfara, I.; Hauer, M.; Doudna, J.A.; Charpentier, E. A Programmable Dual-RNA–Guided DNA Endonuclease in Adaptive Bacterial Immunity. *Science* **2012**, *337*, 816–821. [CrossRef]
38. Wang, H.; La Russa, M.; Qi, L.S. CRISPR/Cas9 in Genome Editing and Beyond. *Annu. Rev. Biochem.* **2016**, *85*, 227–264. [CrossRef] [PubMed]
39. Wang, J.Y.; Doudna, J.A. CRISPR Technology: A Decade of Genome Editing Is Only the Beginning. *Science* **2023**, *379*, eadd8643. [CrossRef] [PubMed]
40. Gasiunas, G.; Barrangou, R.; Horvath, P.; Siksnys, V. Cas9–CrRNA Ribonucleoprotein Complex Mediates Specific DNA Cleavage for Adaptive Immunity in Bacteria. *Proc. Natl. Acad. Sci. USA* **2012**, *109*, E2579–E2586. [CrossRef] [PubMed]
41. Makarova, K.S.; Koonin, E.V. Annotation and Classification of CRISPR-Cas Systems. *Methods Mol. Biol.* **2015**, *1311*, 47–75. [CrossRef] [PubMed]
42. Bortesi, L.; Fischer, R. The CRISPR/Cas9 System for Plant Genome Editing and Beyond. *Biotechnol. Adv.* **2015**, *33*, 41–52. [CrossRef] [PubMed]
43. Feng, Z.; Zhang, B.; Ding, W.; Liu, X.; Yang, D.-L.; Wei, P.; Cao, F.; Zhu, S.; Zhang, F.; Mao, Y.; et al. Efficient Genome Editing in Plants Using a CRISPR/Cas System. *Cell Res.* **2013**, *23*, 1229–1232. [CrossRef]
44. Li, J.-F.; Norville, J.E.; Aach, J.; McCormack, M.; Zhang, D.; Bush, J.; Church, G.M.; Sheen, J. Multiplex and Homologous Recombination–Mediated Genome Editing in *Arabidopsis* and *Nicotiana benthamiana* Using Guide RNA and Cas9. *Nat. Biotechnol.* **2013**, *31*, 688–691. [CrossRef]
45. Shan, Q.; Wang, Y.; Li, J.; Zhang, Y.; Chen, K.; Liang, Z.; Zhang, K.; Liu, J.; Xi, J.J.; Qiu, J.-L.; et al. Targeted Genome Modification of Crop Plants Using a CRISPR-Cas System. *Nat. Biotechnol.* **2013**, *31*, 686–688. [CrossRef] [PubMed]
46. Xie, K.; Yang, Y. RNA-Guided Genome Editing in Plants Using a CRISPR–Cas System. *Mol. Plant* **2013**, *6*, 1975–1983. [CrossRef]
47. Hilscher, J.; Bürstmayr, H.; Stoger, E. Targeted Modification of Plant Genomes for Precision Crop Breeding. *Biotechnol. J.* **2017**, *12*, 1600173. [CrossRef] [PubMed]
48. Jiang, W.; Zhou, H.; Bi, H.; Fromm, M.; Yang, B.; Weeks, D.P. Demonstration of CRISPR/Cas9/SgRNA-Mediated Targeted Gene Modification in Arabidopsis, Tobacco, Sorghum and Rice. *Nucleic Acids Res.* **2013**, *41*, e188. [CrossRef]
49. Upadhyay, S.K.; Kumar, J.; Alok, A.; Tuli, R. RNA-Guided Genome Editing for Target Gene Mutations in Wheat. *G3 Genes Genomes Genet.* **2013**, *3*, 2233–2238. [CrossRef] [PubMed]
50. Liang, Z.; Zhang, K.; Chen, K.; Gao, C. Targeted Mutagenesis in *Zea mays* Using TALENs and the CRISPR/Cas System. *J. Genet. Genom.* **2014**, *41*, 63–68. [CrossRef]
51. Jacobs, T.B.; LaFayette, P.R.; Schmitz, R.J.; Parrott, W.A. Targeted Genome Modifications in Soybean with CRISPR/Cas9. *BMC Biotechnol.* **2015**, *15*, 16. [CrossRef]
52. Cai, Y.; Chen, L.; Liu, X.; Sun, S.; Wu, C.; Jiang, B.; Han, T.; Hou, W. CRISPR/Cas9-Mediated Genome Editing in Soybean Hairy Roots. *PLoS ONE* **2015**, *10*, e0136064. [CrossRef]
53. Hsu, P.D.; Lander, E.S.; Zhang, F. Development and Applications of CRISPR-Cas9 for Genome Engineering. *Cell* **2014**, *157*, 1262–1278. [CrossRef]
54. Capdeville, N.; Schindele, P.; Puchta, H. Getting Better All the Time—Recent Progress in the Development of CRISPR/Cas-Based Tools for Plant Genome Engineering. *Curr. Opin. Biotechnol.* **2023**, *79*, 102854. [CrossRef] [PubMed]
55. Waddington, S.N.; Privolizzi, R.; Karda, R.; O'Neill, H.C. A Broad Overview and Review of CRISPR-Cas Technology and Stem Cells. *Curr. Stem. Cell Rep.* **2016**, *2*, 9–20. [CrossRef] [PubMed]
56. Li, C.; Chu, W.; Gill, R.A.; Sang, S.; Shi, Y.; Hu, X.; Yang, Y.; Zaman, Q.U.; Zhang, B. Computational Tools and Resources for CRISPR/Cas Genome Editing. *Genom. Proteom. Bioinform.* 2022, in press. [CrossRef]
57. Gao, X.; Chen, J.; Dai, X.; Zhang, D.; Zhao, Y. An Effective Strategy for Reliably Isolating Heritable and *Cas9* -Free Arabidopsis Mutants Generated by CRISPR/Cas9-Mediated Genome Editing. *Plant Physiol.* **2016**, *171*, 1794–1800. [CrossRef]
58. He, Y.; Mudgett, M.; Zhao, Y. Advances in Gene Editing without Residual Transgenes in Plants. *Plant Physiol.* **2022**, *188*, 1757–1768. [CrossRef]
59. Wada, N.; Osakabe, K.; Osakabe, Y. Expanding the Plant Genome Editing Toolbox with Recently Developed CRISPR–Cas Systems. *Plant Physiol.* **2022**, *188*, 1825–1837. [CrossRef]
60. Abdallah, N.A.; Elsharawy, H.; Abulela, H.A.; Thilmony, R.; Abdelhadi, A.A.; Elarabi, N.I. Multiplex CRISPR/Cas9-Mediated Genome Editing to Address Drought Tolerance in Wheat. *GM Crops Food* **2022**. [CrossRef] [PubMed]

61. Yang, N.; Yan, L.; Zheng, Z.; Zhang, Y.; Zhan, H.; Tian, Y.; Zhang, T.; Li, R.; Gong, X.; Xu, M.; et al. Editing Gene Families by CRISPR/Cas9: Accelerating the Isolation of Multiple Transgene-Free Null Mutant Combinations with Much Reduced Labor-Intensive Analysis. *Plant Biotechnol. J.* **2022**, *20*, 241–243. [CrossRef]
62. Liu, J.L.; Chen, M.M.; Chen, W.Q.; Liu, C.M.; He, Y.; Song, X.F. A CASE Toolkit for Easy and Efficient Multiplex Transgene-Free Gene Editing. *Plant Physiol.* **2022**, *188*, 1843–1847. [CrossRef]
63. Singh, J.; Sharma, D.; Brar, G.S.; Sandhu, K.S.; Wani, S.H.; Kashyap, R.; Kour, A.; Singh, S. CRISPR/Cas Tool Designs for Multiplex Genome Editing and Its Applications in Developing Biotic and Abiotic Stress-Resistant Crop Plants. *Mol. Biol. Rep.* **2022**, *49*, 11443–11467. [CrossRef]
64. Lorenzo, C.D.; Debray, K.; Herwegh, D.; Develtere, W.; Impens, L.; Schaumont, D.; Vandeputte, W.; Aesaert, S.; Coussens, G.; De Boe, Y.; et al. BREEDIT: A Multiplex Genome Editing Strategy to Improve Complex Quantitative Traits in Maize. *Plant Cell* **2023**, *35*, 218–238. [CrossRef]
65. Pan, C.; Li, G.; Malzahn, A.A.; Cheng, Y.; Leyson, B.; Sretenovic, S.; Gurel, F.; Coleman, G.D.; Qi, Y. Boosting Plant Genome Editing with a Versatile CRISPR-Combo System. *Nat. Plants* **2022**, *8*, 513–525. [CrossRef]
66. Gaillochet, C.; Develtere, W.; Jacobs, T.B. CRISPR Screens in Plants: Approaches, Guidelines, and Future Prospects. *Plant Cell* **2021**, *33*, 794–813. [CrossRef] [PubMed]
67. Liu, H.-J.; Jian, L.; Xu, J.; Zhang, Q.; Zhang, M.; Jin, M.; Peng, Y.; Yan, J.; Han, B.; Liu, J.; et al. High-Throughput CRISPR/Cas9 Mutagenesis Streamlines Trait Gene Identification in Maize. *Plant Cell* **2020**, *32*, 1397–1413. [CrossRef]
68. Pan, C.; Li, G.; Bandyopadhyay, A.; Qi, Y. Guide RNA Library-Based CRISPR Screens in Plants: Opportunities and Challenges. *Curr. Opin. Biotechnol.* **2023**, *79*, 102883. [CrossRef]
69. Raman, R. The Impact of Genetically Modified (GM) Crops in Modern Agriculture: A Review. *GM Crops Food* **2017**, *8*, 195–208. [CrossRef] [PubMed]
70. Rao, Y.; Yang, X.; Pan, C.; Wang, C.; Wang, K. Advance of Clustered Regularly Interspaced Short Palindromic Repeats-Cas9 System and Its Application in Crop Improvement. *Front. Plant Sci.* **2022**, *13*, 839001. [CrossRef]
71. Shan, S.; Mavrodiev, E.V.; Li, R.; Zhang, Z.; Hauser, B.A.; Soltis, P.S.; Soltis, D.E.; Yang, B. Application of CRISPR/Cas9 to *Tragopogon* (Asteraceae), an Evolutionary Model for the Study of Polyploidy. *Mol. Ecol. Resour.* **2018**, *18*, 1427–1443. [CrossRef] [PubMed]
72. FAO; ITPS. *Status of the World's Soil Resources (SWSR)—Main Report*; Food and Agriculture Organization of the United Nations; Intergovernmental Technical Panel on Soils: Rome, Italy, 2015.
73. Ma, X.; Zhu, Q.; Chen, Y.; Liu, Y.-G. CRISPR/Cas9 Platforms for Genome Editing in Plants: Developments and Applications. *Mol. Plant* **2016**, *9*, 961–974. [CrossRef]
74. Husaini, A.M. High-Value Pleiotropic Genes for Developing Multiple Stress-Tolerant Biofortified Crops for 21st-Century Challenges. *Heredity* **2022**, *128*, 460–472. [CrossRef]
75. Hartmann, A.; Berkowitz, O.; Whelan, J.; Narsai, R. Cross-Species Transcriptomic Analyses Reveals Common and Opposite Responses in Arabidopsis, Rice and Barley Following Oxidative Stress and Hormone Treatment. *BMC Plant Biol.* **2022**, *22*, 62. [CrossRef]
76. Tan, Q.W.; Lim, P.K.; Chen, Z.; Pasha, A.; Provart, N.; Arend, M.; Nikoloski, Z.; Mutwil, M. Cross-Stress Gene Expression Atlas of Marchantia Polymorpha Reveals the Hierarchy and Regulatory Principles of Abiotic Stress Responses. *Nat. Commun.* **2023**, *14*, 986. [CrossRef]
77. Wu, T.Y.; Goh, H.Z.; Azodi, C.B.; Krishnamoorthi, S.; Liu, M.J.; Urano, D. Evolutionarily Conserved Hierarchical Gene Regulatory Networks for Plant Salt Stress Response. *Nat. Plants* **2021**, *7*, 787–799. [CrossRef]
78. Bashir, S.S.; Hussain, A.; Hussain, S.J.; Wani, O.A.; Zahid Nabi, S.; Dar, N.A.; Baloch, F.S.; Mansoor, S. Plant Drought Stress Tolerance: Understanding Its Physiological, Biochemical and Molecular Mechanisms. *Biotechnol. Biotechnol. Equip.* **2021**, *35*, 1912–1925. [CrossRef]
79. Satoh, Y.; Yoshimura, K.; Pokhrel, Y.; Kim, H.; Shiogama, H.; Yokohata, T.; Hanasaki, N.; Wada, Y.; Burek, P.; Byers, E.; et al. The Timing of Unprecedented Hydrological Drought under Climate Change. *Nat. Commun.* **2022**, *13*, 3287. [CrossRef]
80. Ndehedehe, C.E.; Ferreira, V.G.; Adeyeri, O.E.; Correa, F.M.; Usman, M.; Oussou, F.E.; Kalu, I.; Okwuashi, O.; Onojeghuo, A.O.; Getirana, A.; et al. Global Assessment of Drought Characteristics in the Anthropocene. *Resour. Environ. Sustain.* **2023**, *12*, 100105. [CrossRef]
81. Oguz, M.C.; Aycan, M.; Oguz, E.; Poyraz, I.; Yildiz, M. Drought Stress Tolerance in Plants: Interplay of Molecular, Biochemical and Physiological Responses in Important Development Stages. *Physiologia* **2022**, *2*, 180–197. [CrossRef]
82. Shelake, R.M.; Kadam, U.S.; Kumar, R.; Pramanik, D.; Singh, A.K.; Kim, J.Y. Engineering Drought and Salinity Tolerance Traits in Crops through CRISPR-Mediated Genome Editing: Targets, Tools, Challenges, and Perspectives. *Plant Commun.* **2022**, *3*, 100417. [CrossRef] [PubMed]
83. Razi, K.; Muneer, S. Drought Stress-Induced Physiological Mechanisms, Signaling Pathways and Molecular Response of Chloroplasts in Common Vegetable Crops. *Crit. Rev. Biotechnol.* **2021**, *41*, 669–691. [CrossRef] [PubMed]
84. VanBuren, R.; Wai, C.M.; Giarola, V.; Župunski, M.; Pardo, J.; Kalinowski, M.; Grossmann, G.; Bartels, D. Core Cellular and Tissue-specific Mechanisms Enable Desiccation Tolerance in *Craterostigma*. *Plant J.* **2023**, *114*, 231–245. [CrossRef] [PubMed]

85. Wu, J.; Yan, G.; Duan, Z.; Wang, Z.; Kang, C.; Guo, L.; Liu, K.; Tu, J.; Shen, J.; Yi, B.; et al. Roles of the Brassica Napus DELLA Protein BnaA6.RGA, in Modulating Drought Tolerance by Interacting With the ABA Signaling Component BnaA10.ABF2. *Front. Plant Sci.* **2020**, *11*, 577. [CrossRef]
86. Peng, Y.; Chen, L.; Zhu, L.; Cui, L.; Yang, L.; Wu, H.; Bie, Z. CsAKT1 Is a Key Gene for the CeO 2 Nanoparticle's Improved Cucumber Salt Tolerance: A Validation from CRISPR-Cas9 Lines. *Environ. Sci. Nano* **2022**, *9*, 4367–4381. [CrossRef]
87. Han, J.; Li, X.; Li, W.; Yang, Q.; Li, Z.; Cheng, Z.; Lv, L.; Zhang, L.; Han, D. Isolation and Preliminary Functional Analysis of FvICE1, Involved in Cold and Drought Tolerance in Fragaria Vesca through Overexpression and CRISPR/Cas9 Technologies. *Plant Physiol. Biochem.* **2023**, *196*, 270–280. [CrossRef] [PubMed]
88. Zhong, X.; Hong, W.; Shu, Y.; Li, J.; Liu, L.; Chen, X.; Islam, F.; Zhou, W.; Tang, G. CRISPR/Cas9 Mediated Gene-Editing of GmHdz4 Transcription Factor Enhances Drought Tolerance in Soybean (*Glycine max* [L.] Merr.). *Front. Plant Sci.* **2022**, *13*, 988505. [CrossRef]
89. Wang, K.; Bu, T.; Cheng, Q.; Dong, L.; Su, T.; Chen, Z.; Kong, F.; Gong, Z.; Liu, B.; Li, M. Two Homologous LHY Pairs Negatively Control Soybean Drought Tolerance by Repressing the Abscisic Acid Responses. *New Phytol.* **2021**, *229*, 2660–2675. [CrossRef]
90. Xu, C.; Shan, J.; Liu, T.; Wang, Q.; Ji, Y.; Zhang, Y.; Wang, M.; Xia, N.; Zhao, L. CONSTANS-LIKE 1a Positively Regulates Salt and Drought Tolerance in Soybean. *Plant Physiol.* **2022**, *191*, 2427–2446. [CrossRef] [PubMed]
91. Du, Y.-T.; Zhao, M.-J.; Wang, C.-T.; Gao, Y.; Wang, Y.-X.; Liu, Y.-W.; Chen, M.; Chen, J.; Zhou, Y.-B.; Xu, Z.-S.; et al. Identification and Characterization of GmMYB118 Responses to Drought and Salt Stress. *BMC Plant Biol.* **2018**, *18*, 320. [CrossRef]
92. Yang, C.; Huang, Y.; Lv, P.; Antwi-Boasiako, A.; Begum, N.; Zhao, T.; Zhao, J. NAC Transcription Factor GmNAC12 Improved Drought Stress Tolerance in Soybean. *Int. J. Mol. Sci.* **2022**, *23*, 12029. [CrossRef]
93. Yang, C.; Huang, Y.; Lv, W.; Zhang, Y.; Bhat, J.A.; Kong, J.; Xing, H.; Zhao, J.; Zhao, T. GmNAC8 Acts as a Positive Regulator in Soybean Drought Stress. *Plant Sci.* **2020**, *293*, 110442. [CrossRef]
94. Singer, S.D.; Burton Hughes, K.; Subedi, U.; Dhariwal, G.K.; Kader, K.; Acharya, S.; Chen, G.; Hannoufa, A. The CRISPR/Cas9-Mediated Modulation of Squamosa Promoter-Binding Protein-like 8 in Alfalfa Leads to Distinct Phenotypic Outcomes. *Front. Plant Sci.* **2022**, *12*, 3203. [CrossRef] [PubMed]
95. Li, G.; Ma, Y.; Wang, X.; Cheng, N.; Meng, D.; Chen, S.; Wang, W.; Wang, X.; Hu, X.; Yan, L.; et al. CRISPR/Cas9 Gene Editing of NtAITRs, a Family of Transcription Repressor Genes, Leads to Enhanced Drought Tolerance in Tobacco. *Int. J. Mol. Sci.* **2022**, *23*, 15268. [CrossRef] [PubMed]
96. Xu, L.; Gao, Q.; Feng, J.; Xu, Y.; Jiang, J.; Deng, L.; Lu, Y.; Zeng, W.; Xing, J.; Xiang, H.; et al. Physiological and Phosphoproteomic Analyses Revealed That the NtPOD63 L Knockout Mutant Enhances Drought Tolerance in Tobacco. *Ind. Crops Prod.* **2023**, *193*, 116218. [CrossRef]
97. Gao, Y.; Yang, J.; Duan, W.; Ma, X.; Qu, L.; Xu, Z.; Yang, Y.; Xu, J. NtRAV4 Negatively Regulates Drought Tolerance in Nicotiana Tabacum by Enhancing Antioxidant Capacity and Defence System. *Plant Cell Rep.* **2022**, *41*, 1775–1788. [CrossRef] [PubMed]
98. Liu, J.; Shen, J.; Xu, Y.; Li, X.; Xiao, J.; Xiong, L. Ghd2, a CONSTANS -like Gene, Confers Drought Sensitivity through Regulation of Senescence in Rice. *J. Exp. Bot.* **2016**, *67*, 5785–5798. [CrossRef] [PubMed]
99. Zhao, W.; Wang, X.; Zhang, Q.; Zheng, Q.; Yao, H.; Gu, X.; Liu, D.; Tian, X.; Wang, X.; Li, Y.; et al. H3K36 Demethylase JMJ710 Negatively Regulates Drought Tolerance by Suppressing MYB48-1 Expression in Rice. *Plant Physiol.* **2022**, *189*, 1050–1064. [CrossRef]
100. Yue, E.; Cao, H.; Liu, B. OsmiR535, a Potential Genetic Editing Target for Drought and Salinity Stress Tolerance in *Oryza sativa*. *Plants* **2020**, *9*, 1337. [CrossRef] [PubMed]
101. Zhang, Y.; Wang, X.; Luo, Y.; Zhang, L.; Yao, Y.; Han, L.; Chen, Z.; Wang, L.; Li, Y. OsABA8ox2, an ABA Catabolic Gene, Suppresses Root Elongation of Rice Seedlings and Contributes to Drought Response. *Crop J.* **2020**, *8*, 480–491. [CrossRef]
102. Santosh Kumar, V.V.; Verma, R.K.; Yadav, S.K.; Yadav, P.; Watts, A.; Rao, M.V.; Chinnusamy, V. CRISPR-Cas9 Mediated Genome Editing of Drought and Salt Tolerance (OsDST) Gene in Indica Mega Rice Cultivar MTU1010. *Physiol. Mol. Biol. Plants* **2020**, *26*, 1099–1110. [CrossRef]
103. Ogata, T.; Ishizaki, T.; Fujita, M.; Fujita, Y. CRISPR/Cas9-Targeted Mutagenesis of OsERA1 Confers Enhanced Responses to Abscisic Acid and Drought Stress and Increased Primary Root Growth under Nonstressed Conditions in Rice. *PLoS ONE* **2020**, *15*, e0243376. [CrossRef]
104. Gu, H.; Zhang, K.; Chen, J.; Gull, S.; Chen, C.; Hou, Y.; Li, X.; Miao, J.; Zhou, Y.; Liang, G. OsFTL4, an FT-like Gene, Regulates Flowering Time and Drought Tolerance in Rice (*Oryza sativa* L.). *Rice* **2022**, *15*, 47. [CrossRef]
105. Jiang, M.; Liu, Y.; Li, R.; Li, S.; Tan, Y.; Huang, J.; Shu, Q. An Inositol 1,3,4,5,6-Pentakisphosphate 2-Kinase 1 Mutant with a 33-Nt Deletion Showed Enhanced Tolerance to Salt and Drought Stress in Rice. *Plants* **2021**, *10*, 23. [CrossRef]
106. Wu, Q.; Liu, Y.; Xie, Z.; Yu, B.; Sun, Y.; Huang, J. OsNAC016 Regulates Plant Architecture and Drought Tolerance by Interacting with the Kinases GSK2 and SAPK8. *Plant Physiol.* **2022**, *189*, 1296–1313. [CrossRef] [PubMed]
107. Wang, B.; Wang, Y.; Yu, W.; Wang, L.; Lan, Q.; Wang, Y.; Chen, C.; Zhang, Y. Knocking Out the Transcription Factor OsNAC092 Promoted Rice Drought Tolerance. *Biology* **2022**, *11*, 1830. [CrossRef] [PubMed]
108. Han, M.-L.; Lv, Q.-Y.; Zhang, J.; Wang, T.; Zhang, C.-X.; Tan, R.-J.; Wang, Y.-L.; Zhong, L.-Y.; Gao, Y.-Q.; Chao, Z.-F.; et al. Decreasing Nitrogen Assimilation under Drought Stress by Suppressing DST-Mediated Activation of Nitrate Reductase 1.2 in Rice. *Mol. Plant* **2022**, *15*, 167–178. [CrossRef] [PubMed]

109. Luo, Z.; Xiong, J.; Xia, H.; Wang, L.; Hou, G.; Li, Z.; Li, J.; Zhou, H.; Li, T.; Luo, L. Pentatricopeptide Repeat Gene-Mediated Mitochondrial RNA Editing Impacts on Rice Drought Tolerance. *Front. Plant Sci.* **2022**, *13*, 926285. [CrossRef]
110. Usman, B.; Nawaz, G.; Zhao, N.; Liao, S.; Liu, Y.; Li, R. Precise Editing of the OsPYL9 Gene by RNA-Guided Cas9 Nuclease Confers Enhanced Drought Tolerance and Grain Yield in Rice (*Oryza sativa* L.) by Regulating Circadian Rhythm and Abiotic Stress Responsive Proteins. *Int. J. Mol. Sci.* **2020**, *21*, 7854. [CrossRef] [PubMed]
111. Lim, C.; Kang, K.; Shim, Y.; Yoo, S.-C.; Paek, N.-C. Inactivating Transcription Factor OsWRKY5 Enhances Drought Tolerance through Abscisic Acid Signaling Pathways. *Plant Physiol.* **2022**, *188*, 1900–1916. [CrossRef]
112. Liao, S.; Qin, X.; Luo, L.; Han, Y.; Wang, X.; Usman, B.; Nawaz, G.; Zhao, N.; Liu, Y.; Li, R. CRISPR/Cas9-Induced Mutagenesis of Semi-Rolled Leaf1,2 Confers Curled Leaf Phenotype and Drought Tolerance by Influencing Protein Expression Patterns and ROS Scavenging in Rice (*Oryza sativa* L.). *Agronomy* **2019**, *9*, 728. [CrossRef]
113. Um, T.; Choi, J.; Park, T.; Chung, P.J.; Jung, S.E.; Shim, J.S.; Kim, Y.S.; Choi, I.-Y.; Park, S.C.; Oh, S.-J.; et al. Rice MicroRNA171f/SCL6 Module Enhances Drought Tolerance by Regulation of Flavonoid Biosynthesis Genes. *Plant Direct* **2022**, *6*, e374. [CrossRef]
114. Chung, P.J.; Chung, H.; Oh, N.; Choi, J.; Bang, S.W.; Jung, S.E.; Jung, H.; Shim, J.S.; Kim, J.-K. Efficiency of Recombinant CRISPR/RCas9-Mediated MiRNA Gene Editing in Rice. *Int. J. Mol. Sci.* **2020**, *21*, 9606. [CrossRef]
115. Li, J.; Zhang, M.; Yang, L.; Mao, X.; Li, J.; Li, L.; Wang, J.; Liu, H.; Zheng, H.; Li, Z.; et al. OsADR3 Increases Drought Stress Tolerance by Inducing Antioxidant Defense Mechanisms and Regulating OsGPX1 in Rice (*Oryza sativa* L.). *Crop J.* **2021**, *9*, 1003–1017. [CrossRef]
116. Du, C.; Cai, W.; Lin, F.; Wang, K.; Li, S.; Chen, C.; Tian, H.; Wang, D.; Zhao, Q. Leucine-Rich Repeat Receptor-like Kinase OsASLRK Regulates Abscisic Acid and Drought Responses via Cooperation with S-like RNase OsRNS4 in Rice. *Environ. Exp. Bot.* **2022**, *201*, 104949. [CrossRef]
117. Gao, W.; Li, M.; Yang, S.; Gao, C.; Su, Y.; Zeng, X.; Jiao, Z.; Xu, W.; Zhang, M.; Xia, K. MiR2105 and the Kinase OsSAPK10 Co-Regulate OsbZIP86 to Mediate Drought-Induced ABA Biosynthesis in Rice. *Plant Physiol.* **2022**, *189*, 889–905. [CrossRef] [PubMed]
118. Bang, S.W.; Choi, S.; Jin, X.; Jung, S.E.; Choi, J.W.; Seo, J.S.; Kim, J.-K. Transcriptional Activation of Rice CINNAMOYL-CoA REDUCTASE 10 by OsNAC5, Contributes to Drought Tolerance by Modulating Lignin Accumulation in Roots. *Plant Biotechnol. J.* **2022**, *20*, 736–747. [CrossRef]
119. Huang, L.; Fu, W.; Ji, E.; Tanveer, M.; Shabala, S.; Yu, M.; Jiang, M. A Novel R3H Protein, OsDIP1, Confers ABA-Mediated Adaptation to Drought and Salinity Stress in Rice. *Plant Soil* **2022**, *477*, 501–519. [CrossRef]
120. Yang, L.; Chen, Y.; Xu, L.; Wang, J.; Qi, H.; Guo, J.; Zhang, L.; Shen, J.; Wang, H.; Zhang, F.; et al. The OsFTIP6-OsHB22-OsMYBR57 Module Regulates Drought Response in Rice. *Mol. Plant* **2022**, *15*, 1227–1242. [CrossRef] [PubMed]
121. Xu, W.; Dou, Y.; Geng, H.; Fu, J.; Dan, Z.; Liang, T.; Cheng, M.; Zhao, W.; Zeng, Y.; Hu, Z.; et al. OsGRP3 Enhances Drought Resistance by Altering Phenylpropanoid Biosynthesis Pathway in Rice (*Oryza sativa* L.). *Int. J. Mol. Sci.* **2022**, *23*, 7045. [CrossRef] [PubMed]
122. Jian, L.; Kang, K.; Choi, Y.; Suh, M.C.; Paek, N.-C. Mutation of OsMYB60 Reduces Rice Resilience to Drought Stress by Attenuating Cuticular Wax Biosynthesis. *Plant J.* **2022**, *112*, 339–351. [CrossRef] [PubMed]
123. Wang, B.; Zhong, Z.; Wang, X.; Han, X.; Yu, D.; Wang, C.; Song, W.; Zheng, X.; Chen, C.; Zhang, Y. Knockout of the OsNAC006 Transcription Factor Causes Drought and Heat Sensitivity in Rice. *Int. J. Mol. Sci.* **2020**, *21*, 2288. [CrossRef] [PubMed]
124. Jung, S.E.; Kim, T.H.; Shim, J.S.; Bang, S.W.; Bin Yoon, H.; Oh, S.H.; Kim, Y.S.; Oh, S.-J.; Seo, J.S.; Kim, J.-K. Rice NAC17 Transcription Factor Enhances Drought Tolerance by Modulating Lignin Accumulation. *Plant Sci.* **2022**, *323*, 111404. [CrossRef]
125. Diyang, Q.; Rui, H.; Ji, L.; Ying, L.; Jierong, D.; Kuaifei, X.; Xuhua, Z.; Zhongming, F.; Mingyong, Z. Peptide Transporter OsNPF8.1 Contributes to Sustainable Growth under Salt and Drought Stresses, and Grain Yield under Nitrogen Deficiency in Rice. *Rice Sci.* **2023**, *30*, 113–126. [CrossRef]
126. Yao, L.; Cheng, X.; Gu, Z.; Huang, W.; Li, S.; Wang, L.; Wang, Y.-F.; Xu, P.; Ma, H.; Ge, X. The AWPM-19 Family Protein OsPM1 Mediates Abscisic Acid Influx and Drought Response in Rice. *Plant Cell* **2018**, *30*, 1258–1276. [CrossRef]
127. Qin, Q.; Wang, Y.; Huang, L.; Du, F.; Zhao, X.; Li, Z.; Wang, W.; Fu, B. A U-Box E3 Ubiquitin Ligase OsPUB67 Is Positively Involved in Drought Tolerance in Rice. *Plant Mol. Biol.* **2020**, *102*, 89–107. [CrossRef] [PubMed]
128. Chen, S.; Xu, K.; Kong, D.; Wu, L.; Chen, Q.; Ma, X.; Ma, S.; Li, T.; Xie, Q.; Liu, H.; et al. Ubiquitin Ligase OsRINGzf1 Regulates Drought Resistance by Controlling the Turnover of OsPIP2;1. *Plant Biotechnol. J.* **2022**, *20*, 1743–1755. [CrossRef] [PubMed]
129. Lou, D.; Wang, H.; Liang, G.; Yu, D. OsSAPK2 Confers Abscisic Acid Sensitivity and Tolerance to Drought Stress in Rice. *Front. Plant Sci.* **2017**, *8*, 993. [CrossRef]
130. Lou, D.; Lu, S.; Chen, Z.; Lin, Y.; Yu, D.; Yang, X. Molecular Characterization Reveals That OsSAPK3 Improves Drought Tolerance and Grain Yield in Rice. *BMC Plant Biol.* **2023**, *23*, 53. [CrossRef]
131. Chen, F.; Zhang, H.; Li, H.; Lian, L.; Wei, Y.; Lin, Y.; Wang, L.; He, W.; Cai, Q.; Xie, H.; et al. IPA1 Improves Drought Tolerance by Activating SNAC1 in Rice. *BMC Plant Biol.* **2023**, *23*, 55. [CrossRef]
132. Shi, X.; Tian, Q.; Deng, P.; Zhang, W.; Jing, W. The Rice Aldehyde Oxidase OsAO3 Gene Regulates Plant Growth, Grain Yield, and Drought Tolerance by Participating in ABA Biosynthesis. *Biochem. Biophys. Res. Commun.* **2021**, *548*, 189–195. [CrossRef] [PubMed]

133. Shen, C.; Zhang, Y.; Li, Q.; Liu, S.; He, F.; An, Y.; Zhou, Y.; Liu, C.; Yin, W.; Xia, X. PdGNC Confers Drought Tolerance by Mediating Stomatal Closure Resulting from NO and H_2O_2 Production via the Direct Regulation of PdHXK1 Expression in Populus. *New Phytol.* **2021**, *230*, 1868–1882. [CrossRef]
134. Zhou, Y.; Zhang, Y.; Wang, X.; Han, X.; An, Y.; Lin, S.; Shen, C.; Wen, J.; Liu, C.; Yin, W.; et al. Root-Specific NF-Y Family Transcription Factor, PdNF-YB21, Positively Regulates Root Growth and Drought Resistance by Abscisic Acid-Mediated Indoylacetic Acid Transport in Populus. *New Phytol.* **2020**, *227*, 407–426. [CrossRef] [PubMed]
135. Li, S.; Lin, Y.-C.J.; Wang, P.; Zhang, B.; Li, M.; Chen, S.; Shi, R.; Tunlaya-Anukit, S.; Liu, X.; Wang, Z.; et al. The AREB1 Transcription Factor Influences Histone Acetylation to Regulate Drought Responses and Tolerance in Populus Trichocarpa. *Plant Cell* **2019**, *31*, 663–686. [CrossRef]
136. Wang, P.; Luo, Q.; Yang, W.; Ahammed, G.J.; Ding, S.; Chen, X.; Wang, J.; Xia, X.; Shi, K. A Novel Role of Pipecolic Acid Biosynthetic Pathway in Drought Tolerance through the Antioxidant System in Tomato. *Antioxidants* **2021**, *10*, 1923. [CrossRef] [PubMed]
137. Chen, M.; Zhu, X.; Liu, X.; Wu, C.; Yu, C.; Hu, G.; Chen, L.; Chen, R.; Bouzayen, M.; Zouine, M.; et al. Knockout of Auxin Response Factor SlARF4 Improves Tomato Resistance to Water Deficit. *Int. J. Mol. Sci.* **2021**, *22*, 3347. [CrossRef]
138. Zhao, W.; Huang, H.; Wang, J.; Wang, X.; Xu, B.; Yao, X.; Sun, L.; Yang, R.; Wang, J.; Sun, A.; et al. Jasmonic Acid Enhances Osmotic Stress Responses by MYC2-Mediated Inhibition of Protein Phosphatase 2C1 and Response Regulators 26 Transcription Factor in Tomato. *Plant J.* **2023**, *113*, 546–561. [CrossRef] [PubMed]
139. Wang, X.; Liu, Y.; Li, H.; Wang, F.; Xia, P.; Li, W.; Zhang, X.; Zhang, N.; Guo, Y.-D. SlSNAT2, a Chloroplast-Localized Acetyltransferase, Is Involved in Rubisco Lysine Acetylation and Negatively Regulates Drought Stress Tolerance in Tomato. *Environ. Exp. Bot.* **2022**, *201*, 105003. [CrossRef]
140. Liu, L.; Zhang, J.; Xu, J.; Li, Y.; Guo, L.; Wang, Z.; Zhang, X.; Zhao, B.; Guo, Y.-D.; Zhang, N. CRISPR/Cas9 Targeted Mutagenesis of SlLBD40, a Lateral Organ Boundaries Domain Transcription Factor, Enhances Drought Tolerance in Tomato. *Plant Sci.* **2020**, *301*, 110683. [CrossRef]
141. Wang, L.; Chen, L.; Li, R.; Zhao, R.; Yang, M.; Sheng, J.; Shen, L. Reduced Drought Tolerance by CRISPR/Cas9-Mediated SlMAPK3 Mutagenesis in Tomato Plants. *J. Agric. Food Chem.* **2017**, *65*, 8674–8682. [CrossRef]
142. Li, R.; Liu, C.; Zhao, R.; Wang, L.; Chen, L.; Yu, W.; Zhang, S.; Sheng, J.; Shen, L. CRISPR/Cas9-Mediated SlNPR1 Mutagenesis Reduces Tomato Plant Drought Tolerance. *BMC Plant Biol.* **2019**, *19*, 38. [CrossRef]
143. dos Reis Moreira, J.; Quiñones, A.; Lira, B.S.; Robledo, J.M.; Curtin, S.J.; Vicente, M.H.; Ribeiro, D.M.; Ryngajllo, M.; Jiménez-Gómez, J.M.; Peres, L.E.P.; et al. *SELF PRUNING 3C* Is a Flowering Repressor That Modulates Seed Germination, Root Architecture, and Drought Responses. *J. Exp. Bot.* **2022**, *73*, 6226–6240. [CrossRef]
144. Ramírez Gonzales, L.; Shi, L.; Bergonzi, S.B.; Oortwijn, M.; Franco-Zorrilla, J.M.; Solano-Tavira, R.; Visser, R.G.F.; Abelenda, J.A.; Bachem, C.W.B. Potato cycling Dof factor 1 and Its LncRNA Counterpart StFLORE Link Tuber Development and Drought Response. *Plant J.* **2021**, *105*, 855–869. [CrossRef] [PubMed]
145. Mohr, T.; Horstman, J.; Gu, Y.Q.; Elarabi, N.I.; Abdallah, N.A.; Thilmony, R. CRISPR-Cas9 Gene Editing of the Sal1 Gene Family in Wheat. *Plants* **2022**, *11*, 2259. [CrossRef] [PubMed]
146. He, J.; Li, C.; Hu, N.; Zhu, Y.; He, Z.; Sun, Y.; Wang, Z.; Wang, Y. ECERIFERUM1-6A Is Required for the Synthesis of Cuticular Wax Alkanes and Promotes Drought Tolerance in Wheat. *Plant Physiol.* **2022**, *190*, 1640–1657. [CrossRef] [PubMed]
147. Wang, N.; Chen, J.; Gao, Y.; Zhou, Y.; Chen, M.; Xu, Z.; Fang, Z.; Ma, Y. Genomic Analysis of Isopentenyltransferase Genes and Functional Characterization of TaIPT8 Indicates Positive Effects of Cytokinins on Drought Tolerance in Wheat. *Crop J.* **2023**, *11*, 46–56. [CrossRef]
148. Mao, H.; Jian, C.; Cheng, X.; Chen, B.; Mei, F.; Li, F.; Zhang, Y.; Li, S.; Du, L.; Li, T.; et al. The Wheat ABA Receptor Gene TaPYL1-1B Contributes to Drought Tolerance and Grain Yield by Increasing Water-Use Efficiency. *Plant Biotechnol. J.* **2022**, *20*, 846–861. [CrossRef]
149. Clemens, M.; Faralli, M.; Lagreze, J.; Bontempo, L.; Piazza, S.; Varotto, C.; Malnoy, M.; Oechel, W.; Rizzoli, A.; Dalla Costa, L. VvEPFL9-1 Knock-Out via CRISPR/Cas9 Reduces Stomatal Density in Grapevine. *Front. Plant Sci.* **2022**, *13*, 878001. [CrossRef]
150. Shi, J.; Gao, H.; Wang, H.; Lafitte, H.R.; Archibald, R.L.; Yang, M.; Hakimi, S.M.; Mo, H.; Habben, J.E. ARGOS8 Variants Generated by CRISPR-Cas9 Improve Maize Grain Yield under Field Drought Stress Conditions. *Plant Biotechnol. J.* **2017**, *15*, 207–216. [CrossRef]
151. Feng, X.; Xiong, J.; Zhang, W.; Guan, H.; Zheng, D.; Xiong, H.; Jia, L.; Hu, Y.; Zhou, H.; Wen, Y.; et al. ZmLBD5, a Class-II LBD Gene, Negatively Regulates Drought Tolerance by Impairing Abscisic Acid Synthesis. *Plant J.* **2022**, *112*, 1364–1376. [CrossRef] [PubMed]
152. Zhang, M.; Chen, Y.; Xing, H.; Ke, W.; Shi, Y.; Sui, Z.; Xu, R.; Gao, L.; Guo, G.; Li, J.; et al. Positional Cloning and Characterization Reveal the Role of a MiRNA Precursor Gene ZmLRT in the Regulation of Lateral Root Number and Drought Tolerance in Maize. *J. Integr. Plant Biol.* **2022**, *65*, 772–790. [CrossRef]
153. Guo, Y.; Shi, Y.; Wang, Y.; Liu, F.; Li, Z.; Qi, J.; Wang, Y.; Zhang, J.; Yang, S.; Wang, Y.; et al. The Clade F PP2C Phosphatase ZmPP84 Negatively Regulates Drought Tolerance by Repressing Stomatal Closure in Maize. *New Phytol.* **2023**, *237*, 1728–1744. [CrossRef]
154. Wang, C.; Gao, B.; Chen, N.; Jiao, P.; Jiang, Z.; Zhao, C.; Ma, Y.; Guan, S.; Liu, S. A Novel Senescence-Specific Gene (ZmSAG39) Negatively Regulates Darkness and Drought Responses in Maize. *Int. J. Mol. Sci.* **2022**, *23*, 15984. [CrossRef] [PubMed]

155. Jiao, P.; Liu, T.; Zhao, C.; Fei, J.; Guan, S.; Ma, Y. ZmTCP14, a TCP Transcription Factor, Modulates Drought Stress Response in *Zea mays* L. *Environ. Exp. Bot.* **2023**, *208*, 105232. [CrossRef]
156. Jiao, P.; Jiang, Z.; Wei, X.; Liu, S.; Qu, J.; Guan, S.; Ma, Y. Overexpression of the Homeobox-Leucine Zipper Protein ATHB-6 Improves the Drought Tolerance of Maize (*Zea mays* L.). *Plant Sci.* **2022**, *316*, 111159. [CrossRef] [PubMed]
157. Yang, Y.; Shi, J.; Chen, L.; Xiao, W.; Yu, J. ZmEREB46, a Maize Ortholog of Arabidopsis WAX INDUCER1/SHINE1, Is Involved in the Biosynthesis of Leaf Epicuticular Very-Long-Chain Waxes and Drought Tolerance. *Plant Sci.* **2022**, *321*, 111256. [CrossRef]
158. Gao, H.; Cui, J.; Liu, S.; Wang, S.; Lian, Y.; Bai, Y.; Zhu, T.; Wu, H.; Wang, Y.; Yang, S.; et al. Natural Variations of ZmSRO1d Modulate the Trade-off between Drought Resistance and Yield by Affecting ZmRBOHC-Mediated Stomatal ROS Production in Maize. *Mol. Plant* **2022**, *15*, 1558–1574. [CrossRef] [PubMed]
159. Tian, T.; Wang, S.; Yang, S.; Yang, Z.; Liu, S.; Wang, Y.; Gao, H.; Zhang, S.; Yang, X.; Jiang, C.; et al. Genome Assembly and Genetic Dissection of a Prominent Drought-Resistant Maize Germplasm. *Nat. Genet.* **2023**, *55*, 496–506. [CrossRef] [PubMed]
160. Pan, Z.; Liu, M.; Zhao, H.; Tan, Z.; Liang, K.; Sun, Q.; Gong, D.; He, H.; Zhou, W.; Qiu, F. ZmSRL5 Is Involved in Drought Tolerance by Maintaining Cuticular Wax Structure in Maize. *J. Integr. Plant Biol.* **2020**, *62*, 1895–1909. [CrossRef]
161. Chang, H.; Yi, B.; Ma, R.; Zhang, X.; Zhao, H.; Xi, Y. CRISPR/Cas9, a Novel Genomic Tool to Knock down MicroRNA in Vitro and in Vivo. *Sci. Rep.* **2016**, *6*, 22312. [CrossRef]
162. Shomali, A.; Das, S.; Arif, N.; Sarraf, M.; Zahra, N.; Yadav, V.; Aliniaeifard, S.; Chauhan, D.K.; Hasanuzzaman, M. Diverse Physiological Roles of Flavonoids in Plant Environmental Stress Responses and Tolerance. *Plants* **2022**, *11*, 3158. [CrossRef]
163. Guo, Z.; Kuang, Z.; Wang, Y.; Zhao, Y.; Tao, Y.; Cheng, C.; Yang, J.; Lu, X.; Hao, C.; Wang, T.; et al. PmiREN: A Comprehensive Encyclopedia of Plant MiRNAs. *Nucleic Acids Res.* **2020**, *48*, D1114–D1121. [CrossRef]
164. Bin Rahman, A.N.M.R.; Zhang, J. Flood and Drought Tolerance in Rice: Opposite but May Coexist. *Food Energy Secur.* **2016**, *5*, 76–88. [CrossRef]
165. Tamang, B.G.; Li, S.; Rajasundaram, D.; Lamichhane, S.; Fukao, T. Overlapping and Stress-specific Transcriptomic and Hormonal Responses to Flooding and Drought in Soybean. *Plant J.* **2021**, *107*, 100–117. [CrossRef] [PubMed]
166. Zscheischler, J.; Westra, S.; van den Hurk, B.J.J.M.; Seneviratne, S.I.; Ward, P.J.; Pitman, A.; AghaKouchak, A.; Bresch, D.N.; Leonard, M.; Wahl, T.; et al. Future Climate Risk from Compound Events. *Nat. Clim. Change* **2018**, *8*, 469–477. [CrossRef]
167. Cui, H.; Jiang, S.; Ren, L.; Xiao, W.; Yuan, F.; Wang, M.; Wei, L. Dynamics and Potential Synchronization of Regional Precipitation Concentration and Drought-Flood Abrupt Alternation under the Influence of Reservoir Climate. *J. Hydrol. Reg. Stud.* **2022**, *42*, 101147. [CrossRef]
168. Fukao, T.; Barrera-Figueroa, B.E.; Juntawong, P.; Peña-Castro, J.M. Submergence and Waterlogging Stress in Plants: A Review Highlighting Research Opportunities and Understudied Aspects. *Front. Plant Sci.* **2019**, *10*, 340. [CrossRef] [PubMed]
169. Jackson, M.B.; Colmer, T.D. Response and Adaptation by Plants to Flooding Stress. *Ann. Bot.* **2005**, *96*, 501–505. [CrossRef]
170. Pan, J.; Sharif, R.; Xu, X.; Chen, X. Mechanisms of Waterlogging Tolerance in Plants: Research Progress and Prospects. *Front. Plant Sci.* **2021**, *11*, 627331. [CrossRef]
171. Sun, J.; Zhang, G.; Cui, Z.; Kong, X.; Yu, X.; Gui, R.; Han, Y.; Li, Z.; Lang, H.; Hua, Y.; et al. Regain Flood Adaptation in Rice through a 14-3-3 Protein OsGF14h. *Nat. Commun.* **2022**, *13*, 5664. [CrossRef]
172. Liang, Y.; Biswas, S.; Kim, B.; Bailey-Serres, J.; Septiningsih, E.M. Improved Transformation and Regeneration of Indica Rice: Disruption of SUB1A as a Test Case via CRISPR-Cas9. *Int. J. Mol. Sci.* **2021**, *22*, 6989. [CrossRef]
173. Ye, N.-H.; Wang, F.-Z.; Shi, L.; Chen, M.-X.; Cao, Y.-Y.; Zhu, F.-Y.; Wu, Y.-Z.; Xie, L.-J.; Liu, T.-Y.; Su, Z.-Z.; et al. Natural Variation in the Promoter of Rice Calcineurin B-like Protein10 (OsCBL10) Affects Flooding Tolerance during Seed Germination among Rice Subspecies. *Plant J.* **2018**, *94*, 612–625. [CrossRef]
174. Verslues, P.E.; Bailey-Serres, J.; Brodersen, C.; Buckley, T.N.; Conti, L.; Christmann, A.; Dinneny, J.R.; Grill, E.; Hayes, S.; Heckman, R.W.; et al. Burning Questions for a Warming and Changing World: 15 Unknowns in Plant Abiotic Stress. *Plant Cell* **2022**, *35*, 67–108. [CrossRef]
175. Shrivastava, P.; Kumar, R. Soil Salinity: A Serious Environmental Issue and Plant Growth Promoting Bacteria as One of the Tools for Its Alleviation. *Saudi J. Biol. Sci.* **2015**, *22*, 123–131. [CrossRef] [PubMed]
176. Litalien, A.; Zeeb, B. Curing the Earth: A Review of Anthropogenic Soil Salinization and Plant-Based Strategies for Sustainable Mitigation. *Sci. Total Environ.* **2020**, *698*, 134235. [CrossRef]
177. Qadir, M.; Quillérou, E.; Nangia, V.; Murtaza, G.; Singh, M.; Thomas, R.J.; Drechsel, P.; Noble, A.D. Economics of Salt-Induced Land Degradation and Restoration. *Nat. Resour. Forum* **2014**, *38*, 282–295. [CrossRef]
178. Chen, J.; Mueller, V. Coastal Climate Change, Soil Salinity and Human Migration in Bangladesh. *Nat. Clim. Change* **2018**, *8*, 981–985. [CrossRef]
179. Li, J.; Pu, L.; Han, M.; Zhu, M.; Zhang, R.; Xiang, Y. Soil Salinization Research in China: Advances and Prospects. *J. Geogr. Sci.* **2014**, *24*, 943–960. [CrossRef]
180. Casanova, M.; Salazar, O.; Oyarzún, I.; Tapia, Y.; Fajardo, M. Field Monitoring of 2010-Tsunami Impact on Agricultural Soils and Irrigation Waters: Central Chile. *Water Air Soil Pollut.* **2016**, *227*, 411. [CrossRef]
181. Kumar, K.; Kumar, M.; Kim, S.-R.; Ryu, H.; Cho, Y.-G. Insights into Genomics of Salt Stress Response in Rice. *Rice* **2013**, *6*, 27. [CrossRef] [PubMed]
182. Moradi, F.; Ismail, A.M. Responses of Photosynthesis, Chlorophyll Fluorescence and ROS-Scavenging Systems to Salt Stress During Seedling and Reproductive Stages in Rice. *Ann. Bot.* **2007**, *99*, 1161–1173. [CrossRef] [PubMed]

183. Parihar, P.; Singh, S.; Singh, R.; Singh, V.P.; Prasad, S.M. Effect of Salinity Stress on Plants and Its Tolerance Strategies: A Review. *Environ. Sci. Pollut. Res.* **2015**, *22*, 4056–4075. [CrossRef]
184. Razzaq, A.; Ali, A.; Safdar, L.B.; Zafar, M.M.; Rui, Y.; Shakeel, A.; Shaukat, A.; Ashraf, M.; Gong, W.; Yuan, Y. Salt Stress Induces Physiochemical Alterations in Rice Grain Composition and Quality. *J. Food Sci.* **2020**, *85*, 14–20. [CrossRef]
185. Munns, R. Comparative Physiology of Salt and Water Stress: Comparative Physiology of Salt and Water Stress. *Plant Cell Environ.* **2002**, *25*, 239–250. [CrossRef]
186. Zhu, J.-K. Salt and drought stress signal transduction in plants. *Annu. Rev. Plant Biol.* **2002**, *53*, 247–273. [CrossRef]
187. Hasanuzzaman, M.; Bhuyan, M.H.M.B.; Parvin, K.; Bhuiyan, T.F.; Anee, T.I.; Nahar, K.; Hossen, M.S.; Zulfiqar, F.; Alam, M.M.; Fujita, M. Regulation of ROS Metabolism in Plants under Environmental Stress: A Review of Recent Experimental Evidence. *IJMS* **2020**, *21*, 8695. [CrossRef]
188. Shavrukov, Y. Salt Stress or Salt Shock: Which Genes Are We Studying? *J. Exp. Bot.* **2013**, *64*, 119–127. [CrossRef]
189. Arif, Y.; Singh, P.; Siddiqui, H.; Bajguz, A.; Hayat, S. Salinity Induced Physiological and Biochemical Changes in Plants: An Omic Approach towards Salt Stress Tolerance. *Plant Physiol. Biochem.* **2020**, *156*, 64–77. [CrossRef]
190. Deinlein, U.; Stephan, A.B.; Horie, T.; Luo, W.; Xu, G.; Schroeder, J.I. Plant Salt-Tolerance Mechanisms. *Trends Plant Sci.* **2014**, *19*, 371–379. [CrossRef]
191. van Zelm, E.; Zhang, Y.; Testerink, C. Salt Tolerance Mechanisms of Plants. *Annu. Rev. Plant Biol.* **2020**, *71*, 403–433. [CrossRef] [PubMed]
192. Shabala, S.; Bose, J.; Hedrich, R. Salt Bladders: Do They Matter? *Trends Plant Sci.* **2014**, *19*, 687–691. [CrossRef] [PubMed]
193. Gupta, B.; Huang, B. Mechanism of Salinity Tolerance in Plants: Physiological, Biochemical, and Molecular Characterization. *Int. J. Genom.* **2014**, *2014*, 701596. [CrossRef] [PubMed]
194. Rathinapriya, P.; Pandian, S.; Rakkammal, K.; Balasangeetha, M.; Alexpandi, R.; Satish, L.; Rameshkumar, R.; Ramesh, M. The Protective Effects of Polyamines on Salinity Stress Tolerance in Foxtail Millet (*Setaria italica* L.), an Important C4 Model Crop. *Physiol. Mol. Biol. Plants* **2020**, *26*, 1815–1829. [CrossRef] [PubMed]
195. Tiburcio, A.F.; Altabella, T.; Bitrián, M.; Alcázar, R. The Roles of Polyamines during the Lifespan of Plants: From Development to Stress. *Planta* **2014**, *240*, 1–18. [CrossRef]
196. Wani, S.H.; Kumar, V.; Khare, T.; Guddimalli, R.; Parveda, M.; Solymosi, K.; Suprasanna, P.; Kavi Kishor, P.B. Engineering Salinity Tolerance in Plants: Progress and Prospects. *Planta* **2020**, *251*, 76. [CrossRef]
197. Seki, M.; Kamei, A.; Yamaguchi-Shinozaki, K.; Shinozaki, K. Molecular Responses to Drought, Salinity and Frost: Common and Different Paths for Plant Protection. *Curr. Opin. Biotechnol.* **2003**, *14*, 194–199. [CrossRef] [PubMed]
198. Takagi, H.; Tamiru, M.; Abe, A.; Yoshida, K.; Uemura, A.; Yaegashi, H.; Obara, T.; Oikawa, K.; Utsushi, H.; Kanzaki, E.; et al. MutMap Accelerates Breeding of a Salt-Tolerant Rice Cultivar. *Nat. Biotechnol.* **2015**, *33*, 445–449. [CrossRef]
199. Zhang, A.; Liu, Y.; Wang, F.; Li, T.; Chen, Z.; Kong, D.; Bi, J.; Zhang, F.; Luo, X.; Wang, J.; et al. Enhanced Rice Salinity Tolerance via CRISPR/Cas9-Targeted Mutagenesis of the OsRR22 Gene. *Mol Breed.* **2019**, *39*, 47. [CrossRef] [PubMed]
200. Li, M.; Chen, R.; Jiang, Q.; Sun, X.; Zhang, H.; Hu, Z. GmNAC06, a NAC Domain Transcription Factor Enhances Salt Stress Tolerance in Soybean. *Plant Mol. Biol.* **2021**, *105*, 333–345. [CrossRef] [PubMed]
201. Wang, T.; Xun, H.; Wang, W.; Ding, X.; Tian, H.; Hussain, S.; Dong, Q.; Li, Y.; Cheng, Y.; Wang, C.; et al. Mutation of GmAITR Genes by CRISPR/Cas9 Genome Editing Results in Enhanced Salinity Stress Tolerance in Soybean. *Front. Plant Sci.* **2021**, *12*, 779598. [CrossRef]
202. Tian, H.; Chen, S.; Yang, W.; Wang, T.; Zheng, K.; Wang, Y.; Cheng, Y.; Zhang, N.; Liu, S.; Li, D.; et al. A Novel Family of Transcription Factors Conserved in Angiosperms Is Required for ABA Signalling. *Plant Cell Environ.* **2017**, *40*, 2958–2971. [CrossRef]
203. Sohail, H.; Noor, I.; Nawaz, M.A.; Ma, M.; Shireen, F.; Huang, Y.; Yang, L.; Bie, Z. Genome-Wide Identification of Plasma-Membrane Intrinsic Proteins in Pumpkin and Functional Characterization of CmoPIP1-4 under Salinity Stress. *Environ. Exp. Bot.* **2022**, *202*, 104995. [CrossRef]
204. Dong, L.; Hou, Z.; Li, H.; Li, Z.; Fang, C.; Kong, L.; Li, Y.; Du, H.; Li, T.; Wang, L.; et al. Agronomical Selection on Loss-of-Function of GIGANTEA Simultaneously Facilitates Soybean Salt Tolerance and Early Maturity. *J. Integr. Plant Biol.* **2022**, *64*, 1866–1882. [CrossRef]
205. Fu, L.; Wu, D.; Zhang, X.; Xu, Y.; Kuang, L.; Cai, S.; Zhang, G.; Shen, Q. Vacuolar H+-Pyrophosphatase HVP10 Enhances Salt Tolerance via Promoting Na^+ Translocation into Root Vacuoles. *Plant Physiol.* **2022**, *188*, 1248–1263. [CrossRef] [PubMed]
206. Alam, M.S.; Kong, J.; Tao, R.; Ahmed, T.; Alamin, M.; Alotaibi, S.S.; Abdelsalam, N.R.; Xu, J.-H. CRISPR/Cas9 Mediated Knockout of the OsbHLH024 Transcription Factor Improves Salt Stress Resistance in Rice (*Oryza sativa* L.). *Plants* **2022**, *11*, 1184. [CrossRef] [PubMed]
207. Wang, X.; Ren, P.; Ji, L.; Zhu, B.; Xie, G. OsVDE, a Xanthophyll Cycle Key Enzyme, Mediates Abscisic Acid Biosynthesis and Negatively Regulates Salinity Tolerance in Rice. *Planta* **2021**, *255*, 6. [CrossRef] [PubMed]
208. Teng, Y.; Lv, M.; Zhang, X.; Cai, M.; Chen, T. BEAR1, a BHLH Transcription Factor, Controls Salt Response Genes to Regulate Rice Salt Response. *J. Plant Biol.* **2022**, *65*, 217–230. [CrossRef]
209. Liu, S.; Liu, W.; Lai, J.; Liu, Q.; Zhang, W.; Chen, Z.; Gao, J.; Song, S.; Liu, J.; Xiao, Y. OsGLYI3, a Glyoxalase Gene Expressed in Rice Seed, Contributes to Seed Longevity and Salt Stress Tolerance. *Plant Physiol. Biochem.* **2022**, *183*, 85–95. [CrossRef]

210. Zhang, M.; Zhao, R.; Wang, H.; Ren, S.; Shi, L.; Huang, S.; Wei, Z.; Guo, B.; Jin, J.; Zhong, Y.; et al. OsWRKY28 Positively Regulates Salinity Tolerance by Directly Activating OsDREB1B Expression in Rice. *Plant Cell Rep.* **2022**, *42*, 223–234. [CrossRef]
211. Huang, J.; Liu, F.; Chao, D.; Xin, B.; Liu, K.; Cao, S.; Chen, X.; Peng, L.; Zhang, B.; Fu, S.; et al. The WRKY Transcription Factor OsWRKY54 Is Involved in Salt Tolerance in Rice. *Int. J. Mol. Sci.* **2022**, *23*, 11999. [CrossRef] [PubMed]
212. Shohat, H.; Cheriker, H.; Cohen, A.; Weiss, D. Tomato ABA-IMPORTING TRANSPORTER 1.1 Inhibits Seed Germination under High Salinity Conditions. *Plant Physiol.* **2023**, *191*, 1404–1415. [CrossRef]
213. Ding, F.; Qiang, X.; Jia, Z.; Li, L.; Hu, J.; Yin, M.; Xia, S.; Chen, B.; Qi, J.; Li, Q.; et al. Knockout of a Novel Salt Responsive Gene SlABIG1 Enhance Salinity Tolerance in Tomato. *Environ. Exp. Bot.* **2022**, *200*, 104903. [CrossRef]
214. Tran, M.T.; Doan, D.T.H.; Kim, J.; Song, Y.J.; Sung, Y.W.; Das, S.; Kim, E.; Son, G.H.; Kim, S.H.; Van Vu, T.; et al. CRISPR/Cas9-Based Precise Excision of SlHyPRP1 Domain(s) to Obtain Salt Stress-Tolerant Tomato. *Plant Cell Rep.* **2021**, *40*, 999–1011. [CrossRef] [PubMed]
215. Zhong, M.; Yue, L.; Liu, W.; Qin, H.; Lei, B.; Huang, R.; Yang, X.; Kang, Y. Genome-Wide Identification and Characterization of the Polyamine Uptake Transporter (Put) Gene Family in Tomatoes and the Role of Put2 in Response to Salt Stress. *Antioxidants* **2023**, *12*, 228. [CrossRef] [PubMed]
216. Bradl, H.B. Chapter 1 Sources and Origins of Heavy Metals. In *Interface Science and Technology*; Elsevier: Amsterdam, The Netherlands, 2005; Volume 6, pp. 1–27. ISBN 978-0-12-088381-3.
217. Li, C.; Zhou, K.; Qin, W.; Tian, C.; Qi, M.; Yan, X.; Han, W. A Review on Heavy Metals Contamination in Soil: Effects, Sources, and Remediation Techniques. *Soil Sediment Contam. Int. J.* **2019**, *28*, 380–394. [CrossRef]
218. Zoffoli, H.J.O.; do Amaral-Sobrinho, N.M.B.; Zonta, E.; Luisi, M.V.; Marcon, G.; Tolón-Becerra, A. Inputs of Heavy Metals Due to Agrochemical Use in Tobacco Fields in Brazil's Southern Region. *Environ. Monit Assess* **2013**, *185*, 2423–2437. [CrossRef] [PubMed]
219. Huang, Y.; Wang, L.; Wang, W.; Li, T.; He, Z.; Yang, X. Current Status of Agricultural Soil Pollution by Heavy Metals in China: A Meta-Analysis. *Sci. Total Environ.* **2019**, *651*, 3034–3042. [CrossRef]
220. Chowdhury, R.; Ramond, A.; O'Keeffe, L.M.; Shahzad, S.; Kunutsor, S.K.; Muka, T.; Gregson, J.; Willeit, P.; Warnakula, S.; Khan, H.; et al. Environmental Toxic Metal Contaminants and Risk of Cardiovascular Disease: Systematic Review and Meta-Analysis. *BMJ* **2018**, *362*, k3310. [CrossRef]
221. Guo, G.; Zhang, D.; Wang, Y. Probabilistic Human Health Risk Assessment of Heavy Metal Intake via Vegetable Consumption around Pb/Zn Smelters in Southwest China. *Int. J. Environ. Res. Public Health* **2019**, *16*, 3267. [CrossRef]
222. Meza-Ramírez, V.; Espinoza-Ortiz, X.; Ramírez-Verdugo, P.; Hernández-Lazcano, P.; Rojas Hermosilla, P. Pb-Contaminated Soil from Quintero-Ventanas, Chile: Remediation Using Sarcocornia Neei. *Sci. World J.* **2021**, *2021*, 2974786. [CrossRef]
223. Khanam, R.; Kumar, A.; Nayak, A.K.; Shahid, M.; Tripathi, R.; Vijayakumar, S.; Bhaduri, D.; Kumar, U.; Mohanty, S.; Panneerselvam, P.; et al. Metal(Loid)s (As, Hg, Se, Pb and Cd) in Paddy Soil: Bioavailability and Potential Risk to Human Health. *Sci. Total Environ.* **2020**, *699*, 134330. [CrossRef]
224. Wang, P.; Chen, H.; Kopittke, P.M.; Zhao, F.-J. Cadmium Contamination in Agricultural Soils of China and the Impact on Food Safety. *Environ. Pollut.* **2019**, *249*, 1038–1048. [CrossRef]
225. DalCorso, G.; Manara, A.; Furini, A. An Overview of Heavy Metal Challenge in Plants: From Roots to Shoots. *Metallomics* **2013**, *5*, 1117. [CrossRef]
226. Clemens, S. Toxic Metal Accumulation, Responses to Exposure and Mechanisms of Tolerance in Plants. *Biochimie* **2006**, *88*, 1707–1719. [CrossRef]
227. Kumar, J.; Gaur, S.; Srivastava, P.K.; Mishra, R.K.; Prasad, S.M.; Chauhan, D.K. (Eds.) *Heavy Metals in Plants: Physiological to Molecular Approach*; CRC Press: Boca Raton, FL, USA, 2022; ISBN 978-1-00-311057-6.
228. Ding, Z.; Wu, J.; You, A.; Huang, B.; Cao, C. Effects of Heavy Metals on Soil Microbial Community Structure and Diversity in the Rice (*Oryza sativa* L. Subsp. Japonica, Food Crops Institute of Jiangsu Academy of Agricultural Sciences) Rhizosphere. *Soil Sci. Plant Nutr.* **2017**, *63*, 75–83. [CrossRef]
229. Belykh, E.S.; Maystrenko, T.A.; Velegzhaninov, I.O. Recent Trends in Enhancing the Resistance of Cultivated Plants to Heavy Metal Stress by Transgenesis and Transcriptional Programming. *Mol. Biotechnol.* **2019**, *61*, 725–741. [CrossRef]
230. Williams, L.E.; Mills, R.F. P1B-ATPases—An Ancient Family of Transition Metal Pumps with Diverse Functions in Plants. *Trends Plant Sci.* **2005**, *10*, 491–502. [CrossRef]
231. Milner, M.J.; Seamon, J.; Craft, E.; Kochian, L.V. Transport Properties of Members of the ZIP Family in Plants and Their Role in Zn and Mn Homeostasis. *J. Exp. Bot.* **2013**, *64*, 369–381. [CrossRef] [PubMed]
232. Li, X.; Wu, Y.; Li, B.; He, W.; Yang, Y.; Yang, Y. Genome-Wide Identification and Expression Analysis of the Cation Diffusion Facilitator Gene Family in Turnip Under Diverse Metal Ion Stresses. *Front. Genet.* **2018**, *9*, 103. [CrossRef] [PubMed]
233. Pittman, J.K.; Hirschi, K.D. CAX-Ing a Wide Net: Cation/H^+ Transporters in Metal Remediation and Abiotic Stress Signalling. *Plant Biol. J.* **2016**, *18*, 741–749. [CrossRef]
234. Yuan, M.; Li, X.; Xiao, J.; Wang, S. Molecular and Functional Analyses of COPT/Ctr-Type Copper Transporter-like Gene Family in Rice. *BMC Plant Biol.* **2011**, *11*, 69. [CrossRef] [PubMed]
235. Chen, S.; Han, X.; Fang, J.; Lu, Z.; Qiu, W.; Liu, M.; Sang, J.; Jiang, J.; Zhuo, R. *Sedum alfredii SaNramp6* Metal Transporter Contributes to Cadmium Accumulation in Transgenic *Arabidopsis thaliana*. *Sci. Rep.* **2017**, *7*, 13318. [CrossRef] [PubMed]
236. Yao, X.; Cai, Y.; Yu, D.; Liang, G. BHLH104 Confers Tolerance to Cadmium Stress in *Arabidopsis thaliana*. *J. Integr. Plant Biol.* **2018**, *60*, 691–702. [CrossRef]

237. Luo, J.-S.; Huang, J.; Zeng, D.-L.; Peng, J.-S.; Zhang, G.-B.; Ma, H.-L.; Guan, Y.; Yi, H.-Y.; Fu, Y.-L.; Han, B.; et al. A Defensin-like Protein Drives Cadmium Efflux and Allocation in Rice. *Nat. Commun.* **2018**, *9*, 645. [CrossRef] [PubMed]
238. Verbruggen, N.; Hermans, C.; Schat, H. Molecular Mechanisms of Metal Hyperaccumulation in Plants. *New Phytol.* **2009**, *181*, 759–776. [CrossRef] [PubMed]
239. Ghuge, S.A.; Nikalje, G.C.; Kadam, U.S.; Suprasanna, P.; Hong, J.C. Comprehensive Mechanisms of Heavy Metal Toxicity in Plants, Detoxification, and Remediation. *J. Hazard. Mater.* **2023**, *450*, 131039. [CrossRef] [PubMed]
240. Tang, L.; Dong, J.; Qu, M.; Lv, Q.; Zhang, L.; Peng, C.; Hu, Y.; Li, Y.; Ji, Z.; Mao, B.; et al. Knockout of OsNRAMP5 Enhances Rice Tolerance to Cadmium Toxicity in Response to Varying External Cadmium Concentrations via Distinct Mechanisms. *Sci. Total Environ.* **2022**, *832*, 155006. [CrossRef] [PubMed]
241. Yang, C.; Zhang, Y.; Huang, C. Reduction in Cadmium Accumulation in Japonica Rice Grains by CRISPR/Cas9-Mediated Editing of OsNRAMP5. *J. Integr. Agric.* **2019**, *18*, 688–697. [CrossRef]
242. Chen, H.; Ye, R.; Liang, Y.; Zhang, S.; Liu, X.; Sun, C.; Li, F.; Yi, J. Generation of Low-Cadmium Rice Germplasms via Knockout of OsLCD Using CRISPR/Cas9. *J. Environ. Sci.* **2023**, *126*, 138–152. [CrossRef]
243. Liu, C.-X.; Yang, T.; Zhou, H.; Ahammed, G.J.; Qi, Z.-Y.; Zhou, J. The E3 Ubiquitin Ligase Gene Sl1 Is Critical for Cadmium Tolerance in *Solanum lycopersicum* L. *Antioxidants* **2022**, *11*, 456. [CrossRef]
244. Wang, F.-Z.; Chen, M.-X.; Yu, L.-J.; Xie, L.-J.; Yuan, L.-B.; Qi, H.; Xiao, M.; Guo, W.; Chen, Z.; Yi, K.; et al. OsARM1, an R2R3 MYB Transcription Factor, Is Involved in Regulation of the Response to Arsenic Stress in Rice. *Front. Plant Sci.* **2017**, *8*, 1868. [CrossRef]
245. Zhang, Y.; Chen, K.; Zhao, F.-J.; Sun, C.; Jin, C.; Shi, Y.; Sun, Y.; Li, Y.; Yang, M.; Jing, X.; et al. OsATX1 Interacts with Heavy Metal P1B-Type ATPases and Affects Copper Transport and Distribution. *Plant Physiol.* **2018**, *178*, 329–344. [CrossRef]
246. Nieves-Cordones, M.; Mohamed, S.; Tanoi, K.; Kobayashi, N.I.; Takagi, K.; Vernet, A.; Guiderdoni, E.; Périn, C.; Sentenac, H.; Véry, A.-A. Production of Low-Cs+ Rice Plants by Inactivation of the K+ Transporter OsHAK1 with the CRISPR-Cas System. *Plant J.* **2017**, *92*, 43–56. [CrossRef] [PubMed]
247. Yue, E.; Rong, F.; Liu, Z.; Ruan, S.; Lu, T.; Qian, H. Cadmium Induced a Non-Coding RNA MicroRNA535 Mediates Cd Accumulation in Rice. *J. Environ. Sci.* **2023**, *130*, 149–162. [CrossRef] [PubMed]
248. Songmei, L.; Jie, J.; Yang, L.; Jun, M.; Shouling, X.; Yuanyuan, T.; Youfa, L.; Qingyao, S.; Jianzhong, H. Characterization and Evaluation of OsLCT1 and OsNramp5 Mutants Generated Through CRISPR/Cas9-Mediated Mutagenesis for Breeding Low Cd Rice. *Rice Sci.* **2019**, *26*, 88–97. [CrossRef]
249. Chu, C.; Huang, R.; Liu, L.; Tang, G.; Xiao, J.; Yoo, H.; Yuan, M. The Rice Heavy-Metal Transporter OsNRAMP1 Regulates Disease Resistance by Modulating ROS Homoeostasis. *Plant Cell Environ.* **2022**, *45*, 1109–1126. [CrossRef] [PubMed]
250. Li, Z.; Rao, M.J.; Li, J.; Wang, Y.; Chen, P.; Yu, H.; Ma, C.; Wang, L. CRISPR/Cas9 Mutant Rice Ospmei12 Involved in Growth, Cell Wall Development, and Response to Phytohormone and Heavy Metal Stress. *Int. J. Mol. Sci.* **2022**, *23*, 16082. [CrossRef] [PubMed]
251. Lal, R. Soil Carbon Sequestration to Mitigate Climate Change. *Geoderma* **2004**, *123*, 1–22. [CrossRef]
252. White, P.J.; Brown, P.H. Plant Nutrition for Sustainable Development and Global Health. *Ann. Bot.* **2010**, *105*, 1073–1080. [CrossRef]
253. Lal, R. Crop Residues as Soil Amendments and Feedstock for Bioethanol Production. *Waste Manag.* **2008**, *28*, 747–758. [CrossRef]
254. Tilman, D.; Cassman, K.G.; Matson, P.A.; Naylor, R.; Polasky, S. Agricultural Sustainability and Intensive Production Practices. *Nature* **2002**, *418*, 671–677. [CrossRef]
255. Montgomery, D.R. Soil Erosion and Agricultural Sustainability. *Proc. Natl. Acad. Sci. USA* **2007**, *104*, 13268–13272. [CrossRef]
256. Pretty, J.; Bharucha, Z.P. Sustainable Intensification in Agricultural Systems. *Ann. Bot.* **2014**, *114*, 1571–1596. [CrossRef]
257. Wheeler, T.; von Braun, J. Climate Change Impacts on Global Food Security. *Science* **2013**, *341*, 508–513. [CrossRef]
258. Pereira, P.; Bašić, F.; Bogunovic, I.; Barcelo, D. Russian-Ukrainian War Impacts the Total Environment. *Sci. Total Environ.* **2022**, *837*, 155865. [CrossRef]
259. Pahalvi, H.N.; Rafiya, L.; Rashid, S.; Nisar, B.; Kamili, A.N. Chemical Fertilizers and Their Impact on Soil Health. In *Microbiota and Biofertilizers*; Dar, G.H., Bhat, R.A., Mehmood, M.A., Hakeem, K.R., Eds.; Springer International Publishing: Cham, Switzerland, 2021; Volume 2, pp. 1–20. ISBN 978-3-030-61009-8.
260. Pimentel, D.; Harvey, C.; Resosudarmo, P.; Sinclair, K.; Kurz, D.; McNair, M.; Crist, S.; Shpritz, L.; Fitton, L.; Saffouri, R.; et al. Environmental and Economic Costs of Soil Erosion and Conservation Benefits. *Science* **1995**, *267*, 1117–1123. [CrossRef]
261. Li, S.; Zhang, C.; Li, J.; Yan, L.; Wang, N.; Xia, L. Present and Future Prospects for Wheat Improvement through Genome Editing and Advanced Technologies. *Plant Commun.* **2021**, *2*, 100211. [CrossRef] [PubMed]
262. Liang, C.; Wang, Y.; Zhu, Y.; Tang, J.; Hu, B.; Liu, L.; Ou, S.; Wu, H.; Sun, X.; Chu, J.; et al. OsNAP Connects Abscisic Acid and Leaf Senescence by Fine-Tuning Abscisic Acid Biosynthesis and Directly Targeting Senescence-Associated Genes in Rice. *Proc. Natl. Acad. Sci. USA* **2014**, *111*, 10013–10018. [CrossRef] [PubMed]
263. Chen, J.; Zhang, Y.; Tan, Y.; Zhang, M.; Zhu, L.; Xu, G.; Fan, X. Agronomic Nitrogen-Use Efficiency of Rice Can Be Increased by Driving *OsNRT2.1* Expression with the *OsNAR2.1* Promoter. *Plant Biotechnol. J.* **2016**, *14*, 1705–1715. [CrossRef]
264. Wang, D.; Xu, T.; Yin, Z.; Wu, W.; Geng, H.; Li, L.; Yang, M.; Cai, H.; Lian, X. Overexpression of OsMYB305 in Rice Enhances the Nitrogen Uptake Under Low-Nitrogen Condition. *Front. Plant Sci.* **2020**, *11*, 369. [CrossRef] [PubMed]
265. Liu, Y.; Hu, B.; Chu, C. Toward Improving Nitrogen Use Efficiency in Rice: Utilization, Coordination, and Availability. *Curr. Opin. Plant Biol.* **2023**, *71*, 102327. [CrossRef] [PubMed]

266. Sathee, L.; Jagadhesan, B.; Pandesha, P.H.; Barman, D.; Adavi, B.S.; Nagar, S.; Krishna, G.K.; Tripathi, S.; Jha, S.K.; Chinnusamy, V. Genome Editing Targets for Improving Nutrient Use Efficiency and Nutrient Stress Adaptation. *Front. Genet.* **2022**, *13*, 1427. [CrossRef]
267. Aluko, O.O.; Kant, S.; Adedire, O.M.; Li, C.; Yuan, G.; Liu, H.; Wang, Q. Unlocking the Potentials of Nitrate Transporters at Improving Plant Nitrogen Use Efficiency. *Front. Plant Sci.* **2023**, *14*, 1074839. [CrossRef] [PubMed]
268. Lu, Y.; Zhu, J.-K. Precise Editing of a Target Base in the Rice Genome Using a Modified CRISPR/Cas9 System. *Mol. Plant* **2017**, *10*, 523–525. [CrossRef]
269. Hu, B.; Wang, W.; Ou, S.; Tang, J.; Li, H.; Che, R.; Zhang, Z.; Chai, X.; Wang, H.; Wang, Y.; et al. Variation in NRT1.1B Contributes to Nitrate-Use Divergence between Rice Subspecies. *Nat. Genet.* **2015**, *47*, 834–838. [CrossRef]
270. Zhang, J.; Zhang, H.; Li, S.; Li, J.; Yan, L.; Xia, L. Increasing Yield Potential through Manipulating of an *ARE1* Ortholog Related to Nitrogen Use Efficiency in Wheat by CRISPR/Cas9. *J. Integr. Plant Biol.* **2021**, *63*, 1649–1663. [CrossRef] [PubMed]
271. Karunarathne, S.D.; Han, Y.; Zhang, X.-Q.; Li, C. CRISPR/Cas9 Gene Editing and Natural Variation Analysis Demonstrate the Potential for HvARE1 in Improvement of Nitrogen Use Efficiency in Barley. *J. Integr. Plant Biol.* **2022**, *64*, 756–770. [CrossRef]
272. Shen, C.; Li, Q.; An, Y.; Zhou, Y.; Zhang, Y.; He, F.; Chen, L.; Liu, C.; Mao, W.; Wang, X.; et al. The Transcription Factor GNC Optimizes Nitrogen Use Efficiency and Growth by Up-Regulating the Expression of Nitrate Uptake and Assimilation Genes in Poplar. *J. Exp. Bot.* **2022**, *73*, 4778–4792. [CrossRef]
273. Yang, X.; Nong, B.; Chen, C.; Wang, J.; Xia, X.; Zhang, Z.; Wei, Y.; Zeng, Y.; Feng, R.; Wu, Y.; et al. OsNPF3.1, a Member of the NRT1/PTR Family, Increases Nitrogen Use Efficiency and Biomass Production in Rice. *Crop J.* **2023**, *11*, 108–118. [CrossRef]
274. Varshney, P.; Mikulic, P.; Vonshak, A.; Beardall, J.; Wangikar, P.P. Extremophilic Micro-Algae and Their Potential Contribution in Biotechnology. *Bioresour. Technol.* **2015**, *184*, 363–372. [CrossRef]
275. Fernández-Marín, B.; Gulías, J.; Figueroa, C.M.; Iñiguez, C.; Clemente-Moreno, M.J.; Nunes-Nesi, A.; Fernie, A.R.; Cavieres, L.A.; Bravo, L.A.; García-Plazaola, J.I.; et al. How Do Vascular Plants Perform Photosynthesis in Extreme Environments? An Integrative Ecophysiological and Biochemical Story. *Plant J.* **2020**, *101*, 979–1000. [CrossRef]
276. Barnard, D.; Casanueva, A.; Tuffin, M.; Cowan, D. Extremophiles in Biofuel Synthesis. *Environ. Technol.* **2010**, *31*, 871–888. [CrossRef]
277. Chien, A.; Edgar, D.B.; Trela, J.M. Deoxyribonucleic Acid Polymerase from the Extreme Thermophile Thermus Aquaticus. *J. Bacteriol.* **1976**, *127*, 1550. [CrossRef]
278. Marasco, R.; Rolli, E.; Ettoumi, B.; Vigani, G.; Mapelli, F.; Borin, S.; Abou-Hadid, A.F.; El-Behairy, U.A.; Sorlini, C.; Cherif, A.; et al. A Drought Resistance-Promoting Microbiome Is Selected by Root System under Desert Farming. *PLoS ONE* **2012**, *7*, e48479. [CrossRef] [PubMed]
279. Acuña-Rodríguez, I.S.; Hansen, H.; Gallardo-Cerda, J.; Atala, C.; Molina-Montenegro, M.A. Antarctic Extremophiles: Biotechnological Alternative to Crop Productivity in Saline Soils. *Front. Bioeng. Biotechnol.* **2019**, *7*, 22. [CrossRef] [PubMed]
280. Jorquera, M.A.; Graether, S.P.; Maruyama, F. Editorial: Bioprospecting and Biotechnology of Extremophiles. *Front. Bioeng. Biotechnol.* **2019**, *7*, 204. [CrossRef] [PubMed]
281. Cavieres, L.A.; Sáez, P.; Sanhueza, C.; Sierra-Almeida, A.; Rabert, C.; Corcuera, L.J.; Alberdi, M.; Bravo, L.A. Ecophysiological Traits of Antarctic Vascular Plants: Their Importance in the Responses to Climate Change. *Plant Ecol.* **2016**, *217*, 343–358. [CrossRef]
282. Morales, M.; Munné-Bosch, S. Oxidative Stress: A Master Regulator of Plant Trade-Offs? *Trends Plant Sci.* **2016**, *21*, 996–999. [CrossRef]
283. Orellana, R.; Macaya, C.; Bravo, G.; Dorochesi, F.; Cumsille, A.; Valencia, R.; Rojas, C.; Seeger, M. Living at the Frontiers of Life: Extremophiles in Chile and Their Potential for Bioremediation. *Front. Microbiol.* **2018**, *9*, 2309. [CrossRef]
284. Oh, D.-H.; Dassanayake, M.; Bohnert, H.J.; Cheeseman, J.M. Life at the Extreme: Lessons from the Genome. *Genome Biol.* **2013**, *13*, 241. [CrossRef]
285. Barak, S.; Farrant, J.M. Extremophyte Adaptations to Salt and Water Deficit Stress. *Funct. Plant Biol.* **2016**, *43*, v–x. [CrossRef]
286. Lindgren, A.R.; Buckley, B.A.; Eppley, S.M.; Reysenbach, A.L.; Stedman, K.M.; Wagner, J.T. Life on the Edge—The Biology of Organisms Inhabiting Extreme Environments: An Introduction to the Symposium. *Integr. Comp. Biol.* **2016**, *56*, 493–499. [CrossRef]
287. Bechtold, U.; Field, B. Molecular Mechanisms Controlling Plant Growth during Abiotic Stress. *J. Exp. Bot.* **2018**, *69*, 2753–2758. [CrossRef] [PubMed]
288. Waqas, M.A.; Kaya, C.; Riaz, A.; Farooq, M.; Nawaz, I.; Wilkes, A.; Li, Y. Potential Mechanisms of Abiotic Stress Tolerance in Crop Plants Induced by Thiourea. *Front. Plant Sci.* **2019**, *10*, 1336. [CrossRef] [PubMed]
289. Ostria-Gallardo, E.; Larama, G.; Berríos, G.; Fallard, A.; Gutiérrez-Moraga, A.; Ensminger, I.; Bravo, L.A. A Comparative Gene Co-Expression Analysis Using Self-Organizing Maps on Two Congener Filmy Ferns Identifies Specific Desiccation Tolerance Mechanisms Associated to Their Microhabitat Preference. *BMC Plant Biol.* **2020**, *20*, 56. [CrossRef] [PubMed]
290. Costa-Silva, J.; Domingues, D.; Lopes, F.M. RNA-Seq Differential Expression Analysis: An Extended Review and a Software Tool. *PLoS ONE* **2017**, *12*, e0190152. [CrossRef]
291. Ali, A.; Cheol Park, H.; Aman, R.; Ali, Z.; Yun, D.-J. Role of HKT1 in Thellungiella Salsugine *a*, a Model Extremophile Plant. *Plant Signal. Behav.* **2013**, *8*, e25196. [CrossRef] [PubMed]

292. Ali, A.; Khan, I.U.; Jan, M.; Khan, H.A.; Hussain, S.; Nisar, M.; Chung, W.S.; Yun, D.-J. The High-Affinity Potassium Transporter EpHKT1;2 From the Extremophile Eutrema Parvula Mediates Salt Tolerance. *Front. Plant Sci.* **2018**, *9*, 1108. [CrossRef] [PubMed]
293. Wang, W.-Y.; Liu, Y.-Q.; Duan, H.-R.; Yin, X.-X.; Cui, Y.-N.; Chai, W.-W.; Song, X.; Flowers, T.J.; Wang, S.-M. SsHKT1;1 Is Coordinated with SsSOS1 and SsNHX1 to Regulate Na^+ Homeostasis in Suaeda Salsa under Saline Conditions. *Plant Soil* **2020**, *449*, 117–131. [CrossRef]
294. Boulc'h, P.-N.; Caullireau, E.; Faucher, E.; Gouerou, M.; Guérin, A.; Miray, R.; Couée, I. Abiotic Stress Signalling in Extremophile Land Plants. *J. Exp. Bot.* **2020**, *71*, 5771–5785. [CrossRef]
295. Flowers, T.J.; Colmer, T.D. Salinity Tolerance in Halophytes. *New Phytol.* **2008**, *179*, 945–963. [CrossRef]
296. Qiu, Q.; Ma, T.; Hu, Q.; Liu, B.; Wu, Y.; Zhou, H.; Wang, Q.; Wang, J.; Liu, J. Genome-Scale Transcriptome Analysis of the Desert Poplar, Populus Euphratica. *Tree Physiol.* **2011**, *31*, 452–461. [CrossRef] [PubMed]
297. Vu, T.V.; Sivankalyani, V.; Kim, E.-J.; Doan, D.T.H.; Tran, M.T.; Kim, J.; Sung, Y.W.; Park, M.; Kang, Y.J.; Kim, J.-Y. Highly Efficient Homology-Directed Repair Using CRISPR/Cpf1-Geminiviral Replicon in Tomato. *Plant Biotechnol. J.* **2020**, *18*, 2133–2143. [CrossRef]
298. Zhao, H.; Wang, L.; Zhao, F.-J.; Wu, L.; Liu, A.; Xu, W. SpHMA1 Is a Chloroplast Cadmium Exporter Protecting Photochemical Reactions in the Cd Hyperaccumulator Sedum Plumbizincicola. *Plant Cell Environ.* **2019**, *42*, 1112–1124. [CrossRef]
299. Biswas, P.; Anand, U.; Ghorai, M.; Pandey, D.K.; Jha, N.K.; Behl, T.; Kumar, M.; Chauhan, R.; Shekhawat, M.S.; Dey, A. Unraveling the Promise and Limitations of CRISPR/Cas System in Natural Product Research: Approaches and Challenges. *Biotechnol. J.* **2022**, *17*, 2100507. [CrossRef] [PubMed]
300. Ahmad, S.; Wei, X.; Sheng, Z.; Hu, P.; Tang, S. CRISPR/Cas9 for Development of Disease Resistance in Plants: Recent Progress, Limitations and Future Prospects. *Brief. Funct. Genom.* **2020**, *19*, 26–39. [CrossRef]
301. Chen, J.; Li, S.; He, Y.; Li, J.; Xia, L. An Update on Precision Genome Editing by Homology-Directed Repair in Plants. *Plant Physiol.* **2022**, *188*, 1780–1794. [CrossRef]
302. Gong, Z.; Cheng, M.; Botella, J.R. Non-GM Genome Editing Approaches in Crops. *Front. Genome Ed.* **2021**, *3*, 40. [CrossRef] [PubMed]
303. Chen, Z.; Debernardi, J.M.; Dubcovsky, J.; Gallavotti, A. Recent Advances in Crop Transformation Technologies. *Nat. Plants* **2022**, *8*, 1343–1351. [CrossRef] [PubMed]
304. Lowe, K.; Wu, E.; Wang, N.; Hoerster, G.; Hastings, C.; Cho, M.-J.; Scelonge, C.; Lenderts, B.; Chamberlin, M.; Cushatt, J.; et al. Morphogenic Regulators *Baby Boom* and *Wuschel* Improve Monocot Transformation. *Plant Cell* **2016**, *28*, 1998–2015. [CrossRef]
305. Cao, X.; Xie, H.; Song, M.; Lu, J.; Ma, P.; Huang, B.; Wang, M.; Tian, Y.; Chen, F.; Peng, J.; et al. Cut–Dip–Budding Delivery System Enables Genetic Modifications in Plants without Tissue Culture. *Innovation* **2023**, *4*, 100345. [CrossRef]
306. Lee, T.G.; Hutton, S.F. Field Evaluation of CRISPR-Driven Jointless Pedicel Fresh-Market Tomatoes. *Agronomy* **2021**, *11*, 1957. [CrossRef]
307. Neequaye, M.; Stavnstrup, S.; Harwood, W.; Lawrenson, T.; Hundleby, P.; Irwin, J.; Troncoso-Rey, P.; Saha, S.; Traka, M.H.; Mithen, R.; et al. CRISPR-Cas9-Mediated Gene Editing of MYB28 Genes Impair Glucoraphanin Accumulation of Brassica Oleracea in the Field. *CRISPR J.* **2021**, *4*, 416–426. [CrossRef]
308. Shabbir, R.; Singhal, R.K.; Mishra, U.N.; Chauhan, J.; Javed, T.; Hussain, S.; Kumar, S.; Anuragi, H.; Lal, D.; Chen, P. Combined Abiotic Stresses: Challenges and Potential for Crop Improvement. *Agronomy* **2022**, *12*, 2795. [CrossRef]
309. Metje-Sprink, J.; Sprink, T.; Hartung, F. Genome-Edited Plants in the Field. *Curr. Opin. Biotechnol.* **2020**, *61*, 1–6. [CrossRef]
310. Faure, J.-D.; Napier, J.A. Europe's First and Last Field Trial of Gene-Edited Plants? *eLife* **2018**, *7*, e42379. [CrossRef]
311. Raffan, S.; Oddy, J.; Mead, A.; Barker, G.; Curtis, T.; Usher, S.; Burt, C.; Halford, N.G. Field Assessment of Genome-edited, Low Asparagine Wheat: Europe's First CRISPR Wheat Field Trial. *Plant Biotechnol. J.* **2023**. [CrossRef] [PubMed]
312. Medvedieva, M.O.; Blume, Y.B. Legal Regulation of Plant Genome Editing with the CRISPR/Cas9 Technology as an Example. *Cytol. Genet.* **2018**, *52*, 204–212. [CrossRef]
313. Kuzma, J. Social Concerns and Regulation of Cisgenic Crops in North America. In *Cisgenic Crops: Safety, Legal and Social Issues*; Chaurasia, A., Kole, C., Eds.; Concepts and Strategies in Plant Sciences; Springer International Publishing: Cham, Switzerland, 2023; pp. 179–194. ISBN 978-3-031-10721-4.
314. Ahmad, A.; Munawar, N.; Khan, Z.; Qusmani, A.T.; Khan, S.H.; Jamil, A.; Ashraf, S.; Ghouri, M.Z.; Aslam, S.; Mubarik, M.S.; et al. An Outlook on Global Regulatory Landscape for Genome-Edited Crops. *Int. J. Mol. Sci.* **2021**, *22*, 11753. [CrossRef] [PubMed]
315. Gatica-Arias, A. The Regulatory Current Status of Plant Breeding Technologies in Some Latin American and the Caribbean Countries. *Plant Cell Tiss. Organ Cult.* **2020**, *141*, 229–242. [CrossRef]
316. Sprink, T.; Wilhelm, R.; Hartung, F. Genome Editing around the Globe: An Update on Policies and Perceptions. *Plant Physiol.* **2022**, *190*, 1579–1587. [CrossRef]
317. Wunderlich, S.; Gatto, K.A. Consumer Perception of Genetically Modified Organisms and Sources of Information. *Adv. Nutr.* **2015**, *6*, 842–851. [CrossRef]
318. Ortega, D.L.; Lin, W.; Ward, P.S. Consumer Acceptance of Gene-Edited Food Products in China. *Food Qual. Prefer.* **2022**, *95*, 104374. [CrossRef]
319. da Silva Santos, C.R.; Teixeira, S.M.; Cruz, J.E.; Bron, P.C. Perception of Producers and Consumers on the Adoption of Genetically Modified Food: The Case of the Transgenic Bean BRSFC401 RMD. *Rev. Econ. Sociol. Rural* **2023**, *61*, e25027. [CrossRef]

320. Bearth, A.; Kaptan, G.; Kessler, S.H. Genome-Edited versus Genetically-Modified Tomatoes: An Experiment on People's Perceptions and Acceptance of Food Biotechnology in the UK and Switzerland. *Agric. Hum. Values* **2022**, *39*, 1117–1131. [CrossRef]
321. Menz, J.; Modrzejewski, D.; Hartung, F.; Wilhelm, R.; Sprink, T. Genome Edited Crops Touch the Market: A View on the Global Development and Regulatory Environment. *Front. Plant Sci.* **2020**, *11*, 586027. [CrossRef]

plants

MDPI

Article

Genome and Transcriptome Identification of a Rice Germplasm with High Cadmium Uptake and Translocation

Jin-Song Luo [1,2,*], Bao Guo [1,2], Yiqi He [1,2], Chun-Zhu Chen [3], Yong Yang [1,2] and Zhenhua Zhang [1,2,*]

1 College of Resources, Hunan Agricultural University, Changsha 410128, China
2 Hunan Provincial Key Laboratory of Farmland Pollution Control and Agricultural Resources Use, Hunan Provincial Key Laboratory of Nutrition in Common University, National Engineering Laboratory on Soil and Fertilizer Resources Efficient Utilization, Changsha 410128, China
3 3D Medicines, Block A, Building 2, No.158 Xinjunhuan Road, Pujiang Town, Minhang District, Shanghai 201210, China
* Correspondence: 0609020317@163.com (J.-S.L.); zhzh1468@163.com (Z.Z.)

Abstract: The safe production of food on Cd-polluted land is an urgent problem to be solved in South China. Phytoremediation or cultivation of rice varieties with low Cd are the main strategies to solve this problem. Therefore, it is very important to clarify the regulatory mechanism of Cd accumulation in rice. Here, we identified a rice variety with an unknown genetic background, YSD, with high Cd accumulation in its roots and shoots. The Cd content in the grains and stalks were 4.1 and 2.8 times that of a commonly used japonica rice variety, ZH11, respectively. The Cd accumulation in the shoots and roots of YSD at the seedling stage was higher than that of ZH11, depending on sampling time, and the long-distance transport of Cd in the xylem sap was high. Subcellular component analysis showed that the shoots, the cell wall, organelles, and soluble fractions of YSD, showed higher Cd accumulation than ZH11, while in the roots, only the cell wall pectin showed higher Cd accumulation. Genome-wide resequencing revealed mutations in 22 genes involved in cell wall modification, synthesis, and metabolic pathways. Transcriptome analysis in Cd-treated plants showed that the expression of pectin methylesterase genes was up-regulated and the expression of pectin methylesterase inhibitor genes was down-regulated in YSD roots, but there were no significant changes in the genes related to Cd uptake, translocation, or vacuole sequestration. The yield and tiller number per plant did not differ significantly between YSD and ZH11, but the dry weight and plant height of YSD were significantly higher than that of ZH11. YSD provides an excellent germplasm for the exploration of Cd accumulation genes, and the cell wall modification genes with sequence- and expression-level variations provide potential targets for phytoremediation.

Keywords: cadmium accumulation; cell wall; pectin methylesterase; rice; phytoremediation; multi-omics

1. Introduction

Owing to the rapid development of industrialization in both urban and rural areas of China, the Cd pollution levels in some farmlands exceed China's national standard of 0.2 mg kg^{-1} (GB2762-2005), leading to Cd rice production, which poses a potential threat to human health. Cd in the human body can affect human reproduction and cause Itai-itai Disease [1]. Breeding low-Cd crop varieties and the phytoremediation of Cd-contaminated farmland are two important coping strategies [2]. Therefore, it is of great practical significance to elucidate the accumulation mechanism of Cd in rice.

Phytoremediation, in which metal pollutants of soil, such as Cd, can be absorbed by natural or artificially bred special plants through their roots to remove Cd or reduce their bioavailability in soil [3,4], which mainly includes plant extraction, plant stabilization, plant filtration, and plant stimulation [5,6]. Such methods are economically feasible, environmentally friendly, and highly applicable.

At present, Cd uptake and transport in rice is mainly thought to be conducted by the transporters of some essential metal elements. For example, the NRAMP (natural resistance-associated macrophage protein) family, NRAMP5, which is responsible for Mn uptake [7], NRAMP1, which is responsible for Fe and Mn uptake [8,9], and Zip5/9, which is responsible for Zn uptake, all contribute to the uptake of Cd [10,11]. Under the action of the P1B-type ATPases, known as heavy metal ATPases (HMAs), HMA3, some of the Cd taken up by the roots is stored in vacuoles [12,13]. Under the action of HMA2, some of it is transported to the shoots via long-distance transport in the xylem sap [14,15]. The Cd in the shoots is distributed from the nodes to the grains under the action of a low-affinity cation transporter, LCT1 [16]. Recently, OsCD1, belonging to the major facilitator superfamily, was found to be involved in root Cd uptake and contribute to grain accumulation in rice. Natural variation in OsCd1 with a missense mutation, Val449Asp, is responsible for the divergence of rice grain Cd accumulation between indica and japonica [17]. The rice defencin gene, CAL1, can chelate cytoplasmic Cd and efflux it into the xylem sap for long-distance transport to the leaves for storage [18].

The cell wall plays an important role in the detoxification and hyperaccumulation of Cd in *Sedum plumbizincicola* [19]. It has been reported that the cell wall can bind Cd, preventing Cd from entering the cytoplasm and reducing the toxicity of Cd in Arabidopsis [20]. When pectin in the cell wall is demethylated, it exposes the Cd ions to more carboxyl binding, playing a major role in Cd detoxification in *Brassica napus* [21]. The application of boron can decrease Cd content in plant tissues and improve the antioxidative system whilst increasing the Cd content in the cell wall, thus alleviating Cd and oxidation stress in rice [22]. Salicylic acid remarkably decreased Cd concentrations in roots and shoots of rice seedlings, and increased the distribution ratio of Cd in the root cell wall fraction through regulating root cell wall composition via nitric oxide signaling [23]. This study aimed to identify the physiological and molecular mechanisms of a rice germplasm YSD with high Cd uptake and translocation, which potentially provide valuable materials and genetic resources for phytoremediation.

2. Results

2.1. YSD Accumulated More Cd in the Straw and Grains Than ZH11

At the harvest stage of the rice, the Cd content in the rice straw and grains of YSD was significantly higher than that of ZH11, and the Cd accumulation in the rice straw and grains of YSD was 2.8 times and 4.1 times that of ZH11, respectively (Figure 1A,B). However, there were no significant differences in other elements in rice straw and grains, except for Ca, which was 54% lower in the YSD straw than in the ZH11 straw (Figure 1C,D).

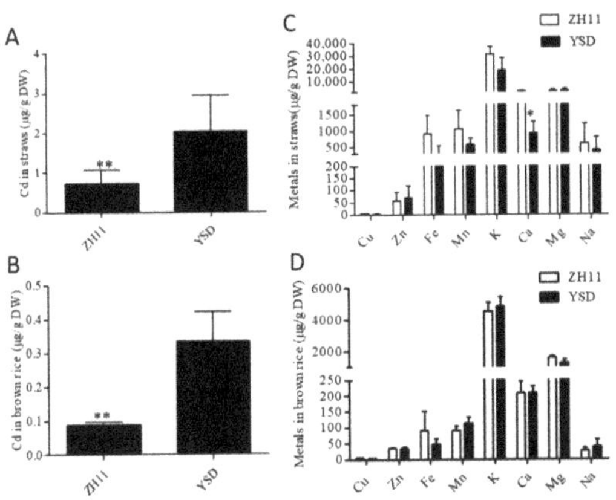

Figure 1. Cd content in YSD straw and grain was significantly higher than that in ZH11. Determination of Cd and other metal elements in YSD and ZH11 straw (**A**,**C**) and brown rice (**B**,**D**) at ripening stage

stage grow in 0.4 mg/kg Cd-contaminated field. Data are presented as the means (±SD), n = 5. Significant differences were determined using the Student's t-test: * $p < 0.05$, ** $p < 0.01$.

2.2. Time-Dependent Higher Cd Accumulation in the Shoots and Roots of YSD

To further determine the phenotype of the high Cd accumulation in YSD, we treated rice with 10 μM Cd for 1, 3, and 5 days. The results showed that the Cd accumulation in the shoots and roots of YSD was significantly and time-dependently higher than that of ZH11. The Cd accumulation of YSD was 1.4, 1.6, and 1.7 times that of ZH11 in the shoots after Cd treatment for 1, 3, and 5 days, respectively (Figure 2A); it was 1.2, 1.2, and 1.5 times that of ZH11 in the roots (Figure 2B). We detected the Cd content in xylem sap treated with 10 μM Cd for 1 day and found that the concentration of Cd in the xylem sap of YSD was significantly higher (2.1 times higher) than that of ZH11 (Figure 2C). In addition, the Cd content at the junction of the root and stem of YSD was 25% higher than that of ZH11 (Figure 2D).

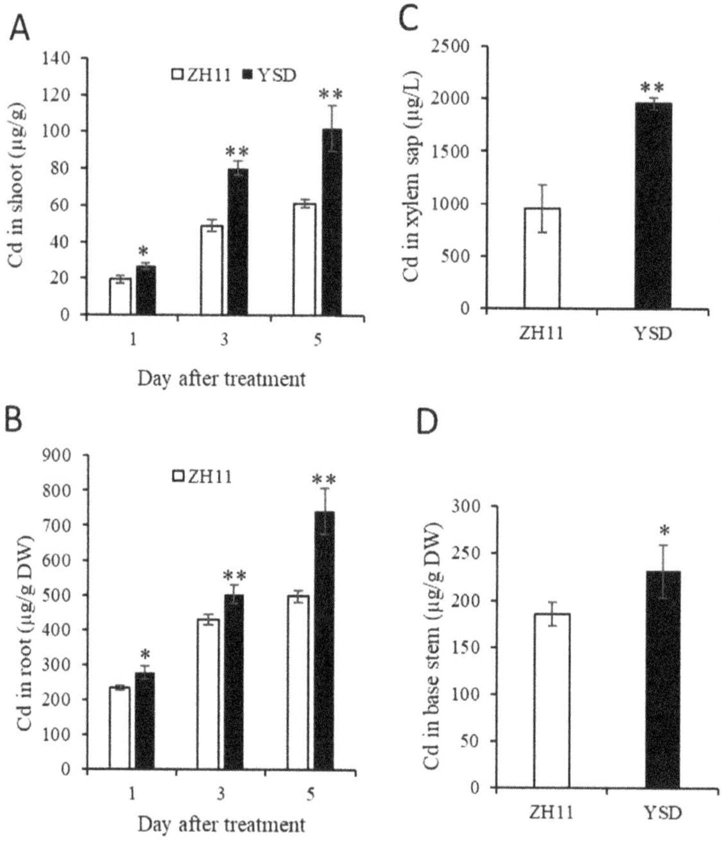

Figure 2. Time-dependent higher Cd accumulation in shoots and roots of YSD than ZH11. After 4 weeks of hydroponics, rice seedlings were treated with 10 μM Cd for 1, 3, and 5 days, Cd content in the shoot (**A**), root (**B**), basal stem (**C**) and xylem sap (**D**) were determined by ICP-MS. Data are presented as the means (±SD), n = 5. Significant differences were determined using the Student's t-test: * $p < 0.05$, ** $p < 0.01$.

2.3. Cd Content in Various Subcellular and Cell Wall Components of YSD and ZH11

The Cd content in the shoot cell wall, organelle, and soluble fractions of YSD was 2.3, 1.7, and 1.5 times higher, respectively, than that of ZH11 (Figure 3A–C), representing a significant difference. The Cd content in the YSD roots was 59.8% higher than that in the ZH11 roots for the cell wall fraction, but there were no significant differences in the Cd content in the organelle and soluble fractions (Figure 3A–C). When we detected the Cd content in different components of the cell wall, the results showed that the Cd content in the shoot cellulose, hemicellulose, and pectin of YSD was 3.4, 1.9, and 3.6 times higher, respectively, than that of ZH11, representing a significant difference (Figure 3D–F). The Cd content in the pectin in the YSD roots was 223% higher than that in ZH11 roots, but there were no significant differences in the Cd content in the cellulose or hemicellulose in the roots (Figure 3D–F). These results indicate that the cell wall, organelles, and soluble fractions of YSD shoots all possess characteristics of high Cd accumulation, while the high Cd accumulation in the roots may be mainly determined by the cell wall pectin.

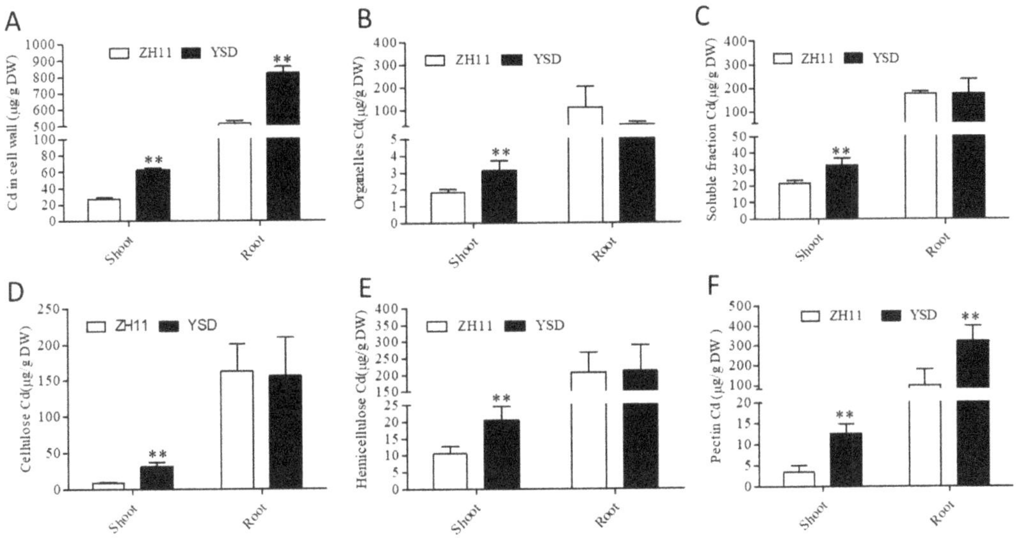

Figure 3. Cd content in different subcellular and cell wall components of YSD and ZH11. After 4 weeks of hydroponics, rice seedlings were treated with 10 μM Cd for 3 days, Cd content in cell wall (**A**), organelles (**B**), soluble part (**C**), cellulose (**D**), hemicellulose (**E**), and pectin (**F**) of the shoot and root were determined by ICP-MS. Data are presented as the means (±SD), n = 3. Significant differences were determined using the Student's *t*-test: ** $p < 0.01$.

2.4. Genome-Level Variation of YSD Compared with ZH11

To study the effect of YSD genome-level variation on Cd accumulation, genome resequencing was conducted. The results showed that, compared with ZH11, there were 968,795 SNP sites and 185,084 InDel sites in the whole YSD genome (Figure 4A). These SNPs resulted in non-synonymous mutations in the coding regions of 5006 genes, the premature termination of 363 genes, the termination codon loss of 112 genes, and synonymous mutations in the coding regions of 3844 genes (Figure 4B). The InDel loci resulted in frameshift mutations in the coding regions of 1498 genes, non-frameshift mutations in 1258 genes, the premature termination of 84 genes, and the loss of termination codons in 12 genes (Figure 4C). KEGG pathway analysis of these affected genes revealed that 156 are carbohydrate-metabolism-related genes, 92 are lipid-metabolism-related genes, and 53 are membrane transport- and catalysis-related genes (Figure 4D). GO enrichment found 22 mutated genes related to cell wall synthesis and modification (Table 1).

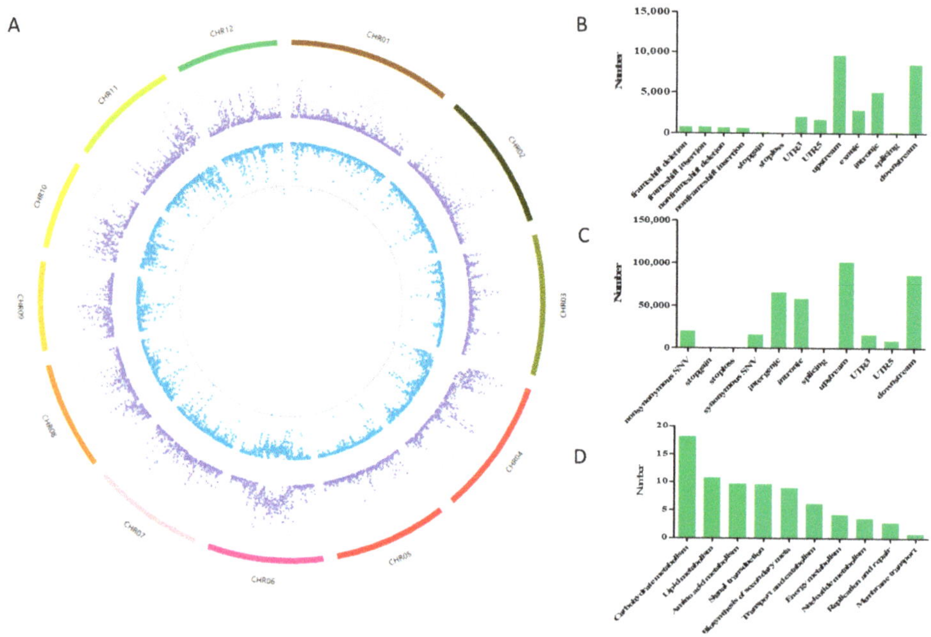

Figure 4. Genome-level variation of YSD compared to ZH11. Fresh leaves of 10-day-old plants were sampled for isolation of genomic DNA (gDNA) to perform whole-genome resequencing. (**A**) Genome variation Circos diagram. Circle 1: The chromosome; Circle 2: The purple locus represents the distribution of SNP density in the genome; Circle 3: Blue sites represent fractions of genomic InDel density. (**B**) Statistical map of SNP annotation results at gene level. (**C**) Map of InDel annotation results at gene level, and (**D**) pathway analysis of mutated gene loci.

Table 1. Cell-wall-associated variation in gene annotation.

GO Annotation	Number	Mutant Gene ID
Cell wall biogenesis	4	*Os03g0172000; Os03g0172200; Os03g0172700; Os03g0172100*
Cell wall modification	8	*Os02g0558200; Os01g0160100; Os01g0188400; Os01g0857400; Os03g0584224; Os01g0168600; Os01g0159800; Os04g0600800*
Plant-type cell wall organization	6	*Os02g0156600; Os01g0918400; Os01g0917900; Os02g0816200; Os02g0139300; Os04g0449000*
Cell wall macromolecule catabolic process	4	*Os01g0357800; Os04g0116200; Os04g0166000; Os01g0224000*

2.5. Differentially Expressed Genes between YSD and ZH11 Roots

To reveal the differentially expressed genes in response to Cd in YSD roots, transcriptome sequencing was performed on ZH11 and YSD roots. Principal component analysis showed that the first principal component could explain 84.83% and the second 8.7% of the gene expression differences in the roots of the two cultivars (Figure 5A). There were 602 up-regulated and 297 down-regulated genes in the ZH11 roots, and 584 up-regulated and 297 down-regulated genes in the YSD roots when exposed to Cd treatment (Figure 5B). Compared to ZH11, 972 genes in YSD were up-regulated and 1073 down-regulated under Cd treatment (Figure 5B). Under control treatment conditions, 965 YSD genes were up-regulated and 1207 down-regulated (Figure 5C). A Venn diagram showed that 354 genes responded to the change in Cd in both YSD and ZH11 (Figure 5C), while 545 genes responded only in ZH11, and 527 genes responded only in YSD. Under both, Cd treatment and control conditions, 1419 differentially changed genes were consistent in YSD and ZH11.

There were 778 genes that were differentially expressed only under control conditions, and 625 genes that responded only to Cd treatment. Finally, 24 genes differed in all four comparisons (Figure 5C). Figure 5D shows a heat map of differentially expressed genes under Cd treatment. We conducted GO enrichment analysis on the differentially expressed genes, and the results showed that the differentially expressed genes in the ZH11 roots were mainly enriched in the process of nicotinamide synthesis, ethylene signal transduction, and ammonium transmembrane transport. The differentially expressed genes were mainly located in the extracellular apoplast (Figure 5E). The differentially expressed genes in the YSD roots were mainly enriched in GSH metabolism, iron and superoxide balance, and the differentially expressed genes were mainly located in the extracellular region of the cell wall (Figure 5F).

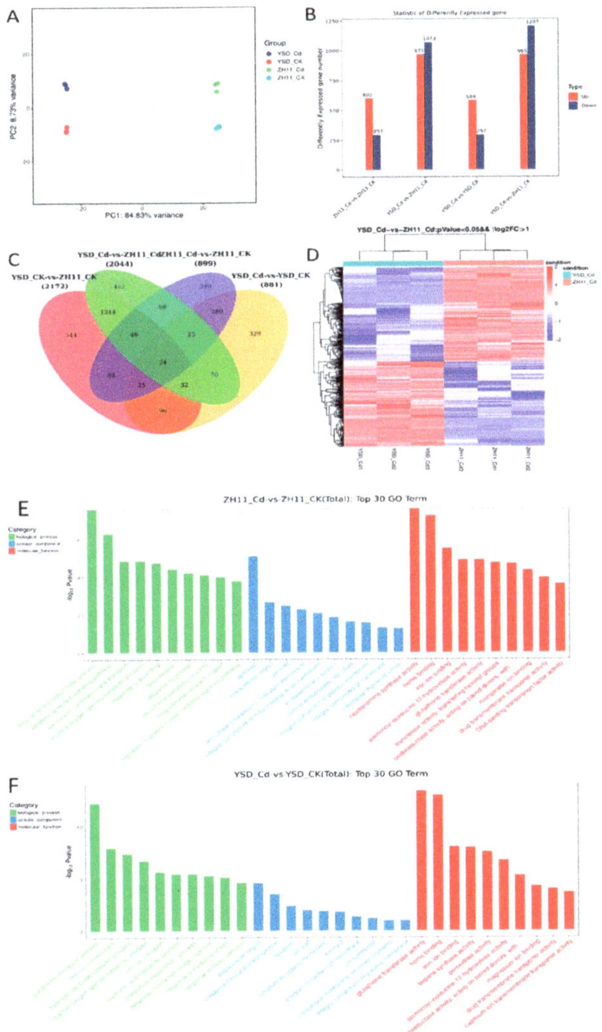

Figure 5. Analysis of transcriptome data from YSD and ZH11 roots. After 4 weeks of hydroponics, rice seedlings were treated with 0, 10 μM Cd for 3 days, and roots were taken for RNA_seq analysis.

(**A**) Principal component analysis. (**B**) Number of up_ and down_regulated differentially expressed genes. (**C**) Venn diagram analysis of differentially expressed genes, (**D**) Heat map of differentially expressed genes. (**E**) GO annotation of differentially expressed genes in response to Cd stress in the roots of ZH11. (**F**) GO annotation of differentially expressed genes in response to Cd stress in the roots of YSD.

2.6. Nicotinamide Promotes Cd Translocation in Rice

According to the GO enrichment results of ZH11 in response to Cd, we found that the expression of the nicotinamide synthesis genes, NAS1 and NAS2, in ZH11 under Cd treatment was 3.5 and 3.6 times higher, respectively, than that in YSD (Supplementary Figure S1A,B). The nicotinamide content in the ZH11 roots and shoots was 37% and 21% higher, respectively, than that in YSD. In ZH11 (Supplementary Figure S1C), the expression of YSL2, which is responsible for NA–Fe transport, was 592 times higher than that of YSD (Supplementary Figure S1D). Based on the Cd accumulation characteristics of ZH11 and YSD, as well as this result, we speculated that nicotinamide synthesis and YSL2-mediated metal chelate transport may be negatively correlated with Cd accumulation. We designed an in vitro experiment involving the addition of 0, 10, and 100 μM nicotinamide to a rice solution, and found that in vitro, 10 and 100 μM nicotinamide treatment increased the Cd accumulation in ZH11 shoots by 175% and 195% (Supplementary Figure S1E), and in YSD shoots by 165% and 175%, respectively (Supplementary Figure S1E). This is contrary to our speculation, indicating that the differential accumulation characteristics of YSD and ZH11 Cd may not be caused by the differences in nicotinamide synthesis.

According to the GO enrichment results of Cd in YSD, we found that GSH transferase gene expression in YSD under Cd treatment was 2–3 times higher than that in ZH11, and the GSH content in the roots and shoots was significantly lower (37% and 30%) than that in ZH11 (Supplementary Figure S2). This result indicates that GSH metabolism may be involved in the different Cd accumulation and tolerance characteristics of YSD and ZH11.

2.7. No Significant Changes in Cd Uptake and Transport-Related Genes

In addition, we analyzed the expression of the currently reported genes responsible for Cd uptake, translocation, and vacuolar storage, and found that the genes responsible for Cd uptake (*NRMP1/NRAMP5/ZIP5*), long-distance transport (*HMA2/CAL1*), and vacuolar storage (*HMA3*) did not differ significantly between YSD and ZH11 (Supplementary Figure S3), except for the significantly down-regulated expression of the Cd uptake gene *ZIP9* in YSD roots (Supplementary Figure S3). These results indicate that the high Cd accumulation characteristics of YSD may not be caused by changes in the expression of these genes.

2.8. Expression of Pectin Modification and Metabolism-Related Genes

Previous studies found that differences in Cd accumulation between YSD and ZH11 roots were mainly determined by cell wall pectin. We analyzed the expression of genes related to pectin modification and catabolism. In YSD roots, the expression levels of PME genes (*PME22/PME35/PME67/PME32*) and pectin-synthesis-related genes (*Os05g0385116/GAUT3/BYM1*) were significantly higher than in ZH11 roots (Figure 6A), while those of PME-inhibitor genes (*PMEI8/PMEI28*) and pectin cleavage-related genes (*Os05g0213900/PAE5*) were significantly lower (Figure 6A). The PME activity was 51% higher in YSD than in ZH11 roots after Cd treatment (Figure 6B). These results reveal that pectin modification plays an important role in determining root Cd accumulation.

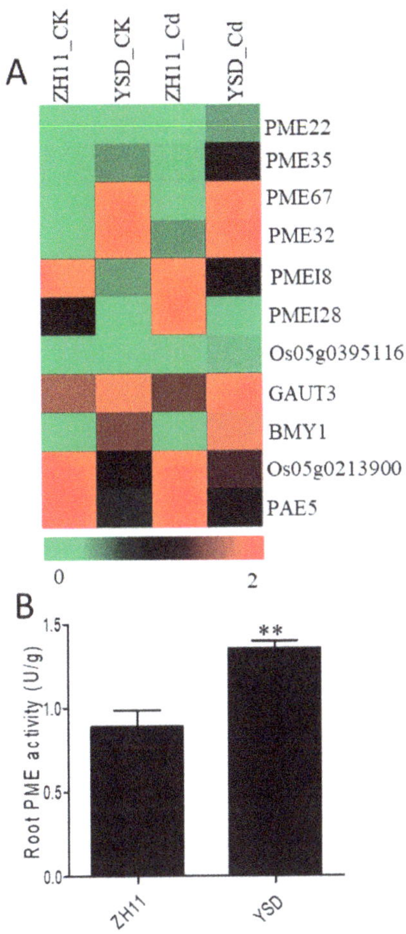

Figure 6. Differentially expressed genes (DEGs) involved in cell wall pectin metabolism and PME activity between YSD and ZH11 root under Cd treatments. (**A**) Transcriptional profiling of pectin methylesterase (PME 22/PME35/PME67/PME32), pectin methylesterase inhibitor protein (PMEI8/PMEI28), pectate lyase (Os05g0213900/PAE5) and pectin biosynthesis (CAUT3/BMY1). (**B**) PME activity in leaves of 30-day-old rice treated with 10 μM $CdCl_2$ for 3 days. Data are presented as the means of 3 independent biological replicates (n = 3) and vertical bars represent the SD., ** indicates significant differences between the cultivars at $p < 0.01$.

2.9. Comparison of Agronomic Characteristics between YSD and ZH11

We compared the agronomic traits of YSD and ZH11. The plant height of YSD was 175% higher than that of ZH11 (Figure 7A) and the straw dry weight of YSD was 3.15 times than that of ZH11 (Figure 7B). There were no significant differences in the tiller number or yield per plant between the two cultivars (Figure 7C,D).

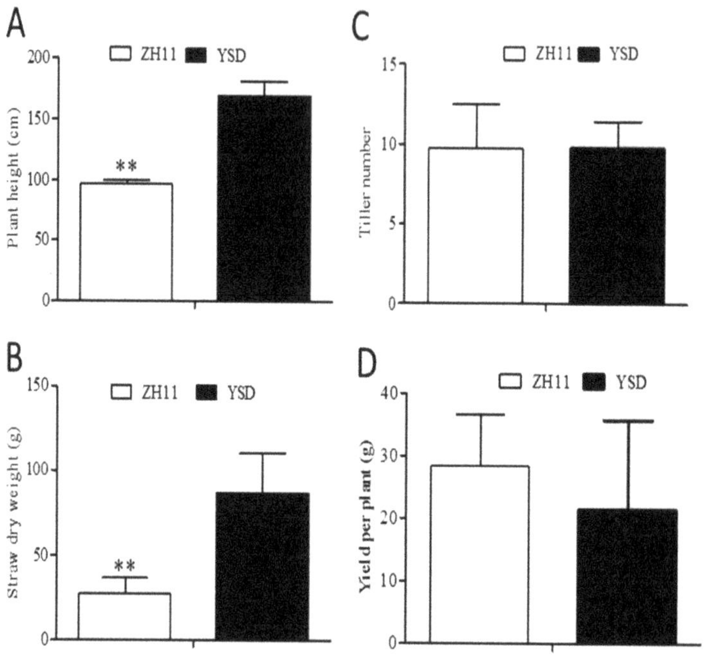

Figure 7. Analysis of agronomic characters of YSD and ZH11. Measurement of rice height (**A**), straw dry weight (**B**), tiller number (**C**) and yield per plant (**D**) at harvest stage. Data are presented as the means (±SD), n = 10. Significant differences were determined using the Student's *t*-test: ** $p < 0.01$.

3. Discussion

A wild rice variety, YSD, with high Cd accumulation at the seedling and harvest stages, was identified (Figures 1 and 2). Agronomic trait analysis showed that the plant height and straw dry weight of YSD were significantly higher than that of ZH11 (Figure 7). Genome resequencing showed that the mutant genes were significantly enriched in genes related to carbon metabolism (Figure 4D), which might be the reason for the higher biomass of YSD.

The cell wall is the first defense barrier for plant cells. Many studies have shown that different components of the cell wall can chelate Cd and reduce the toxicity of Cd to cells [19–21]. Compared to ZH11, the cell wall components, soluble fraction, and organelles of YSD shoots showed obvious high Cd accumulation characteristics, while only the cell walls in the root showed high Cd accumulation (Figure 3). Genome-wide resequencing showed that 22 genes related to the synthesis and modification of YSD cell walls were mutated (Table 1). Pectin-modifying genes in the roots were significantly changed: the expressions of pectin methylesterase genes and pectin synthesis genes were up-regulated, while those of pectin-inhibiting genes and lytic-related genes were down-regulated (Figure 6). These results suggest that YSD cell wall plays an important role in high Cd accumulation, which may be caused by the enhancement of the Cd apoplast transport pathway.

The transcriptomic analysis of genes responsible for Cd uptake, long-distance transport, and vacuolar storage in YSD roots showed no significant changes compared to ZH11 (Supplementary Figure S3), suggesting that the high accumulation of Cd in YSD may not be caused by these reported genes. Transcriptomics revealed that, when YSD responds to Cd stress, the GSH metabolic pathway is significantly enriched, and the GSH content in YSD plants is significantly reduced (Supplementary Figure S2). GSH maintains the redox balance in YSD cells as a precursor of phytochelatin synthesis, and two proteins of the ABC family in Arabidopsis mediate the vacuole storage of the phytochelatin–Cd complex [24–27]. Currently, the gene that mediates PC–Cd transport in rice has not been identified; thus, it may also be mediated by the ABC family. Therefore, YSD provides valuable germplasm resources for the subsequent cloning of new Cd accumulating rice varieties.

Nicotinamide, as a ligand of some metal elements including Fe, Mn, and Cu, mediates the absorption and transport of NA–metal complexes, with the participation of YSL transporters [28]. Our study found that the expression level and content of nicotinamide synthesis genes in YSD were significantly lower than those in ZH11, and the expression level of YSL12 was significantly lower than that in ZH11 (Supplementary Figure S1A–D). These results suggest that the nicotinamide content and YSL12 expression in rice may be negatively correlated with Cd transport. In vitro nicotinamide treatment increased Cd accumulation in rice shoots (Supplementary Figure S1E), but had no effect on Cd uptake in the roots (Supplementary Figure S1F), suggesting that nicotinamide had different effects on Cd uptake and transport in vivo and in vitro.

4. Conclusions

In summary, we identified YSD, a rice variety with high Cd accumulation that provides excellent germplasm resources for the exploration of Cd accumulation-related genes, and the cell wall modification genes with sequence- and expression-level variations that provide potential targets for phytoremediation. Moreover, our findings suggest that the apoplast cell wall may play a key role in the accumulation of Cd.

5. Materials and Methods

5.1. Plant Materials, Analysis of Agronomic Characters and Growth Conditions

Zhonghua 11 (ZH11) is a commonly used japonica rice variety; whereas, YSD is a rice germplasm collected from Huaihua city with unclear genetic background. The rice was grown in paddy fields contaminated with 0.4 mg/kg Cd in Changsha, until the seeds were harvested. At the mature stage, 10 rice plants were randomly selected from YSD and ZH11 varieties, and the plant height, straw dry weight and yield per plant were measured, respectively.

The conditions for the hydroponic cultivation of the rice in the greenhouse were as follows: the seeds were soaked in water at 30 °C for 36 h in the dark, and the seeds with good germination were planted in 96-well bottomless plates. Rice seedlings were hydroponically cultivated, as previously described [18].

5.2. Plant Sampling and Elemental Determination

At 4 weeks, the hydroponically grown rice seedlings were exposed to the Cd treatments indicated in the following section before being sampled. The xylem sap collected from five plants was pooled into one replicate, and a total of three replicates were used for each line, as previously described [18]. The metal content was determined using inductively coupled plasma mass spectrometry (ICP-MS), as previously described [29].

In vitro niacinamide treatment was performed on 4-week-old hydroponic rice seedlings treated with 10 μM Cd, 10 μM Cd + 10 μM niacinamide and 10 μM Cd + 100 μM niacinamide for 3 days, respectively. The content of Cd in the shoots was determined by ICP-MS.

5.3. Extraction of Subcellular and Cell Wall Components and Determination of Cd Content

At 4 weeks, the rice plants were treated with 10 μM $CdCl_2$ for 3 days. The shoots and roots of the plants were then collected. The subcellular components were extracted by differential centrifugation, as previously described [17]. The extracted cell wall and organelle fractions were dried in an oven, and after drying, the cell wall, organelle, and soluble fractions were digested in 70% HNO_3 (v/v), before the Cd concentration was determined by ICP-MS.

Pectin, cellulose, and hemicellulose were extracted according to the methods, as previously described [21]. The Cd concentrations were determined by ICP-MS after dilution. Pectin methylesterase (PME) activity was determined, as previously described [20].

5.4. Whole-Genome Resequencing

Fresh leaves from 10-day-old plants were sampled for the isolation of genomic DNA. An Illumina HiSeq 4000 system (read length 350–500 bp, paired end) belonging to the OE Biotech company (Shanghai, China) was used to perform whole-genome resequencing to distinguish variations in the genomic DNA [30]. A total of 17G of data was generated, covering 40 times the rice genome. Genome-wide single-nucleotide polymorphisms (SNPs) and insertions/deletions (InDels) were identified and characterized between 'YSD' and 'ZH11' by the OE Biotech company (Shanghai, China).

5.5. RNA-SEQ

The roots of ZH11 and YSD were cultured for 28 days and sampled after 10 μM Cd treatment for 3 days. Three plant roots were sampled as replicates and quickly submerged in liquid nitrogen and sent to the OE Biotech company (Shanghai, China) for RNA sequencing analysis. Each replicated sample produced an average of 6G data. Three biological replicates were set. Heat maps of the differentially expressed genes were constructed using the MeV software (http://www.tm4.org/, 8 July 2021).

5.6. Determination of Glutathione and Nicotinamide Content

The glutathione (GSH) and nicotinamide content were determined by Suzhou Comin Biotechnology Co., Ltd. Specifically, the GSH in the roots and shoots was measured using a kit (GSH-1-W), with the o-phthalaldehyde fluorescence derivatization method. Nicotinamide was detected as described, with minor modification [31]. First, approximately 0.1 g of the sample was weighed out and 1 mL of mobile phase A was added. The mixture was then homogenized in an ice bath and an ice-bath ultrasound was conducted for 30 min. The sample was centrifuged at 8000× *g* for 10 min and the supernatant was collected, filtered using a needle filter, and subjected to high-performance liquid chromatography. A Rigol L-3000 system with a CST Daiso C18 (250 mm × 4.6 mm, 5 μm) column was used. Mobile phase A was 0.05 mol/L sodium acetate aqueous solution (acetic acid was used to adjust the pH to 4.5) and mobile phase B was methanol. The ratio of A to B was 90:10. The injection volume was 10 μL, the flow rate was 1 mL/min, the column temperature was 35 °C, the sample departure time was 30 min, and a UV detector wavelength of 261 nm was used for the method group.

5.7. Statistical Analysis and Data Availability

The data were analyzed using minimum differential multiple-range comparisons with the SPSS software, and each experiment was carried out with at least three biological replicates. $p < 0.05$ was considered to indicate a significant difference, and $p < 0.01$ was considered to indicate a highly significant difference. Charts were prepared with GraphPad Prism 8. The raw data for the whole-genome resequencing and mRNA transcriptome sequencing are available from the corresponding author on request.

Supplementary Materials: The following supporting information can be downloaded at: https://www.mdpi.com/article/10.3390/plants12061226/s1, Figure S1: Nicotinamide treatment promoted the translocation of cadmium in rice. Figure S2: GST transferase expression level and GSH content in YSD and ZH11. Figure S3: Expression level of Cd uptake and translocation gene betweenYSD and ZH11 root under Cd treatments.

Author Contributions: J.-S.L. designed the experiments and analyzed the data; B.G. and Y.H. performed most of the experiments; C.-Z.C. analyzed genomic and transcriptome data; Y.Y. determined the element content Z.Z. and J.-S.L. wrote and revised the manuscript. All authors have read and agreed to the published version of the manuscript.

Funding: This study was partly supported by the National Natural Science Foundation of Hunan Province (2021RC3086, 2021JJ20001), National Natural Science Foundation of China (32272811), National Key Research and Development Program of China (2022YFD1700103), and the China Agriculture Research System Project (CARS-01-02A).

Institutional Review Board Statement: Not applicable.

Informed Consent Statement: Not applicable.

Data Availability Statement: The data presented in this study are available on request from the corresponding author.

Conflicts of Interest: The authors declare no conflict of interest.

Abbreviations

CDTA: cyclohexane diamine tetraacetic acid; GSH, glutathione; ICP-MS, inductively coupled plasma mass spectrometry; PME, pectin methylesterase; SNP, single-nucleotide polymorphism.

References

1. Kumar, S.; Sharma, A. Cadmium toxicity: Effects on human reproduction and fertility. *Rev. Environ. Health* **2019**, *34*, 327–338. [CrossRef]
2. Clemens, S.; Aarts, M.G.; Thomine, S.; Verbruggen, N. Plant science: The key to preventing slow cadmium poisoning. *Trends Plant Sci.* **2013**, *18*, 92–99. [CrossRef] [PubMed]
3. DalCorso, G.; Fasani, E.; Manara, A.; Visioli, G.; Furini, A. Heavy Metal Pollutions: State of the Art and Innovation in Phytoremediation. *Int. J. Mol. Sci.* **2019**, *20*, 3412. [CrossRef]
4. Ali, H.; Khan, E.; Sajad, M.A. Phytoremediation of heavy metals–concepts and applications. *Chemosphere* **2013**, *91*, 869–881. [CrossRef]
5. Salt, D.E.; Blaylock, M.; Kumar, N.P.; Dushenkov, V.; Ensley, B.D.; Chet, I.; Raskin, I. Phytoremediation: A novel strategy for the removal of toxic metals from the environment using plants. *Nat. Biotechnol.* **1995**, *13*, 468–474. [CrossRef]
6. Jacob, J.M.; Karthik, C.; Saratale, R.G.; Kumar, S.S.; Prabakar, D.; Kadirvelu, K.; Pugazhendhi, A. Biological approaches to tackle heavy metal pollution: A survey of literature. *J. Environ. Manag.* **2018**, *217*, 56–70. [CrossRef] [PubMed]
7. Sasaki, A.; Yamaji, N.; Yokosho, K.; Ma, J.F. Nramp5 is a major transporter responsible for manganese and cadmium uptake in rice. *Plant Cell* **2012**, *24*, 2155–2167. [CrossRef] [PubMed]
8. Chang, J.D.; Huang, S.; Yamaji, N.; Zhang, W.; Ma, J.F.; Zhao, F.J. OsNRAMP1 transporter contributes to cadmium and manganese uptake in rice. *Plant Cell Environ.* **2020**, *43*, 2476–2491. [CrossRef]
9. Takahashi, R.; Ishimaru, Y.; Senoura, T.; Shimo, H.; Ishikawa, S.; Arao, T.; Nakanishi, H.; Nishizawa, N.K. The OsNRAMP1 iron transporter is involved in Cd accumulation in rice. *J. Exp. Bot.* **2011**, *62*, 4843–4850. [CrossRef]
10. Tan, L.; Qu, M.; Zhu, Y.; Peng, C.; Wang, J.; Gao, D.; Chen, C. ZINC TRANSPORTER5 and ZINC TRANSPORTER9 Function Synergistically in Zinc/Cadmium Uptake. *Plant Physiol.* **2020**, *183*, 1235–1249. [CrossRef] [PubMed]
11. Tan, L.; Zhu, Y.; Fan, T.; Peng, C.; Wang, J.; Sun, L.; Chen, C. OsZIP7 functions in xylem loading in roots and inter-vascular transfer in nodes to deliver Zn/Cd to grain in rice. *Biochem. Biophys. Res. Commun.* **2019**, *512*, 112–118. [CrossRef] [PubMed]
12. Miyadate, H.; Adachi, S.; Hiraizumi, A.; Tezuka, K.; Nakazawa, N.; Kawamoto, T.; Katou, K.; Kodama, I.; Sakurai, K.; Takahashi, H.; et al. OsHMA3, a P1B-type of ATPase affects root-to-shoot cadmium translocation in rice by mediating efflux into vacuoles. *New Phytol.* **2011**, *189*, 190–199. [CrossRef] [PubMed]
13. Ueno, D.; Yamaji, N.; Kono, I.; Huang, C.F.; Ando, T.; Yano, M.; Ma, J.F. Gene limiting cadmium accumulation in rice. *Proc. Natl. Acad. Sci. USA* **2010**, *107*, 16500–16505. [CrossRef] [PubMed]
14. Yamaji, N.; Xia, J.; Mitani-Ueno, N.; Yokosho, K.; Feng Ma, J. Preferential delivery of zinc to developing tissues in rice is mediated by P-type heavy metal ATPase OsHMA2. *Plant Physiol.* **2013**, *162*, 927–939. [CrossRef] [PubMed]
15. Takahashi, R.; Ishimaru, Y.; Shimo, H.; Ogo, Y.; Senoura, T.; Nishizawa, N.K.; Nakanishi, H. The OsHMA2 transporter is involved in root-to-shoot translocation of Zn and Cd in rice. *Plant Cell Environ.* **2012**, *35*, 1948–1957. [CrossRef]
16. Uraguchi, S.; Kamiya, T.; Sakamoto, T.; Kasai, K.; Sato, Y.; Nagamura, Y.; Yoshida, A.; Kyozuka, J.; Ishikawa, S.; Fujiwara, T. Low-affinity cation transporter (OsLCT1) regulates cadmium transport into rice grains. *Proc. Natl. Acad. Sci. USA* **2011**, *8*, 20959–20964. [CrossRef]
17. Xu, W.; Xie, J.; Gao, Y.; Wu, L.; Sun, L.; Feng, L.; Chen, X.; Zhang, T.; Dai, C.; Li, T.; et al. Variation of a major facilitator superfamily gene contributes to differential cadmium accumulation between rice subspecies. *Nat. Commun.* **2019**, *10*, 2562.
18. Luo, J.S.; Huang, J.; Zeng, D.L.; Peng, J.S.; Zhang, G.B.; Ma, H.L.; Guan, Y.; Yi, H.Y.; Fu, Y.L.; Han, B.; et al. A defensin-like protein drives cadmium efflux and allocation in rice. *Nat. Commun.* **2018**, *9*, 645. [CrossRef]
19. Peng, J.S.; Wang, Y.J.; Ding, G.; Ma, H.L.; Zhang, Y.J.; Gong, J.M. A Pivotal Role of Cell Wall in Cadmium Accumulation in the Crassulaceae hyperaccumulator *Sedum plumbizincicola*. *Mol. Plant* **2017**, *10*, 771–774. [CrossRef]
20. Xiao, Y.; Wu, X.; Liu, D.; Yao, J.; Liang, G.; Song, H.; Ismail, A.M.; Luo, J.S.; Zhang, Z. Cell Wall Polysaccharide-Mediated Cadmium Tolerance between Two *Arabidopsis thaliana* Ecotypes. *Front. Plant Sci.* **2020**, *11*, 473. [CrossRef]

21. Wu, X.; Song, H.; Guan, C.; Zhang, Z. Boron mitigates cadmium toxicity to rapeseed (*Brassica napus*) shoots by relieving oxidative stress and enhancing cadmium chelation onto cell walls. *Environ. Pollut.* **2020**, *263*, 114546. [CrossRef]
22. Riaz, M.; Kamran, M.; Fang, Y.; Yang, G.; Rizwan, M.; Ali, S.; Zhou, Y.; Wang, Q.; Deng, L.; Wang, Y.; et al. Boron supply alleviates cadmium toxicity in rice (*Oryza sativa* L.) by enhancing cadmium adsorption on cell wall and triggering antioxidant defense system in roots. *Chemosphere* **2021**, *266*, 128938. [CrossRef] [PubMed]
23. Pan, J.; Guan, M.; Xu, P.; Chen, M.; Cao, Z. Salicylic acid reduces cadmium (Cd) accumulation in rice (*Oryza sativa* L.) by regulating root cell wall composition via nitric oxide signaling. *Sci. Total Environ.* **2021**, *797*, 149202. [CrossRef] [PubMed]
24. Brunetti, P.; Zanella, L.; De Paolis, A.; Di Litta, D.; Cecchetti, V.; Falasca, G.; Barbieri, M.; Altamura, M.M.; Costantino, P.; Cardarelli, M. Cadmium-inducible expression of the ABC-type transporter AtABCC3 increases phytochelatin-mediated cadmium tolerance in Arabidopsis. *J. Exp. Bot.* **2015**, *66*, 3815–3829. [CrossRef] [PubMed]
25. Clemens, S.; Kim, E.J.; Neumann, D.; Schroeder, J.I. Tolerance to toxic metals by a gene family of phytochelatin synthases from plants and yeast. *EMBO J.* **1999**, *18*, 3325–3333. [CrossRef] [PubMed]
26. Park, J.; Song, W.Y.; Ko, D.; Eom, Y.; Hansen, T.H.; Schiller, M.; Lee, T.G.; Martinoia, E.; Lee, Y. The phytochelatin transporters AtABCC1 and AtABCC2 mediate tolerance to cadmium and mercury. *Plant J.* **2012**, *69*, 278–288. [CrossRef] [PubMed]
27. Higuchi, K.; Suzuki, K.; Nakanishi, H.; Yamaguchi, H.; Nishizawa, N.K.; Mori, S. Cloning of nicotianamine synthase genes, novel genes involved in the biosynthesis of phytosiderophores. *Plant Physiol.* **1999**, *119*, 471–480. [CrossRef]
28. Ishimaru, Y.; Masuda, H.; Bashir, K.; Inoue, H.; Tsukamoto, T.; Takahashi, M.; Nakanishi, H.; Aoki, N.; Hirose, T.; Ohsugi, R.; et al. Rice metal-nicotianamine transporter, OsYSL2, is required for the long-distance transport of iron and manganese. *Plant J.* **2010**, *62*, 379–390. [CrossRef]
29. Gong, J.M.; Lee, D.A.; Schroeder, J.I. Long-distance root-to-shoot transport of phytochelatins and cadmium in Arabidopsis. *Proc. Natl. Acad. Sci. USA* **2003**, *100*, 10118–10123. [CrossRef]
30. McKenna, A.; Hanna, M.; Banks, E.; Sivachenko, A.; Cibulskis, K.; Kernytsky, A.; Garimella, K.; Altshuler, D.; Gabriel, S.; Daly, M.; et al. The Genome Analysis Toolkit: A MapReduce framework for analyzing next-generation DNA sequencing data. *Genome Res.* **2010**, *20*, 1297–1303. [CrossRef]
31. Lang, R.; Yagar, E.F.; Eggers, R.; Hofmann, T. Quantitative investigation of trigonelline, nicotinic acid, and nicotinamide in foods, urine, and plasma by means of LC-MS/MS and stable isotope dilution analysis. *J. Agric. Food Chem.* **2008**, *56*, 11114–11121. [CrossRef] [PubMed]

plants

MDPI

Review

Plant Tolerance to Drought Stress with Emphasis on Wheat

Sarah Adel [1] and Nicolas Carels [2,*]

[1] Genetic Department, Faculty of Agriculture, Ain Shams University, Cairo 11241, Egypt; sarah_adel@agr.asu.edu.eg
[2] Laboratory of Biological System Modeling, Center of Technological Development for Health (CDTS), Oswaldo Cruz Foundation (Fiocruz), Rio de Janeiro 21040-361, Brazil
* Correspondence: nicolas.carels@fiocruz.com.br

Abstract: Environmental stresses, such as drought, have negative effects on crop yield. Drought is a stress whose impact tends to increase in some critical regions. However, the worldwide population is continuously increasing and climate change may affect its food supply in the upcoming years. Therefore, there is an ongoing effort to understand the molecular processes that may contribute to improving drought tolerance of strategic crops. These investigations should contribute to delivering drought-tolerant cultivars by selective breeding. For this reason, it is worthwhile to review regularly the literature concerning the molecular mechanisms and technologies that could facilitate gene pyramiding for drought tolerance. This review summarizes achievements obtained using QTL mapping, genomics, synteny, epigenetics, and transgenics for the selective breeding of drought-tolerant wheat cultivars. Synthetic apomixis combined with the msh1 mutation opens the way to induce and stabilize epigenomes in crops, which offers the potential of accelerating selective breeding for drought tolerance in arid and semi-arid regions.

Keywords: ChIP; climate change; epigenetic; genomics; histone code; QTL; transcription factors; transgenic crops

1. Introduction

Agriculture is an essential economic activity, as well as a source of food and jobs, which makes its sustainability mandatory [1]. However, the quantity and quality of cultivated crops depend on local weather variables [2] and particularly on water availability. The global warming that we now observe [3] makes it difficult to predict the type of challenges that selective breeders will face to guarantee food security in the near future [4]. Nonetheless, selective breeding for increased tolerance to drought is vital [5] to mitigate inevitable pressures on water supply [6]. With the deficit of water progressively increasing [7–9], its value is being assessed more carefully [10]. In any case, the consequences of global warming suggest that societies will have to adapt their behavior to reach sustainability [11] since water shortages and/or unpredictable supply from one year to the next will remain a recurrent problem for farmers from now on [12,13].

In the mid-XX century, agriculture changed the course of its evolution through the so-called Green Revolution, i.e., the introduction of cultivation protocols involving high-yielding varieties, chemical fertilizers, pesticides, irrigation, and mechanization. Today, together with maize, rice, and soybean, wheat represents a staple food for humanity [14]. However, the pressure for the continuous increase of food production by agriculture is becoming critical, despite the necessity of biodiversity preservation and climate change mitigation. Because wheat is a primary source of calories and proteins, biotic and abiotic stresses may constitute bottlenecks for the human population, whose size continues to grow. In particular, rust disease, a fungal pathogen of cereals, is seen as a foremost threat to the sustainable growth of wheat production [15,16]. Indeed, agricultural practices will need to keep pace with the intensification of sustainable food production to face the challenge of feeding a world population estimated to reach over nine billion by 2050 [17,18].

Since drought is a major hurdle that limits crop yields, further investigations of wheat drought tolerance-related processes are critical for the genetic improvement of drought tolerance in this crop [19,20].

A document from the Intergovernmental Panel on Climate Change [21] indicated that dangers emerging from climate change could worsen the risks, vulnerability, and unpredictability facing humans and ecosystems. To address these threats, the European Union issued policies in 2021 to "*adapt to climate change in a consistent manner in all policy areas*" and to "focus, in particular, on the most vulnerable and impacted populations and sectors" [22]. Taking Egypt as an example, people rely exclusively on irrigation to produce about 9 million of the 20 million tons of wheat consumed annually and import the other half. Because Egypt is currently the largest wheat-importing country in the world, it tries to become less reliant on imports by increasing its production [23]. The authorities promote the increase of wheat production via (i) vertical expansion by way of offering the wheat farmers all their needs including fertilizers, pesticides, advice, laboratory support, and first-class cultivars, and (ii) horizontal expansion by cultivating wheat in new and reclaimed lands. Hopefully, given its political effort, Egypt may have witnessed in 2019 a ranking of 5th among the regions of high grain productivity [24] with the benefit of introducing new cultivars [25]. Because rain precipitation may decrease in the future, the wheat yield of this country could fall by as much as 9% in 2030 and by 20% in 2060 [3,26].

Because we are concerned about plant adaptation for tolerance to drought-stress, with particular emphasis on wheat production challenges in the era of climate change, we focused this review on selective breeding through genetics and epigenetics in an effort to integrate data and concepts. We concluded that synthetic apomixis combined with the msh1 mutation opens the way to induce and stabilize epigenomes in crops, which offers the potential for accelerating selective breeding of drought-tolerant crops for arid and semi-arid regions.

2. QTL Mapping for Drought Stress Tolerance in Plants

Drought-tolerance refers to polygenic traits or quantitative trait loci (QTL) identified in crop plants (Table S1a,b). Conceptually, drought tolerance occurred during the transition of plants from water to land, and the role of drought tolerance was to allow this transition [27]. Some of the drought-related-QTLs are associated with root architecture, plant biomass, water soluble carbohydrates, membrane stability index, and grain yield. Some of these QTLs are linked to agronomic or physiological traits associated with drought-stress and can contribute up to 20% phenotypic variation [28]. Examples of QTLs involved in drought stress tolerance and their related candidate genes in model plants are given in Table S2 [29]. For some other interesting references on QTLs of drought tolerance in crops of economical importance, see [30–48].

Genome-wide association study (GWAS) of wheat has been used to map QTLs associated with seedling drought tolerance [49]. Multiple significant QTLs associated with seedling drought tolerance were identified on chromosomes 1B, 2A, 2B, 2D, 3A, 3B, 3D, 4B, 5A, 5B, 6B, and 7B. Twelve stable QTLs responded to drought stress for various traits; among them were shoot length and leaf chlorophyll fluorescence, which were good indicators of drought stress tolerance. Some QTLs of wheat detected by Maulana et al. [49] co-localized with previously reported QTLs (i) for root and shoot traits at the seedling stage and (ii) for canopy temperature at the grain-filling stage. Some significant single-nucleotide polymorphisms (SNPs) were identified in candidate genes involved in plant abiotic stress responses, which will allow marker-assisted selective breeding for drought tolerance at the seedling stage [50].

The combination of GWAS and QTL mapping at seedling stage has revealed candidate genes and SNP networks controlling traits of recovery and tolerance associated with drought in seedlings of winter wheat [51]. Mathew et al. [52] found that drought tolerance and drought-recovery were only weakly correlated. Interestingly, most QTLs associated with drought-recovery were different from those associated with drought tolerance. These authors also colocalized SNPs with 14 genes, as shown in Table S1a. The markers complete the genetic map of QTLs for drought tolerance that was developed by Hussain et al. [53] by mapping SNPs obtained through crossing drought-tolerant (Harry) and drought-susceptible (Wesley) lines (Figure 1).

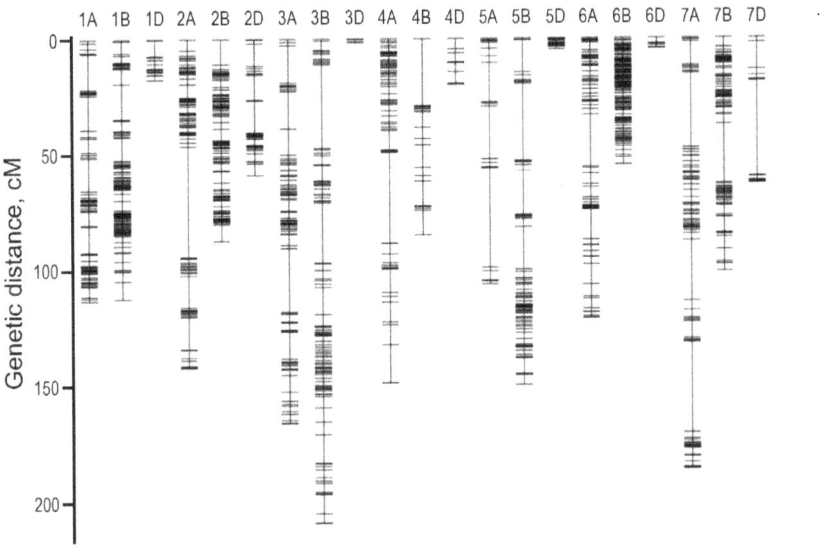

Figure 1. Linkage map constructed from genotyping-by-sequencing derived SNPs in a recombinant inbred population obtained from a cross between drought-tolerant (Harry) and drought-susceptible (Wesley) lines of wheat (adapted with permission from ref. [53], Copyright 2017 owned by Nature. More details on "Copyright and Licensing" are available via: https://creativecommons.org/licenses/by/4.0/).

The genetic map by Hussain et al. [53] clearly shows a distortion of the physical map of drought-associated SNPs as reported by Sallam et al. [51] (Figure 2). The biased distribution of transcriptionally active regions along cereal chromosomes was also reported by other authors [54,55]. The same observation was made for QTLs associated with other traits [48]. Despite regular Giemsa banding along *Hordeum vulgare* [56], Figure 2 clearly shows that pericentromeric areas are generally empty of drought-associated SNPs, which tend to be more frequent at the distal chromosome extremities. This trend has been recognized since the decade of the 1990s [57–60] and suggests a distal preference for transcriptionally active genes along plant chromosomes [61] correlating with a higher crossing-over frequency [60,62].

The high rate of synteny in Poaceae [63–66] is expected to help to find orthologous genes in other cereals [67]. Thus, it is interesting to look at genes for drought tolerance in other plant crops of this group and even in other plant models [48,68–72]. In this sense, a methodology was developed to identify key drought-adaptive genes and the mechanisms in which they are involved, to test their evolutionary conservation. Empirically defined filtering criteria were used to facilitate a robust integration of microarray experiments from Arabidopsis, rice, wheat, and barley available in public databases [73].

Sorghum (*Sorghum bicolor* L. Moench) is a C4 cereal crop adapted to different environmental conditions, including drought-prone areas where many other staple crops fail [74]. It is among the most drought-adapted cereal crops, but its adaptation is not yet well understood. Transcriptome analysis of drought tolerance at the seedling stage revealed that about half (n = 70) of the up-regulated genes in response to drought were novel with no known function and the remainder were TFs, signaling and stress-related proteins implicated in drought tolerance in other crops [75].

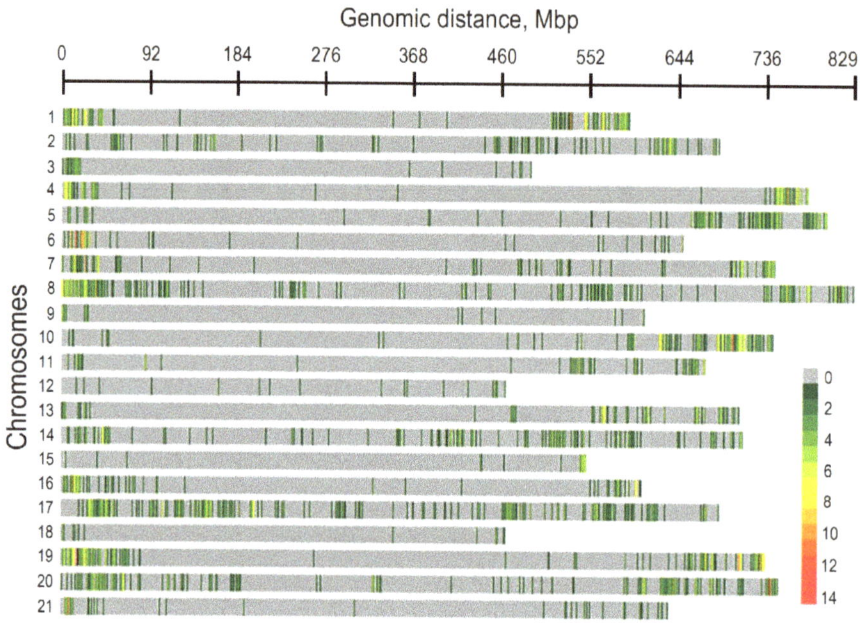

Figure 2. Distribution of SNPs on wheat chromosomes in a biparental mapping population derived from crossing drought-tolerant (Harry) and drought-susceptible (Wesley) inbred lines (adapted with permission from ref. [51], Copyright 2022 owned by Elsevier. More details on "Copyright and Licensing" are available via: https://creativecommons.org/licenses/by/4.0/).

In sorghum, premature leaf death occurs when water starts to be limiting during the grain filling stage. Premature leaf senescence, in turn, leads to charcoal rot, stalk lodging, and significant yield loss. Most sorghum cultivars have pre-flowering drought tolerance, but many do not have any significant post-flowering drought tolerance. Stay-green is one form of drought-tolerance mechanism giving sorghum a tolerance to premature senescence under soil drought-stress during the post-flowering stage. The stay-green trait results in larger functional photosynthetic leaf area during grain filling and even after physiological maturity [76]. In sorghum, four genomic regions associated with the stay-green trait were identified in all field trials and accounted for 53.5% of the phenotypic variance [77]. In wheat, deciphering stay-green revealed variations in the exons of *CaO* and *RCCR* associated with a significant difference in the regulation of *CaO* and *Cab* at seven days after anthesis under terminal heat stress [78].

3. Genomics

As outlined above, plant species may use various drought adaptive mechanisms depending on their biological setup, which reflects an adaptive evolutionary genome strategy, given their environment. Still, for most plant species, the eco-physiological and genetic mechanisms underlying variation for drought adaptation in the field are not known. Thus, deciphering the mechanisms of drought adaptation remains a foremost challenge in plant biology and breeding [79–81].

3.1. Genome Phenotype and Genome Strategy

In vertebrates, the genome was shown to be organized into two major components with specific functionalities [82] according to guanine plus cytosine (GC) content. The base changes during vertebrate evolution were such that the GC-poor compartment remained poor in cold-blooded vertebrates (fishes, amphibians, reptiles), but their GC-rich compartment became even richer in GC during their transition to warm-blooded vertebrates (mammals and birds). This compositional transition led to a range of functional adaptations at the molecular level [82]. The effect of this major transition [83] on the genome structure of modern species of both groups can still be observed through GC composition at the regional level, which led Bernardi to coin the term isochores in 2007 for DNA stretches larger than 300 Kbp that do not vary by more than an average standard deviation of ~2% GC [84].

The concept of genome phenotype correlates with that of genome strategy and this last concept was recognized through codon usage analyses by Grantham et al. [85] and extended to plant genomes by Carels [86]. Indeed, as noted by Grantham et al. [85] "Systematic third base choices can be used to establish a kind of genetic distance, which reflects differences in coding strategy. The patterns of codon choice we find seem compatible with the idea that the genome and not the individual gene is the unit of selection. Each gene in a genome tends to conform to its species' usage of the codon catalog". By extension, Grantham et al. [85] suggested that selection acts on genomes by shaping coding sequences through their codon usage, which would comprise an adaptive strategy given an evolutionary history resulting in particular features of the genome phenotype. Carels [86] recognized the difference in the organization of intergenic sequences in humans and maize despite their similarity in genome size and bias of GC composition of codons. This is particularly clear when comparing the compartmentalization in the maize gene space to the compartmentalization in the human genome, which shows that the compositional regression line of GC in the 3rd position of codons with intergenic sequences is too steep in maize to have a complete separation between GC-poor and GC-rich compartments. The slope of the maize orthogonal regression line is about four times steeper than that of the human one, which leads to about seven times the contraction of the gene space compared to humans. As a result, only ~3% GC separates the GC-poor and GC-rich compartments. Interestingly, zein genes, which are expressed only during the embryonic phase, are located in the GC-poor compartment [86]. This situation mimics the expression of embryonic genes in the vertebrate GC-poor compartment. More recently, the concept of genome strategy as a response to adaptation to environmental challenges was recognized in bacteria [87]. Surprisingly, almost all the interval of GC variation observed among bacterial species [88], the so-called universal correlation, is found within the genomes of species of warm-blooded vertebrates. This implies that genes are spread over GC-poor as well as GC-rich contexts offering a richer range of promoter sequences for interaction with TFs than might be the case with only one GC-poor compartment. Therefore, one may see in genome compartmentalization according to regional composition an opportunity for a more diverse genetic code exploration on which environmental selection may operate. A type of driving adaptation that has been proposed is the maximization of energy use in challenging conditions [89]. Another functional correlation that could also operate is a trend for larger use of beta-sheets in proteins from GC-poor genes and of turns in proteins from GC-rich genes [89].

Considering plants, it is interesting to note here that a compositional transition occurred in Poaceae compared to other monocots and dicots [90], which is very clear when considering the third positions of codons [91,92]. However, when considering the intergenic sequences, the transition occurred in the shifting mode [93] and an isochore structure typical of warm-blooded vertebrates could not be found [86]. By contrast, the existence of the GC-rich compartment inside the gene space of maize indicates that some long-term process of regional GC enrichment mechanism has played a role in the genome compartmentalization at sequences of a size around ~100 kb, in the past history of maize. This compartmentalization that results from the correlation between the compositional distribution of GC-poor, GC-rich genes, and ~100 kb sequences is striking since it fits with the compositional distribution and interspersion rate of retrotranspon families in the gene space. The organization in GC-poor and GC-rich compartments still exists with the gene space preferentially within the GC-rich compartment [86]. The absence of an isochore structure may reflect the different chromosome and nucleus organization between plants and vertebrates.

Vertebrates have a complex Giemsa banding that correlates with isochore distribution [94,95] and nucleus spatial organization [95,96] following the fractal globule model of chromosome distribution [97,98]. The degree of DNA condensation of human chromosomes is relatively low compared to that of plants for which there may be more associated proteins on their surface; hence a longer digestion time is needed [56,99] for karyotype preparation. Since transcriptionally active regions show a distal preference for gene location along plant chromosomes (especially in the cases of large genomes) as in vertebrates [54,55,61,84,100], one may conclude that pericentromeric euchromatin islands are richer in GC-poor genes [101,102], which are known to be less active transcriptionally [61,103] and under other regulation processes [104].

In contrast to warm-blooded vertebrates where centromeric regions and heterochromatine are packed on the nucleus membrane, while euchromatin loops toward its center [95], interphase plant cells present two types of spatial organization [105]: (i) Rosette, in Arabidopsis, i.e., heavily methylated interspersed heterochromatin segments of chromosomes that interact with their centromers (B compartment) to form chromocenters, while euchromatin (A compartment) oozes from chromocenters as loops spanning 0.2–2 Mbp. Euchromatin loops are rich in acetylated histones, whereas chromocenters contain less acetylated histones [106]. (ii) Rabbi, in cereals, i.e., telomeres and centromeres of chromosomes cluster at two different poles in the nucleus [107]. The interphase chromosome distribution in Arabidopsis follows the fractal globule distribution as in vertebrates [108], but the 3D distribution of Rabbi chromosomes is quite different since they have a parallel spatial arrangement [109]. These different chromosome distributions in interphase cells of vertebrates and plants are another layer of genome phenotype corresponding to a genome strategy coding for a molecular strategy pervading the whole cell machinery [89,91,110]. Such a biological framework cannot be modified within a short time, which means that it can be, eventually, negatively selected and lead to extinction in case of low fitness for drastic environmental changes since it implies a long evolutionary history correlated to evolutionary inertia to produce them.

It is clear from the discussion above that Poaceae occupy a special place in the spectrum of genome organization in angiosperms. Even if intergenic sequences are relatively homogeneous in composition, GC-rich genes have CpG islands in their promoter regions, while this is not the case for GC-poor genes [93]. Since recombination appears to be redirected to gene promoters and CpG islands [111] in humans, it is likely to be the case in plants as well because of the similarity of chromatin features between both biological systems. As outlined above, these regulatory domains may indicate adaptative peculiarities in Poaceae compared to other dicots and monocots, such as a potential for higher tolerance to various abiotic stresses.

3.2. *Wheat Genome*

During GFAR [112], it was stated that scientists from the International Maize and Wheat Improvement Center (CIMMYT) would sequence the genomes of 15 wheat varieties from breeding programs worldwide. This initiative enabled scientists and breeders to identify much faster the key genes able to improve yield as well as tolerance to heat, drought, pest, and other important crop traits.

Wheat is the largest crop genome decoded to date [113]. For sake of comparison, the largest plant genome to date is the Japanese andromeda (*Paris japonica* (Franch. and Sav.) Franch.) with 149 Gbp and the smallest is *Arabidopsis thaliana* L. with 125 Mbp. The wheat haploid genome (~17 Gbp) is about five to six times larger than the human one, so it is a complex structure comprised of three independent genomes. Each sub-genome accounts for ~5.5 Gbp [114]. The high proportion of repetitive DNA (retrotransposons covering >80% of the genome) [115] makes it difficult to decrypt the genome structure of this crop. However, the mapping of markers such as (i) expressed sequence tags (EST), (ii) SNPs, and (iii) contigs (see: https://wheat.pw.usda.gov/cgi-bin/GG3/browse.cgi, accessed on 14 January 2023) is helping in this task and a complete genome sequence is being regularly updated by Ensembl (see: https://plants.ensembl.org/Triticum_aestivum/Info/Index, accessed on 14 January 2023). Until now, more than 1.07 million wheat ESTs were sequenced [116,117], and there are BAC libraries for each of the three sub-genomes (A, B, and D) covering each specific chromosome or chromosome arm. According to the IWGSC Annual Report from 2021, the "WGSC Annotation v2.1 contains 266,753 genes, comprising 106,913 high confidence (HC) genes and 159,840 low confidence (LC) genes" (https://www.wheatgenome.org/content/download/4298/40943/version/1/file/2021_IWGSC_AnnualReport.pdf, accessed on 14 January 2023). BLAST searchable wheat genomic sequences are now available at https://wheat.pw.usda.gov/cgi-bin/seqserve/blast_wheat.cgi (accessed on 14 January 2023).

When investigating the genome of *T. aestivum* cv. 'Chinese Spring', *T. urartu*, *Aegilops speltoides*, *T. turgidum* cv. Cappelli, *T. turgidum* cv. Strongfield by whole genome high-throughput shotgun sequencing, class I retroelements were confirmed to be the most abundant. Compared to B, class II mobile elements (DNA transposons) were more frequent in the A and D genomes, respectively [115].

The diploid wheat A and D genome were sequenced first to detect their homologous genes. Shotgun sequencing of wheat genome was initially performed with SOLiD, 454, and Illumina [118]. However, the large quantity of repeat DNA made it difficult to assemble the complete genome. To overcome this difficulty, sequencing was combined with SNP markers, physical and genetic mapping [119–121]. GWAS combined with Targeted Induced Local Lesions in Genome (TILLING) was used to annotate the tetraploid and hexaploid wheat gene function as well as to characterize the genetic diversity and ancestral relations between populations [122]. In drought conditions, a larger number of genes differentially expressed were mapped on the B genome than on the other subgenomes, particularly on chromosomes 3B, 5B and 2B. These genes were involved in 116 different pathways [123].

3.3. Genomic Contribution to Understanding the Molecular Bases of Wheat Response to Stress

Plants offer a living context for a wide range of molecular interactions to occur among different stress factors. The large number of genes controlling abiotic stress [124] is a challenge that obliged the adoption of multi-omics strategies [125].

Since map-based cloning of genes for tolerance to abiotic stresses is still difficult with forward genetic strategies in wheat (by seeking the genetic basis of a phenotype—typically by inducing mutations), reverse genetic methods (by seeking to learn which phenotypes are controlled by particular genetic sequences) have been widely used to identify heat-responsive genes in wheat. Transcriptome analysis including microarray and RNA-seq are high-throughput strategies to detect differentially expressed genes that were used to analyze the response to heat stress [126]. For instance, differentially expressed genes (DEGs) in wheat grain and flag leaf enabled the identification of three genes, *TaFBR1*, *TaFBR2*, and *TaFBR3*, which may be crucial for wheat's ability to increase its thermotolerance [127]. The resistance of genes to harmful environmental conditions that could be involved in flavonoid synthesis has been extensively investigated [128]. For instance, a prior study showed that *TaFSL1* improved transgenic Arabidopsis' ability to tolerate salinity by causing primary root extension when compared to control plants [129]. In another transcriptome investigation, *TabZIP60*, a transcription factor encoding gene, was up-regulated as a result of heat stress and controlled the expression patterns of 1104 genes [130].

Because of their sessile standing, plants are continuously challenged by various biotic and abiotic stresses during their life cycle [131]. To cope with this challenge, plants have evolved complex molecular mechanisms to adapt to these abiotic stresses. As has been shown, DNA methylation and non-coding RNAs (ncRNA) are involved in the post-transcriptional regulation of wheat response to heat [130]. The discovery of miRNAs and their regulatory mechanisms for controlling genes involved in various developmental, biological, and stress responses has advanced our understanding of gene regulation in plants [132]. miRNA appears to target many genes. For instance, in the case of the tolerance to salinity in roots, 75 miRNAs were identified by in silico investigation to possibly target 861 mRNA. The genetic development of wheat cultivars may be aided by this information, which might promote our understanding of how miRNAs mediate the molecular mechanisms of tolerance to abiotic stresses [133].

LncRNAs are a diversified class of RNAs generally defined as transcripts larger than 200 nucleotides that are not translated into protein. LncRNAs include intergenic lincRNAs, intronic ncRNAs as well as sense and antisense lncRNAs involved in the control of transcriptional regulation and genome imprinting in processes such as plant development, disease resistance and nutrient acquisition, through chromatin remodeling, histone modification, pri-mRNA alternative splicing, or acting as 'target mimicry' (see refs in [131]). The analysis of differentially expressed lncRNAs, miRNAs, and genes has revealed regulatory networks of lncRNA-miRNA-mRNA modules involved in response to drought stress in wheat [134].

4. Epigenetic Modifications

There is growing evidence indicating that plants implement sophisticated epigenetic mechanisms to fine-tune their responses to environmental stresses. Epigenetic processes involving DNA methylation, histone modification, chromatin remodeling, and ncRNAs make up wheat's response to drought stress. Changes in chromatin, histone, and DNA mainly serve the purpose of memorizing past stress events and enable plants to deal with climate challenges and thrive in stressful environments.

Epigenetics is the study of heritable phenotypic changes that do not involve alteration of the genetic code itself. Epigenetics includes DNA methylation [135], histone modifications, histone variants, and ncRNA modifications that influence the structure and accessibility of chromatin [136] to transcription machinery and alter gene expression [137]. An increased number of investigations have shown the participation of epigenetic mechanisms in the response of plants to abiotic stresses [138]. Therefore, deciphering the epigenetic codes of plant stress response could be of great significance for the selective breeding of drought-tolerant crops [139].

Plants respond through various short-term and long-term strategies depending on whether the stress is permanent or transitory. Short-term strategies include alteration in plant homeostasis while long-term strategies include trans-generational changes involving the development of heritable gene expression changes. This mechanism consists of creating new epigenetic marks while erasing old ones as well as increasing the expression of some genes while silencing the expression of others [140]. Epigenetic reprogramming in response to various environmental challenges contributes to phenotypic diversity as well as tolerance to these challenges [141]. According to Priyanka et al. [141], the mechanism of an epigenetic process can be divided into three stages: (i) Epigenator, which is a trigger, such as foods, toxins, radiation, or hormones, that alters the cells' environment to produce an epigenetic phenotype. These triggering cues are transient but last long enough to initiate the epigenetic process. (ii) Epigenetic initiator, which translates the epigenator signal into an epigenetic modification of chromatin. An epigenetic initiator is primed by the epigenator and determines the location on a chromosome that should be marked. (iii) Epigenetic maintainer, which sustains the chromatin environment in the current and succeeding generations. Persistence of the chromatin configuration may require cooperation between the initiator and maintainer.

4.1. DNA Methylation

As proven by the analysis of Methylation Sensitive Amplification Polymorphism (MSAP), plants can remember their past environmental experience and retrieve their associated information [142,143]. DNA demethylation is an active DNA repair-related mechanism. For example, the Decreased DNA Methylation 1 (DDM1), a nucleosome remodeler involved in the maintenance of DNA methylation, and the Repressor of Transcriptional Silencing 1 (ROS1), a primary factor required for active DNA demethylation, were determined to be involved in UV-B DNA damage repair. A short-term memory induced by a single stress on the parent resulted in epigenetic modifications that were also observed in their first offspring and were reprogrammed in further generations. By contrast, multiple stresses led to long-term memory, and parent modifications were transferred to several generations [144]. Short-term memory allows plants to remain tolerant to certain stresses for up to about 10 days. By contrast, long-term stress memory is regulated by epigenetic modifications and can potentially last for the whole life of plants suffering from continuous stress; ultimately, it may be transferred to the offspring. The transgenerational epigenetic memory of meiotic prophase chromatin features may be inherited from interphase somatic imprinting since plants do not have germline. This long-term stress memory plays an important role in the adaptation and evolution of plants [145]. Epigenetic regulation of genes depends on the types of epigenetic marks and their position on genes [146]. In plants, DNA methylation happens in diverse types of cytosine contexts, including CpG, CpHpG, and CpHpH (where H represents A, T, or C); methylation at these sites is catalyzed by METHYLTRANSFERASE 1 (MET1), CHROMOMETHYLASE 3 (CMT3), and DOMAINS REARRANGED METHYLTRANSFERASE 2 (DRM2), respectively. In Arabidopsis, DNA methylation can be erased by members of the DNA glycosylase family of demethylases, including DEMETER (DME), ROS1, DEMETER-LIKE 1 (DML1), DML2, and DML3.

4.2. Histone Code

The N-terminal extremity of histone is known as the *tail* and is rich in basic amino acids such as lysine and arginine. In addition to DNA methylation, covalent modifications to histone tails, such as methylation, acetylation, phosphorylation, ubiquitination, sumoylation, glycosylation, and ADP-ribosylation constitute another type of epigenetic code, which is conserved across different kingdoms. Modifications to histone tails alter chromatin packaging with the consequence that condensed chromatin structure, such as in the B compartment, results in transcription inhibition because the transcription machinery is unable to access DNA, while loose chromatin, such as in the A compartment, results in transcription activation [147].

According to the histone code hypothesis, the transcription of genetic information encoded in DNA is in part regulated by chemical modifications carried by histone proteins on their unstructured ends (tail) [148]. Unlike acetylation, methylation shows a diverse pattern; several different arginine and lysine residues (R3 of H2A, R3, K20 of H4 and K4, K9, K27, K36, R2, and R17 of H3, etc.) can undergo various types of methylation such as mono, di or tri-methyl residues (arginine undergoes mono and di-methylation only while lysine can undergo mono, di and tri methylation). Depending on the type of methylation, histones can either activate or inhibit the expression of a gene. For example, H3K4 trimethylation activates transcription but K9 and K27 dimethylation in H3 acts as a transcription repressor. Methylation does not alter the net charge of histone-like acetylation but it affects hydrophobicity by the addition of methyl groups and hence may change histone DNA interactions or may create a binding site for various proteins that promote or inhibit the transcription machinery. Histone lysine methyl transferases (HKMT) and protein arginine methyl transferases (PRMT) are responsible for lysine and arginine methylation, respectively. As an example, CAM plants switch from the C3-photosynthetic cycle to the CAM pathway to reduce water loss in case of drought stress [149]. This switching process from the C3 to CAM pathway is coupled with the activation of genomic methylation and hypermethylation of satellite DNA [150].

4.3. Histone Modifications

ChIP-seq and spectrophotometer analyses of the four histone modifications, histone H3 lysine4 trimethylation (H3K4me2), H3K4me3, H3K9me2, or H3K27me3 in seedlings of Arabidopsis plants under salt and drought stress, revealed that the priming treatment (under which seeds are rehydrated and then dried back just before germination) altered the epigenomic landscape. After re-watering, leaf water potential, membrane stability, photosynthetic processes, reactive oxygen species (ROS) generation, anti-oxidative activities, lipid peroxidation, and osmotic potential completely recover in the case of moderately stressed plants but do not fully recover in severely stressed plants [151–153]. When rewatered, plants recover by growing new shoots. The extent of recovery due to rewatering strongly depends on drought intensity, duration, and plant genetics. In wheat, higher photosynthetic rates during drought and rapid recovery after rewatering produce less-pronounced yield declines in tolerant cultivars than in susceptible ones, which suggests that their ability to maintain functions during drought-stress and to quickly recover after rewatering are important factors determining their final productivity [65].

The output of gene transcription is influenced by the input of stress signals due to the well-documented trans-cis interaction between TFs and target DNA regions. However, there are still gaps between the transcription of genes and the anchoring of TFs, and these gaps in chromatin accessibility may be of fundamental importance. Histone methylation is one of the most functionally diverse epigenetic mechanisms because it can either exert activation or repression on target genes depending on the location of lysine residues, in contrast to DNA methylation or histone acetylation, which mark gene repression and activation, respectively [154]. Sani et al. [155] showed that priming-induced changes also depended on the specific lysine residues that were methylated. Despite being small, the changes were confirmed by semi-quantitative-PCR; they varied in number and preferentially targeted TFs. Several genes with priming-induced differences in H3K27me3 showed altered transcriptional responsiveness to a second stress treatment but the number of location-specific changes of H3K27me3 decreased, suggesting that the memory fades over time. In a genome-wide investigation of the gene expression profile in rice under drought-stress, differential H3K4me3 methylation was found to be significantly and positively correlated with transcript level only for a subset of genes under drought stress conditions. Moreover, for the H3K4me3-regulated stress-related genes, the H3K4me3 modification level was mainly increased in genes with low expression and decreased in genes with high expression under drought-stress [156].

Indeed, histone modifications, such as H3K4me3, H3K9ac, H3K9me2, H3K23ac, H3K27ac, H3K27me3, and H4ac, along with DNA methylation could be correlated with gene expression in response to abiotic stresses [157]. Some of the histone modifications occur quickly in response to environmental variations, while others occur gradually according to the challenges imposed on gene expression to control physiological homeostasis and development under environmental stress. Since histone acetylation tends to induce gene activation its removal can lead to gene repression and silencing. Although changes in histone modifications can be correlated with gene activity, the molecular mechanisms through which the chemical modifications influence the chromosomal structure and the accessibility of TFs are still not fully understood.

Plants memorize abiotic stress when exposed to it and can better tolerate it when experiencing it again later on [158]. The memorization process induced by DNA and histone modifications occurs through (i) the up-regulation of small RNAs (micro-RNAs or miRNAs) and short interfering RNAs (siRNAs), (ii) the down-regulation of repressors (specific proteins inhibiting transcription), and (iii) the down-regulation of specific siRNAs, which induce the up-regulation of activators required for the regulation of plant hormone and transcription factors. In Arabidopsis, in addition to siRNAs, it has been shown that lncRNA also participates in the regulation of plant stress-responsive gene expression via DNA de novo cytosine methylation through the RNA-directed DNA methylation (RdDM) pathway. RNAPIV-generated siRNAs could be loaded to Argonaute 4 (AGO4) and interact with lncRNAs generated by RNAPII to constitute a siRNA–AGO4–lncRNA silencing complex, which subsequently recruits the DMT domains rearranged methyltransferase 2 (DRM2) to mediate DNA de novo cytosine methylation. Mutants deficient in NRPD2, an essential subunit of RNAPIV, were hypersensitive to heat stress, suggesting that the RdDM pathway is essential to the regulation of plant stress responses (see refs in [159]).

Interestingly, the family of SWI2/SNF2 chromatin remodeling complexes can regulate the methylation of both DNA and histone marks for gene activation under environmental stresses [160], as was shown for drought tolerance in rice [161]. Examples of histone modifications and chromatin remodeling are (i) rice SWI/SNF2 ATPase BRHIS1 that regulates the expression of disease defense-related *OsPBZc* and *OsSIRK1* genes through specific interaction with mono-ubiquitinated *H2A.Xa/H2A.Xb/H2A.3* and *H2B.7* variants at those gene loci, (ii) Arabidopsis BRM that represses the expression of heat-activated *HSFA3* and *HSP101* genes by removing H4K16ac at their chromatin loci through interaction with HD2C, and (iii) PWR proteins that form a chromatin-remodeling complex with HDA9 and ABI4 to repress the drought-responsive *CYP707A1/2* genes in Arabidopsis. Histone modifiers and transcription regulators also coordinate chromatin dynamics and nucleosome configurations at transcription sites to modulate gene expression [162].

The interplay between histone modifications and DNA methylation provides plants with a multifaceted and robust regulatory circuitry for transcriptional reprogramming in response to stress. For example, DNA methylation changes and various histone modifications, such as H3K4me3, H3K9ac, and H3K9me2 are concertedly regulated for transcriptional activation or repression of the salt-responsive genes such as *Glyma08g41450*, *Glyma11g02400*, *Glyma20g30840* in soybean as well as *SUVH2/5/8*, *ROS1*, *MSH6*, *APUM3*, *MOS6*, and *DRB2* in Arabidopsis. The rice transcriptional complex SUVH7-BAG4-MYB106, consisting of a DNA methylation reader, a chaperone regulator, and a TF, activates *OsHKT1*. In addition, the expression of the Arabidopsis pathogenesis-related *SNC1* gene is cooperatively regulated by the chromatin-remodeling proteins CHR5 and DDM1/SYD for nucleosome occupancy and DNA methylation, along with the histone modifiers HUB1/2 and ATXR7/MOS9 for H2Bub and H3K4me3, respectively, in plant immune responses [163].

The HDA9-PWR-ABI4 complex plays a central role in the regulation of the molecular response to drought stress through the inhibition of ABA catabolism to promote the accumulation of active ABA, thereby protecting plants against dehydration. ABA regulates several physiological processes, including plant growth and development, flowering time as well as leaf size and morphology. Further investigations are required to understand how this complex modulates genes activated by ABI4 at a chromatin level [164].

5. Chromatin Structure

Chromatin immunoprecipitation or ChIP is a method used to investigate the dynamic interactions of specific proteins and DNA sites. These interactions have a significant role in various cellular processes such as replication, transcription, DNA damage repair, genome stability, and gene regulation [165]. This technique allows the study of various cellular mechanisms inside cells according to protein-DNA interactions. As the name chromatin immunoprecipitation suggests, this method allows the identification of DNA-protein interactions by immunoprecipitation. ChIP is being extensively used to depict TFs, variants of histone, chromatin-modifying enzymes, and histone post-translational modifications [166,167]. ChIP followed by high-throughput sequencing (ChIP-seq) is used to study genome-wide protein-DNA interactions to understand gene regulation in native chromatin. ChIP-seq is used to identify, map, and characterize the specific DNA fragments that interact with proteins in vivo [168].

ChIP assays have been successful in identifying the histone modifications responsible for epigenetic regulation. However, the results obtained by ChIP assays provide only indirect candidate residues that may be targets of modifications [136]. Thus, gathering direct or indirect pieces of evidence that enable correlation of the functions of histone modifications with the plant epigenetic landscape is still a challenge, as shown in Figure 3 [169].

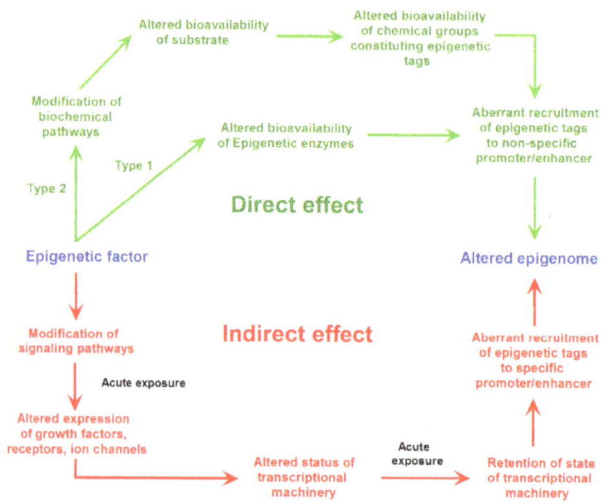

Figure 3. Ways that an epigenetic factor might affect the epigenome and change how genes are expressed. Direct and indirect influences on the epigenome might result from epigenetic changes produced by an external source or intrinsic environment. Red is for epigenetic factors with indirect impacts, while green is for factors with direct effects. Here, epigenetic factor is understood as any molecule, such as ncRNAs, metabolites, or phytohormones induced by endogenous or environmental stresses, able to induce global changes affecting the DNA methylation of multiple genes or modify the expression of specific genes (adapted from ref. [169], Copyright 2014 owned by Frontiers. More details on "Copyright and Licensing" are available via: https://creativecommons.org/licenses/by/4.0/).

The mapping of chromatin by CHIP-seq in seedlings of allohexaploid wheat enabled detection of distinct chromatin architectural features surrounding various functional elements including genes, promoters, enhancer-like elements, and transposons [170]. Thousands of new gene regions, trans- and cis-regulatory elements were identified based on the combinatorial pattern of chromatin features. As expected, the subset of genome active regulatory elements that include promoters and enhancer-like elements (roughly 1.5%) is characterized by a high degree of chromatin openness, histone acetylation, abundance of CpG islands, and low DNA methylation levels since they are associated with genes, which are found in their majority in the A compartment [147]. A comparison across sub-genomes revealed that negative selection is targeting sequence and chromatin features involved in gene regulation. The divergent enrichment of cis-elements between enhancer-like sequences and promoters implies that these functional elements are targeted by different TFs [170].

In mammals, chromatin analyses revealed that: (i) low recombination domains and regions of elevated linkage disequilibrium (LD) tend to coincide with topologically associated domains (TAD) and isochores; (ii) double-strand break (DSB) and recombination frequencies increase in the short loops of GC-rich TADs, whereas recombination cold spots are typical of lamina-associated domains (LAD), which mediate chromatin tethering to the lamina of the nuclear envelope; and (iii) binding and loading of proteins, which are critical for DSB and meiotic recombination, are higher in GC-rich TADs. Recombination is favored or suppressed in specific mega-base-sized regions [171] according to linkage disequilibrium (LD), which is positively correlated to GC%, with strong LD being typical of GC-poor regions [172]. In humans, hotspots of recombination are associated with local GC spikes (1 to 2 kb in size) and with an increase of the local mutation rate resulting in G or C nucleotides, but without an effect on substitution rate or divergence [173], which means that only allele interchanges between sister chromatin are affected. Thus, GC-rich R-bands favor DSB due to lower chromatin fiber stiffness [174]. As a result, TADs and LADs are characterized by a non-random association of chromosomal recombination profile and chromatin architecture. TADs represent an assembly of chromatin loops (with boundaries of 0.2–2 Mb in size [175], which can be resolved into contact domains, of 185 Kb in median size [176]. LADs have a median size of 500 Kb, are GC-poor, and are distributed over the whole genome [177–179]. DSBs correlate with H3K4me3 [180,181] and typically take place in GC-rich nucleosome-depleted loop sequences with the consequence that GC-rich domains are more recombinogenic compared to GC-poor ones. GC-rich regions are typically (i) highly transcriptionally active, (ii) characterized by H3K4me3, H3K9ac, and H3K36me3 marks, and (iii) harbor shorter loops than GC-poor domains [179]. The correspondence between LD-regions, isochores, and TADs boundaries strongly suggests the existence of a common genomic code of chromatin architecture in mammalian meiotic and mitotic cells [182].

In plants, chromatin functionalities occurred through DNA mechanical anchoring via nucleolus-associated domains (NADs), which are chromatin regions interacting with the nucleolus, and LADs. Chromosomal A and B compartments can be further segmented as TADs of intertwined loops in the range of 0.1–1.0 Mb of relatively independent chromatin regions, which promote a larger rate of interaction between cis and trans-regulatory sequences (enhancers and promoters) as well as gene kissing [183,184]. TADs were only observed in plant species having genome sizes larger than 400 Mb [170,185,186]. Chromatin loops connect the distant regulatory elements to their target loci by physical approximation. The regulatory function of chromatin loops occurs through the formation of (i) a repressive loop by the intermediation of histone modifiers—H3K27me3—polycomb protein—lncRNAs complex, such as in heterochromatin, which is characterized by a low gene density, low transcriptional activity, high repressive epigenetic modifications, and higher transposon density (B compartment); (ii) a silencing loop by the intermediation of H3K9me2-reader (ADCP1)—ncRNAs complex, and (iii) a transcriptional loop hub formed by the intermediation of the H3K4me3 modifiers-RNA Pol-II-eRNAs complex and characterized by high gene density, activating epigenetic modifications, and active transcriptional activity (A compartment) [161,187].

A general model for the role of epigenetic elements in stress responses of wheat and barley has been proposed by Kong et al. [159] where DNA methylation, histone modifications, and chromatin remodeling are regulated through siRNA and lncRNA epigenetic processes, including RdDM and histone modification as well as genome topology changes, affecting gene expression in response to the environmental stresses.

6. Transcription Factors

TFs help plants to respond to abiotic stress through priming stimuli by inducing the accumulation of inactive transcription factors, and inactive signaling compounds. A primed state is a kind of state of readiness for another stress event by which plants are capable of quicker and more effective activation of stress-protective responses (see refs in [188]). TFs are the endpoints of signal transduction networks and may control numerous pathways simultaneously, which has prompted practical strategies for engineering plants with improved stress tolerance [189,190]. Many key genes and transcription regulators governing morpho-physiological traits were found to control root architecture and stomatal development for soil moisture extraction and its retention, which justifies their use as targets for molecular breeding strategies of selective breeding for drought tolerance [191]. For example, *DRO1* in rice (*Oryza sativa* L.) and *ERECTA* in Arabidopsis and rice were identified as enhancers of drought tolerance via regulation of root traits and leaf transpiration. Tools such as (i) the reference genome sequence for wheat, (ii) functional reverse genetics, and (iii) genome editing technologies are expected to aid in deciphering the functional roles of genes and regulatory networks underlying adaptive phenological traits as well as the development of drought-tolerant cultivars [192]. Another example is *DUO-B1*, which is a gene encoding an APETALA2/ethylene response factor (AP2/ERF) protein, one of the largest families of transcription factors that have been linked to a variety of biological processes. Based on the DNA binding domain (DBD), AP2/ERFs were classified into AP2, RAV (related to *Abscisic acid insensitive3/Viviparous1*), DREB (subgroup A1–A6), ERF (subgroup B1–B6), and others [193]. The members of the RAV (one of the most abundant transcription factor families in plants) and ethylene response (ERF) subfamily are frequently engaged in phytohormone signaling, disease resistance, and abiotic stress response. CRISPR-Cas9 genome editing showed that the wild gene controls meristem activity and branching in its orthologues from *Brachypodium distachyon* (L.) P. Beauv., bread wheat, and Arabidopsis. CRISPR-induced mutations of *DUO-B1* led to mild supernumerary spikelets, increased grain number per spike, and increased yield under field conditions without affecting other major agronomic traits. Pyramiding this gene with loci that are responsible for improved photosynthesis and nitrogen use efficiency could broaden the utilization of *DUO-B1* alleles to increase yield potential [194].

It has been found that genes involved in drought tolerance were regulated by TFs such as: (i) Basic leucine zipper (bZIP), which controls various biological functions involved in pathogen defense, seed maturation, light and stress signaling, and flowering [195]; (ii) Basic helix-loop-helix (bHLH) TFs, which act as transcriptional regulators involved in stress response, phytochrome signaling, anthocyanin biosynthesis, fruit ripening, carpel and epidermal development [195]; (iii) WRKY, which has a ~60 amino acid domain whose WRKYGQK sequence is completely conserved and followed by a zinc finger motif, participating in signaling networks for plant defense (some WRKY family members may play important roles in the maturation of root cells, senescence, dormancy, fruit maturity, tannin synthesis in the seed coat, and embryo development [196]); (iv) Myeloblastosis (MYB), which is involved in the regulation of primary and secondary metabolism, the control of cell development and the cell cycle as well as in the participation in defense and response to various biotic and abiotic stresses, hormone synthesis, and signal transduction [195]; and (v) NACs, which have a variety of significant functions in plant development as well as in responses to biotic (defense to parasites) and abiotic stresses. The establishment of shoot apical meristems, lateral root development, senescence, cell wall construction, and secondary metabolism are other processes that NACs control in plants [195].

7. Transgenic Approach

The main response mechanisms to drought stress were uncovered by studying *A. thaliana*, and multiple drought tolerance genes that are highly conserved among plants are being engineered into crops [197]. These genes relate to drought escape, control of flowering time, stomatal responses, T6P pathways, and some root traits. So far, most plants with enhanced drought resistance have displayed reduced crop yield. However, the uncoupling of drought tolerance and plant growth seems possible at least in Arabidopsis since it was shown that increasing brassinosteroid receptors in vascular plant tissues confers tolerance to drought without penalizing growth [198]. Arabidopsis is a good model to test drought-responsive strategies that may have interesting agronomic potential and might be translated into crops by transformation. Of course, this issue must be carefully addressed since genes from different plant species may not behave in the same way from one system to the other. One obvious example is that the GC content of an Arabidopsis gene will be below that of the grass genome context and would probably end up methylated. The question here is not merely to transfer a gene as it is but rather to transfer the functional information it carries.

Plants exposed to abiotic stress exhibit enhanced levels of ROS, which affect chlorophyll biosynthesis, photosynthetic capacity, carbohydrate, protein, lipid, and antioxidant enzyme activities. Thus, transgenic breeding offers a tempting platform to mitigate these negative effects [199].

However, pyramidizing genes by transgeny may be difficult because stress tolerance phenotypes are QTLs. They must first be dissected to exploit their use by accelerating genetic introgression using molecular markers or site-directed mutagenesis [200].

Genetic improvement for drought-tolerant wheat cultivars is a major aim of wheat breeders as indicated in Table S3. This table summarizes the current genes used to improve the drought tolerance of wheat through genetic engineering [201].

8. Epigenetics and Crop Improvement

As seen above, epigenetics is another layer of information enabling an organism to adapt to environmental conditions such as biotic and abiotic stresses. This adaptation occurs by modulating massive gene expression through hundreds of molecular players including ncRNAs [202,203]. In this way, epigenetic changes act as a buffer mechanism to alleviate short term stressful effects from the environment by adapting the organism's gene expression response [204]. When stress conditions last for a long time, the gene expression profile may, eventually, be passed on to the next generation, but the benefits could be at the expense of undesirable phenotypic traits as well [205], which means that the mechanism involved must be well understood. Both short and long-term memory has specific advantages and drawbacks that favor selective breeding to produce so-called smart-crops [206]. However, a difficulty with epigenetics is that the effect it induces can be confused with the genetic trait that supports it. Hofmeister et al. [207] developed an epigenotyping procedure that enabled uncoupling this type of interaction, which allows a better understanding of patterns of epiallele inheritance. In Arabidopsis, exposure to abiotic stress during several generations was observed to induce heritable phenotypic changes, but the observed effects depended on plant genotype, which suggested an interaction between genetic background and inheritance of induced epigenetic patterns [208]. Since the experiment involved several generations, allele shuffling must have occurred from one generation to the next and even under the hypothesis that the epigenetic pattern would be the same, different phenotypes would result because of different allele combinations being expressed. Nonetheless, there is no reason to think that it is not possible to select alleles for the epigenetic (epialleles) process itself and this could result in different epigenetic patterning. Surveys have shown that stable epialleles can accumulate, segregate in populations, and be selected for [209–211]. Stable epialleles can therefore be linked to quantitative trait loci (epiQTL). To identify epiQTLs, epigenetic recombinant inbred lines (epiRILs) are created using genetically identical parental (isogenic) plants, which differ in levels of DNA methylation [207]. Epigenetic traits that could contribute to improving drought tolerance were reviewed in [206] and could be obtained by recurrent epi-selection, hybrid mimics, epigenomic selection, epigenome editing, and stress priming mechanisms.

Recently, the msh1 mutant was discovered to induce epigenetic reprogramming. The MSH1 is a plant-specific gene which encodes a protein that targets plastids. Disruption of MSH1 function leads to variation in plant growth rate, flowering time, response to short day length, leaf morphology, variegation, and stress response, which are phenotypes that are reproducible across a range of plant species [212]. The msh1 mutant induces genome-wide DNA methylation repatterning, changes in siRNA expression, and heritable nongenetic memory. Gene networks affected by methylation repatterning include auxin-related pathways, so that altered expression of auxin-response genes contributes to the increased plant vigor phenotype in a way similar to heterosis [213].

As an example of an MSH1 system application for the induction of epigenetic variation in crops, Raju et al. [214] developed epi-lines by crossing a wild type with an isogenic transgenic soybean line for msh1. The progeny showed a wide variation both in greenhouse and field trials and a low epitype–environment interaction, which indicated yield stability and low effect of environmental constraints.

One may conclude from the experiments with msh1 that involvement of a significant amount of individual RNA and proteins seems to be essential for manifesting phenotypic diversity and phenotypic expression of heterosis and hybrid vigor. According to Bharti Thapa and Shrestha [215], the best hybrids seem to have a significantly larger number of gene expressed compared to the best parent at different stages. Concerning methylation level, it was found that (i) it is lower in heterotic hybrids than in related non-heterotic hybrids, (ii) it is higher in old, low yielding inbred lines, and (iii) it is lower in modern inbred lines, especially those selected for high yield [215].

As seen above, plant germ lines develop from somatic cells [216], which means that epigenetic information can be transferred by vegetative cloning, graft, or through gametogenesis, but epigenetic states can often be unstable in this last case. Vegetative cloning and graft are not feasible in cereals, but apomixis seems to be a viable option by enabling F1 hybrids with superior traits to be clonally propagated. Engineering synthetic apomixis was performed in inbred rice by converting the meiotic division into mitotic division and triggering parthenogenesis. A hybrid apomictic clone that produces more than 95% of clonal seeds across at least three generations was produced from a commercial F1 rice hybrid by T-DNA transformation of embryo-derived callus [217]. This achievement combined with the msh1 mutation opens the way to induce and stabilize epigenomes in crops, which offers the potential of accelerating selective breeding for drought tolerance in arid and semi-arid regions.

9. Conclusions

Understanding the physiological processes that lead to increased drought tolerance in plant models such as Arabidopsis, rice, and maize is helping to map the genes that can be validated as functional for selective breeding purposes. Drought tolerance is a QTL, but some of the genes involved in that QTL usually explain a larger proportion of the detected phenotype. Genetic and physical mapping through SNP association and GWAS helps to detect these genes, which can then be proven to be involved in drought tolerance by functional assays such as knockdown. Such genes can then be transferred by transgenesis or by selective breeding. According to the classical approach, in vitro culture and transgenesis help to produce clones carrying specific traits that can then be crossed with elite lines for pyramiding genes. Nonetheless, improving the understanding of epigenetics and stress memory will also contribute to the selective breeding of complex agronomic traits. The uncovering of DNA-binding sites of particular binding proteins, such as transcription factors or chromatin-associated proteins is enabling the exploration of the genome epigenetic landscape for such purposes. However, epigenomes are known to be unstable from one generation to the next, which makes their industrial exploitation difficult. Recently the discovery of the msh1 mutation has made it possible to produce epigenetic variability in several crops with similar effects to heterosis. New techniques of synthetic apomixis are being developed that permit clonal propagation accessions for several generations, which enable us to glimpse the future contribution of epigenetics to selective breeding including for drought tolerance.

Supplementary Materials: The following supporting information can be downloaded at: https://www.mdpi.com/article/10.3390/plants12112170/s1, References [218–260] are cited in the supplementary materials. Table S1a: QTLs of agronomic traits identified for drought tolerance in cereals (updated from [29]); Table S1b: QTLs of agronomic traits identified in soybean for drought tolerance (updated from [228]); Table S2: Drought stress and their related candidate genes in model plants (updated from [29]); Table S3: Improving drought tolerance of cereals through engineering genes (updated from [201]).

Author Contributions: Conceptualization, S.A. and N.C.; formal analysis, S.A. and N.C.; investigation, S.A. and N.C.; resources, S.A. and N.C.; data curation, S.A. and N.C.; writing—original draft preparation, S.A. and N.C.; writing—review and editing, S.A. and N.C.; visualization, S.A. and N.C.; supervision, N.C.; project administration, S.A. and N.C.; funding acquisition, N.C. All authors have read and agreed to the published version of the manuscript.

Funding: The publication charges of this review were provided by Fiocruz (process number 25380.000890/2023-58).

Data Availability Statement: Not applicable.

Acknowledgments: Distribution of original figures were performed for (i) Figure 1 by Sci. Rep. (Nature) under CC BY 4.0, (ii) Figure 2 by Genomics (Elsevier) under CC BY 4.0, (iv) Figure 3 by Front. Cell. Dev. Biol. (Frontiers) under CC BY.

Conflicts of Interest: The authors declare no conflict of interest.

References

1. Savickienė, J.; Miceikienė, A. Sustainable economic development assessment model for family farms. *Agric. Econ. - Czech* **2018**, *64*, 527–535. [CrossRef]
2. Stoddard, F.L.; Mäkelä, P.S.A.; Puhakainen, T. Adaptation of boreal field crop production to climate change. In *Climate Change—Research and Technology for Adaptation and Mitigation*; Blanco, J., Kheradmand, H., Eds.; IntechOpen Limited: London, UK, 2011; Chapter 21; Available online: https://www.intechopen.com/chapters/18730 (accessed on 10 March 2023). [CrossRef]
3. Demelash, T.; Amou, M.; Gyilbag, A.; Tesfay, G.; Xu, Y. Adaptation potential of current wheat cultivars and planting dates under the changing climate in Ethiopia. *Agronomy* **2022**, *12*, 37. [CrossRef]
4. Meza, I.; Siebert, S.; Döll, P.; Kusche, J.; Herbert, C.; Rezaei, E.E.; Nouri, H.; Gerdener, H.; Popat, E.; Frischen, J.; et al. Global-scale drought risk assessment for agricultural systems. *Nat. Hazards Earth Syst. Sci.* **2020**, *20*, 695–712. [CrossRef]
5. Mohammadi, R.; Golkari, S. Genetic resources for enhancing drought tolerance from a mini-core collection of spring bread wheat (*Triticum aestivum* L.). *Acta Sci. Agron.* **2022**, *44*, e56129. [CrossRef]
6. Bocci, M.; Smanis, T.; Gaaloul, N. *Assessment of the Impacts of Climate Change on the Agriculture Sector in the Southern Mediterranean: Foreseen Developments and Policy Measures*; Union for the Mediterranean, European Union: Barcelona, Spain, 2019; 36p, Available online: https://ufmsecretariat.org/wp-content/uploads/2019/04/Climate-Change-impact-on-Agriculture.pdf (accessed on 14 January 2023).
7. Motiee, H.; Salamat, A.; Mc Bean, E.E. Drought as a water related disaster; a case study of Oroomieh lake. *Aqua-Lac* **2012**, *4*, 7–18. Available online: https://unesdoc.unesco.org/ark:/48223/pf0000227014 (accessed on 14 January 2023).
8. Vatter, J.; Wagnitz, P.; Schmiester, J.; Hernandez, E. *Drought Risk: The Global Thirst for Water in the Era of Climate Crisis*; Hartwig, L., Ed.; WWF Germany: Berlin, Germany, 2019; 27p, Available online: https://wwfint.awsassets.panda.org/downloads/drought_risk__wwf_.pdf (accessed on 14 January 2023).
9. Siswanto, S.; Wardani, K.K.; Purbantoro, B.; Rustanto, A.; Zulkarnain, F.; Anggraheni, E.; Dewanti, R.; Nurlambang, T.; Dimyati, M. Satellite-based meteorological drought indicator to support food security in Java Island. *PLoS ONE* **2022**, *17*, e0260982. [CrossRef]
10. WWDR. *The United Nations World Water Development Report. Valuing Water*; UNESCO: Paris, France, 2021; 206p, Available online: https://unesdoc.unesco.org/ark:/48223/pf0000375724 (accessed on 14 January 2023).
11. Williamson, K.; Satre-Meloy, A.; Velasco, K.; Green, K. *Climate Change Needs Behavior Change: Making the Case for Behavioral Solutions to Reduce Global Warming*; Rare: Arlington VA, USA, 2018; 56p, Available online: https://rare.org/wp-content/uploads/2019/02/2018-CCNBC-Report.pdf (accessed on 14 January 2023).
12. Velimirovic, A.; Jovovic, Z.; Przulj, N. From neolithic to late modern period: Brief history of wheat. *Genetika* **2021**, *53*, 407–417. [CrossRef]
13. Chaves, M.C.; Martinelli, J.A.; Guterres, C.W.; Graichen, F.A.S.; Brammer, S.P.; Scagliusi, S.M.; Da Silva, P.R.; Wiethölter, P.; Torres, G.A.M.; Lau, E.Y.; et al. The importance for food security of maintaining rust resistance in wheat. *Food Secur.* **2013**, *5*, 157–176. [CrossRef]
14. Bhardwaj, S.C.; Singh, G.P.; Gangwar, O.P.; Prasad, P.; Kumar, S. Status of wheat rust research and progress in rust management-indian context. *Agronomy* **2019**, *9*, 892. [CrossRef]
15. Gu, D.; Andreev, K.; Dupre, M.E. Major trends in population growth around the world. *China CDC Wkly.* **2021**, *3*, 604–613. [CrossRef]
16. Buettner, T. World Population Prospects—A Long View. *Econ. Stat. /Econ. Stat.* **2020**, *520–521*, 9–27. [CrossRef]
17. Diffenbaugh, N.S.; Singh, D.; Mankin, J.S.; Rajaratnam, B. Quantifying the influence of global warming on unprecedented extreme climate events. *Proc. Natl. Acad. Sci. USA* **2017**, *114*, 4881–4886. [CrossRef]
18. Ornes, S. How does climate change influence extreme weather? Impact attribution research seeks answers. *Proc. Natl. Acad. Sci. USA* **2018**, *115*, 8232–8235. [CrossRef]
19. Mahpara, S.; Hussain, S.T.; Farooq, J. Drought tolerance studies in wheat (*Triticum aestivum* L.). *Cercet. Agron. Mold.* **2014**, *47*, 133–140. [CrossRef]
20. Bapela, T.; Shimelis, H.; Tsilo, T.J.; Mathew, I. Genetic improvement of wheat for drought tolerance: Progress, challenges and opportunities. *Plants* **2022**, *11*, 1331. [CrossRef]
21. IPCC. Climate Change 2014: Synthesis Report. In *Contribution of Working Groups I, II and III to the Fifth Assessment Report of the Intergovernmental Panel on Climate Change*; Core Writing Team, Pachauri, R.K., Meyer, L.A., Eds.; IPCC: Geneva, Switzerland; 151p, Available online: https://www.ipcc.ch/site/assets/uploads/2018/02/AR5_SYR_FINAL_Front_matters.pdf (accessed on 14 January 2023).
22. Climate-ADAPT. 2022. Available online: https://climate-adapt.eea.europa.eu/en/observatory/policy-context/european-policy-framework/climate-adaptation-and-health (accessed on 14 January 2023).

23. Asseng, S.; Kheir, A.M.S.; Kassie, B.T.; Hoogenboom, G.; Abdelaal, A.I.N.; Haman, D.Z.; Ruane, A.C. Can Egypt become self-sufficient in wheat? *Environ. Res. Lett.* **2018**, *13*, 094012. [CrossRef]
24. Heiba, S.A.A.; Osman, S.A.; Eldessouky, S.E.I.; Haiba, A.A.A.; Ali, R.T. Genetic and biochemical studies on some Egyptian wheat genotypes under drought stress. *Bull. Natl. Res. Cent.* **2021**, *45*, 151. [CrossRef]
25. Sayed, M.A.; El-sadek, A.N.; Bakry, B.A.; Ali, M.B.; Leon, J.; Salem, E.M. QTL analysis in Barley Across Environments in Egypt. *Egypt. J. Agron.* **2017**, *39*, 53–70. [CrossRef]
26. Ata, R.M.; Fahmy, A.A. Wheat Growers' (WGs) practices to adapt with climate change in the Nile Delta(ND) of Egypt. *J. Agric. Sci.* **2021**, *3*, 317–326. [CrossRef]
27. Bowles, A.M.C.; Paps, J.; Bechtold, U. Evolutionary origins of drought tolerance in spermatophytes. *Front. Plant Sci.* **2021**, *12*, 655924. [CrossRef]
28. Bennani, S.; Birouk, A.; Jlibene, M.; Sanchez-Garcia, M.; Nsarellah, N.; Gaboun, F.; Tadesse, W. Drought-tolerance QTLs associated with grain yield and related traits in spring bread wheat. *Plants* **2022**, *11*, 986. [CrossRef]
29. Andleeb, T.; Shah, T.; Nawaz, R.; Munir, I.; Munsif, F.; Jalal, A. QTL mapping for drought stress tolerance in plants. In *Salt and Drought Stress Tolerance in Plants*; Signaling and Communication in Plants, Hasanuzzaman, M., Tanveer, M., Eds.; Springer: Cham, Switzerland, 2020; pp. 383–403.
30. Brensha, W.; Kantartzi, S.K.; Meksem, K.; Grier IV, R.L.; Barakat, A.; Lightfoot, D.A.; Kassem, M.A. Genetic analysis of root and shoot traits in the 'Essex' by 'Forrest' Recombinant Inbred Line (RIL) population of soybean [*Glycine max* (L.) Merr.]. *Plant Genet. Genom. Biotechnol.* **2012**, *1*, 1–9. [CrossRef]
31. Gupta, P.K.; Balyan, H.S.; Gahlaut, V. QTL analysis for drought tolerance in wheat: Present status and future possibilities. *Agronomy* **2017**, *7*, 5. [CrossRef]
32. Jiang, Y.; Wang, X.; Yu, X.; Zhao, X.; Luo, N.; Pei, Z.; Liu, H.; Garvin, D.F. Quantitative trait loci associated with drought tolerance in *Brachypodium distachyon*. *Front. Plant Sci.* **2017**, *8*, 811. [CrossRef]
33. Gudys, K.; Guzy-Wrobelska, J.; Janiak, A.; Dziurka, M.A.; Ostrowska, A.; Hura, K.; Jurczyk, B.; Żmuda, K.; Grzybkowska, D.; Śróbka, J.; et al. Prioritization of candidate genes in QTL regions for physiological and biochemical traits underlying drought response in barley (*Hordeum vulgare* L.). *Front. Plant Sci.* **2018**, *9*, 769. [CrossRef] [PubMed]
34. Dossa, K.; Li, D.; Zhou, R.; Yu, J.; Wang, L.; Zhang, Y.; You, J.; Liu, A.; Mmadi, M.A.; Fonceka, D.; et al. The genetic basis of drought tolerance in the high oil crop *Sesamum indicum*. *Plant Biotechnol. J.* **2019**, *17*, 1788–1803. [CrossRef] [PubMed]
35. Pham, A.T.; Maurer, A.; Pillen, K.; Brien, C.; Dowling, K.; Berger, B.; Eglinton, J.K.; March, T.J. Genome-wide association of barley plant growth under drought stress using a nested association mapping population. *BMC Plant Biol.* **2019**, *19*, 134. [CrossRef] [PubMed]
36. Rodziewicz, P.; Chmielewska, K.; Sawikowska, A.; Marczak, L.; Łuczak, M.; Bednarek, P.; Mikołajczak, K.; Ogrodowicz, P.; Kuczyńska, A.; Krajewski, P.; et al. Identification of drought responsive proteins and related proteomic QTLs in barley. *J. Exp. Bot.* **2019**, *70*, 2823–2837. [CrossRef] [PubMed]
37. Diouf, I.; Albert, E.; Duboscq, R.; Santoni, S.; Bitton, F.; Gricourt, J.; Causse, M. Integration of QTL, transcriptome and polymorphism studies reveals candidate genes for water stress response in tomato. *Genes* **2020**, *11*, 900. [CrossRef]
38. Moursi, Y.S.; Thabet, S.G.; Amro, A.; Dawood, M.F.A.; Baenziger, P.S.; Sallam, A. Detailed genetic analysis for identifying QTLs associated with drought tolerance at seed germination and seedling stages in barley. *Plants* **2020**, *9*, 1425. [CrossRef]
39. Raman, H.; Raman, R.; Mathews, K.; Diffey, S.; Salisbury, P. QTL mapping reveals genomic regions for yield based on an incremental tolerance index to drought stress and related agronomic traits in canola. *Crop Pasture Sci.* **2020**, *71*, 562–577. [CrossRef]
40. Verma, A.; Niranjana, M.; Jha, S.K.; Mallick, N.; Agarwal, P. QTL detection and putative candidate gene prediction for leaf rolling under moisture stress condition in wheat. *Sci. Rep.* **2020**, *10*, 18696. [CrossRef] [PubMed]
41. Hu, X.; Wang, G.; Du, X.; Zhang, H.; Xu, Z.; Wang, J.; Chen, G.; Wang, B.; Li, X.; Chen, X.; et al. QTL analysis across multiple environments reveals promising chromosome regions associated with yield-related traits in maize under drought conditions. *Crop J.* **2021**, *9*, 759–766. [CrossRef]
42. Magwanga, R.O.; Lu, P.; Kirungu, J.N.; Cai, X.; Zhou, Z.; Agong, S.G.; Wang, K.; Liu, F. Identification of QTLs and candidate genes for physiological traits associated with drought tolerance in cotton. *J. Cotton Res.* **2020**, *3*, 3. [CrossRef]
43. Gad, M.; Chao, H.; Li, H.; Zhao, W.; Lu, G.; Li, M. QTL mapping for seed germination response to drought stress in *Brassica napus*. *Front. Plant Sci.* **2021**, *11*, 629970. [CrossRef] [PubMed]
44. Liang, J.; Sun, J.; Ye, Y.; Yan, X.; Yan, T.; Rao, Y.; Zhou, H.; Le, M. QTL mapping of PEG-induced drought tolerance at the early seedling stage in sesame using whole genome re-sequencing. *PLoS ONE* **2021**, *16*, e0247681. [CrossRef]
45. Ren, Y.; Liu, J.; Zhang, J.; Dreisigacker, S.; Xia, X.; Geng, H. QTL mapping of drought tolerance at germination stage in wheat using the 50 K SNP array. *Plant Genet. Resour. Characterisation Util.* **2021**, *19*, 453–460. [CrossRef]
46. Selamat, N.; Nadarajah, K.K. Meta-analysis of quantitative traits loci (QTL) identified in drought response in rice (*Oryza sativa* L.). *Plants* **2021**, *10*, 716. [CrossRef] [PubMed]
47. Ouyang, W.; Chen, L.; Ma, J.; Liu, X.; Chen, H.; Yang, H.; Guo, W.; Shan, Z.; Yang, Z.; Chen, S.; et al. Identification of Quantitative Trait Locus and Candidate Genes for Drought Tolerance in a Soybean Recombinant Inbred Line Population. *Int. J. Mol. Sci.* **2022**, *23*, 10828. [CrossRef]

48. Saini, D.K.; Srivastava, P.; Pal, N.; Gupta, P.K. Meta-QTLs, ortho-meta-QTLs and candidate genes for grain yield and associated traits in wheat (*Triticum aestivum* L.). *Theor. Appl. Genet.* **2022**, *135*, 1049–1081. [CrossRef]
49. Maulana, F.; Huang, W.; Anderson, J.D.; Ma, X.F. Genome-wide association mapping of seedling drought tolerance in winter wheat. *Front. Plant Sci.* **2020**, *11*, 573786. [CrossRef]
50. El Gataa, Z.; Samir, K.; Tadesse, W. Genetic dissection of drought tolerance of elite bread wheat (*Triticum aestivum* L.) genotypes using genome wide association study in Morocco. *Plants* **2022**, *11*, 2705. [CrossRef] [PubMed]
51. Sallam, A.; Eltaher, S.; Alqudah, A.M.; Belamkar, V.; Baenziger, P.S. Combined GWAS and QTL mapping revealed candidate genes and SNP network controlling recovery and tolerance traits associated with drought tolerance in seedling winter wheat. *Genomics* **2022**, *114*, 110358. [CrossRef] [PubMed]
52. Mathew, I.; Shimelis, H.; Shayanowako, A.I.T.; Laing, M.; Chaplot, V. Genome-wide association study of drought tolerance and biomass allocation in wheat. *PLoS ONE* **2019**, *14*, e0225383. [CrossRef] [PubMed]
53. Hussain, W.; Baenziger, P.S.; Belamkar, V.; Guttieri, M.J.; Venegas, J.P.; Easterly, A.; Sallam, A.; Poland, J. Genotyping-by-sequencing derived high-density linkage map and its application to QTL mapping of flag leaf traits in bread wheat. *Sci. Rep.* **2017**, *7*, 16394. [CrossRef] [PubMed]
54. Soderlund, C.; Descour, A.; Kudrna, D.; Bomhoff, M.; Boyd, L.; Currie, J.; Angelova, A.; Collura, K.; Wissotski, M.; Ashley, E.; et al. Sequencing, mapping, and analysis of 27,455 maize full-length cDNAs. *PLoS Genet.* **2009**, *5*, e1000740. [CrossRef]
55. Pingault, L.; Choulet, F.; Alberti, A.; Glover, N.; Wincker, P.; Feuillet, C.; Paux, E. Deep transcriptome sequencing provides new insights into the structural and functional organization of the wheat genome. *Genome Biol.* **2015**, *16*, 29. [CrossRef]
56. Chen, R.; Song, W.; Li, X.; An, Z. Chromosome G-banding in plants by inducing with trypsin and urea. *Cell Res.* **1994**, *4*, 79–87. [CrossRef]
57. Gill, K.S.; Gill, B.S.; Endo, T.R. A chromosome region-specific mapping strategy reveals gene-rich telomeric ends in wheat. *Chromosoma* **1993**, *102*, 374–381. [CrossRef]
58. Gill, K.S.; Gill, B.S.; Endo, T.R.; Boyko, E.V. Identification and high-density mapping of gene-rich regions in chromosome group 5 of wheat. *Genetics* **1996**, *143*, 1001–1012. [CrossRef] [PubMed]
59. Gill, K.S.; Gill, B.S.; Endo, T.R.; Taylor, T. Identification and high-density mapping of gene-rich regions inchromosome group 1 of wheat. *Genetics* **1996**, *144*, 1883–1891. [CrossRef]
60. Sidhu, D.; Gill, K.S. Distribution of genes and recombination in wheat and other eukaryotes. *Plant Cell Tissue Organ Cult.* **2005**, *79*, 257–270. [CrossRef]
61. Wu, J.; Maehara, T.; Shimokawa, T.; Yamamoto, S.; Harada, C.; Takazaki, Y.; Ono, N.; Mukai, Y.; Koike, K.; Yazaki, J.; et al. A comprehensive rice transcript map containing 6591 expressed sequence tag sites. *Plant Cell* **2002**, *14*, 525–535. [CrossRef] [PubMed]
62. Kianian, P.M.A.; Wang, M.; Simons, K.; Ghavami, F.; He, Y.; Dukowic-Schulze, S.; Sundararajan, A.; Sun, Q.; Pillardy, J.; Mudge, J.; et al. High-resolution crossover mapping reveals similarities and differences of male and female recombination in maize. *Nat. Commun.* **2018**, *9*, 2370. [CrossRef] [PubMed]
63. Moore, G.; Devos, K.M.; Wang, Z.; Gale, M.D. Grasses, line up and form a circle. *Curr. Biol.* **1995**, *5*, 737–739. [CrossRef]
64. Moore, G.; Foote, T.; Helentjaris, T.; Devos, K.; Kurata, N.; Gale, M. Was there a single ancestral cereal chromosome? *Trends Genet.* **1995**, *11*, 81–82. [CrossRef] [PubMed]
65. Gale, M.D.; Devos, K.M. Comparative genetics in the grasses. *Proc. Natl. Acad. Sci. USA* **1998**, *95*, 1971–1974. [CrossRef]
66. Visendi, P.; Batley, J.; Edwards, D. Next generation characterisation of cereal genomes for marker discovery. *Biology* **2013**, *2*, 1357–1377. [CrossRef]
67. Faye, J.M.; Akata, E.A.; Sine, B.; Diatta, C.; Cisse, N.; Fonceka, D.; Morris, G.P. Quantitative and population genomics suggest a broad role of stay-green loci in the drought adaptation of sorghum. *Plant Genome* **2022**, *15*, e20176. [CrossRef]
68. Feuillet, C.; Keller, B. High gene density is conserved at syntenic loci of small and large grass genomes. *Proc. Natl. Acad. Sci. USA* **1999**, *96*, 8265–8270. [CrossRef]
69. Paterson, A.H.; Lan, T.-H.; Reischmann, K.P.; Chang, C.; Lin, Y.-R.; Liu, S.-C.; Burow, M.D.; Kowalski, S.P.; Katsar, C.S.; DelMonte, T.A.; et al. Toward a unified genetic map of higher plants, transcending the monocot-dicot divergence. *Nat. Genet.* **1996**, *14*, 380–382. [CrossRef] [PubMed]
70. Salse, J.; Feuillet, C. Palaeogenomics in cereals: Modeling of ancestors for modern species improvement. *Comptes Rendus Biol.* **2011**, *334*, 205–211. [CrossRef] [PubMed]
71. Zhao, T.; Schranz, M.E. Network-based microsynteny analysis identifies major differences and genomic outliers in mammalian and angiosperm genomes. *Proc. Natl. Acad. Sci. USA* **2019**, *116*, 2165–2174. [CrossRef]
72. Zhao, T.; Zwaenepoel, A.; Xue, J.-Y.; Kao, S.-M.; Li, Z.; Schranz, M.E.; Van de Peer, Y. Whole-genome microsynteny-based phylogeny of angiosperms. *Nat. Commun.* **2021**, *12*, 3498. [CrossRef]
73. Shaar-Moshe, L.; Hübner, S.; Peleg, Z. Identification of conserved drought-adaptive genes using a cross-species meta-analysis approach. *BMC Plant Biol.* **2015**, *15*, 111. [CrossRef]
74. Maina, F.; Harou, A.; Hamidou, F.; Morris, G.P. Ecophysiological genomics identifies a pleiotropic locus mediating drought tolerance in sorghum. *bioRxiv* **2021**. [CrossRef]
75. Abdel-Ghany, S.E.; Ullah, F.; Ben-Hur, A.; Reddy, A.S.N. Transcriptome analysis of drought-resistant and drought-sensitive sorghum (*Sorghum bicolor*) genotypes in response to PEG-induced drought stress. *Int. J. Mol. Sci.* **2020**, *21*, 772. [CrossRef]

76. Galyuon, I.K.A.; Gay, A.; Borrell, A.K.; Howarth, C.J. Physiological and biochemical basis for stay-green trait in sorghum. *Afr. Crop Sci. J.* **2019**, *7*, 653–677. [CrossRef]
77. Sanchez, A.C.; Subudhi, P.K.; Rosenow, D.T.; Nguyen, H.T. Mapping QTLs associated with drought resistance in sorghum (*Sorghum bicolor* L. Moench). *Plant Mol. Biol.* **2002**, *48*, 713–726. [CrossRef] [PubMed]
78. Latif, S.; Wang, L.; Khan, J.; Ali, Z.; Sehgal, S.K.; Babar, M.A.; Wang, J.; Quraishi, U.M. Deciphering the role of stay-green trait to mitigate terminal heat stress in bread wheat. *Agronomy* **2020**, *10*, 1001. [CrossRef]
79. Tuberosa, R. Phenotyping for drought tolerance of crops in the genomics era. *Front. Physiol.* **2012**, *3*, 347. [CrossRef] [PubMed]
80. Mickelbart, M.V.; Hasegawa, P.M.; Bailey-Serres, J. Genetic mechanisms of abiotic stress tolerance that translate to crop yield stability. *Nat. Rev. Genet.* **2015**, *16*, 237–251. [CrossRef] [PubMed]
81. Tardieu, F.; Simonneau, T.; Muller, B. The physiological basis of drought tolerance in crop plants: A scenario-dependent probabilistic approach. *Annu. Rev. Plant Biol.* **2018**, *69*, 733–759. [CrossRef]
82. Bernardi, G. The "genomic code": DNA pervasively moulds chromatin structures leaving no room for "Junk". *Life* **2021**, *11*, 342. [CrossRef]
83. Bernardi, G.; Hughes, S.; Mouchiroud, D. The major compositional transitions in the vertebrate genome. *J. Mol. Evol.* **1997**, *44*, S44–S51. [CrossRef] [PubMed]
84. Costantini, M.; Clay, O.; Auletta, F.; Bernardi, G. An isochore map of human chromosomes. *Genome Res.* **2006**, *16*, 536–541. [CrossRef]
85. Grantham, R.; Gautier, G.; Gouy, M.; Mercier, R.; Paré, A. Codon catalog usage and the genome hypothesis. *Nucleic Acids Res.* **1980**, *8*, r49–r62. [CrossRef]
86. Carels, N. The maize gene space is compositionally compartimentalized. *FEBS Lett.* **2005**, *579*, 3867–3871. [CrossRef]
87. Wu, H.; Zhang, Z.; Hu, S.; Yu, J. On the molecular mechanism of GC content variation among eubacterial genomes. *Biol. Direct* **2012**, *7*, 2. [CrossRef]
88. D'Onofrio, G.; Jabbari, K.; Musto, H.; Bernardi, G. The correlation of protein hydropathy with the base composition of coding sequences. *Gene* **1999**, *238*, 3–14. [CrossRef]
89. Ponce de Leon, M.; de Miranda, A.B.; Alvarez-Valin, F.; Carels, N. The purine bias of coding sequences is encoded by physico-chemical constraints on proteins. *Bioinform. Biol. Insights* **2014**, *8*, 93–108. [CrossRef]
90. Salinas, J.; Matassi, G.; Montero, L.M.; Bernardi, G. Compositionnal compartmentalization and compositionnal patterns in the nuclear genomes of plants. *Nucleic Acids Res.* **1988**, *16*, 4269–4285. [CrossRef] [PubMed]
91. Carels, N.; Hatey, P.; Jabbari, K.; Bernardi, G. Compositional properties of homologous coding sequences from plants. *J. Mol. Evol.* **1998**, *46*, 45–53. [CrossRef] [PubMed]
92. Serres-Giardi, L.; Belkhir, K.; David, J.; Glémin, S. Patterns and evolution of nucleotide landscapes in seed plants. *Plant Cell* **2012**, *24*, 1379–1397. [CrossRef] [PubMed]
93. Carels, N. The genome organization of angiosperms. In *Recent Research Developments in Plant Science*; Pandalai, S.G., Ed.; Research Signpost: Trivandrum, India, 2005; pp. 129–194.
94. Federico, C.; Andreozzi, L.; Saccone, S.; Bernardi, G. Gene density in the Giemsa bands of human chromosomes. *Chromosome Res.* **2000**, *8*, 737–746. [CrossRef] [PubMed]
95. Saccone, S.; Federico, C.; Bernardi, G. Localization of the gene-richest and the gene-poorest isochores in the interphase nuclei of mammals and birds. *Gene* **2002**, *300*, 169–178. [CrossRef]
96. Cremer, T.; Cremer, C. Chromosome territories, nuclear architecture and gene regulation in mammalian cells. *Nat. Rev. Genet.* **2001**, *2*, 292–301. [CrossRef]
97. Lieberman-Aiden, E.; van Berkum, N.L.; Williams, L.; Imakaev, M.; Ragoczy, T.; Telling, A.; Amit, I.; Lajoie, B.R.; Sabo, P.J.; Dorschner, M.O.; et al. Comprehensive mapping of long-range interactions reveals folding principles of the human genome. *Science* **2009**, *326*, 289–293. [CrossRef]
98. Knoch, T.A.; Wachsmuth, M.; Kepper, N.; Lesnussa, M.; Abuseiris, A.; Ali Imam, A.M.; Kolovos, P.; Zuin, J.; Kockx, C.E.M.; Brouwer, R.W.W.; et al. The detailed 3D multi-loop aggregate/rosette chromatin architecture and functional dynamic organization of the human and mouse genomes. *Epigenetics Chromatin* **2016**, *9*, 58. [CrossRef]
99. Ansari, H.A.; Ellison, N.W.; Bassett, S.A.; Hussain, S.W.; Bryan, G.T.; Williams, W.M. Fluorescence chromosome banding and FISH mapping in perennial ryegrass, *Lolium perenne* L. *BMC Genom.* **2016**, *17*, 977. [CrossRef]
100. Pavlícek, A.; Jabbari, K.; Paces, J.; Paces, V.; Hejnar, J.V.; Bernardi, G. Similar integration but different stability of Alus and LINEs in the human genome. *Gene* **2001**, *276*, 39–45. [CrossRef] [PubMed]
101. Carels, N.; Bernardi, G. The compositional organization and the expression of the Arabidopsis genome. *FEBS Lett.* **2000**, *472*, 302–306. [CrossRef] [PubMed]
102. Carels, N. A history of genomic structures: The big picture. In *Plant Biology and Biotechnology: Plant Genomics and Biotechnology*; Bahadur, B., Venkat Rajam, M., Sahijram, L., Krishnamurthy, K.V., Eds.; Springer: New Delhi, India, 2015; Volume 2, pp. 131–178. [CrossRef]
103. Liu, J.; Zhang, L.; Quan, H.; Tian, H.; Meng, L.; Yang, L.; Feng, H.; Gao, Y.Q. From 1D sequence to 3D chromatin dynamics and cellular functions: A phase separation perspective. *Nucleic Acids Res.* **2018**, *46*, 9367–9383. [CrossRef]
104. Carels, N.; Bernardi, G. Two classes of genes in plants. *Genetics* **2000**, *154*, 1819–1825. [CrossRef]

105. Tiang, C.L.; He, Y.; Pawlowski, W.P. Chromosome organization and dynamics during interphase, mitosis, and meiosis in plants. *Plant Physiol.* **2011**, *158*, 26–34. [CrossRef]
106. Fransz, P.; De Jong, J.H.; Lysak, M.; Castiglione, M.R.; Schubert, I. Interphase chromosomes in Arabidopsis are organized as well defined chromocenters from which euchromatin loops emanate. *Proc. Natl. Acad. Sci. USA* **2002**, *99*, 14584–14589. [CrossRef]
107. Schubert, I.; Shaw, P. Organization and dynamics of plant interphase chromosomes. *Trends Plant Sci.* **2011**, *16*, 273–281. [CrossRef] [PubMed]
108. Pecinka, A.; Schubert, V.; Meister, A.; Kreth, G.; Klatte, M.; Lysak, M.A.; Fuchs, J.; Schubert, I. Chromosome territory arrangement and homologous pairing in nuclei of Arabidopsis thaliana are predominantly random except for NOR-bearing chromosomes. *Chromosoma* **2004**, *113*, 258–269. [CrossRef]
109. Concia, L.; Veluchamy, A.; Ramirez-Prado, J.S.; Martin-Ramirez, A.; Huang, Y.; Perez, M.; Domenichini, S.; Rodriguez Granados, N.Y.; Kim, S.; Blein, T.; et al. Wheat chromatin architecture is organized in genome territories and transcription factories. *Genome Biol.* **2020**, *21*, 104. [CrossRef]
110. Carels, N.; Ponce de Leon, M. An interpretation of the ancestral codon from Miller's amino acids and nucleotide correlations in modern coding sequences. *Bioinform. Biol. Insights* **2015**, *9*, 37–47. [CrossRef]
111. Morgan, A.P.; Gatti, D.M.; Najarian, M.L.; Keane, T.M.; Galante, R.J.; Pack, A.I.; Mott, R.; Churchill, G.A.; de Villena, F.P. Structural variation shapes the landscape of recombination in mouse. *Genetics* **2017**, *206*, 603–619. [CrossRef]
112. Millere, M. The Global Forum on Agricultural Research and Innovation. *Tangible Agricultural Solutions Shine at First Online AGRF.* 2020. Available online: https://www.cimmyt.org/news/tangible-agricultural-solutions-shine-at-first-online-agrf/ (accessed on 14 January 2023).
113. Shi, X.; Ling, H.-Q. Current advances in genome sequencing of common wheat and its ancestral species. *Crop J.* **2018**, *6*, 15–21. [CrossRef]
114. Marcussen, T.; Sandve, S.R.; Heier, L.; Spannagl, M.; Pfeifer, M.; International Wheat Genome Sequencing Consortium; Jakobsen, K.S.; Wulff, B.B.; Steuernagel, B.; Mayer, K.F.; et al. Ancient hybridizations among the ancestral genomes of bread wheat. *Science* **2014**, *345*, 1250092. [CrossRef]
115. Muthamilarasan, M.; Prasad, M. An overview of wheat genome sequencing and its implications for crop improvement. *J. Genet.* **2014**, *93*, 619–622. [CrossRef] [PubMed]
116. Krasileva, K.V.; Buffalo, V.; Bailey, P.; Pearce, S.; Ayling, S.; Tabbita, F.; Soria, M.; Wang, S.; IWGS Consortium; Akhunov, E.; et al. Separating homeologs by phasing in the tetraploid wheat transcriptome. *Genome Biol.* **2013**, *14*, R66. [CrossRef]
117. Krasileva, K.V.; Vasquez-Gross, H.A.; Howell, T.; Bailey, P.; Paraiso, F.; Clissold, L.; Simmonds, J.; Ramirez-Gonzalez, R.H.; Wang, X.; Borrill, P.; et al. Uncovering hidden variation in polyploid wheat. *Proc. Natl. Acad. Sci. USA* **2017**, *114*, E913–E921. [CrossRef]
118. Brenchley, R.; Spannagl, M.; Pfeifer, M.; Barker, G.L.; D'Amore, R.; Allen, A.M.; McKenzie, N.; Kramer, M.; Kerhornou, A.; Bolser, D.; et al. Analysis of the bread wheat genome using whole-genome shotgun sequencing. *Nature* **2012**, *491*, 705–710. [CrossRef]
119. Cui, F.; Zhang, N.; Fan, X.L.; Zhang, W.; Zhao, C.-H.; Yang, L.-J.; Pan, R.-Q.; Chen, M.; Han, J.; Zhao, X.-Q.; et al. Utilization of a Wheat660K SNP array-derived high-density genetic map for high-resolution mapping of a major QTL for kernel number. *Sci. Rep.* **2017**, *7*, 3788. [CrossRef] [PubMed]
120. Wen, W.; He, Z.; Gao, F.; Liu, J.; Jin, H.; Zhai, S.; Qu, Y.; Xia, X. A high-density consensus map of common wheat integrating four mapping populations scanned by the 90K SNP Array. *Front. Plant Sci.* **2017**, *8*, 1389. [CrossRef]
121. Qu, P.; Wang, J.; Wen, W.; Gao, F.; Liu, J.; Xia, X.; Peng, H.; Zhang, L. Construction of consensus genetic map with applications in gene mapping of wheat (*Triticum aestivum* L.) using 90K SNP array. *Front. Plant Sci.* **2021**, *12*, 727077. [CrossRef]
122. Jia, M.; Guan, J.; Zhai, Z.; Geng, S.; Zhang, X.; Mao, L.; Li, A. Wheat functional genomics in the era of next generation sequencing: An update. *Crop J.* **2018**, *6*, 7–14. [CrossRef]
123. Poersch-Bortolon, L.B.; Pereira, J.F.; Nhani, A., Jr.; Gonzáles, H.H.; Torres, G.A.; Consoli, L.; Arenhart, R.A.; Bodanese-Zanettini, M.H.; Margis-Pinheiro, M. Gene expression analysis reveals important pathways for drought response in leaves and roots of a wheat cultivar adapted to rainfed cropping in the Cerrado biome. *Genet. Mol. Biol.* **2016**, *39*, 629–645. [CrossRef] [PubMed]
124. Mantri, N.; Patade, V.; Penna, S.; Ford, R.; Pang, E. Abiotic stress responses in plants present and future. In *Abiotic Stress Responses in Plants Metabolism, Productivity and Sustainability*; Ahmad, P., Prasad, M., Eds.; Springer: New York, NY, USA, 2012. [CrossRef]
125. Farooqi, M.Q.U.; Nawaz, G.; Wani, S.H.; Choudhary, J.R.; Rana, M.; Sah, R.P.; Afzal, M.; Zahra, Z.; Ganie, S.A.; Razzaq, A.; et al. Recent developments in multi-omics and breeding strategies for abiotic stress tolerance in maize (*Zea mays* L.). *Front. Plant Sci.* **2022**, *13*, 965878. [CrossRef] [PubMed]
126. Qin, D.; Wu, H.; Peng, H.; Yao, Y.; Ni, Z.; Li, Z.; Zhou, C.; Sun, Q. Heat stress-responsive transcriptome analysis in heat susceptible and tolerant wheat (*Triticum aestivum* L) by using Wheat Genome Array. *BMC Genom.* **2008**, *9*, 432. [CrossRef] [PubMed]
127. Su, P.; Jiang, C.; Qin, H.; Hu, R.; Feng, J.; Chang, J.; Yang, G.; He, G. Identification of Potential Genes Responsible for Thermotolerance in Wheat under High Temperature Stress. *Genes* **2019**, *10*, 174. [CrossRef]
128. Nguyen, N.H.; Kim, J.H.; Kwon, J.; Jeong, C.Y.; Lee, W.; Lee, D.; Hong, S.W.; Lee, H. Characterization of *Arabidopsis thaliana* FLAVONOL SYNTHASE 1 (FLS1) -overexpression plants in response to abiotic stress. *Plant Physiol. Biochem.* **2016**, *103*, 133–142. [CrossRef]
129. Wang, M.; Qin, L.; Xie, C.; Li, W.; Yuan, J.; Kong, L.; Yu, W.; Xia, G.; Liu, S. Induced and constitutive DNA methylation in a salinity-tolerant wheat introgression line. *Plant Cell Physiol.* **2014**, *55*, 1354–1365. [CrossRef]

130. Sun, L.; Wen, J.; Peng, H.; Yao, Y.; Hu, Z.; Ni, Z.; Sun, Q.; Xin, M. The genetic and molecular basis for improving heat stress tolerance in wheat. *Abiotech* **2021**, *3*, 25–39. [CrossRef]
131. Jha, U.C.; Nayyar, H.; Jha, R.; Khurshid, M.; Zhou, M.; Mantri, N.; Siddique, K.H.M. Long non-coding RNAs: Emerging players regulating plant abiotic stress response and adaptation. *BMC Plant Biol.* **2020**, *20*, 466. [CrossRef]
132. Swarup, R.; Denyer, T. miRNAs in plant development. *Annu. Plant Rev.* **2019**, 2. [CrossRef]
133. Zeeshan, M.; Qiu, C.-W.; Naz, S.; Cao, F.; Wu, F. Genome-wide discovery of miRNAs with differential expression patterns in responses to salinity in the two contrasting wheat cultivars. *Int. J. Mol. Sci.* **2021**, *22*, 12556. [CrossRef] [PubMed]
134. Li, N.; Liu, T.; Guo, F.; Yang, J.; Shi, Y.; Wang, S.; Sun, D. Identification of long non-coding RNA-microRNA-mRNA regulatory modules and their potential roles in drought stress response in wheat (*Triticum aestivum* L.). *Front. Plant Sci.* **2022**, *13*, 1011064. [CrossRef]
135. Deng, J.; Kou, S.; Zhang, C.; Zou, Q.; Li, P.; Zhang, C.; Yuan, P. DNA methylation and plant stress responses. *J. Plant Physiol. Pathol.* **2018**, *6*, 4. [CrossRef]
136. Kim, J.-M.; Sasaki, T.; Ueda, M.; Sako, K.; Seki, M. Chromatin changes in response to drought, salinity, heat, and cold stresses in plants. *Front. Plant Sci.* **2015**, *6*, 114. [CrossRef] [PubMed]
137. Duan, C.G.; Zhu, J.K.; Cao, X. Retrospective and perspective of plant epigenetics in China. *J. Genet. Genom.* **2018**, *45*, 621–638. [CrossRef]
138. Singroha, S.; Sharma, P. Epigenetic modifications in plants under abiotic stress. In *Epigenetics*; Meccariello, R., Ed.; IntechOpen: London, UK, 2019. [CrossRef]
139. Chang, Y.N.; Zhu, C.; Jiang, J.; Zhang, H.; Zhu, J.K.; Duan, C.G. Epigenetic regulation in plant abiotic stress responses. *J. Integr. Plant Biol.* **2020**, *62*, 563–580. [CrossRef]
140. Dhar, M.K.; Vishal, P.; Sharma, R.; Kaul, S. Epigenetic dynamics: Role of epimarks and underlying machinery in plants exposed to abiotic stress. *Int. J. Genom.* **2014**, *2014*, 187146. [CrossRef]
141. Priyanka, V.; Goel, N.; Dhaliwal, I.; Sharma, M.; Kumar, R.; Kaushik, P. Epigenetics: A key to comprehending biotic and abiotic stress tolerance in family Poaceae. *Preprints* **2021**, 2021070135. [CrossRef]
142. Fortes, A.M.; Gallusci, P. Plant stress responses and phenotypic plasticity in the epigenomics era: Perspectives on the grapevine scenario, a model for perennial crop plants. *Front. Plant Sci.* **2017**, *8*, 82. [CrossRef]
143. Zhang, X.; Li, C.; Tie, D.; Quan, J.; Yue, M.; Liu, X. Epigenetic memory and growth responses of the clonal plant Glechoma longituba to parental recurrent UV-B stress. *Funct. Plant Biol.* **2021**, *48*, 827–838. [CrossRef]
144. Zheng, X.; Chen, L.; Li, M.; Lou, Q.; Xia, H.; Wang, P.; Li, T.; Liu, H.; Luo, L. Transgenerational variations in DNA methylation induced by drought stress in two rice varieties with distinguished difference to drought resistance. *PLoS ONE* **2013**, *8*, e80253. [CrossRef] [PubMed]
145. Sun, C.; Ali, K.; Yan, K.; Fiaz, S.; Dormatey, R.; Bi, Z.; Bai, J. Exploration of epigenetics for improvement of drought and other stress resistance in crops: A review. *Plants* **2021**, *10*, 1226. [CrossRef] [PubMed]
146. Lucibelli, F.; Valoroso, M.C.; Aceto, S. Plant DNA methylation: An epigenetic mark in development, environmental interactions, and evolution. *Int. J. Mol. Sci.* **2022**, *23*, 8299. [CrossRef] [PubMed]
147. Dong, P.; Tu, X.; Chu, P.-Y.; Lü, P.; Zhu, N.; Grierson, D.; Du, B.; Li, P.; Zhong, S. 3D Chromatin architecture of large plant genomes determined by local A/B compartments. *Mol. Plant* **2017**, *10*, 1497–1509. [CrossRef] [PubMed]
148. Talbert, P.B.; Henikoff, S. The Yin and Yang of Histone marks in transcription. *Annu. Rev. Genom. Hum. Genet.* **2021**, *22*, 147–170. [CrossRef] [PubMed]
149. Saraswat, S.; Yadav, A.K.; Sirohi, P.; Singh, N.K. Role of epigenetics in crop improvement: Water and heat stress. *J. Plant Biol.* **2017**, *60*, 231–240. [CrossRef]
150. Dyachenko, O.V.; Zakharchenko, N.S.; Shevchuk, T.V.; Bohnert, H.J.; Cushman, J.C.; Buryanov, Y.I. Effect of hypermethylation of CCWGG sequences in DNA of *Mesembryanthemum crystallinum* plants on their adaptation to salt stress. *Biochemistry* **2006**, *71*, 461–465. [CrossRef]
151. Zhang, X.; Lei, L.; Lai, J.; Zhao, H.; Song, W. Effects of drought stress and water recovery on physiological responses and gene expression in maize seedlings. *BMC Plant Biol.* **2018**, *18*, 68. [CrossRef]
152. Gessler, A.; Bottero, A.; Marshall, J.; Arend, M. The way back: Recovery of trees from drought and its implication for acclimation. *New Phytol.* **2020**, *228*, 1704–1709. [CrossRef]
153. Abid, M.; Ali, S.; Qi, L.K.; Zahoor, R.; Tian, Z.; Jiang, D.; Snider, J.L.; Dai, T. Physiological and biochemical changes during drought and recovery periods at tillering and jointing stages in wheat (*Triticum aestivum* L.). *Sci. Rep.* **2018**, *8*, 4615. [CrossRef]
154. Xiao, M.; Wang, J.; Xu, F. Methylation hallmarks on the histone tail as a linker of osmotic stress and gene transcription. *Front. Plant Sci.* **2022**, *13*, 967607. [CrossRef]
155. Sani, E.; Herzyk, P.; Perrella, G.; Colot, V.; Amtmann, A. Hyperosmotic priming of Arabidopsis seedlings establishes a long-term somatic memory accompanied by specific changes of the epigenome. *Genome Biol.* **2013**, *14*, R59. [CrossRef]
156. Zhong, W.; Zhong, X.; You, J.; Xiong, L. Genome-wide profiling of histone H3K4-tri-methylation and gene expression in rice under drought stress. *Plant Mol. Biol.* **2013**, *81*, 175–188. [CrossRef]
157. Igolkina, A.A.; Zinkevich, A.; Karandasheva, K.O.; Popov, A.A.; Selifanova, M.V.; Nikolaeva, D.; Tkachev, V.; Penzar, D.; Nikitin, D.M.; Buzdin, A. H3K4me3, H3K9ac, H3K27ac, H3K27me3 and H3K9me3 histone tags suggest distinct regulatory evolution of open and condensed chromatin landmarks. *Cells* **2019**, *8*, 1034. [CrossRef] [PubMed]

158. Bandurska, H. Drought stress responses: Coping strategy and resistance. *Plants* **2022**, *11*, 922. [CrossRef] [PubMed]
159. Kong, L.; Liu, Y.; Wang, X.; Chang, C. Insight into the role of epigenetic processes in abiotic and biotic stress response in wheat and barley. *Int. J. Mol. Sci.* **2020**, *21*, 1480. [CrossRef] [PubMed]
160. Jégu, T.; Veluchamy, A.; Ramirez-Prado, J.S.; Rizzi-Paillet, C.; Perez, M.; Lhomme, A.; Latrasse, D.; Coleno, E.; Vicaire, S.; Legras, S.; et al. The Arabidopsis SWI/SNF protein BAF60 mediates seedling growth ontrol by modulating DNA ccessibility. *Genome Biol.* **2017**, *18*, 114. [CrossRef] [PubMed]
161. Zhao, L.; Wang, S.; Cao, Z.; Ouyang, W.; Zhang, Q.; Xie, L.; Zheng, R.; Guo, M.; Ma, M.; Hu, Z.; et al. Chromatin loops associated with active genes and heterochromatin shape rice genome architecture for transcriptional regulation. *Nat. Commun.* **2019**, *10*, 3640. [CrossRef]
162. Kim, J.-H. Multifaceted chromatin structure and transcription changes in plant stress response. *Int. J. Mol. Sci.* **2021**, *22*, 2013. [CrossRef]
163. Kim, T.H. Mechanism of ABA signal transduction: Agricultural highlights for improving drought tolerance. *J. Plant Biol.* **2014**, *57*, 1–8. [CrossRef]
164. Ali, A.; Yun, D.J. Chromatin remodeling complex HDA9-PWR-ABI4 epigenetically regulates drought stress response in plants. *Plant Signal. Behav.* **2020**, *15*, 1803568. [CrossRef]
165. Nikitaki, Z.; Holá, M.; Donà, M.; Pavlopoulou, A.; Michalopoulos, I.; Angelis, K.J.; Georgakilas, A.G.; Macovei, A.; Balestrazzi, A. Integrating plant and animal biology for the search of novel DNA damage biomarkers. *Mutat. Res. Rev. Mutat. Res.* **2018**, *775*, 21–38. [CrossRef]
166. Ranawaka, B.; Tanurdzic, M.; Waterhouse, P.; Naim, F. An optimised chromatin immunoprecipitation (ChIP) method for starchy leaves of *Nicotiana benthamiana* to study histone modifications of an allotetraploid plant. *Mol. Biol. Rep.* **2020**, *47*, 9499–9509. [CrossRef] [PubMed]
167. Sharma, K.; Subash, C.S.; Mahilkar, S. Chromatin Immunoprecipitation (Chip). In *Protocols Used in Molecular Biology*; Kumar, S., Kumar, D., Eds.; Bentham Science Publishers: Singapore, 2019.
168. Soyer, J.L.; Möller, M.; Schotanus, K.; Connolly, L.R.; Galazka, J.M.; Freitag, M.; Stukenbrock, E.H. Chromatin analyses of *Zymoseptoria tritici*: Methods for chromatin immunoprecipitation followed by high-throughput sequencing (ChIP-seq). *Fungal Genet. Biol.* **2015**, *79*, 63–70. [CrossRef] [PubMed]
169. Kanherkar, R.R.; Bhatia-Dey, N.; Csoka, A.B. Epigenetics across the human lifespan. *Front. Cell Dev. Biol.* **2014**, *2*, 49. [CrossRef] [PubMed]
170. Li, Z.; Wang, M.; Lin, K.; Xie, Y.; Guo, J.; Ye, L.; Zhuang, Y.; Teng, W.; Ran, X.; Tong, Y.; et al. The bread wheat epigenomic map reveals distinct chromatin architectural and evolutionary features of functional genetic elements. *Genome Biol.* **2019**, *20*, 139. [CrossRef]
171. Smagulova, F.; Gregoretti, I.V.; Brick, K.; Khil, P.; Camerini-Otero, R.D.; Petukhova, G.V. Genome-wide analysis reveals novel molecular features of mouse recombination hotspots. *Nature* **2011**, *472*, 375–378. [CrossRef]
172. Smith, A.V.; Thomas, D.J.; Munro, H.M.; Abecasis, G.R. Sequence features in regions of weak and strong linkage disequilibrium. *Genome Res.* **2005**, *15*, 1519–1534. [CrossRef]
173. Spencer, C.C.; Deloukas, P.; Hunt, S.; Mullikin, J.; Myers, S.; Silverman, B.; Donnelly, P.; Bentley, D.; McVean, G. The influence of recombination on human genetic diversity. *PLoS Genet.* **2006**, *2*, e148. [CrossRef]
174. Kleckner, N. Chiasma formation: Chromatin/axis interplay and the role(s) of the synaptonemal complex. *Chromosoma* **2006**, *115*, 175–194. [CrossRef]
175. Dixon, J.R.; Selvaraj, S.; Yue, F.; Kim, A.; Li, Y.; Shen, Y.; Hu, M.; Liu, J.S.; Ren, B. Topological domains in mammalian genomes identified by analysis of chromatin interactions. *Nature* **2012**, *485*, 376–380. [CrossRef] [PubMed]
176. Rao, S.S.; Huntley, M.H.; Durand, N.C.; Stamenova, E.K.; Bochkov, I.D.; Robinson, J.T.; Sanborn, A.L.; Machol, I.; Omer, A.D.; Lander, E.S.; et al. A 3D map of the human genome at kilobase resolution reveals principles of chromatin looping. *Cell* **2014**, *159*, 1665–1680. [CrossRef]
177. Meuleman, W.; Peric-Hupkes, D.; Kind, J.; Beaudry, J.B.; Pagie, L.; Kellis, M.; Reinders, M.; Wessels, L.; van Steensel, B. Constitutive nuclear lamina-genome interactions are highly conserved and associated with A/T-rich sequence. *Genome Res.* **2013**, *23*, 270–280. [CrossRef] [PubMed]
178. Kind, J.; Pagie, L.; de Vries, S.S.; Nahidiazar, L.; Dey, S.S.; Bienko, M.; Zhan, Y.; Lajoie, B.; de Graaf, C.A.; Amendola, M.; et al. Genome-wide maps of nuclear lamina interactions in single human cells. *Cell* **2015**, *163*, 134–147. [CrossRef] [PubMed]
179. Jabbari, K.; Bernardi, G. An isochore ramework underlies chromatin architecture. *PLoS ONE* **2017**, *12*, e0168023. [CrossRef]
180. Hayashi, K.; Yoshida, K.; Matsui, Y.A. histone H3 methyltransferase controls epigenetic events required for meiotic prophase. *Nature* **2005**, *438*, 374–378. [CrossRef] [PubMed]
181. Brick, K.; Smagulova, F.; Khil, P.; Camerini-Otero, R.D.; Petukhova, G.V. Genetic recombination is directed away from functional genomic elements in mice. *Nature* **2012**, *485*, 642–645. [CrossRef]
182. Jabbari, K.; Wirtz, J.; Rauscher, M.; Wiehe, T. A common genomic code for chromatin architecture and recombination landscape. *PLoS ONE* **2019**, *14*, e0213278. [CrossRef]
183. Abranches, R.; Santos, A.P.; Wegel, E.; Williams, S.; Castilho, A.; Christou, P.; Shaw, P.; Stoger, E. Widely separated multiple transgene integration sites in wheat chromosomes are brought together at interphase. *Plant J.* **2000**, *24*, 713–723. [CrossRef]

184. Zuckerkandl, E.; Cavalli, G. Combinatorial epigenetics, "junk DNA", and the evolution of complex organisms. *Gene* **2007**, *390*, 232–242. [CrossRef]
185. Dogan, E.S.; Liu, C. Three-dimensional chromatin packing and positioning of plant genomes. *Nat. Plants* **2018**, *4*, 521–529. [CrossRef] [PubMed]
186. Grob, S. Three-dimensional chromosome organization in flowering plants. *Brief. Funct. Genom.* **2020**, *19*, 83–91. [CrossRef] [PubMed]
187. Kumar, S.; Kaur, S.; Seem, K.; Kumar, S.; Mohapatra, T. Understanding 3D genome organization and its effect on transcriptional gene regulation under environmental stress in plant: A chromatin perspective. *Front. Cell Dev. Biol.* **2021**, *9*, 774719. [CrossRef] [PubMed]
188. Wiszniewska, A. Priming strategies for benefiting plant performance under toxic trace metal exposure. *Plants* **2021**, *10*, 623. [CrossRef]
189. Ahmed, R.F.; Irfan, M.; Shakir, H.A.; Khan, M.; Chen, L. Engineering drought tolerance in plants by modification of transcription and signalling factors. *Biotechnol. Biotechnol. Equip.* **2020**, *34*, 781–789. [CrossRef]
190. Manna, M.; Thakur, T.; Chirom, O.; Mandlik, R.; Deshmukh, R.; Salvi, P. Transcription factors as key molecular target to strengthen the drought stress tolerance in plants. *Physiol. Plant.* **2021**, *172*, 847–868. [CrossRef]
191. Wang, J.; Li, C.; Li, L.; Reynolds, M.; Mao, X.; Jing, R. Exploitation of drought tolerance-related genes for crop improvement. *Int. J. Mol. Sci.* **2021**, *22*, 10265. [CrossRef] [PubMed]
192. Kulkarni, M.; Soolanayakanahally, R.; Ogawa, S.; Uga, Y.; Selvaraj, M.G.; Kagale, S. Drought response in wheat: Key genes and regulatory mechanisms controlling root system architecture and transpiration efficiency. *Front. Chem.* **2017**, *5*, 106. [CrossRef]
193. Sakuma, Y.; Liu, Q.; Dubouzet, J.G.; Abe, H.; Shinozaki, K.; Yamaguchi-Shinozaki, K. DNA-binding specificity of the ERF/AP2 domain of Arabidopsis DREBs, transcription factors involved in dehydration- and cold-inducible gene expression. *Biochem. Biophys. Res. Commun.* **2002**, *290*, 998–1009. [CrossRef]
194. Wang, Y.; Du, F.; Wang, J.; Wang, K.; Tian, C.; Qi, X.; Lu, F.; Liu, X.; Ye, X.; Jiao, Y. Improving bread wheat yield through modulating an unselected *AP2/ERF* gene. *Nat. Plants* **2022**, *8*, 930–939. [CrossRef]
195. Hrmova, M.; Hussain, S.S. Plant transcription factors involved in drought and associated stresses. *Int. J. Mol. Sci.* **2021**, *22*, 5662. [CrossRef] [PubMed]
196. Okay, S.; Derelli, E.; Unver, T. Transcriptome-wide identification of bread wheat WRKY transcription factors in response to drought stress. *Mol. Genet. Genom.* **2014**, *289*, 765–781. [CrossRef] [PubMed]
197. Arya, H.; Singh, M.B.; Bhalla, P.L. Towards developing drought-smart soybeans. *Front. Plant Sci.* **2021**, *12*, 750664. [CrossRef] [PubMed]
198. Martignago, D.; Rico-Medina, A.; Blasco-Escámez, D.; Fontanet-Manzaneque, J.B.; Caño-Delgado, A.I. Drought resistance by engineering plant tissue-specific responses. *Front. Plant Sci.* **2020**, *10*, 1676. [CrossRef]
199. Anwar, A.; Kim, J.K. Transgenic breeding approaches for improving abiotic stress tolerance: Recent progress and future perspectives. *Int. J. Mol. Sci.* **2020**, *21*, 2695. [CrossRef] [PubMed]
200. Villalobos-López, M.A.; Arroyo-Becerra, A.; Quintero-Jiménez, A.; Iturriaga, G. Biotechnological advances to improve abiotic stress tolerance in crops. *Int. J. Mol. Sci.* **2022**, *23*, 12053. [CrossRef] [PubMed]
201. Khan, S.; Anwar, S.; Yu, S.; Sun, M.; Yang, Z.; Gao, Z.Q. Development of drought-tolerant transgenic wheat: Achievements and limitations. *Int. J. Mol. Sci.* **2019**, *20*, 3350. [CrossRef]
202. Liu, H.; Able, A.J.; Able, J.A. Water-deficit stress-responsive microRNAs and their targets in four durum wheat genotypes. *Funct. Integr. Genom.* **2017**, *17*, 237–251. [CrossRef]
203. Gallusci, P.; Agius, D.R.; Moschou, P.N.; Dobránszki, J.; Kaiserli, E.; Martinelli, F. Deep inside the epigenetic memories of stressed plants. *Trends Plant Sci.* **2023**, *28*, 142–153. [CrossRef]
204. Crisp, P.A.; Ganguly, D.; Eichten, S.R.; Borevitz, J.O.; Pogson, B.J. Reconsidering plant memory: Intersections between stress recovery, RNA turnover, and epigenetics. *Sci. Adv.* **2016**, *2*, e1501340. [CrossRef]
205. Chinnusamy, V.; Zhu, J.K. Epigenetic regulation of stress responses in plants. *Curr. Opin. Plant Biol.* **2009**, *12*, 133–139. [CrossRef] [PubMed]
206. Varotto, S.; Tani, E.; Abraham, E.; Krugman, T.; Kapazoglou, A.; Melzer, R.; Radanović, A.; Miladinović, D. Epigenetics: Possible applications in climate-smart crop breeding. *J. Exp. Bot.* **2020**, *71*, 5223–5236. [CrossRef] [PubMed]
207. Hofmeister, B.T.; Lee, K.; Rohr, N.A.; Hall, D.W.; Schmitz, R.J. Stable inheritance of DNA methylation allows creation of epigenotype maps and the study of epiallele inheritance patterns in the absence of genetic variation. *Genome Biol.* **2017**, *18*, 155. [CrossRef] [PubMed]
208. Suter, L.; Widmer, A. Environmental heat and salt stress induce transgenerational phenotypic changes in *Arabidopsis thaliana*. *PLoS ONE* **2013**, *8*, e60364. [CrossRef]
209. Noshay, J.M.; Springer, N.M. Stories that can't be told by SNPs; DNA methylation variation in plant populations. *Curr. Opin. Plant Biol.* **2021**, *61*, 101989. [CrossRef] [PubMed]
210. Schmitz, R.J.; He, Y.; Valdés-López, O.; Khan, S.M.; Joshi, T.; Urich, M.A.; Nery, J.R.; Diers, B.; Xu, D.; Stacey, G.; et al. Epigenomewide inheritance of cytosine methylation variants in a recombinant inbred population. *Genome Res.* **2013**, *23*, 1663–1674. [CrossRef]

211. Turcotte, H.; Hooker, J.; Samanfar, B.; Parent, J.-S. Can epigenetics guide the production of better adapted cultivars? *Agronomy* **2022**, *12*, 838. [CrossRef]
212. Virdi, K.S.; Wamboldt, Y.; Kundariya, H.; Laurie, J.D.; Keren, I.; Kumar, K.R.S.; Block, A.; Basset, G.; Luebker, S.; Elowsky, C.; et al. MSH1 is a plant organellar DNA binding and thylakoid protein under precise spatial regulation to alter development. *Mol. Plant* **2016**, *9*, 245–260. [CrossRef]
213. Kundariya, H.; Yang, X.; Morton, K.; Sanchez, R.; Axtell, M.J.; Hutton, S.F.; Fromm, M.; Mackenzie, S.A. MSH1-induced heritable enhanced growth vigor through grafting is associated with the RdDM pathway in plants. *Nat. Commun.* **2020**, *11*, 5343. [CrossRef]
214. Raju, S.K.K.; Shao, M.R.; Sanchez, R.; Xu, Y.Z.; Sandhu, A.; Graef, G.; Mackenzie, S. An epigenetic breeding system in soybean for increased yield and stability. *Plant Biotechnol. J.* **2018**, *16*, 1836–1847. [CrossRef]
215. Thapa, B.; Shrestha, A. Epigenetic mechanisms and its role in plant growth and development. *J. Plant Biochem. Physiol.* **2020**, *8*, 255. Available online: https://www.longdom.org/open-access/epigenetic-mechanisms-and-its-role-in-plant-growth-and-development.pdf (accessed on 12 March 2023).
216. Feng, S.; Jacobsen, S.E.; Reik, W. Epigenetic reprogramming in plant and animal development. *Science* **2010**, *330*, 622–627. [CrossRef]
217. Vernet, A.; Meynard, D.; Lian, Q.; Mieulet, D.; Gibert, O.; Bissah, M.; Rivallan, R.; Autran, D.; Leblanc, O.; Meunier, A.C.; et al. High-frequency synthetic apomixis in hybrid rice. *Nat. Commun.* **2022**, *13*, 7963. [CrossRef] [PubMed]
218. Rabbi, S.M.H.A.; Kumar, A.; Mohajeri Naraghi, S.; Sapkota, S.; Alamri, M.S.; Elias, E.M.; Kianian, S.; Seetan, R.; Missaoui, A.; Solanki, S.; et al. Identification of main-effect and environmental interaction QTL and their candidate genes for drought tolerance in a wheat RIL population between two elite spring cultivars. *Front. Genet.* **2021**, *12*, 656037. [CrossRef] [PubMed]
219. Yadav, S.; Sandhu, N.; Singh, V.K.; Catolos, M.; Kumar, A. Genotyping-by-sequencing based QTL mapping for rice grain yield under reproductive stage drought stress tolerance. *Sci. Rep.* **2019**, *9*, 14326. [CrossRef]
220. Rajurkar, A.B.; Muthukumar, C.; Bharathi, A.; Thomas, H.B.; Babu, R.C. Saturation mapping of consistent QTLs for yield and days to flowering under drought using locally adapted landrace in rice (*Oryza sativa* L.). *NJAS-Wagening. J. Life Sci.* **2019**, *88*, 66–75. [CrossRef]
221. Sabar, M.; Shabir, G.; Shah, S.M.; Aslam, K.; Naveed, S.A.; Arif, M. Identification and mapping of QTLs associated with drought tolerance traits in rice by a cross between Super Basmati and IR55419-04. *Breed. Sci.* **2019**, *69*, 169–178. [CrossRef]
222. Zhao, X.; Peng, Y.; Zhang, J.; Fang, P.; Wu, B. Identification of QTLs and meta-QTLs for seven agronomic traits in multiple maize populations under well-watered and water-stressed conditions. *Crop Sci.* **2018**, *58*, 507–520. [CrossRef]
223. Arriagada, O.; Mora, F.; Quitral, Y.; Del Pozo, A. Identification of QTL underlying agronomic, morphological and physiological traits in barley under rainfed conditions using SNP markers. *Acta Sci. Agron.* **2017**, *39*, 321–329. [CrossRef]
224. Makhtoum, S.; Sabouri, H.; Gholizadeh, A.; Ahangar, L.; Katouzi, M. QTLs controlling physiological and morphological traits of barley (*Hordeum vulgare* L.) seedlings under salinity, drought, and normal conditions. *BioTech* **2022**, *11*, 26. [CrossRef]
225. Li, C. QTL analysis of drought tolerance traits in rice during vegetative growth period. *Preprint* **2022**. [CrossRef]
226. Zhao, X.; Zhang, J.; Fang, P.; Peng, Y. Comparative QTL analysis for yield components and morphological traits in maize (*Zea mays* L.) under water-stressed and well-watered conditions. *Breed. Sci.* **2019**, *69*, 621–632. [CrossRef] [PubMed]
227. Nikolic, A.; Andjelkovic, V.; Dodig, D.; Mladenovi, S.; Kravi, N.; Ignjatovic-Micic, D. Identification of QTL-s for drought tolerance in maize, II: Yield and yield components. *Genetika* **2013**, *45*, 341–350. [CrossRef]
228. Dhungana, S.K.; Park, J.-H.; Oh, J.-H.; Kang, B.-K.; Seo, J.-H.; Sung, J.-S.; Kim, H.-S.; Shin, S.-O.; Baek, I.-Y.; Jung, C.-S. Quantitative trait locus mapping for drought tolerance in soybean recombinant inbred line population. *Plants* **2021**, *10*, 1816. [CrossRef]
229. Ye, H.; Song, L.; Schapaugh, W.T.; Ali, M.L.; Sinclair, T.R.; Riar, M.K.; Mutava, R.N.; Li, Y.; Vuong, T.; Valliyodan, B.; et al. The importance of slow canopy wilting in drought tolerance in soybean. *J. Exp. Bot.* **2019**, *71*, 642–652. [CrossRef] [PubMed]
230. Abou-Elwafa, S.F. Identification of genes associated with drought tolerance in barley. *Biol. Plant.* **2018**, *62*, 299–306. [CrossRef]
231. Hoang, G.T.; Dinh, L.V.; Nguyen, T.T.; Ta, N.K.; Gathignol, F.; Mai, C.D.; Jouannic, S.; Tran, K.D.; Khuat, T.H.; Do, V.N.; et al. Genome-wide association study of a panel of Vietnamese rice landraces reveals new QTLs for tolerance to water deficit during the vegetative phase. *Rice* **2019**, *12*, 4. [CrossRef]
232. Khan, S.U.; Zheng, Y.; Chachar, Z.; Zhang, X.; Zhou, G.; Zong, N.; Leng, P.; Zhao, J. Dissection of maize drought tolerance at the flowering stage using genome-wide association studies. *Genes* **2022**, *13*, 564. [CrossRef]
233. Shahzad, A.; Qian, M.; Sun, B.; Mahmood, U.; Li, S.; Fan, Y.; Chang, W.; Dai, L.; Zhu, H.; Li, J.; et al. Genome-wide association study identifies novel loci and candidate genes for drought stress tolerance in rapeseed. *Oil Crop Sci.* **2021**, *6*, 12–22. [CrossRef]
234. Liu, W.; Li, S.; Zhang, C.; Jin, F.; Li, W.; Li, X. Identification of candidate genes for drought tolerance at maize seedlings using genome-wide association. *Iran. J. Biotechnol.* **2021**, *19*, e2637. [CrossRef]
235. Harb, A.; Simpson, C.; Guo, W.; Govindan, G.; Kakani, V.G.; Sunkar, R. The effect of drought on transcriptome and hormonal profiles in barley genotypes with contrasting drought tolerance. *Front. Plant Sci.* **2020**, *11*, 618491. [CrossRef]
236. Chen, H.; Feng, H.; Zhang, X.; Zhang, C.; Wang, T. An Arabidopsis E3 ligase HUB2 increases histone H2B monoubiquitination and enhances drought tolerance in transgenic cotton. *Plant Biotechnol. J.* **2019**, *17*, 556–568. [CrossRef] [PubMed]
237. Lin, L.; Wu, J.; Jiang, M.; Wang, Y. Plant mitogen-activated protein kinase cascades in environmental stresses. *Int. J. Mol. Sci.* **2021**, *22*, 1543. [CrossRef]

238. Chen, X.; Wang, J.; Zhu, M.; Jia, H.; Liu, D.; Hao, L.; Guo, X. A cotton Raf-like MAP3K gene, GhMAP3K40, mediates reduced tolerance to biotic and abiotic stress in *Nicotiana benthamiana* by negatively regulating growth and development. *Plant Sci.* **2015**, *240*, 10–24. [CrossRef]
239. Wang, N.-N.; Zhao, L.; Lu, R.; Li, Y.; Li, X.-B. Cotton mitogen-activated protein kinase4 (GhMPK4) confers the transgenic Arabidopsis hypersensitivity to salt and osmotic stresses. *Plant Cell Tissue Organ Cult.* **2015**, *123*, 619–632. [CrossRef]
240. Zhang, J.; Zou, D.; Li, Y.; Sun, X.; Wang, N.N.; Gong, S.Y.; Zheng, Y.; Li, X.B. GhMPK17, a cotton mitogen- activated protein kinase, is involved in plant response to high salinity and osmotic stresses and ABA signaling. *PLoS ONE* **2014**, *9*, e95642. [CrossRef]
241. Long, L.; Gao, W.; Xu, L.; Liu, M.; Luo, X.; He, X.; Yang, X.; Zhang, X.; Zhu, L. GbMPK3, a mitogen-activated protein kinase from cotton, enhances drought and oxidative stress tolerance in tobacco. *Plant Cell Tissue Organ Cult.* **2014**, *116*, 153–162. [CrossRef]
242. Maghsoudi, K.; Emam, Y.; Niazi Azfar, K.; Pessarakli, M.; Arvin, M. P5CS expression level and proline accumulation in the sensitive and tolerant wheat cultivars under control and drought stress conditions in the presence/absence of silicon and salicylic acid. *J. Plant Interact.* **2018**, *13*, 461–471. [CrossRef]
243. Bi, H.; Shi, J.; Kovalchuk, N.; Luang, S.; Bazanova, N.; Chirkova, L.; Zhang, D.; Shavrukov, Y.; Stepanenko, A.; Tricker, P.; et al. Overexpression of the TaSHN1 transcription factor in bread wheat leads to leaf surface modifications, improved drought tolerance, and no yield penalty under controlled growth conditions. *Plant Cell Environ.* **2018**, *41*, 2549–2566. [CrossRef] [PubMed]
244. Xu, Y.; Zou, S.; Zeng, H.; Wang, W.; Wang, B.; Wang, H.; Tang, D. A NAC transcription factor TuNAC69 Contributes to ANK-NLR-WRKY NLR-mediated stripe rust resistance in the diploid wheat *Triticum urartu*. *Int. J. Mol. Sci.* **2022**, *23*, 564. [CrossRef]
245. Terashima, A.; Takumi, S. Allopolyploidization reduces alternative splicing efficiency for transcripts of the wheat DREB2 homolog, WDREB2. *Genome* **2009**, *52*, 100–105. [CrossRef] [PubMed]
246. Punchkhon, C.; Plaimas, K.; Buaboocha, T.; Siangliw, J.L.; Toojinda, T.; Comai, L.; De Diego, N.; Spíchal, L.; Chadchawan, S. Drought-tolerance gene identification using genome comparison and co-expression network analysis of chromosome substitution lines in rice. *Genes* **2020**, *11*, 1197. [CrossRef] [PubMed]
247. Feng, X.; Liu, W.; Qiu, C.W.; Zeng, F.; Wang, Y.; Zhang, G.; Chen, Z.H.; Wu, F. HvAKT2 and HvHAK1 confer drought tolerance in barley through enhanced leaf mesophyll H+ homoeostasis. *Plant Biotechnol. J.* **2020**, *18*, 1683–1696. [CrossRef] [PubMed]
248. Tarczynski, M.C.; Jensen, R.G.; Bohnert, H.J. Expression of a bacterial mtlD gene in transgenic tobacco leads to production and accumulation of mannitol. *Proc. Natl. Acad. Sci. USA* **1992**, *89*, 2600–2604. [CrossRef]
249. Koag, M.C.; Wilkens, S.; Fenton, R.D.; Resnik, J.; Vo, E.; Close, T.J. The K-segment of maize DHN1 mediates binding to anionic phospholipid vesicles and concomitant structural changes. *Plant Physiol.* **2009**, *150*, 1503–1514. [CrossRef]
250. Jia, H.; Li, M.; Li, W.; Liu, L.; Jian, Y.; Yang, Z.; Shen, X.; Ning, Q.; Du, Y.; Zhao, R.; et al. A serine/threonine protein kinase encoding gene KERNEL NUMBER PER ROW6 regulates maize grain yield. *Nat. Commun.* **2020**, *11*, 988. [CrossRef]
251. Luo, Y.; Zhang, M.; Liu, Y.; Liu, J.; Li, W.; Chen, G.; Peng, Y.; Jin, M.; Wei, W.; Jian, L.; et al. Genetic variation in YIGE1 contributes to ear length and grain yield in maize. *New Phytol.* **2021**, *234*, 513–526. [CrossRef]
252. Pei, Y.; Deng, Y.; Zhang, H.; Zhang, Z.; Liu, J.; Chen, Z.; Cai, D.; Li, K.; Du, Y.; Zang, J.; et al. EAR APICAL DEGENERATION1 regulates maize ear development by maintaining malate supply for apical inflorescence. *Plant Cell.* **2022**, *34*, 2222–2241. [CrossRef]
253. Chen, Z.; Liu, Y.; Yin, Y.; Liu, Q.; Li, N.; Li, X.; He, W.; Hao, D.; Liu, X.; Guo, C. Expression of *AtGA2ox1* enhances drought tolerance in maize. *Plant Growth Regul.* **2019**, *89*, 203–215. [CrossRef]
254. Zhang, Q.; Liu, H.; Wu, X.; Wang, W. Identification of drought tolerant mechanisms in a drought-tolerant maize mutant based on physiological, biochemical and transcriptomic analyses. *BMC Plant Biol.* **2020**, *20*, 315. [CrossRef]
255. Li, H.; Han, X.; Liu, X.; Zhou, M.; Ren, W.; Zhao, B.; Ju, C.; Liu, Y.; Zhao, J. A leucine-rich repeat-receptor-like kinase gene *SbER2-1* from sorghum (*Sorghum bicolor* L.) confers drought tolerance in maize. *BMC Genom.* **2019**, *20*, 737. [CrossRef]
256. Ning, Q.; Jian, Y.; Du, Y.; Li, Y.; Shen, X.; Jia, H.; Zhao, R.; Zhan, J.; Yang, F.; Jackson, D.; et al. An ethylene biosynthesis enzyme controls quantitative variation in maize ear length and kernel yield. *Nat. Commun.* **2021**, *12*, 5832. [CrossRef] [PubMed]
257. Wang, B.; Li, Z.; Ran, Q.; Li, P.; Peng, Z.; Zhang, J. ZmNF-YB16 Overexpression improves drought resistance and yield by enhancing photosynthesis and the antioxidant capacity of maize plants. *Front. Plant Sci.* **2018**, *9*, 709. [CrossRef] [PubMed]
258. He, Z.; Zhong, J.; Sun, X.; Wang, B.; Terzaghi, W.; Dai, M. The maize ABA receptors ZmPYL8, 9, and 12 facilitate plant drought resistance. *Front. Plant Sci.* **2018**, *9*, 422. [CrossRef]
259. Liu, S.; Li, C.; Wang, H.; Wang, S.; Yang, S.; Liu, X.; Yan, J.; Li, B.; Beatty, M.; Zastrow-Hayes, G.; et al. Mapping regulatory variants controlling gene expression in drought response and tolerance in maize. *Genome Biol.* **2020**, *21*, 163. [CrossRef]
260. Zhou, S.F.; Chen, X.Y.; Xue, X.N.; Zhang, X.G.; Li, Y.X. Physiological and growth responses of tomato progenies harboring the betaine alhyde dehydrogenase gene to salt stress. *J. Integr. Plant Biol.* **2007**, *49*, 628–637. [CrossRef]

plants

MDPI

Article

Identification of Drought Tolerant Rice (*Oryza Sativa* L.) Genotypes with Asian and African Backgrounds

Cyprien Ndikuryayo [1,2,3,*], Alexis Ndayiragije [2], Newton Lwiyiso Kilasi [1] and Paul Kusolwa [1]

[1] Department of Crop Science and Horticulture, Sokoine University of Agriculture, Morogoro P.O. Box 3001, Tanzania
[2] International Rice Research Institute (IRRI), Bujumbura P.O. Box 5132, Burundi
[3] Burundi Institute of Agricultural Sciences (ISABU), Avenue de la Cathédrale, Bujumbura P.O. Box 795, Burundi
* Correspondence: ndikuryayocyprien@yahoo.fr; Tel.: +257-79749151

Abstract: Drought is among the major abiotic stresses on rice production that can cause yield losses of up to 100% under severe drought conditions. Neither of the rice varieties currently grown in Burundi can withstand very low and irregular precipitation. This study identified genotypes that have putative quantitative trait loci (QTLs) associated with drought tolerance and determined their performance in the field. Two hundred and fifteen genotypes were grown in the field under both drought and irrigated conditions. Genomic deoxyribonucleic acid (DNA) was extracted from rice leaves for further genotypic screening. The results revealed the presence of the QTLs qDTY12.1, qDTY3.1, qDTY2-2_1, and qDTY1.1 in 90%, 85%, 53%, and 22% of the evaluated genotypes, respectively. The results of the phenotypic evaluation showed a significant yield reduction due to drought stress. Yield components and other agronomic traits were also negatively affected by drought. Genotypes having high yield best linear unbiased predictions (BLUPs) with two or more major QTLs for drought tolerance, including IR 108044-B-B-B-3-B-B, IR 92522-45-3-1-4, and BRRI DHAN 55 are of great interest for breeding programs to improve the drought tolerance of lines or varieties with other preferred traits.

Keywords: soil moisture content; quantitative trait loci; irrigated; yield; improvement; Burundi

1. Introduction

Rice is the principal food grain consumed by more than half of the world's population [1]. However, rice (*Oryza sativa* L.) production faces biotic and abiotic constraints worldwide [2]. The most common biotic constraints include blast, sheath rot, and brown spot in Burundi [3,4]. Among abiotic constraints, drought stress is the major one in rain-fed ecologies [2]. For numerous soils, at least two weeks without rainfall induces noticeable negative differences in drought sensitivity during the vegetative stage, and at least seven days without rainfall causes severe drought damage during the reproductive stage [5]. Drought can cause yield losses of up to 21% under mild drought, up to 51% under moderate drought, and up to 90.6% in severe cases [6], depending on the grown variety, growth stage, degree, and duration of the stress.

Reduced grain yield is a result of morphological responses such as increases in leaf rolling, stomata closure, and leaf tip drying; molecular responses that include changes in gene expression (up/down regulation of transcripts) and the activation of relevant transcription factors and signaling pathways; and physiological and biochemical responses such as reductions in transpiration, photosynthesis, chlorophyll content, membrane stability, stomatal conductance, and increases in osmoprotectants [7]. Drought stress reduces the performance of rice varieties that are grown worldwide [8–10].

In East Africa, rice-growing areas are exposed to severe drought [10]. In Burundi, the major constraints on rice production include inputs, flooding, and drought, which accounted for 41%, 30%, and 29%, respectively [11]. In Burundi, irrigated lowland rice is grown in Moso and mostly in the Imbo region. Most farmers in Imbo do not regularly obtain water for irrigation due to insufficient or destroyed infrastructures. Imbo is generally semiarid with low and irregular rainfall that can reach up to 500 mm per year [12]. Neither of the current rice varieties grown in Burundi can withstand such a complicated rainfall pattern. Thus, screening rice varieties suitable for Burundi becomes a priority using an appropriate approach.

A modified conventional breeding approach was suggested to integrate phenotyping, genotyping, and a strategy to screen many lines among which selection can be made [13]. The approach improved the assessment of plant responses to drought stress.

Three levels of drought stress corresponding to 5% (m^3/m^3), 10.6% (m^3/m^3), and 16% (m^3/m^3) soil moisture content, representing severe, moderate, and low drought, respectively, were used by Singh et al. [14] to assess the response of rice cultivars to early-season drought stress. Different methods of screening rice genotypes for drought tolerance have been used by researchers through classical markers or DNA/molecular markers [15,16]. Depending on the specific objective, each type of marker may present advantages and/or disadvantages. Currently, single nucleotides polymorphic markers (SNPs) are mostly used due to their high frequency, low mutation rates, and high-throughput nature [17]. Molecular markers have been a very useful tool mostly in Asian countries (Philippines, India, Nepal, Malaysia, etc.) where different quantitative trait loci (QTLs) for drought tolerance have been identified [16,18].

Polygenic architectures were reported for many traits under both drought and irrigated environments. Previous studies have demonstrated that conditional neutrality is more common than antagonistic pleiotropy [19]. This provides an explanation as to why rice breeders successfully developed drought tolerant rice lines and varieties without a yield penalty in irrigated environments [20]. Shamsudin et al. [21] found a positive interaction between qDTY2.2 and qDTY12.1 in the developed rice lines through marker assisted breeding.

According to IRRI, QTLs with large effects that are qDTY1.1, qDTY2.2, qDTY3.1, qDTY3.2, and qDTY12.1 may be used to improve rice varieties for grain yield under reproductive-stage drought in lowland areas [18]. Introgression or pyramiding of some of these QTLs was successfully achieved, especially in Asia, where drought-tolerant varieties have been released [18,22].

The International Rice Research Institute (IRRI) released genotypes IR 86781-3-3-1-1 and IR 81412-B-B-82-1 in the Philippines, and IR 82077-B-B-71-1, IR 82589-B-B-84-3, and IR 83388-B-B-108-3 in Malawi, Bangladesh, and Nepal, respectively, for drought tolerance in lowland ecosystems. IR 79913-B-176-B-4 and IR 55423-01 are upland varieties that were released in the Philippines and India, respectively [13]. However, there is no information about the use of these genotypes in rice improvement for drought tolerance. Furthermore, there is no report on the release of drought-tolerant rice varieties in Burundi [8].

Therefore, there is a need to effectively utilize the identified QTLs for drought tolerance in developing drought-tolerant rice lines to meet the preferences of producers in Burundi. To enhance breeding efforts, potentially drought-tolerant genotypes were collected from Asia and Eastern and Southern African countries. This study aimed to identify genotypes that have putative QTLs associated with drought tolerance.

2. Results

In the current study, the phenotypic data from a sample of 10 plants and genotypic data from two leaves (two plants) of each of the 215 genotypes were subjected to REML analysis or nonparametric testing. The association between phenotype and genotype, which was assessed through Chi-square test of independence, provided insightful results.

2.1. Evolution of Drought Stress Symptoms and Nonparametric Test for Scores

Before the drought stress, the tested plants had no symptoms of leaf rolling and drying. The appearance of leaf rolling symptoms started between two and three weeks while symptoms of leaf drying appeared between three and four weeks after stress initiation. The drought stress was observed at the vegetative and reproductive stage. At maturity, with two cycles of drought stress, it was easy to differentiate between stressed and control plants.

The Kruskal-Wallis nonparametric test showed no significant differences between genotypes for both leaf rolling and leaf drying. Significant differences were detected between genotypes by the Kruskal-Wallis nonparametric test for plant phenotypic acceptability, panicle phenotypic acceptability, seed phenotypic acceptability, panicle exertion, and the severity of brown spot under both drought stress and irrigated conditions. Differences in the incidence of sheath rot were only significant between the evaluated genotypes under irrigated conditions (Table 1).

Table 1. Kruskal-Wallis rank sum test for phenotypic acceptability and diseases cores for field experiments.

Drought Experiment						
Change	**PPA**	**PaPA**	**SPA**	**PE**	**ISR**	**BS**
Chi-Square	303.18	278.15	290.15	288.03	244.90	315.80
d.f.	214	214	214	214	214	214
Pr > Chi-Square	0.0001	0.0021	0.0004	0.0005	0.0700	0.0000
Irrigated (control)						
Chi-Square	285.54	324.65	340.54	324.89	280.46	262.60
d.f.	214	214	214	214	214	214
Pr > Chi-Square	0.0008	0.000	0.000	0.000	0.002	0.013

PPA = plant phenotypic acceptability, PaPA = panicle phenotypic acceptability, SPA = seed phenotypic acceptability, PE = panicle exertion, ISR = incidence of sheath rot and BS = severity of brown spot, d.f. = degree of freedom.

2.2. *Restricted Maximum Likelihood Analysis for Yield, Yield Components and Other Agronomic Traits*

Linear mixed model analysis revealed highly significant ($p \leq 0.001$) differences in plant height, number of total tillers, days to 50% flowering, days to maturity, number of panicles per plant, panicle length, number of filled grains per panicle, one thousand grain weight, and grain yield of screened genotypes under drought stress and irrigated conditions. Significant differences ($p \leq 0.01$) were detected among the tested genotypes for spikelet fertility (Table 2).

Table 2. Restricted Maximum Likelihood analysis for yield and other agronomic traits for field experiments.

Drought Experiment											
Source of Variation	**d.f.**	**PH**	**TT**	**DFl**	**DM**	**PP**	**PL**	**SF**	**NFGP**	**TGW**	**Yield**
Rep	1	1.48 ns	93.39 *	1771.74 ***	1857.27 ***	75.55 *	26.83 *	23.82 ns	1013.10 ns	1.66 ns	1.48 ns
Rep/block	9	69.99 ***	11.15 ***	71.71 ***	103.36 ***	9.31 ***	4.70 ***	237.77 **	648.80 **	10.63 **	2.93 ***
Genotype [a]	213/214	222.07 ***	9.85 ***	261.53 ***	225.60 ***	8.83 ***	4.37 ***	170.07 **	586.78 ***	24.42 ***	1.95 ***
Residual [b]	181.03–187.08	22.21	2.72	18.86	19.44	2.44	1.36	96.42	282.10	4.48	0.77
LEE [b]	181.03–187.08	24.98	3.10	21.04	21.82	2.79	1.54	106.30	310.46	4.93	0.87
Cv%		6.73	16.56	4.55	3.62	16.93	6.17	28.68	21.05	9.10	34.33
SED		5.00	1.76	4.59	4.67	1.67	1.24	10.31	17.62	2.22	0.93
Irrigated experiment (control)											
Rep	1	251.35 ns	30.13 ns	422.89 *	205.09 ns	30.25 ns	144.07 ***	1096.10 ***	1832.30 ns	55.83 *	1.07 ns
Rep/block	9	143.94 **	18.03 ***	42.41 ***	42.54 ***	14.38 ***	2.52 *	c	1038.20 ***	7.40 ns	1.44 ns
Genotype	214	288.90 ***	6.30 ***	131.02 ***	118.35 ***	5.38 ***	2.95 ***	59.56 **	798.65 ***	31.64 ***	2.95 ***
Residual [b]	185.28–193.01	55.19	2.66	4.96	7.64	2.48	1.10	25.97	254.10	4.17	1.065
LEE [b]	185.28–193.01	59.69	2.93	5.45	8.35	2.73	1.19	c	280.23	4.48	1.1535
CV%		7.96	15.38	2.48	2.34	15.98	4.80	6.08	13.70	7.69	21.083
s.e.d.		7.73	1.71	2.34	2.89	1.65	1.09	5.18	16.74	2.12	1.074

*** Significant at $p \leq 0.001$, ** significant at $p \leq 0.01$, * significant at $p \leq 0.05$, ns = nonsignificant at $p \leq 0.05$, [a] = the degree of freedom of genotype varies because of missing data, [b] = the degree of freedom of residual and LEE varies because of the nature of the lattice layout during analysis, LEE = lattice effective error, CV = coefficient of variation, s.e.d. = standard error of difference, d.f. = degree of freedom, PH = plant height, TT = number of total tillers, DFl = days to flowering, DM = days to maturity, PP = number of panicles per plant, PL = panicle length, SF = spikelet fertility, NFGP = number of filled grains per panicle, TGW = one thousand grain weight, [c] = analysis was performed according to randomized complete block design because blocks were not significant in alpha lattice design for variable spikelet fertility.

The mean yield was 2.71 t/ha, the minimum yield was 0.08 t/ha, and the maximum yield was 5.72 t/ha for screened genotypes under drought stress. The BLUP for yield under drought varied between 1.24 and 3.97 t/ha. For the irrigated experiment, the mean yield was 5.10 t/ha, the minimum yield was 1.58 t/ha, and the maximum yield was 9.12 t/ha. BLUP for yield varied between 2.93 and 7.54 under irrigated conditions. The grand mean of yield reduction was 2.33 t/ha, corresponding to 46.15%. More details on individual genotype performance are provided in the Table 3, Tables S2 and S3 of Supplementary Material.

Table 3. Summary of means and genotypic information for the ten best and ten worst genotypes and checks based on yield BLUPs under drought stress.

DESIGNATION	YBs	Ys	Yns	RY	STIY	PPAs	PPAns	SM	qDTY12.1	qDTY2.2	qDTY3.1	qDTY1.1
IR 108044-B-B-B-3-B-B	3.97	5.06 ab	6.20	18.27	0.82	1.04	4.11	11.98	+:+	+:+	-:-	+:+
IR 108031-B-B-B-2-B-B	3.91	5.03 ab	6.08	17.30	0.83	4.06	2.36	13.34	+:+	+:+	+:+	-:-
MUSESEKARA	3.73	4.50 abcd	5.34	15.88	0.84	3.03	2.99	15.79	+:+	-:-	-:-	-:-
IR 92522-45-3-1-4	3.64	4.42 abcd	6.76	34.68	0.65	3.99	3.21	10.28	+:+	+:+	+:+	-:-
IR 97011-7-4-1-3-B	3.57	4.42 abcd	5.83	26.80	0.73	1.94	2.35	13.73	+:+	-:-	+:+	-:-
YASIMIN AROMATIC	3.54	4.11 abcd	5.62	26.96	0.73	3.69	3.05	16.59	+:+	+:+	+:+	-:-
BRRI DHAN 55	3.53	4.08 abcd	4.70	13.20	0.87	2.90	3.87	10.65	+:+	-:-	+:+	+:+
IR 103421-B-B-5-3	3.51	3.94 abcd	5.17	23.87	0.76	3.97	2.88	8.47	?	*	+:+	+:-
IR 112671-126-1:4-B RGA-B RGA-1	3.48	4.14 abcd	7.10	41.58	0.58	4.02	3.09	13.06	+:+	-:-	+:+	-:-
BASMATI	3.45	3.96 abcd	6.58	40.62	0.59	3.97	2.95	12.94	+:+	-:-	+:+	+:+
IR 106172:496-2007-23-3-6	1.90	1.44 abcd	4.83	70.15	0.30	7.13	4.92	9.21	+:+	+:+	+:+	-:-
SUPA DE NYANZA–LAC	1.87	1.15 abcd	5.69	79.86	0.20	6.99	5.09	14.02	+:+	-:-	+:+	-:-
JAMBO TWENDE	1.81	1.03 bcd	4.27	75.84	0.24	6.99	7.23	13.96	+:+	+:+	+:+	+:+
IR 107015-18-3-1-B	1.77	0.29 d	2.90	89.99	0.10	9.00	9.14	12.07	+:+	-:-	+:+	-:-
NERICA 10	1.74	1.08 abcd	3.11	65.14	0.35	6.03	6.36	13.83	+:+	+:+	+:+	+:+
EDIGET (WAB189-B-B-B-HB)	1.68	0.95 bcd	2.33	59.41	0.41	5.07	7.33	15.21	+:+	+:+	+:+	+:+
LINE-8A-2	1.68	0.99 abcd	2.62	62.36	0.38	6.01	5.78	14.60	+:+	+:+	+:+	+:+
MKIA WA NYUMBU	1.68	0.81 cd	3.29	75.34	0.25	8.94	6.76	7.79	+:+	+:+	+:+	-:-
NERICA 4	1.60	0.83 cd	2.56	67.48	0.33	5.98	3.62	14.71	+:+	+:+	+:+	+:+
FRX 472	1.24	0.08 cd	4.89	98.43	0.02	9.02	6.41	11.07	+:+	-:-	+:+	-:-
IR 86781-3-3-1-1+	3.30	3.74 abcd	5.00	25.22	0.75	3.97	3.80	14.82	+:+	+:+	-:-	-:-
IR 64+	3.02	3.45 abcd	6.10	43.50	0.56	8.06	4.18	8.98	+:+	-:-	+:+	-:-

+ = Check; a,b,c,d = means that share the same letter are not significantly different, they belong to the same group; YBs = yield best linear unbiased prediction under drought stress, Ys = yield under drought stress, Yns = yield under non stress conditions, RY = percentage of reduction in yield, STIY = stress tolerance index for yield, PPAs = plant phenotypic acceptability under drought stress, PPAns = plant phenotypic acceptability under irrigated conditions, SM = soil moisture, qDTY = quantitative trait loci for drought tolerance, +:+ = homozygote, +:- = heterozygote, -:- = negative for targeted QTL, ? = absent in one sample and present in the second sample for qDTY 12.1, * = was homozygote in one sample and heterozygote in other samples for qDTY2.2.

2.3. Association between Phenotypic Data and Targeted QTLs for Drought Tolerance

Most of the screened genotypes had at least two major quantitative trait loci (QTLs) for drought tolerance. Genotypes that had four, three, two, and one of the targeted QTLs had a mean yield BLUP of 2.38 t/ha, 2.75 t/ha, 2.65 t/ha, and 2.82 t/ha, respectively. Furthermore, the mean stress tolerance index for yield was 0.52, 0.56, 0.53, and 0.56 for genotypes with four, three, two, and one of the targeted QTLs, respectively. The genotypic results showed the presence of the QTLs qDTY12.1, qDTY3.1, qDTY2.2, and qDTY1.1 in 194, 183, 114, and 48 genotypes corresponding to 90%, 85%, 53%, and 22% of the evaluated genotypes, respectively (Figure 1a). The QTL qDTY12.1 was more observed in the genotypes from Asia (Figure 1b) compared to those from Africa (Figure 1c); and the opposite occurred for qDTY1.1.

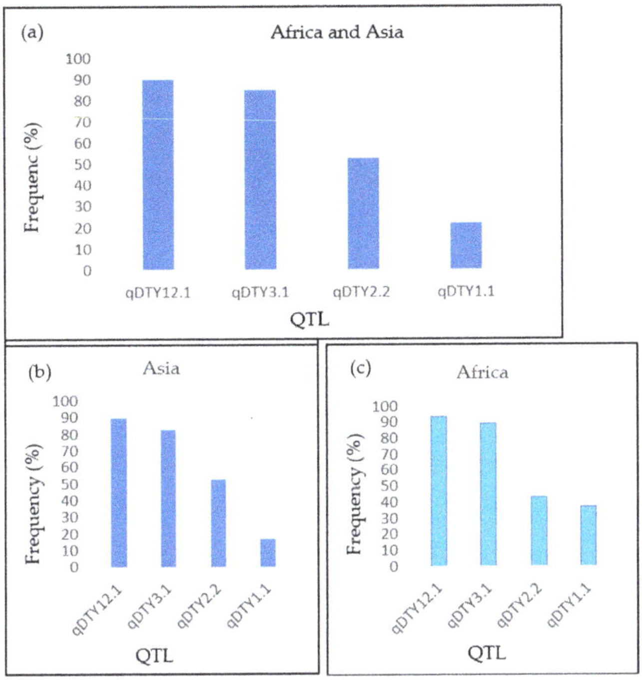

Figure 1. Frequency distribution of targeted QTLs in the evaluated genotypes; (**a**) the percentage of each QTL in the genotypes from both Africa and Asia; (**b**) the occurrence of each QTL in genotypes from Asia alone; (**c**) the frequency of each QTL in genotypes from Africa alone.

All the evaluated genotypes formed a total of eight clusters, where the genotypes from Africa and these from Asia shared six clusters and the other two clusters were owned by the genotypes from Asia. The graphical view of the diversity of these genotypes is globally presented in the neighbor-joining tree (Figure 2) and detailed in the dendrogram (Figure A2 of the Appendix A).

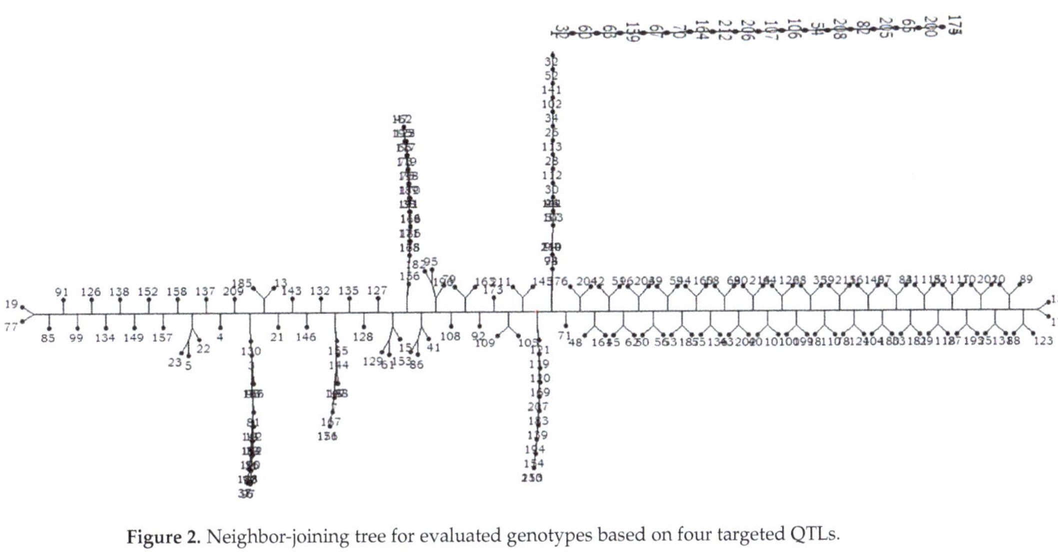

Figure 2. Neighbor-joining tree for evaluated genotypes based on four targeted QTLs.

The likelihood ratio chi-square showed a significant association between qDTY2.2 and all the phenotypic data under drought stress. The QTL qDTY1.1 was only significantly associated with the yield best linear unbiased predictions under drought stress. Other QTLs were not significantly associated with the phenotypic data (Table 4).

Table 4. Chi-Square test of independence for phenotypic and genotypic data.

Variable	d.f.[1]	qDTY2.2 LRC	qDTY3.1 LRC	qDTY12.1 LRC	qDTY1.1 LRC
YBs	132	171.90 *	132.64 ns	106.16 ns	161.08 *
Ys	214	297.83 ***	191.04 ns	141.97 ns	228.33 ns
DFl	213	295.05 ***	188.26 ns	139.20 ns	228.33 ns
DM	213	295.05 ***	191.04 ns	141.97 ns	228.33 ns
SF	214	297.83 ***	191.04 ns	141.97 ns	228.33 ns
TT	210	296.37 ***	190.68 ns	141.75 ns	225.31 ns
PH	213	296.37 ***	190.68 ns	141.75 ns	225.31 ns
PP	213	296.37 ***	190.68 ns	141.75 ns	225.31 ns
PL	209	293.60 ***	187.91 ns	138.98 ns	225.31 ns
TGW	214	297.83 ***	191.04 ns	141.97 ns	228.33 ns
NFGP	214	297.83 ***	191.04 ns	141.97 ns	228.33 ns
PPA	176	252.05 ***	161.22 ns	107.29 ns	183.60 ns
PaPA	168	242.01 ***	159.49 ns	116.65 ns	180.15 ns
SPA	169	234.74 ***	162.26 ns	113.20 ns	195.06 ns
PE	172	234.37 **	155.67 ns	124.29 ns	199.56 ns
LD1	109	146.27 **	109.16 ns	88.19 ns	131.91 ns
LD2	143	195.55 **	141.44 ns	105.56 ns	168.01 ns
LD3	168	241.33 ***	140.76 ns	110.43 ns	179.10 ns
LD4	177	259.69 ***	147.35 ns	114.92 ns	195.06 ns
LR1	137	192.73 **	141.44 ns	103.47 ns	159.33 ns
LR2	175	236.46 **	166.76 ns	130.88 ns	194.69 ns
LR3	172	234.74 ***	158.44 ns	122.56 ns	191.24 ns
LR4	179	250.69 ***	172.31 ns	133.65 ns	187.42 ns
BS	197	278.42 ***	185.49 ns	130.88 ns	217.24 ns
ISR	193	260.74 ***	174.40 ns	124.29 ns	214.47 ns

[1] = the degrees of freedom vary because of variation in cells with expected frequency, YBs = yield best linear unbiased prediction under drought stress, Ys = yield under drought stress, DFl = days to flowering, DM = days to maturity, SF = spikelet fertility, TT = number of total tillers, PH = plant height, PP = number of panicles per plant, PL = panicle length, TGW = one thousand grain weight, NFGP = number of filled grains per panicle, PPA = plant phenotypic acceptability, PaPA = panicle phenotypic acceptability, SPA = seed phenotypic acceptability, PE = panicle exertion, LD = leaf drying, LR = leaf rolling, BS = severity of brown spot, ISR = incidence of sheath rot, d.f. = degree of freedom, qDTY = quantitative trait loci for drought tolerance, LRC = likelihood ratio Chi-square, *** Significant at $p \leq 0.001$, ** significant at $p \leq 0.01$, * significant at $p \leq 0.05$, ns = nonsignificant at $p \leq 0.05$.

2.4. Correlation Analysis for Drought Traits, Yield, and Other Agronomic Traits of Genotypes Evaluated in 2020 at Gihanga Research Station

Correlations between the yield, the number of panicles per plant, and the number of filled grains per panicle were positive and highly significant. Negative and significant correlations were found between yield, leaf rolling, leaf drying, and plant phenotypic acceptability. Strong, positive, and highly significant correlations were detected between the yield, the yield BLUP, and the STI for yield. Correlations between yield and one thousand grain weight, plant height, panicle exertion, and severity of brown spot were negative and non-significant. Strong, positive, and highly significant correlation was detected between the panicle length and the plant height. Highly significant and negative correlations were detected between the plant height, one thousand grain weight, and the number of panicles per plant (Table 5).

Table 5. Correlations between traits of evaluated rice genotypes for the field drought experiment.

	YIELD	DFl	LD1	LD4	LR4	NFGP	TGW	PH	PP	PPA	BS	STIY
DFl	0.08 ns											
LD1	−0.20 **	0.07 ns										
LD4	−0.15 *	0.12 ns	0.32 ***									
LR4	−0.20 **	0.10 ns	0.18 *	0.50 ***								
NFGP	0.41 ***	−0.26 ***	−0.03 ns	−0.04 ns	−0.18 **							
TGW	−0.11 ns	−0.20 **	0.07 ns	−0.09 ns	−0.41 ***	−0.05 ns						
PH	−0.11 ns	−0.07 ns	0.12 ns	−0.03 ns	−0.21 **	0.18 **	0.40 ***					
PP	0.43 ***	0.51 ***	−0.04 ns	0.02 ns	0.18 **	−0.19 **	−0.48 ***	−0.51 ***				
PPA	−0.57 ***	0.05 ns	0.12 ns	0.19 **	0.30 ***	−0.24 ***	−0.07 ns	−0.09 ns	−0.12 ns			
BS	−0.07 ns	−0.17 *	0.06 ns	0.12 ns	0.14 *	−0.07 ns	−0.10 ns	−0.11 ns	0.05 ns	0.30 ***		
STIY	0.75 ***	0.15 *	0.21 **	0.28 ***	0.19 **	0.36 ***	0.05 ns	0.01 ns	0.20 **	0.48 ***	0.03 ns	
YdB	0.95 ***	0.10 ns	0.22 **	0.19 **	0.22 **	0.38 ***	0.15 *	0.11 ns	0.44 ***	0.57 ***	0.09 ns	0.72 ***

*** Significant at $p \leq 0.001$, ** significant at $p \leq 0.01$, * significant at $p \leq 0.05$, ns nonsignificant at $p \leq 0.05$, DFl = days to flowering, LD = leaf drying, LR = leaf rolling, NFGP = number of filled grains per panicle, TGW = one thousand grain weight, PH = plant height, PP = number of panicles per plant, PPA = plant phenotypic acceptability, BS = severity of brown spot, STIY = stress tolerance index for yield, YdB = yield best linear unbiased prediction.

3. Discussion

This study detected significant differences in leaf rolling and leaf drying among genotypes under drought stress. Furthermore, differences in plant height, number of panicles per plant, one thousand grain weight, and grain yield were significant, suggesting genetic diversity among the tested genotypes. This implies the possibility of selecting most drought-tolerant lines for their further use by farmers or by the breeding program. Comparable results were reported by Mohd Ikmal et al. [23] in BC1 F4 lines. Spikelet fertility and yield components were significantly reduced by drought stress. Similarly, all cultivars subjected to drought stress exhibited a significant grain yield reduction in a study conducted by Adhikari et al. [9]. The effects of drought on morphological and agronomic traits, including leaf area, panicle length, plant height, tillering ability, and efficiency, results in decreased yield [24].

The performance of the evaluated genotypes was better under irrigated conditions than under drought stress conditions. The current results agree with the findings of previous studies on drought where the best cultivars under nonstress conditions exhibited poor performance under stress conditions [9,18]. The reduction in performance of a given genotype increases with drought intensity [14]. This validates the significant efforts that breeders and geneticists have put into coping with water scarcity by finding QTLs and genes for drought tolerance and deploying them in genotypes with different backgrounds [13,18,22]. Therefore, growing drought-tolerant rice varieties is an alternative genetic adaptive strategy to increase rice yield and production in areas where farmers have limited access to water for irrigation [8].

A positive correlation between yield and the number of filled grains per panicle and the number of panicles per hill indicates that a higher number of panicles per plant and a higher number of filled grains per panicle lead to higher yield. Abd Allah et al. [25] reported that the number of panicles per plant, the number of filled grains per panicle, and 100 grain weight are key traits in improving yield under both irrigated and drought stress conditions. The negative correlation between yield and plant phenotypic acceptability is due to the nature of the scale for scoring phenotypic acceptability where higher scores correspond to poor performance [5]. The results imply that the phenotypically desirable genotypes also had higher yields.

Strong, positive, and highly significant correlations between the yield, the yield BLUP, and the STI for yield suggest that high yielding genotypes can be selected based on BLUP or STI. However, the results of this study showed that the correlation coefficient between the yield and the BLUP was greater than the one between yield and STI for yield under drought stress. Furthermore, some genotypes with high STI had low yield under both drought and irrigated conditions (Table 3 and Tables S2 and S3).

Best linear unbiased prediction (BLUP) was reported to be the most efficient prediction method among the commonly used methods for selection [26]. Breeding values imply the ability to perform well in crosses, and they have been recommended to select genotypes with high performance in most of the desirable traits [13].

A significantly negative correlation between the number of days to flowering and the number of filled grains per panicle shows that the longer the cycle of the genotypes, the fewer the number of filled grains per panicle due to the increase in drought intensity at the reproductive stage. Guimarães et al. [27] stated that late-flowering genotypes had high spikelet sterility. The negative and nonsignificant correlation between yield and brown spot shows that yield was slightly affected by this disease. Severe cases of this disease were reported with a yield loss of 50–90% in Bengal [17].

Promising lines were found among screened genotypes during the current study. The genotype IR 108044-B-B-B-3-B-B was classified as the best in the field based on yield BLUP. This genotype has three of the targeted QTLs for drought tolerance, including qDTY2.2, which was significantly associated with all the phenotypic data, and qDTY1.1, which was found in a few genotypes and was significantly associated with the yield BLUPs. In the same way, the genotype IR 92522-45-3-1-4 was ranked fourth based on the yield BLUP from the field where it was under severe drought stress and had three of the targeted QTLs for drought tolerance, including qDTY2.2. Similarly, the genotype BRRI DHAN 55 was under severe drought stress but was ranked seventh in yield BLUP and had three of the major targeted QTLs for drought tolerance, including qDTY1.1. The BLUPs and high yield of these genotypes under drought can enable them to be considered parents for drought tolerance, which can be used to improve existing rice varieties. Dhawan et al. [28] used Nagina 22 as a drought-tolerant parent, for which the yield was 1.77 t/ha under drought stress.

The yield of genotypes IR 97013-8-1-3-2-B, IR 13240-108-2-2-3, WAHIWAHI, and IR 97013-19-1-3-1-B was reduced by rodents that strongly attacked them a few days before pesticide application. The high scores of leaf rolling and leaf drying from the field experiment indicate the presence of a high intensity of drought during this study. High leaf drying induced a reduced yield, confirmed by the negative correlation between these traits. Similar results were obtained by Bocco et al. [24] where more plants with high leaf drying provided lower yields.

The majority of evaluated genotypes had at least two QTLs for drought tolerance providing some yield advantage under drought stress. Appropriate QTL combinations is a good approach for improving drought tolerance [16,21,29]. In the current study, some genotypes with all the four major QTLs were among the worst genotypes under drought stress. Indeed, most of these genotypes with low yield under drought stress were low-yielding even under no stress conditions. Researchers have reported that a high yield potential under no stress is a good indicator of a high yield advantage under drought stress [21]. Another reason could be the interaction between these QTLs, even if conditional neutrality was reported to be more common than antagonistic pleiotropy [19]. The QTLs qDTY2.2 and qDTY3.1 pyramided with qDTY12.1 significantly increased the yield of lines having qDTY12.1 in the study of Shamsudin et al. [21]. Therefore, pyramiding the best combination of alleles with favorable interactions is the best strategy to improve the performance of rice varieties under drought stress [13]. In this study, the QTLs qDTY12.1, qDTY3.1, and qDTY2.2 were present in more than 50% of the evaluated genotypes. Shamsudin et al. [21] found qDTY12.1, qDTY3.1, and qDTY2.2 in 82%, 36%, and 18% of selected pyramided lines, respectively. The analysis of genetic diversity of 60 rice genotypes detected qDTY12.1 and qDTY2.2 in 43.3% and 6.67% of evaluated genotypes, respectively [30].

During the current study, only qDTY2.2 was significantly associated with all the phenotypic data. This indicates that qDTY2.2 is a major QTL for drought tolerance that can be used for the improvement of preferred varieties in Burundi. However, qDTY1.1 was also significantly associated with the yield BLUPs only; further study using other genotypes in Burundi shall help to confirm our findings. Kadam et al. [31] demonstrated the complexity of yield traits under drought stress by detecting very many different QTLs between years and treatments, even by comparing with previous studies using the same genotypes. Overlapping transcriptions between the water-use efficiency and the days to flowering revealed a genetic basis for a trade-off between drought avoidance and drought escape in rice [19]. Therefore, a statistical power analysis accompanying a proper genotypic and phenotypic sampling is very important for QTL studies [32,33]. Drought tolerance is a complex trait that is characterized by low heritability, genotype-by-environment interactions, genetic interactions, and polygenic effects [16]. Furthermore, drought and heat are reported to often occur together. The genes regulating tolerance to these stresses are different but share some signaling pathways [34]. Several genetic management approaches have been suggested by researchers to increase rice production in a changing climate [34,35]. Rice improvement for drought tolerance may continue through pyramiding major QTLs for drought tolerance by crossing elite x elite cultivars or by marker-assisted backcrossing [21,29] involving landraces or wild rice as a source of drought tolerance [36] followed by multi-environmental trials to select for a specific environment or location [16].

4. Materials and Methods

4.1. Experimental Plant Materials and Study Area

A total of 215 rice genotypes with diverse origins, including potentially aromatic and potentially drought-tolerant genotypes, were screened. Based on information from previous reports, IR 64 was used as a drought susceptible check [37,38] while IR 86781-3-3-1-1 [13] was used as drought tolerant check. These genotypes were provided by the research institutions ISABU and IRRI in accordance with the national and international regulations of plant materials exchange. More details related to the parentage of evaluated genotypes and their geographical origin can be found in the Supplementary Table S1.

The experiments were set in the field at Gihanga Central Imbo in Burundi, which is located at 29°2′14.3″ E and 3°10′23.9″ S. With an elevation of 839 m above sea level, the annual mean temperature is 24 °C. During this study, the total rainfall was 141.63 mm in five months and half. More information on weather data is provided in the Appendix A (Figure A1).

4.2. Field Experimental Design, Agricultural Practices, and Drought Treatment

Tested genotypes were grown in an alpha lattice design with 2 replications for both irrigated and non-irrigated experiments. Seeding was done on 17 July 2019 and transplanting was performed three weeks after seeding. Each genotype had five rows and occupied 5.4 m^2 with only one seedling per hill. At transplanting time, fertilizers were applied according to the formula NPK 75-30-30 at a rate of 65 kg of DAP, 29 kg of urea, and 50 kg of K_2O per ha, as recommended by the Ministry of Environment, Agriculture, and Livestock [39]. Drought stress was initiated at 28 days after the last date of transplanting by draining the field of the drought experiment [38]. Other agricultural practices were performed as recommended by the MINEAGRIE [39].

Soil samples were taken from the field before plowing and were analyzed for further field and drought management. The results of soil analysis are presented in the Appendix B (Table A1). Soil classification into texture classes was performed according to Moormann and Van Breemen [40] using Texture AutoLookup (TAL) 42 software. The permanent wilting point was then determined [41]. Two polyvinyl chloride (PVC) pipes measuring 1.2-m-long × 2-inch diameter with small perforations at the bottom were installed 1 m below the soil surface in different replicates of the field for water table measurements.

Another PVC pipe measuring 0.52 m long × 2 inches in diameter was installed 0.5 m below the soil surface. Measurement of the water table was performed using a meter stick from one week after draining the field until harvesting time. Using the Hand Held (HH2, version 4.3) soil moisture meter, the soil moisture content was recorded once the genotypes had started showing symptoms of leaf rolling. Re-irrigation was performed once the soil moisture content was almost at the permanent wilting point. Water was removed from the plots after 6 h to initiate the second cycle of drought stress [38]. Experimental plants underwent two cycles of drought stress.

4.3. Genotyping Procedure

Two leaf samples were harvested from each genotype at six weeks after transplanting before booting. These samples were taken to the laboratory at IRRI-Burundi, where they were punched by an EP100 machine and kept in the wells of plates at −80 °C for 24 h. Samples in plates were later transferred to a lyophilizer for 48 h. To check the presence or absence of the major QTLs for drought tolerance, these samples were subjected first to genomic deoxyribonucleic acid (DNA) extraction and then to the Kompetitive Allele Specific Polymerase chain reaction (KASP) method in the INTERTEK laboratory according to the method described by Kanyange [3]. The SNP markers targeting major QTLs for drought tolerance are provided in the Table 6.

Table 6. SNP markers and targeted QTLs for drought tolerance.

SNP ID	QTL	Favorable Allele	Unfavorable Allele
snpOS0085	qDTY3.1	A	G
snpOS0091	qDTY12.1	T	C
snpOS00400	qDTY1.1	G	C
snpOS00412	qDTY2.2	C	A

SNP = single nucleotide polymorphisms, QTL = quantitative trait loci.

4.4. Data Collection and Analysis

The drought traits of leaf rolling and leaf drying were recorded at most twice a week once some genotypes showed symptoms and at the end of the stress cycle before reirrigation [38]. The mean leaf rolling scores were obtained using the IRRI standard evaluation system for rice [5], where: 0 = leaves healthy, 1 = leaves start to fold (shallow), 3 = leaves folding (deep V-shape), 5 = leaves fully cupped (U-shape), 7 = leaf margins touching (0-shape), and 9 = leaves tightly rolled.

In the same way, mean leaf drying scores were obtained using the IRRI standard evaluation system for rice [5] where: 0 = no symptoms, 1 = slight tip drying, 3 = tip drying extended up to ¼ length in most leaves, 5 = one-fourth to ½ of all leaves dried, 7 = more than 2/3 of all leaves fully dried, and 9 = all plants apparently dead. The plant height, the number of days to 50% flowering, and the number of days to maturity were recorded.

At the maturity period, data were recorded from ten hills in each plot for the number of tillers per plant, number of panicles per plant, panicle length, number of filled grains per panicle, and 1000 filled grain weight [5]. Data collected from the whole plot included phenotypic acceptability of the plant, phenotypic acceptability of panicle, phenotypic acceptability of seeds, panicle exertion, and grain yield. The filled grains from each plot were weighed using a high-accuracy electronic scale, and grain yield (t/ha, 13%) for each genotype was computed using Formula (1) [42]:

$$\text{Grain yield (t/ha)} = (\text{Plot grain weight (Kg/plot)} \times 10{,}000 \times (100 - \text{GMC}))/((100 - 13) \times (\text{harvested plot area}) \times 1000) \quad (1)$$

where t/ha is tons per hectare and GMC is grain moisture content (%).

The percentage reduction in grain yield [28] and the stress tolerance index (STI) [9] were calculated using Formulas (2) and (3):

$$\text{Yield reduction (\%)} = (Y_{ins} - Y_{is}) \times 100/Y_{ins} \quad (2)$$

$$\text{STI} = (Y_{ins}\ Y_{is})/((Y_{ins})^2) \quad (3)$$

where, Y_{ins} is the Yield of ith genotype under non-stress condition and Y_{is} represents the yield of ith genotype under stress condition. Other data collected from the whole plot included diseases that were present in many plots or that had high severity, such as sheath rot and brown spot.

The collected data was subjected to restricted maximum likelihood (ReML) and mixed linear model analysis using Genstat14. Genotypes were attributed fixed effects, while replicates and blocks had random effects. Means were separated using Tukey's test at the 5% level of significance [43] after detecting significant differences. Scores for drought traits, phenotypic acceptability, and diseases were subjected to Kruskal–Wallis nonparametric tests [33] using the Statistical Tool for Agricultural Research (STAR).

The genotypic data underwent a numerical scoring method by assigning one to a positive allele and zero to a negative allele. To test the association between phenotypic and genotypic data, a chi-square test of independence was performed using STAR. To display the genetic dissimilarity of tested genotypes, a weighted neighbor-joining tree was constructed in DARwin 6.0.21 [44]. To generate a dendrogram, the genotypic data was subjected to hierarchical clustering with 1000 bootstrap *p*-values in KDCompute 1.5.2.beta [45].

Through multivariate analysis, a biplot was generated and helped to reduce the number of traits to consider for correlation analysis. To determine genotypes that can be considered as potential parents for drought tolerance improvement in rice, best linear unbiased predictions (BLUPs) were calculated using R statistical software.

5. Conclusions

The current study demonstrated that drought stress significantly reduced yield for all tested genotypes. The intensity and duration of drought stress may be considered when selecting drought-tolerant rice lines. Genotypes having high yield best linear unbiased predictions (BLUPs) with two or more major QTLs for drought tolerance, including IR 108044-B-B-B-3-B-B, IR 92522-45-3-1-4, and BRRI DHAN 55, are of great interest for drought tolerance improvement in Burundi. However, further studies using other genotypes, including segregating populations, are needed in Burundi to confirm the effectiveness of qDTY 2.2 and qDTY 1.1 in controlling drought tolerance and their interaction with other potentially putative QTLs. This will enable the breeding program to successfully use these putative QTLs for drought tolerance improvement of locally grown varieties in a changing climate. Future research shall provide a list of genes within the QTLs recommended by IRRI for grain yield under drought stress, their interactions, and their mode of action in the light of ABA-mediated drought tolerance pathways [46], ERECTA-mediated drought tolerance [47], and DREB-based ABA-independent drought tolerance responses [48].

Supplementary Materials: The following supporting information can be downloaded at: https://www.mdpi.com/article/10.3390/plants12040922/s1, Table S1: Origin of genotypes, Table S2: Means for agronomic traits including yield and yield components, drought traits and diseases under drought stress in the field. Table S3: Means for agronomic traits including yield and yield components and diseases under irrigated conditions in the field.

Author Contributions: Conceptualization, C.N., A.N., N.L.K. and P.K.; methodology, C.N., A.N., N.L.K. and P.K.; software, C.N.; validation, A.N., N.L.K. and P.K.; formal analysis C.N.; investigation, A.N., N.L.K. and P.K.; resources, C.N. and A.N. and N.L.K.; data curation, C.N.; writing—original draft preparation, C.N.; writing—review and editing, C.N., A.N., N.L.K. and P.K.; visualization, C.N.; supervision, A.N., N.L.K. and P.K.; project administration, A.N. and N.L.K.; funding acquisition, A.N. and N.L.K. All authors have read and agreed to the published version of the manuscript.

Funding: "This research was funded by the World Bank through the Regional Integrated Agricultural Development Project in the Great Lakes, grant number A-2018-87/15-320-14857" and supported by the International Rice Research Institute (IRRI).

Data Availability Statement: The data supporting the conclusions of this article are provided within the article, in the appendices and in the Supplementary Materials. If more information such as row data is needed, it will be provided by the corresponding author on reasonable request.

Acknowledgments: Authors gratefully acknowledge the support provided by the staff of the International Rice Research Institute (IRRI) during data collection and analysis, especially the calculation of Best Linear Unbiased Predictions.

Conflicts of Interest: The authors declare no conflict of interest. The funders had no role in the design of the study; in the collection, analyses, or interpretation of data; in the writing of the manuscript; or in the decision to publish the results.

Appendix A

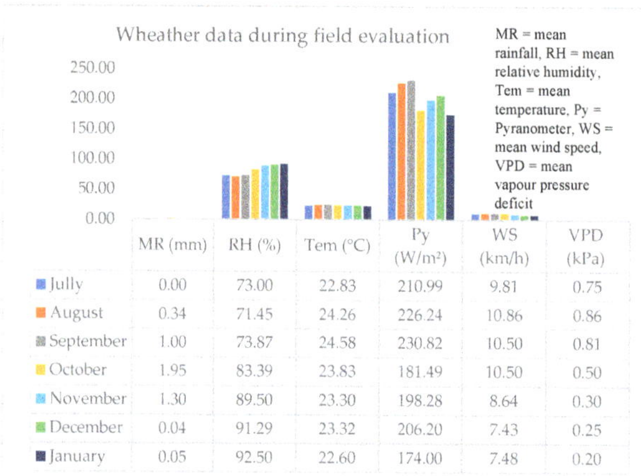

	MR (mm)	RH (%)	Tem (°C)	Py (W/m^2)	WS (km/h)	VPD (kPa)
Jully	0.00	73.00	22.83	210.99	9.81	0.75
August	0.34	71.45	24.26	226.24	10.86	0.86
September	1.00	73.87	24.58	230.82	10.50	0.81
October	1.95	83.39	23.83	181.49	10.50	0.50
November	1.30	89.50	23.30	198.28	8.64	0.30
December	0.04	91.29	23.32	206.20	7.43	0.25
January	0.05	92.50	22.60	174.00	7.48	0.20

Figure A1. Weather data in the field at Gihanga from mid-July 2020 to early January 2021.

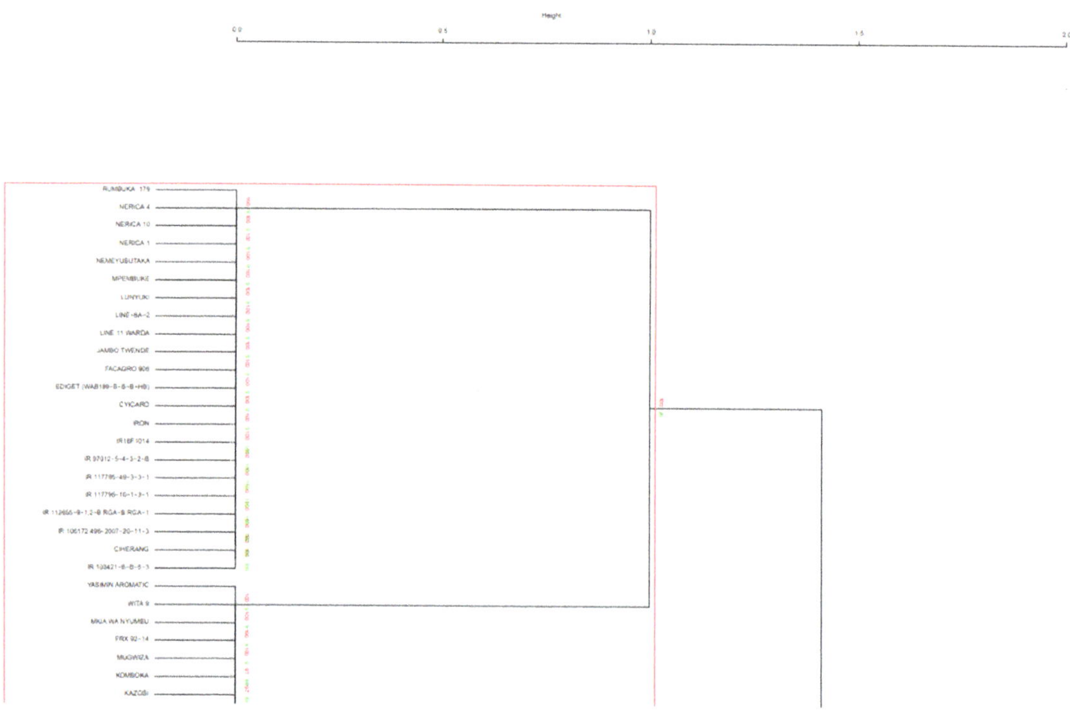

Figure A2. *Cont.*

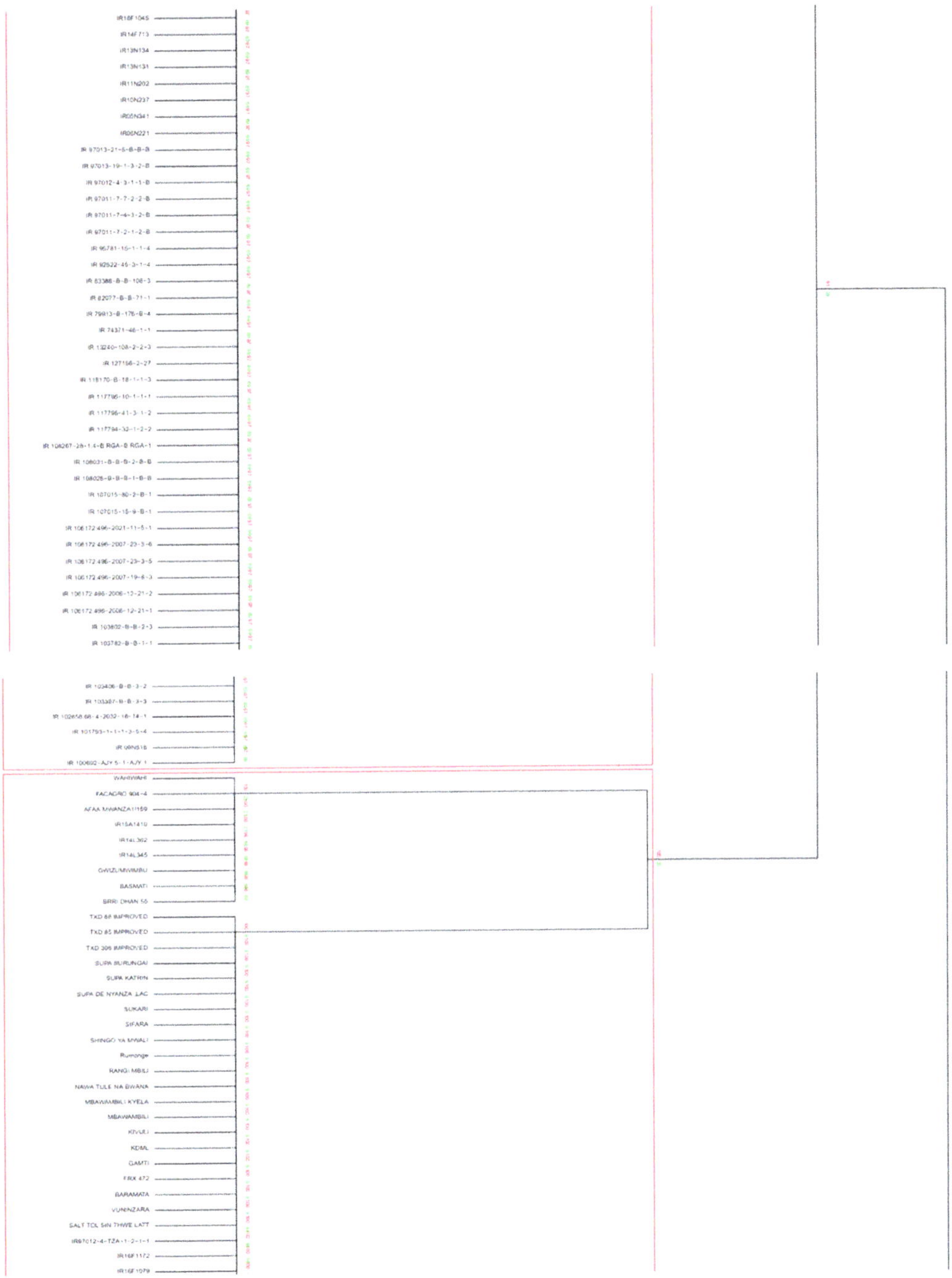

Figure A2. *Cont.*

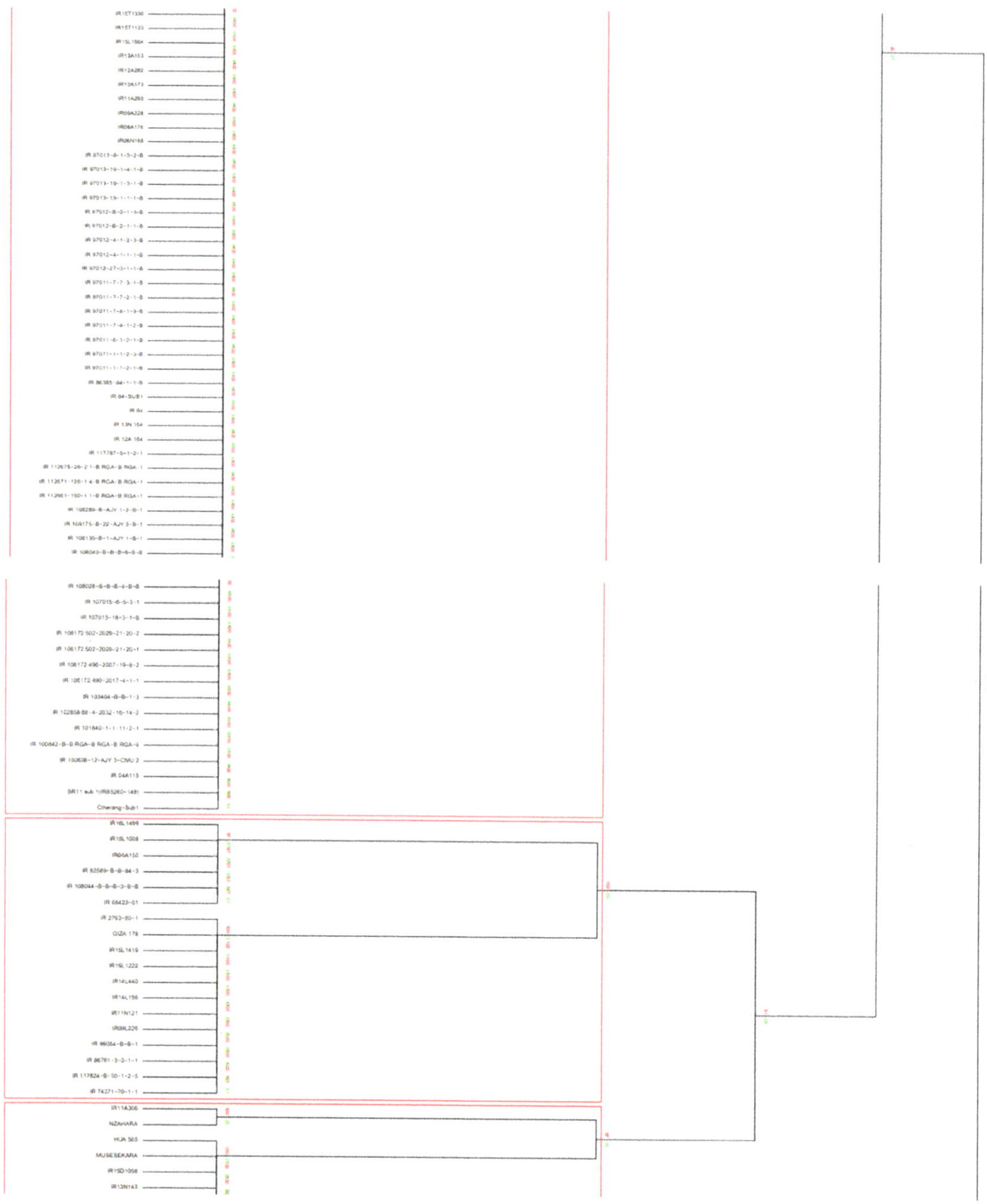

Figure A2. *Cont.*

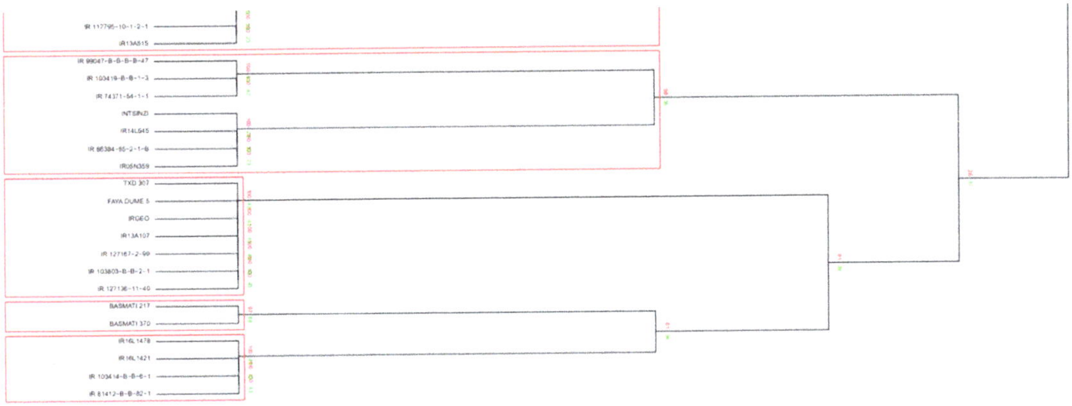

Figure A2. Dendrogram of 215 genotypes based on four targeted QTLs for drought tolerance.

Appendix B

Table A1. Results of soil analysis for Gihanga site in Imbo Lowland.

			Soil Texture			
Rep	**Block**	**Sand (%)**	**Silt (%)**	**Clay (%)**	**Texture**	**PWP (%)**
1	1	85.8	5.81	12.41	loamy sand	8.5
1	2 and 3	62.25	22.54	13.41	sandy loam	9.1
1	4 and 5	81.91	5.4	11.5	loamy sand	8.5
1	6 and 7	84.36	5.94	7.06	loamy sand	8.5
1	8 and 9	83.14	5.67	6.19	loamy sand	8.5
2	1 and 2	83.28	4.85	10.35	loamy sand	8.5
2	3 and 4	79.42	5.91	11.76	sandy loam	9.1
2	5 and 6	76.74	5.3	15.02	sandy loam	9.1
2	7 and 8	74.2	4.62	14.04	sandy loam	9.1
2	9	79.19	4.97	12.83	sandy loam	9.1
			pH and some nutrients content			
pH H_2O	% N	P (mg/kg)	K(mEq/100 g)	Na(mEq/100 g)	Zn(mg/kg)	Fe(mg/kg)
5.5	0.09	32.4	0.33	0.14	0.37	443

PWP = permanent wilting point, pH = hydrogen ion concentration, N = nitrogen, P = phosphorus, K = potassium, Na = sodium, Zn = zinc, Fe = iron, % = percent, mg = milligram, kg = kilogram, mEq = milliequivalent.

References

1. Rezvi, H.U.A.; Tahjib-Ul-Arif, M.; Azim, M.A.; Tumpa, T.A.; Tipu, M.M.H.; Najnine, F.; Dawood, M.F.; Skalicky, M.; Brestič, M. Rice and food security: Climate change implications and the future prospects for nutritional security. *Food Energy Secur.* **2022**, *12*, e430. [CrossRef]
2. Thomas, H.B.; Vangapandu, T.; Ayyenar, B.; Sellamuthu, R. Identification and Mapping of QTLS for Drought Resistance in Rice. Centre for Plant Molecular Biology and Biotechnology, Coimbatore, Tamil Nadu, India. *Int. J. Curr. Microbiol. Appl. Sci.* **2017**, *6*, 1703–1710. [CrossRef]
3. Kanyange, L.; Kamau, J.; Ombori, O.; Ndayiragije, A.; Muthini, M. Genotyping for blast (Pyricularia oryzae) resistance genes in F2 population of supa aromatic rice (*Oryza sativa* L.). *Int. J. Genom.* **2019**, *2019*, 5246820. [CrossRef]
4. Samson, M.; Bez, C.; Georges, H.; Joseph, B.; Venturi, V. Characterization of bacterial strains from bacterial culture collection of rice sheath in Burundi highlights an Alcaligenes species strain with antibacterial activity against Pseudomonas fuscovaginae rice pathogen. *Afr. J. Microbiol. Res.* **2021**, *15*, 497–511.
5. IRRI. *Standard Evaluation System for Rice*, 5th ed.; IRRI: Los Baños, Philippines, 2013; pp. 1–52.
6. Zhang, J.; Zhang, S.; Cheng, M.; Jiang, H.; Zhang, X.; Peng, C.; Lu, X.; Zhang, M.; Jin, J. Effect of drought on agronomic traits of rice and wheat: A meta-analysis. *Int. J. Environ. Res. Public Health* **2018**, *15*, 839. [CrossRef]
7. Ndjiondjop, M.N.; Wambugu, P.W.; Sangare, J.R.; Gnikoua, K. The effects of drought on rice cultivation in sub-Saharan Africa and its mitigation: A review. *Afr. J. Agric. Res.* **2018**, *13*, 1257–1271. [CrossRef]

8. Ndikuryayo, C.; Ndayiragije, A.; Kilasi, N.; Kusolwa, P. Breeding for Rice Aroma and Drought Tolerance: A Review. *Agronomy* **2022**, *12*, 1726. [CrossRef]
9. Adhikari, M.; Adhikari, N.R.; Sharma, S.; Gairhe, J.; Bhandari, R.R.; Paudel, S. Evaluation of drought tolerant rice cultivars using drought tolerant indices under water stress and irrigated condition. *Am. J. Clim. Chang.* **2019**, *8*, 228–236. [CrossRef]
10. Kilimo, T. *Expanding Rice Markets in the East African Community. Regional Solution to Local Problems*; Kilimo Trust, Head Quarters: Kampala, Uganda, 2018; p. 78.
11. Gahungu, A. Socio-Economic and Financial Profitability Analysis of Rice Seed Production by Women Groups « Nawenuze » in the Framework of «Win Win» Programm Implemented by Care International in Burundi. CARE-CERDA Partnership. 2017. Available online: https://www.careevaluations.org/wp-content/uploads/Rice-Cost-effectiveness-Study_CARE_CERDA.pdf (accessed on 25 June 2022).
12. Furaha, M.G. Analyse Comparée des Chaines de Valeur du riz dans la Plaine de la Ruzizi de la Communauté Economique des pays des Grands Lacs (CEPGL). Ph.D. Thesis, Université de Liège, Liège, Belgique, 2017.
13. Sandhu, N.; Kumar, A. Bridging the rice yield gaps under drought: QTLs, genes, and their use in breeding programs. *Agronomy* **2017**, *7*, 27. [CrossRef]
14. Singh, B.; Reddy, K.R.; Redoña, E.D.; Walker, T. Screening of rice cultivars for morpho-physiological responses to early-season soil moisture stress. *Rice Sci.* **2017**, *24*, 322–335. [CrossRef]
15. Gaballah, M.M.; Ghoneim, A.M.; Rehman, H.U.; Shehab, M.M.; Ghazy, M.I.; El-Iraqi, A.S.; Mohamed, A.E.; Waqas, M.; Shamsudin, N.A.A.; Chen, Y. Evaluation of morpho-physiological traits in rice genotypes for adaptation under irrigated and water-limited environments. *Agronomy* **2022**, *12*, 1868. [CrossRef]
16. Sahebi, M.; Hanafi, M.M.; Rafii, M.Y.; Mahmud, T.M.M.; Azizi, P.; Osman, M.; Abiri, R.; Taheri, S.; Kalhori, N.; Shabanimofrad, M.; et al. Improvement of drought tolerance in rice (*Oryza sativa* L.): Genetics, genomic tools, and the WRKY gene family. *Biomed. Res. Int.* **2018**, *18*, 20. [CrossRef] [PubMed]
17. Arif, M.; Jan, T.; Riaz, M.; Fahad, S.; Arif, M.S.; Shakoor, M.B.; Rasul, F. Advances in Rice Research for Abiotic Stress Tolerance: Agronomic Approaches to Improve Rice Production under Abiotic Stress. In *Advances in Rice Research for Abiotic Stress Tolerance*; Woodhead Publishing: Cambridge, UK, 2019; pp. 585–614.
18. Kumar, A.; Dixit, S.; Ram, T.; Yadav, R.B.; Mishra, K.K.; Mandal, N.P. Breeding high-yielding drought-tolerant rice: Genetic variations and conventional and molecular approaches. *J. Exp. Bot.* **2014**, *65*, 6265–6278. [CrossRef] [PubMed]
19. Ćalić, I.; Groen, S.C.; Choi, J.Y.; Joly-Lopez, Z.; Hamann, E.; Natividad, M.A.; Dorph, K.; Cabral, C.L.U.; Torres, R.O.; Vergara, G.V.; et al. The influence of genetic architecture on responses to selection under drought in rice. *Evol. Appl.* **2022**, *15*, 1670–1690. [CrossRef] [PubMed]
20. Kumar, A.; Raman, A.; Yadav, S.; Verulkar, S.B.; Mandal, N.P.; Singh, O.N.; Swain, P.; Ram, T.; Badri, J.; Dwivedi, J.L.; et al. Genetic gain for rice yield in rainfed environments in India. *Field Crops Res.* **2021**, *260*, 107977. [CrossRef] [PubMed]
21. Shamsudin, N.A.A.; Swamy, B.P.; Ratnam, W.; Cruz, S.; Teressa, M.; Sandhu, N.; Raman, A.K.; Kumar, A. Pyramiding of drought yield QTLs into a high-quality Malaysian rice cultivar MRQ74 improves yield under reproductive stage drought. *Rice* **2016**, *9*, 21. [CrossRef]
22. Selamat, N.; Nadarajah, K.K. Meta-Analysis of Quantitative Traits Loci (QTL) Identified in Drought Response in Rice (*Oryza sativa* L.). *Plants* **2021**, *10*, 716. [CrossRef]
23. MohdIkmal, A.; Noraziyah, A.A.S.; Wickneswari, R. Incorporating Drought and Submergence Tolerance QTL in Rice (*Oryza sativa* L.)—The Effects under Reproductive Stage Drought and Vegetative Stage Submergence Stresses. *Plants* **2021**, *10*, 225. [CrossRef]
24. Bocco, R.; Lorieux, M.; Seck, P.A.; Futakuchi, K.; Manneh, B.; Baimey, H.; Ndjiondjop, M.N. Agro-morphological characterization of a population of introgression lines derived from crosses between IR 64 (*Oryza sativa* indica) and TOG 5681 (*Oryza glaberrima*) for drought tolerance. *Plant Sci.* **2012**, *183*, 65–76. [CrossRef]
25. Abd Allah, A.A.; Ammar, M.H.; Badawi, A.T. Screening rice genotypes for drought resistance in Egypt. *J. Plant Breed. Crop Sci.* **2010**, *2*, 205–215.
26. Wang, S.; Wei, J.; Li, R.; Qu, H.; Chater, J.M.; Ma, R.; Li, Y.; Xie, W.; Jia, Z. Identification of optimal prediction models using multi-omic data for selecting hybrid rice. *Heredity* **2019**, *123*, 395–406. [CrossRef] [PubMed]
27. Guimarães, C.M.; de Castro, A.P.; Stone, L.F.; de Oliveira, J.P. Drought tolerance in upland rice: Identification of genotypes and agronomic characteristics. *Acta Sci. Agron.* **2016**, *38*, 201–206. [CrossRef]
28. Dhawan, G.; Kumar, A.; Dwivedi, P.; Gopala Krishnan, S.; Pal, M.; Vinod, K.K.; Bhowmick, P.K.; Bollinedi, H.; Ellur, R.K.; Ravikiran, K.T. Introgression of qDTY1. 1 Governing Reproductive Stage Drought Tolerance into an Elite Basmati Rice Variety "Pusa Basmati 1" through Marker Assisted Backcross Breeding. *Agronomy* **2021**, *11*, 202. [CrossRef]
29. Singh, U.M.; Dixit, S.; Alam, S.; Yadav, S.; Prasanth, V.V.; Singh, A.K.; Venkateshwarlu, C.; Abbai, R.; Vipparla, A.K.; Badri, J.; et al. Marker-assisted forward breeding to develop a drought-, bacterial-leaf-blight-, and blast-resistant rice cultivar. *Plant Genome* **2022**, *15*, 20170. [CrossRef]
30. Anupam, A.; Sinha, S.K.; Banerjee, A.; Roy, S.; Mandal, N.P. Analysis of Genetic Diversity and Survey of QTLs for Grain Yield under Drought Stress in Drought Tolerant Rice Landraces Using DTY QTL-linked Markers. *IJPGR* **2022**, *35*, 250–256. [CrossRef]
31. Kadam, N.N.; Struik, P.C.; Rebolledo, M.C.; Yin, X.; Jagadish, S.V. Genome wide association provides novel genomic loci controlling rice grain yield and its component traits under water-deficit stress during the reproductive stage. *J. Exp. Bot.* **2018**, *69*, 4017–4032. [CrossRef]

32. Liang, F.; Zhan, W.; Hu, G.; Liu, H.; Xing, Y.; Li, Z.; Han, Z. Five plants per RIL for phenotyping traits of high or moderate heritability ensure the power of QTL mapping in a rice MAGIC population. *Mol. Breed.* **2022**, *42*, 28. [CrossRef]
33. Elliott, A.C.; Hynan, L.S. A SAS® macro implementation of a multiple comparison post hoc test for a Kruskal–Wallis analysis. *Comput. Methods Programs Biomed.* **2011**, *102*, 75–80. [CrossRef] [PubMed]
34. Yang, Y.; Yu, J.; Qian, Q.; Shang, L. Enhancement of Heat and Drought Stress Tolerance in Rice by Genetic Manipulation: A Systematic Review. *Rice* **2022**, *15*, 67. [CrossRef]
35. Kilasi, N.L.; Singh, J.; Vallejos, C.E.; Ye, C.; Jagadish, S.K.; Kusolwa, P.; Rathinasabapathi, B. Heat stress tolerance in rice (*Oryza sativa* L.): Identification of quantitative trait loci and candidate genes for seedling growth under heat stress. *Front. Plant Sci.* **2018**, *9*, 1578. [CrossRef] [PubMed]
36. Cortés, A.J.; López-Hernández, F. Harnessing crop wild diversity for climate change adaptation. *Genes* **2021**, *12*, 783. [CrossRef] [PubMed]
37. Borah, P.; Sharma, E.; Kaur, A.; Chandel, G.; Mohapatra, T.; Kapoor, S.; Khurana, J.P. Analysis of drought-responsive signalling network in two contrasting rice cultivars using transcriptome-based approach. *Sci. Rep.* **2017**, *7*, 42131. [CrossRef] [PubMed]
38. Cruz, M.T.S.; Ramos, J.; Anumalla, M.; Catolos, M.; Khanna, A.; Bhosale, S.; Hussain, W. *Standard Operational Procedures for Drought, Salinity, and Submergence Phenotypic Screening Protocols*; IRRI: Los Baños, Philippines, 2020; pp. 1–13.
39. Ministère de l'Environnement, de l'Agriculture et de l'Elevage (MINEAGRIE). *Fiche Technique Harmonisée Pour la Riziculture au Burundi*, 3rd ed.; MINEAGRIE: Bujumbura, Burundi, 2019; pp. 1–2.
40. Moormann, F.R.; Van Breemen, N. *Rice: Soil, Water, Land*; IRRI: Los Baños, Philippines, 1978; pp. 1–185.
41. Rab, M.A.; Chandra, S.; Fisher, P.D.; Robinson, N.J.; Kitching, M.; Aumann, C.D.; Imhof, M. Modelling and prediction of soil water contents at field capacity and permanent wilting point of dryland cropping soils. *Soil Res.* **2011**, *49*, 389–407. [CrossRef]
42. Ndikuryayo, C.; Ochwo-Ssemakula, M.; Gibson, P.; Lamo, J. Resistance to Rice yellow mottle virus and performance of selected improved rice genotypes in central Uganda. *Crop Prot.* **2020**, *129*, 105041. [CrossRef]
43. Rafter, J.A.; Abell, M.L.; Braselton, J.P. Multiple comparison methods for means. *Siam Rev.* **2002**, *44*, 259–278. [CrossRef]
44. Thant, A.A.; Zaw, H.; Kalousova, M.; Singh, R.K.; Lojka, B. Genetic Diversity and Population Structure of Myanmar Rice (*Oryza sativa* L.) Varieties Using DArTseq-Based SNP and SilicoDArT Markers. *Plants* **2021**, *10*, 564. [CrossRef]
45. Badji, A.; Machida, L.; Kwemoi, D.B.; Kumi, F.; Okii, D.; Mwila, N.; Agbahoungba, S.; Ibanda, A.; Bararyenya, A.; Nghituwamhata, S.N.; et al. Factors influencing genomic prediction accuracies of tropical maize resistance to fall armyworm and weevils. *Plants* **2021**, *10*, 29. [CrossRef]
46. Cortés, A.J.; Chavarro, M.; Madriñán, S.; This, D.; Blair, M.W. Molecular ecology and selection in the drought-related Asr gene polymorphisms in wild and cultivated common bean (Phaseolus vulgaris L.). *BMC Genet.* **2012**, *13*, 58. [CrossRef]
47. Blair, M.W.; Cortés, A.J.; This, D. Identification of an ERECTA gene and its drought adaptation associations with wild and cultivated common bean. *Plant Sci.* **2016**, *242*, 250–259. [CrossRef]
48. Cortés, A.J.; This, D.; Chavarro, C.; Madriñán, S.; Blair, M.W. Nucleotide diversity patterns at the drought-related DREB2 encoding genes in wild and cultivated common bean (*Phaseolus vulgaris* L.). *Theor. Appl. Genet.* **2012**, *125*, 1069–1085. [CrossRef]

plants

Article

Comparative Transcriptome Analysis of Tolerant and Sensitive Genotypes of Common Bean (*Phaseolus vulgaris* L.) in Response to Terminal Drought Stress

Mayavan Subramani [1,*], Carlos A. Urrea [2], Rasheed Habib [1], Ketaki Bhide [3], Jyothi Thimmapuram [3] and Venu Kalavacharla [1]

[1] Molecular Genetics and Epigenomics Laboratory, College of Agriculture, Science and Technology (CAST), Delaware State University, Dover, DE 19901, USA
[2] Panhandle Research and Extension Center, University of Nebraska, 4502 Avenue I, Scottsbluff, NE 69361, USA
[3] Bioinformatics Core, Purdue University, West Lafayette, IN 47907, USA
* Correspondence: msubramani@desu.edu

Abstract: We conducted a genome-wide transcriptomic analysis of three drought tolerant and sensitive genotypes of common bean to examine their transcriptional responses to terminal drought stress. We then conducted pairwise comparisons between the root and leaf transcriptomes from the resulting tissue based on combined transcriptomic data from the tolerant and sensitive genotypes. Our transcriptomic data revealed that 491 (6.4%) DEGs (differentially expressed genes) were upregulated in tolerant genotypes, whereas they were downregulated in sensitive genotypes; likewise, 396 (5.1%) DEGs upregulated in sensitive genotypes were downregulated in tolerant genotypes. Several transcription factors, heat shock proteins, and chaperones were identified in the study. Several DEGs in drought DB (data Base) overlapped between genotypes. The GO (gene ontology) terms for biological processes showed upregulation of DEGs in tolerant genotypes for sulfate and drug transmembrane transport when compared to sensitive genotypes. A GO term for cellular components enriched with upregulated DEGs for the apoplast in tolerant genotypes. These results substantiated the temporal pattern of root growth (elongation and initiation of root growth), and ABA-mediated drought response in tolerant genotypes. KEGG (kyoto encyclopedia of genes and genomes) analysis revealed an upregulation of MAPK (mitogen activated protein kinase) signaling pathways and plant hormone signaling pathways in tolerant genotypes. As a result of this study, it will be possible to uncover the molecular mechanisms of drought tolerance in response to terminal drought stress in the field. Further, genome-wide transcriptomic analysis of both tolerant and sensitive genotypes will assist us in identifying potential genes that may contribute to improving drought tolerance in the common bean.

Keywords: transcriptomics; common bean; root; leaf; terminal drought stress; tolerant genotypes; sensitive genotypes

1. Introduction

The common bean (*Phaseolus vulgaris* L.) is grown extensively in various regions, from lowland tropical to semi-arid environments, either as a monoculture or intercropped. As a legume, it is an abundant source of total protein, micronutrients, and energy [1]. In the event of terminal or intermittent drought, common bean yield can be negatively affected by up to 60%, a loss that is only further worsened when soil moisture is reduced to 60–70% during grain filling [2–4]. Plants employ three key strategies to adapt to drought conditions and respond appropriately. Drought escape through changes in molecular mechanisms that allow them to adapt to the environment, drought avoidance through modifications to morphological and physiological characteristics, and drought tolerance by protein stabilization and osmotic adaptations to cope with dehydration [5]. As a result, a combination of morphological, physiological, biochemical, and molecular changes is more than likely to be induced by changes in the upregulation of several regulatory and functional genes to sense and respond to drought stress [6].

Studying signal perception, gene expression and regulation, and metabolic pathways followed during drought stress is crucial for understanding how plants respond to drought stress [7]. Furthermore, analysis of gene enrichment in metabolic pathways under drought stress will aid in screening for potential genes and drought response mechanisms in diverse plants [8]. It is essential to understand plants' common and specific gene expression and regulation during drought stress [9]. Due to metabolic and biological processes differing between tissues above and below ground [5], energy production in the leaves is fine-tuned based on the water availability in the roots. Understanding how plants respond to drought stress can be improved by investigating differential gene expression patterns between the tissues above and below ground [10]. Such a study may lead to the discovery of a coordinated biological process, or a distinct pattern of processes followed by each type of tissue. Therefore, a better understanding of drought-specific biological processes, and crosstalk between the regulatory mechanisms of different tissues may contribute significantly towards advancing knowledge of the general molecular mechanisms underlying drought response [11].

Next-generation sequencing, such as RNA sequencing (RNA-Seq), has become increasingly common in analyzing plants' transcriptome under drought stress. The Andean [12] and Mesoamerican [13] common bean genomes have been sequenced, with a total of 587 megabase pairs (Mbp) and 549.6 Mbp, respectively. These studies provide insight into the genetic basis of biotic and abiotic stress responses [12]. Therefore, genomic information favors the identification of genes that are drought responsive through the use of transcriptomic analysis [2].

Different genotypes of common bean have been studied for differences in gene expression, co-expression of genes, and the relationship between several pathways and the biological functions of specific genes during drought stress [2,14,15]. Genome-wide gene expression analysis in the contrasting genotypes supports the identification of genetic interactions responsible for diverse drought response patterns, genotype-specific drought responses, and possible candidate genes for breeding drought tolerance [16–18]. Studying drought responses in field conditions is likely to contribute to a better understanding of the molecular mechanisms underlying drought responses and facilitate the development of effective drought mitigation strategies [19].

Tolerant chickpea (*Cicer arietinum*) genotypes were found to be able to conserve water during terminal drought stress, which was later utilized in reproductive stages. Yet there were no significant differences in root traits between tolerant and sensitive genotypes. Interestingly, roots temporal pattern of water uptake relates to drought tolerance [19,20]. Another study showed that the root-to-shoot ratio increased when plants were exposed to the terminal and intermittent drought stress [21]. Thus, root and shoot growth are highly coordinated in water deficit conditions [22]. In a comparison of differential gene expression in response to the onset of water stress between three hybrid *Brachiaria* genotypes, it has been suggested that faster root growth can offer advantages to the plant in terms of extracting water from soil during a short period of rain [23].

In leaves of the signal grass (*Brachiaria* (Trin.) Griseb.) genotypes, a GO term for carbohydrate and cell wall metabolism was enriched by upregulated DEGs compared to downregulated DEGs in tolerant genotypes. Alternatively, DEGs for apoplastic peroxidase activities, which involved lignification and elasticity of secondary cell walls, were significantly downregulated in the roots of all genotypes, suggesting the pattern of gene expression and the root structure may be influential in response to water stress [10]. In the present study, it is anticipated that root DEGs derived from tolerant genotypes might be enriched with respect to the temporal pattern of root activities, which include variation in cellular response and signaling pathways when responding to drought stress, specifically terminal drought stress in field conditions.

Our study provides genome-wide transcriptomic analysis of six genotypes, three tolerant and three sensitive, grown under terminal drought stress to drought conditions. Our previous study tested these genotypes at two locations for broad temperate and tropical adaptations [24]. Terminal drought stress (after flowering) was imposed on all genotypes in the field. The transcriptomes of both leaves and roots have been generated, and comparisons have been made between tolerant roots and leaves and sensitive roots and leaves. It is anticipated that the described approach will lead to greater insight into the molecular mechanisms of drought tolerance in common bean genotypes.

2. Materials and Methods

2.1. Field Experiment

Six genotypes of common bean (*Phaseolus vulgaris* L.) were used in the present study. Tolerant genotypes: Merlot'//05F-5055-1/98020-3-1-6-2 (SB-DT3), and USPT-ANT// 'Matterhorn'/98078-5-15-1(SB-DT2) [24], Matterhorn [25], and three sensitive genotypes: Sawtooth [26], Merlot [27], and Stampede [28]. The field experiment was conducted as described previously [29] (Figure 1). A field of silt loam soil (Typic Ustorthents) was used for the cultivation of the common bean genotypes (41°56.6′ N, 103°41.9′ W, 1240 m elevation). There was 75% silt, 15% sand, and 10% clay in the soil. In addition, it had a cation exchange capacity of 17 meq/100 g, a pH of 7.8, and an organic matter content of 14 mg/g. Since the field contained 20.5 kg of residual nitrogen and the manure credit contained 25.9 kg of nitrogen, no additional nitrogen was applied to the field. Most plants reached anthesis (flowering) before the drought was imposed. Until then, the plants were irrigated by drip irrigation. The precipitation totaled 135.1 mm before flowering. This includes irrigation of the plants twice (101.6 mm) and precipitation of 33.5 mm. The rainfall after flowering totaled 16.3 mm, and the plants were not irrigated. Upon reaching almost physiological maturity, the leaves and roots of the plants were collected and frozen in liquid nitrogen. The samples were collected from three sensitive and three tolerant genotypes grown under terminal drought stress in three replicates, resulting in 18 samples [$(3 \times 3) + (3 \times 3)$]. Agronomic characters of the tolerant and sensitive genotypes as well as their starch and fat contents were analyzed in our previous study [29], Subramani et al., 2022. For each genotype, RNA isolation and library preparation were performed separately.

Figure 1. Experiments conducted under stress and non-stress conditions in the field.

2.2. RNA Isolation and cDNA Synthesis

Total RNA was isolated from leaves and roots using the Spectrum Total Plant RNA kit (Sigma-Aldrich, St Louis, MO, USA) according to the manufacturer's instructions. Genomic DNA contamination was eliminated by using DNase I (Invitrogen, Waltham, MA, USA). RNA concentration and purity were determined by the Nanodrop 2000 spectrophotometer (Thermo Scientific, Wilmington, DE, USA). Most RNA samples had a 260/280 ratio between 2 and 2.1. The RNA integrity was then assessed with agarose gel electrophoresis and

Agilent Bioanalyzer 2100 (Agilent Technologies, Santa Clara, CA, USA). The complementary DNA (cDNA) was synthesized from 1µg of total RNA according to the manufacturer's instructions using the ProtoScript II First Strand cDNA Synthesis kit (New England Biolabs, Ipswich, MA, USA).

2.3. Library Preparation and Sequencing

The RNA-Seq libraries were prepared using the Illumina TruSeq Stranded mRNA Sample Preparation Kit (Illumina Inc., San Diego, CA, USA). Briefly, Poly(A) tail mRNA was enriched for first-strand c-DNA synthesis. Following the second strand synthesis, multiple washes were carried out for the end repairs. A single A nucleotide was then added to the 3′ ends. PCR was performed to enrich both ends of the shorter fragments with adapters. The cDNA fragment pools were then loaded, and a paired-end read of 150 bp sequencing was performed on the Illumina HiSeq^{TM} 2500 at the Delaware Biotechnology Institute in Newark, DE, USA.

2.4. Pre-Processing RNA- Seq Data

Phaseolus vulgaris v2.1, downloaded from phytozome v13, was used for genome and gene references. The following URLs were accessed on 8 September 2021. ASTQC (v0.11.9; http://www.bioinformatics.babraham.ac.uk/projects/fastqc/) was used to assess sequence quality, and Trim Galore (v 0.6.5; http://www.bioinformatics.babraham.ac.uk/projects/trim_galore/) was used to trim adapters and reads based on Phred33 score 30, and reads of at least 75 bases were retained. The quality-trimmed reads were mapped to the STAR-indexed genome using STAR (v2.7.9a) [30]. The STAR mapping results, and annotation files (GTF/GFF) were used for reference genome as input for the HTSeq package (version 0.13.5) [31] to calculate the read counts for each gene feature for each sample. For htseq-count, -m union and the -s reverse were used. A read count matrix for all samples was generated by merging counts from all samples with custom Perl scripts.

2.5. Differentially Expressed Genes (DEGs) Identification

DEGs (differentially expressed genes) were identified using DESeq2 (v 1.32.0) [32]. Pairwise comparisons were made between Tolerant root vs leaves and Sensitive root vs leaves. A read count matrix generated using HTSeq was used as input for DESeq2. Genes with a non-zero read count in at least one sample were selected. Then zeros in the matrix were replaced by ones to avoid infinite values being calculated for fold change. DESeq2 uses the negative binomial distribution based on the data model and performs specific estimate variance–mean tests. DESeq2 determines significant DEGs using the Wald test. *p*-value and adjusted *p*-values of false discovery rate (FDR) to correct for multiple tests were added using DESeq2 based on the Wald test.

2.6. Functional Analysis of DEGs

ClusterProfiler (Version 4.0.5) [33], and R package (R Version 4.1.0) were used to perform Gene Ontology (GO) and Kyoto Encyclopedia of Genes and Genomes (KEGG) enrichment of significant differentially expressed genes in all pairwise comparisons (padj < 0.05 and log2 FC +/−2). Only the DEGs lists were included for analysis. To perform GO analysis of corresponding DEGs, the Log_2FC was included as input for agriGO v2 [34] using Singular Enrichment Analysis (SEA) and parametric Analysis of Gene Set Enrichment (PAGE) tools. We selected *Phaseolus vulgaris* species and custom GO database of *P. vulgaris* v 2.1 genome (downloaded from Phytozome v13) (accessed on 8 September 2021) and default parameters such as complete GO database, significance cutoff 0.05, and Hochberg (FDR) method as multiple test correction methods to perform agriGO. The statistically significant GO terms associated with DEGs were derived.

2.7. Volcano Plots

Volcano plots were generated to illustrate a pattern of significant DE genes using the "EnhancedVolcano" R package (v1.10.0) (R version 4.1.0) (accessed on 16 November 2021) using default options/parameters. The pattern and number of significant DE genes were selected based on log2FC and padj < 0.05.

2.8. Quantitative Real-Time (qRT-PCR)

QRT-PCR was performed to validate the RNA-Seq results. The C-DNA was synthesized from the root RNA samples. The primers were designed for the up and down-regulated DEGs. The list of primers and the corresponding genes were provided in Supplementary Table S6. The primers were designed using the primerQuest™ Tool (Integrated DNA Technologies (IDT), Coralville, IA, USA). The primers were validated by conventional PCR before being used in qRT-PCR. The qRT-PCR was performed on the ABI 7500 real-time PCR (Applied Biosystems, Foster City, CA, USA). Each reaction was carried out in a 25 µL master mix containing 100 ng of C-DNA, 10 µM each of forward and reverse primers, and 12.5 µL of SYBR Green PCR Master Mix (Germantown, MD, USA). The reaction was run at an initial denaturation of 95 °C for 10 min, followed by 40 cycles at 95 °C for 15 s and a final extension at 60 °C for 1 min. The samples were run in triplicate, and the actin gene [35] was used to normalize qPCR samples and as an internal control. The comparative CT method $2^{-\Delta CT}$ [36] was used to calculate relative expressions.

3. Results

3.1. Data Quality and Summary of Reads

A total of 4615 million reads were obtained from the tolerant and sensitive root and leaf samples. There were approximately 129 million reads from each sample (Table 1). More than 92% of reads were mapped to the reference genome (*Phaseolus vulgaris* v2.1). These confirmed the high data quality and could be used for further analysis. In order to visualize the variation in RNA sequences between root and leaf samples, principal component analysis was performed with DESeq2. There was a clear distinction between root and leaf replicates from tolerant and sensitive genotypes on the PCA scatter plot (Supplementary Figure S1). In addition, this will provide further confirmation that there is great variation in gene expression between the roots and leaves.

Table 1. Summary of reads with average read counts for the individual samples and the cumulative totals of both roots and leaves.

Tolerant Root		**Tolerant Leaves**		**Sensitive Root**		**Sensitive Leaves**	
Matt_7R	139,640,906	Matt_10L	107,711,956	Swa_4R	130,624,430	Swa_2L	130,454,386
Matt_7R	138,965,052	Matt_14L	116,817,638	Swa_5R	133,482,394	Swa_6L	110,347,824
Matt_9R	154,185,446	Matt_17L	117,538,996	Swa_6R	152,247,298	Swa_9L	125,654,646
Total	**432,791,404**	**Total**	**342,068,590**	**Total**	**416,354,122**	**Total**	**366,456,856**
Average	**144,263,801**	**Average**	**114,022,863**	**Average**	**138,784,707**	**Average**	**122,152,285**
MerX_10R	174,025,346	MerX_11L	162,408,546	Mer_1R	104,058,518	Mer_1L	160,053,034
MerX_11R	119,788,128	MerX_15L	119,504,092	Mer_2R	120,540,136	Mer_5L	144,966,788
MerX_12R	106,557,634	MerX_16L	110,960,258	Mer_3R	112,068,328	Mer_7L	128,432,570
Total	**400,371,108**	**Total**	**392,872,896**	**Total**	**336,666,982**	**Total**	**433,452,392**
Average	**133,457,036**	**Average**	**130,957,632**	**Average**	**112,222,327**	**Average**	**144,484,131**
USPT_13R	84,249,968	USPT_12L	168,923,136	Stam_16R	115,711,432	Stam_3L	110,748,340
USPT_14R	128,161,580	USPT_13L	107,542,292	Stam_17R	102,825,568	Stam_4L	140,862,648
USPT_15R	121,624,450	USPT_18L	138,913,792	Stam_18R	144,821,904	Stam_8L	129,873,270
Total	**334,035,998**	**Total**	**415,379,220**	**Total**	**363,358,904**	**Total**	**381,484,258**
Average	**111,345,333**	**Average**	**138,459,740**	**Average**	**121,119,635**	**Average**	**127,161,419**

Matt—Matterhorn; MerX—MerlotX; Swa—Sawtooth; Mer—Merlot; Stam—Stampede.

3.2. Specific Gene Expression within Tolerant and Sensitive Genotypes

Using the two pairwise comparisons (Tolerant genotypes roots vs. leaves, Sensitive genotypes roots vs. leaves), the total number of responsive genes identified with the restriction to FDR < 0.05 was 15,685 for the tolerant and 16,226 for the sensitive genotypes (Supplementary Tables S1 and S2). With FDR < 0.05 and log 2FC ± 2, 491 (6.4%) upregulated DEGs were unique to the tolerant genotype. Similarly, 396 (5.1%) were enriched in sensitive genotypes (Figure 2). These DEGs are likely to play an important role in response to terminal drought. Furthermore, the scatter plot was constructed based on the log2fold change and $-\log_{10}$ padj to determine the upregulated and downregulated genes by restricting FDR < 0.05 and log2fold change ±2. In total, 3319 DEGs were upregulated, and 3561 genes were downregulated in the tolerant genotypes comparisons. In the sensitive genotypes comparisons, 3224 DEGs were upregulated, and 3652 were downregulated (Figure 3).

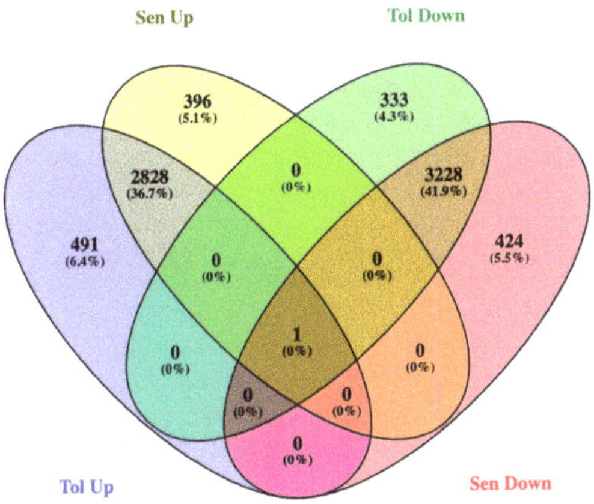

Figure 2. Venn diagram shows the potential upregulated and downregulated DEGs in tolerant and sensitive genotypes. DEGs were quantified based on log2 fold changes ±2 and FDR < 0.05.

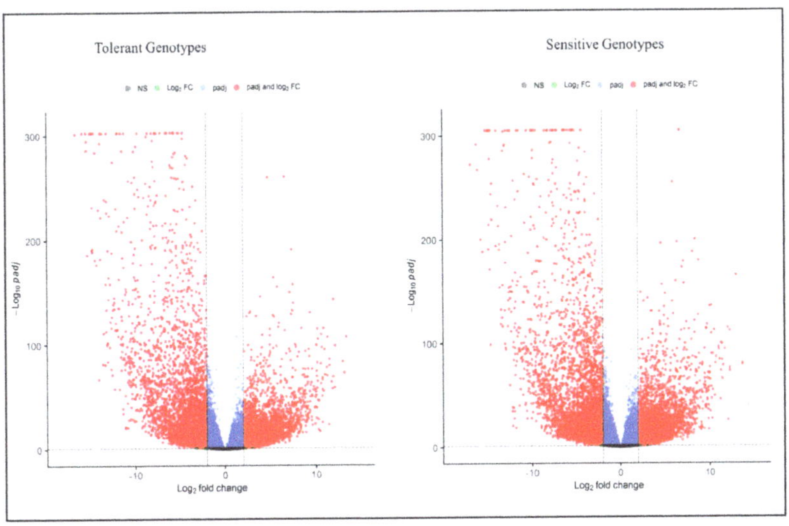

Figure 3. Volcano plots illustrate the up and downregulation of DEGs in tolerant and sensitive genotypes. The red color indicates the up and down regulation of DEGs based on the log2 fold change and padj < 0.05.

3.3. GO Enrichment Analysis

A GO annotation analysis was performed to analyze DEGs based on padj < 0.05, associated with cellular components, molecular functions, and biological processes. In both the genotypes comparisons, biological process enrichment revealed that DEGs enriched in photosynthesis, light harvesting, response to oxidative stress, recognition of pollen, defense response, lipid biosynthetic process, response to biotic stimulus, and response to auxin (Figure 4). Among the selected DEGs 2726 of the tolerant genotypes comparisons, 345 genes were involved in diverse biological processes. In contrast, out of 2706 DEGs, 303 genes were involved in biological processes for the sensitive genotypes. It was found that sulfate transport and drug transmembrane transport were the two biological processes enriched specifically in tolerant genotype comparisons (Figure 4). The gene ratios were almost identical between the two genotypes for each biological process (Supplementary Table S3). Among 3576 DEGs, 133 and 92 were enriched for cellular component function for tolerant and sensitive genotype comparisons, respectively. The DEGs specific to cellular components, such as the apoplast, were enriched in the tolerant genotypes compared to the sensitive genotypes. In each cellular component function, the gene ratios were similar. However, the tolerant genotypes were more enriched than the sensitive genotypes for the molecular function category. These include xenobiotic transmembrane transporter activity, antiporter activity, secondary active sulfate transmembrane transporter activity, sulfate transmembrane transporter activity, electron transfer activity, and xyloglucan:xyloglucosyl transferase activity. Also, specific molecular functions enriched for sensitive genotypes were hydrolase activity, acting on ester bonds, and two iron, two sulfur cluster binding. Higher DEGs ratios for heme binding were 189 out of 3448 and 186 out of 3424 for the tolerant and sensitive genotypes, respectively. The differences in the DEGs enrichment may indicate the possible biological functional differences in the tolerant and sensitive genotypes in response to drought stress.

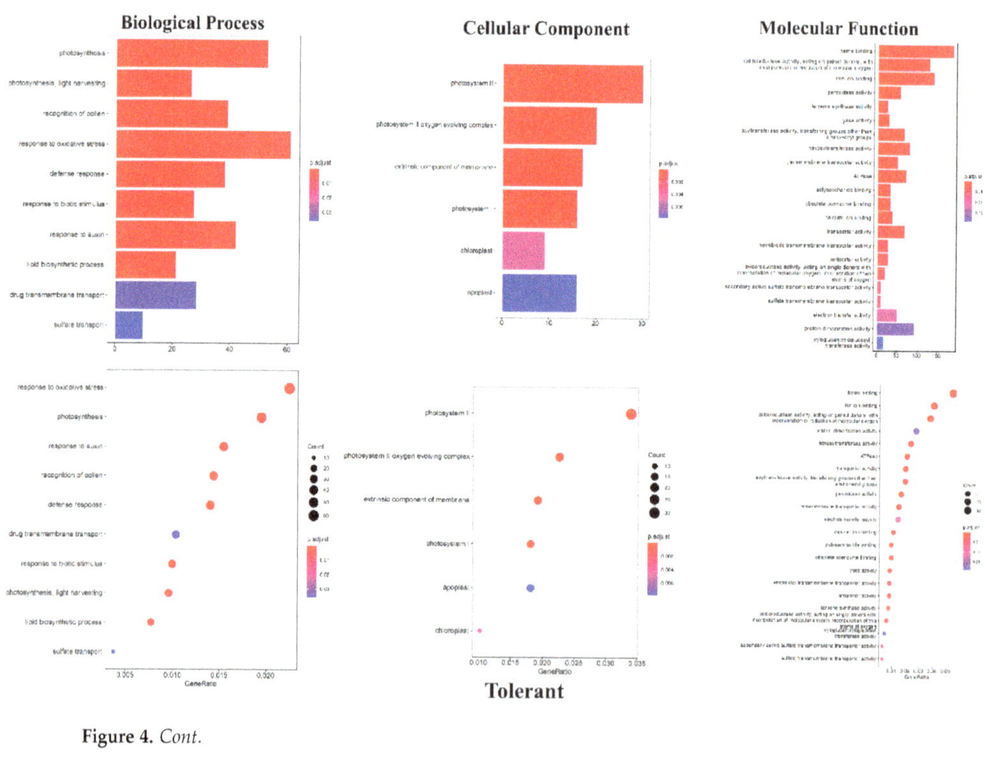

Figure 4. *Cont.*

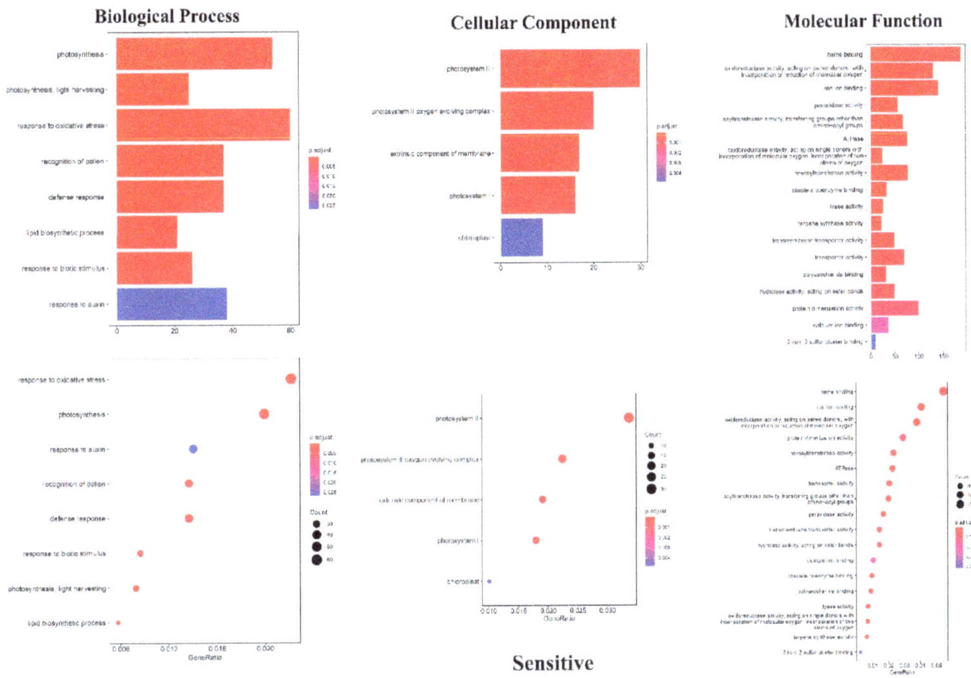

Figure 4. Based on gene ontology (GO) annotation, the root and leaf comparisons of tolerant and sensitive genotypes reveal DEG enrichment for biological processes, cellular components, and molecular functions. Gene ratios for enriched DEGs are included. The number of genes involved are plotted against each GO terms (Biological process, cellular components, molecular function). The DEGs were annotated against the GO database. The GO enrichment was based on the padj < 0.05.

3.4. *KEGG and DEGs*

In addition, KEGG enrichment function analyses were performed for both the tolerant and sensitive genotype comparisons. In both the tolerant and sensitive genotype comparisons, the number of genes enriched for the plant hormone signal transduction pathway followed by phenylpropanoid biosynthesis was higher than the other pathways (Figure 5). MAPK signaling pathway was specific to tolerant genotype comparisons, while monoterpenoid biosynthesis, cyanoamino acid metabolism, zeatin biosynthesis, ABC transporters, and tyrosine metabolism were exclusive to sensitive genotype comparisons. More pathways in sensitive than tolerant genotype comparisons were identified. DEGs for starch and sucrose metabolism were comparatively upregulated in sensitive genotypes. Differences in the pathways may be associated with specific functional roles in response to drought stress.

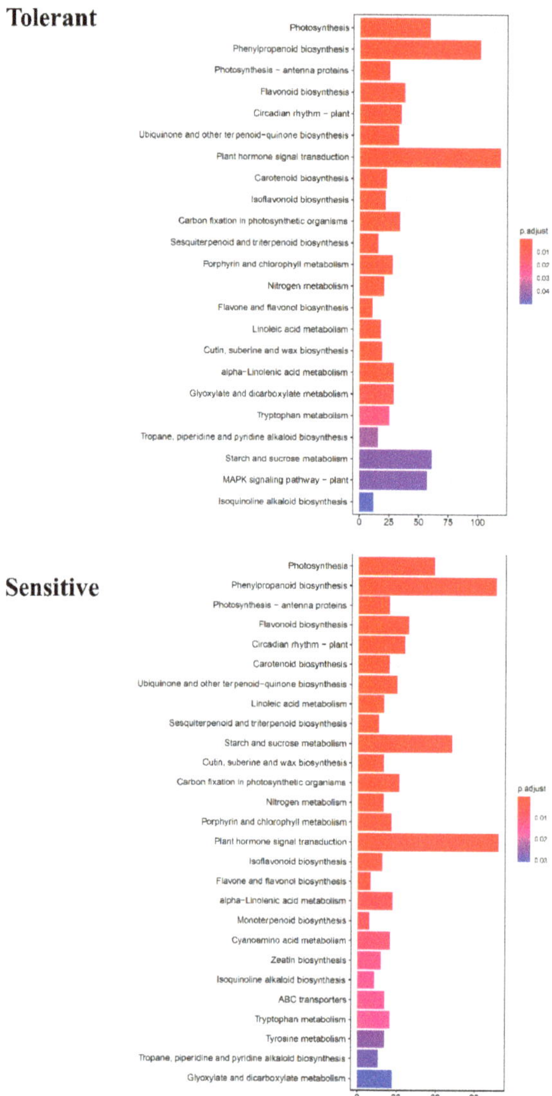

Figure 5. DEG enrichment of the metabolic pathway. There was a difference between the tolerant and sensitive genotypes in the number of genes enriched for the particular pathways.

3.5. DEGs in Drought DB (Data Base)

It was found that 155 and 154 DEGs from tolerant and sensitive genotype comparisons, respectively, had best hits to *Arabidopsis* genes in the drought database. These lists of genes, their names, and descriptions, identified from phytozome, along with the corresponding log2FC, are included in an Excel file (Supplementary Table S4). Among the DEGs, Phvul.002G107100 and Phvul.001G166500 were the two genes with higher expression (Log2FC > 7) in the tolerant and sensitive genotypes comparisons (Figure 6). The Phvul.002G107100 gene is homologous to AHK1 in *Arabidopsis*. The AHK1 (Histidine kinase 1) gene was reported to regulate drought and salt responses through ABA (abscisic acid) independent and dependent signaling pathways [37]. Similarly, the sensitive genotypes contain the homologous gene Phvul.001G166500 to *Arabidopsis* ABCG40. The *atabcg40* mutants exhibited reduced drought tolerance due to impaired lateral root formation and loss of stomatal function [38]. Upregulated DEGs might play an important role in drought responses associated with ABA in both genotypes' comparisons.

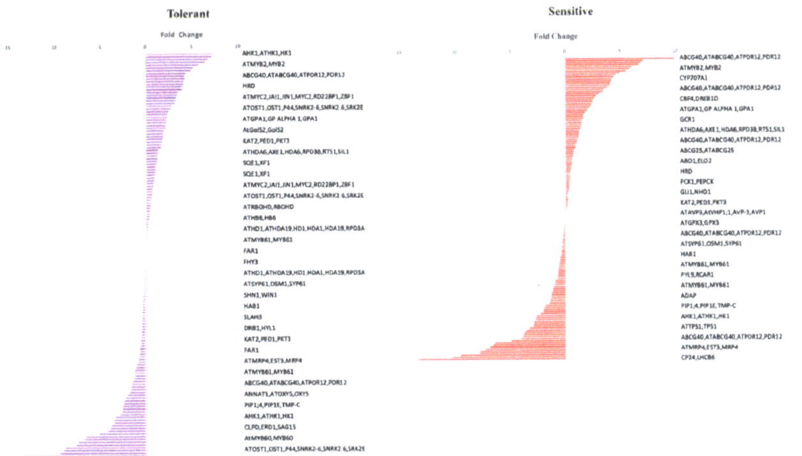

Figure 6. DEGs identified in the tolerant and sensitive genotype comparisons have the best hits to Arabidopsis genes in drought DB. DEGs are shown with fold changes.

3.6. Transcription Factors (TFs)

Out of 25,419 genes, 1418 TFs were identified in the tolerant genotype comparisons. Similarly, out of 25,466 genes in the sensitive genotypes, 1416 TFs were identified (Supplementary Table S5). The C2H2 family protein of TFs was associated with the upregulated DEGs Phvul.002G075900 and Phvul.002G075900 with Log2FC > 10 in tolerant and sensitive genotypes, respectively. Transcription factors zinc finger protein 7 and LOB domain-containing protein 27 were identified specifically in the tolerant genotypes' comparisons. In the sensitive genotypes, myb domain protein 82, myb domain protein 21, WOX family protein, homeodomain GLABROUS 12 AGAMOUS-like 58, myb domain protein 101, myb domain protein 64 and zinc finger protein 3 were identified specifically.

3.7. Validation of Differentially Expressed Genes

To confirm the results of RNA-Seq, we analyzed the expression patterns of randomly selected up and downregulated differentially expressed transcripts using qRT-PCR. QRT-PCR results for selected DEGs were similar to those obtained by RNA-Seq. Comparisons of the expression patterns revealed a strong correlation ($r^2 = 0.8693$) (Supplementary Figure S2) between both techniques.

4. Discussion

The incidence of terminal drought stress in common bean is most common in bean growing areas. Drought stress adversely impacts the quality and yield of common bean, particularly during the pod development and seed filling stages [4,39]. Therefore, improving the plant's ability to withstand this stress is considered a key objective of breeding programs [4,40]. The complete genome sequence of the common bean is now available. This makes it possible to discover transcription regulation patterns and enrich gene databases by exploring transcriptome maps and related pathways among tolerant and sensitive genotypes. In addition to improving drought tolerance, this may also help identify the molecular mechanisms of abiotic stress [2]. Several studies have examined the transcriptional response of common bean to drought stress under controlled conditions [2,15,41–44].

However, studies that developed and utilized transcriptome data associated with terminal drought stress imposed on field grown common bean genotypes are limited. The primary objective of this study is to identify transcriptional changes and molecular mechanisms underlying the terminal drought stress induced on sensitive and tolerant genotypes of field-grown common beans.

4.1. Potential DEGs in Tolerant and Sensitive Genotypes

In total, 486 DEGs that were upregulated in tolerant genotypes were found to be downregulated in sensitive genotypes. It is likely that these DEGs contribute to drought tolerance. Among these genes, Phvul.010G050500, an ethylene-responsive element binding factor 13, has been reported as a drought candidate gene in solanaceous plants and is involved in gene regulation of metabolic pathways [45]. Phvul.010G088500 (S-locus lectin protein kinase family protein) is also known to be expressed highly in rice (*Oryza sativa*) and *Arabidopsis* as a response to stress conditions such as drought, salt, and other biotic stresses [46]. The log2FC of these genes was >5 in tolerant genotypes. Also, the C2H2 and C2HC zinc fingers superfamily protein (Phvul.002G256900) and the ethylene response factor 1 (Phvul.001G160200) both display log2FC > 4 and are upregulated specifically in tolerant genotypes. These transcription factors regulate gene expression in abiotic stresses such as drought and salt genes in *Arabidopsis* [47,48]. Similarly, 392 DEGs upregulated in sensitive genotypes were downregulated in tolerant genotypes. Among the DEGs, Phvul.011G163300, annotated as Ankyrin repeat family protein, had a log2FC > 2. Ankyrin repeat protein (DRA1) is reported to negatively regulate drought tolerance in *Arabidopsis* [49]. The volcano plot analysis indicated that more genes were up and downregulated in the tolerant genotypes than in the sensitive genotypes. In the tolerant genotypes, Phvul.011G182500 had a log2FC > 13 and padj < 0.05. This gene is associated with an MLP-like protein 43 that is highly expressed in roots and cotyledons. It is a positive regulator of drought stress response and regulates ABA-responsive gene expression [50]. Additionally, Phvul.003G285900 is upregulated (Log2Fc > 7) with best hits in drought DB in drought-tolerant genotypes, while its homologous (PCK1) has been reported to influence drought tolerance in *Arabidopsis* by reducing stomatal conductance [51]. Similarly, in sensitive genotypes, Phvul.001G166500 showed upregulation with log2FC > 9, and its homologous PDR12 (pleiotropic drug resistance 12) was reported to improve drought tolerance mediated through efficient ABA transport in *Arabidopsis* [38]. DEGs such as Phvul.008G147800 and Phvul.L002632 were upregulated in both tolerant and sensitive genotypes with log2FC > 13. They belong to the RmlC-like cupins superfamily protein, which is known to function in the biological process of small molecules (metabolites) involved in oxidoreductase and disomerase activities. It has been reported that these metabolites, such as amines and sulfur amino acids, contribute to maintaining osmotic potential and preventing water loss in plants during stress conditions [52]. In addition, a similar gene was identified (Phvul.002G107100) that has the highest hits in drought DB as well as a higher expression level, i.e., log2FC > 7, in both the tolerant and sensitive genotypes. The homologous version of this gene (HK1) has enhanced drought tolerance and indicates ABA-dependent drought response in *Arabidopsis* [37,53].

4.2. GO Enrichment

GO enrichment was carried out for the selected DEGs to identify the preferred GO terms based on padj < 0.05. The analysis revealed enrichment in DEGs in response to various stresses or inducers, such as photosystem, oxidative stress, signaling, hormone, biotic stimulus, metabolite transport, defense response, and enzymes. In common bean, these GO terms are primarily enriched during drought stress [2], indicating that DEG enrichment is drought-specific and responsive to terminal drought stress. Among GO terms for biological processes, the DEGs upregulated in the tolerant genotypes specifically are related to sulfate transport and drug transmembrane transport compared to sensitive genotypes. Similarly, Pereira et al., 2020 [2] reported DEG enrichment associated with the formation of sulfur-containing compounds in drought-tolerant genotypes of common bean. Sulfate transport is an important component of abiotic stress responses, such as drought. When plant roots are exposed to abiotic stresses, such as drought, sulfate transport plays a critical role in producing sufficient ABA and glutathione compounds [54]. As part of the ABA biosynthesis process, cysteine is utilized as a sulfur donor via sulfate transport. Additionally, sulfate accumulation in the root system increases cysteine and glutathione compounds, which contribute to root growth during drought stress [55]. Therefore, it is possible that upregulation of sulfate transport in tolerant genotypes is critical for drought tolerance through an increase in ABA synthesis, glutathione, and root growth during terminal drought stress. In addition, DEGs' enrichment for drug transmembrane transport is elevated in drug-tolerant genotypes. The upregulation of DEGs for drug transmembrane transport has been reported in other plant species such as *Phormium tenax* (New Zealand Flax), *Pinus massoniana* (Masson Pine), and *Boea hygrometrica* [56–58] under drought stress as well. In contrast, both genotypes exhibited higher enrichment of gene ratios related to the GO term "response to oxidative stress".

In the GO term for cellular components, DEG enrichment for apoplast was upregulated in the tolerant genotypes, but downregulated in the sensitive genotypes. In drought-stressed plants, the alkaline nature of apoplast aids in the accumulation of anionic form of ABA to initiate stomata closure [59]. During drought stress, apoplasts are also reported to participate in nutrient transfer from roots and ROS (reactive oxygen species)-mediated cell signaling [60,61]. Moreover, apoplastic ROS can enhance root elongation via loosening cell walls and protect the root from ROS-induced damage during water stress [62]. Thus, apoplasts play a role in endogenous signaling as well as in the initiation of root growth in tolerant genotypes during terminal drought stress. For the molecular function category, DEGs' enrichment was found to be upregulated for more GO terms such as xenobiotic transmembrane transporter activity, antiporter activity, secondary active sulfate transmembrane transporter activity, sulfate transmembrane transporter activity, electron transfer activity, and xyloglucan:xyloglucosyl transferase activity in tolerant genotypes than in sensitive genotypes. Their role in enhancing drought tolerance has been reported [63–65].

4.3. KEGG

KEGG pathway enrichment was conducted for significant DEGs in the tolerant and sensitive genotype comparisons. In both genotypes, the number of DEGs associated with plant hormone signal transduction pathways is high, followed by phenylpropanoid biosynthesis. Plant hormones play a key role in regulating plant growth, development, and response to biotic and abiotic stresses. Several plant hormones, including ABA, jasmonic acids (JA), ethylene (ET), and salicylic acids (SA), may act as mediators in the prevention or response to drought stress [66]. The hormones work through a coordinated network of signal transduction and cross-talk between hormones to facilitate the switchover of pathways to cope with drought stress [67]. Similarly, Wang et al. (2020) [68] report that the shared DEGs in response to drought stress indicate overlap and crosstalk between hormone signal transduction pathways in Tobacco (*Nicotiana tabacum*).

Phenylpropanoids, such as flavonoids, isoflavonoids, plant hormones, anthocyanins, and lignins, play an important role in biotic and abiotic stress [69]. DEGs associated with the phenylpropanoid pathway were enriched in tolerant genotypes in this study. A similar enrichment of DEGs for the phenylpropanoids pathway has been reported in foxtail millet (*Setaria italica* (L.) P. Beauv.) under drought stress [70].

However, DEGs' enrichment for the MAPK signaling pathway was upregulated in tolerant genotypes. Mitogen-activated protein kinases (MAPKs) are signal transduction modules that transmit extracellular signals into cells. They regulate gene expression by phosphorylating many transcription factors. A number of MAPK components, including MAPK kinase and MAPK kinase kinase, contribute to the development of cells, hormonal activities, and biotic and abiotic stresses [71,72]. A drought-induced MAPK pathway enrichment with upregulated DEGs has been reported for pearl millet (*Pennisetum glaucum*

(L.) [73]. There is also a possibility that in this study, the interaction of enriched pathways, such as the plant hormone signaling pathway and MAPK pathways, can positively impact drought tolerance in tolerant genotypes, as they share common byproducts [74].

DEGs enrichment for monoterpenoid biosynthesis was upregulated in sensitive genotypes. Plant terpenes are synthesized in the cytosol as well as plastid through the mevalonate (MVA) and 2C-methyl-D-erythritol-4-phosphate pathway (MEP) [75]. Terpenoid metabolites have been reported to protect plants against biotic and abiotic stresses [76]. Field drought stress conditions were found to modulate monoterpenoid biosynthesis. Long-term and severe field drought stress resulted in the downregulation of DEGs that encode structural enzymes for monoterpenoid biosynthesis [77]. This may explain the downregulation of DEGs for monoterpenoid biosynthesis in tolerant genotypes in this study. Similarly, Wan et al. (2022) [78] reported that DEGs downregulated for monoterpenoid biosynthesis in alfalfa (*Medicago sativa* L.) under drought stress.

5. Conclusions

We analyzed comparative transcriptomes of roots and leaves of tolerant as well as sensitive genotypes of field-grown common beans under terminal drought stress. DEG regulation was significantly different between tolerant and sensitive genotypes, with downregulated DEGs significantly higher in sensitive genotypes than upregulated DEGs. DEGs overlapping between genotypes were also found in the drought database. Several transcription factors, heat shock proteins, and chaperones were identified in both genotype comparisons. ABA-regulated drought stress responses were associated with DEGs with higher expression in both genotypes. It was found that DEGs in the tolerant genotype were involved in signal transduction, oxidative stress damage, and transportation. The enrichment of DEGs for biological and cellular components' GO terms in tolerant genotypes may further explain the temporal pattern of root growth and the ABA-dependent drought tolerance. KEGG pathway analysis indicates that drought-tolerant genotypes exhibit crosstalk between pathways. It is likely that transcription factors associated with pathways, including ERF, BHLH, EIL, and bZIP, contribute to drought tolerance in response to terminal drought stress. The pairwise transcriptomic approaches revealed molecular signatures such as upregulated DEGs enriched with cellular components and biological processes for apoplasts, sulfate, and drug membrane transport in tolerant genotypes, compared to sensitive genotypes. KEGG analysis revealed MAPK and plant hormone signaling pathways interaction in tolerant genotypes. These molecular signatures may contribute to the development of genotypes' tolerance against terminal drought stress. In the future, this study may also serve as a reference for understanding the drought stress transcriptomes of other legume species.

Supplementary Materials: The following are available online at https://www.mdpi.com/article/10.3390/plants12010210/s1, Supplementary Figure S1: Principle component analysis indicates differences between the replicates of leaves and roots, as well as root samples differing from leaf samples from each genotype; Supplementary Figure S2: (a) Comparison of qRT-PCR and RNA-Seq results for the selected DEGs. Data shown are the mean value of triplicates ±SD (b) Correlation between qRT-PCR and RNA-Seq techniques; Supplementary Table S1 and Table S2: The table shows the up or downregulated DEGs in tolerant and sensitive genotypes compared between roots and leaves with FDR < 0.05; Supplementary Table S3: Comparison of the tolerant and sensitive genotypes in root vs. leaf shows the enrichment of DEGs on biological processes, cellular components, and molecular functions; Supplementary Table S4: A summary of DEG enrichment in the droughtDB is presented in the table along with the identification of the best hit Arabidopsis ID. Most DEGs of tolerant and sensitive genotypes in the drought database overlap; Supplementary Table S5: An analysis of transcription factors and their family of tolerant and sensitive genotypes is presented in the table. DEGs with higher log2FC are similar between the two genotypes; Supplementary Table S6: List of up and downregulated genes and their primer sequences for the validation of RNA-Seq by qRT-PCR.

Author Contributions: V.K. designed the study. M.S. and R.H. performed the experiments. M.S. drafted the manuscript. C.A.U. conducted the field experiment, provided the plant samples, and edited the manuscript. K.B. and J.T. analyzed the data, assisted in writing, and edited the manuscript. All authors have read and agreed to the published version of the manuscript.

Funding: This research was supported by the National Institute of Food and Agriculture (NIFA), USDA, under a Capacity Building Grant (CBG) 2019-38821-29054. Funding bodies were not involved in the design, collection, analysis, or interpretation of data or the preparation of the manuscript.

Institutional Review Board Statement: Not applicable.

Informed Consent Statement: Not applicable.

Data Availability Statement: Data is contained within the article or supplementary material.

Conflicts of Interest: The authors declared no conflict of interest.

References

1. Petry, N.; Boy, E.; Wirth, J.P.; Hurrell, R.F. Review: The Potential of the Common Bean (*Phaseolus vulgaris*) as a Vehicle for Iron Biofortification. *Nutrients* **2015**, *7*, 1144. [CrossRef] [PubMed]
2. Pereira, W.J.; Melo, A.T.D.O.; Coelho, A.S.G.; Rodrigues, F.A.; Mamidi, S.; de Alencar, S.A.; Lanna, A.C.; Valdisser, P.A.M.R.; Brondani, C.; Do Nascimento-Júnior, I.R.; et al. Genome-Wide Analysis of the Transcriptional Response to Drought Stress in Root and Leaf of Common Bean. *Genet. Mol. Biol.* **2020**, *43*, 1–16. [CrossRef] [PubMed]
3. White, J.W.; Singh, S.P. Sources and Inheritance of Earliness in Tropically Adapted Indeterminate Common Bean. *Euphytica* **1991**, *55*, 15–19. [CrossRef]
4. Beebe, S.E.; Rao, I.M.; Blair, M.W.; Acosta-Gallegos, J.A. Phenotyping Common Beans for Adaptation to Drought. *Front. Physiol.* **2013**, *4*, 35. [CrossRef] [PubMed]
5. Kang, Z.; Babar, M.A.; Khan, N.; Guo, J.; Khan, J.; Islam, S.; Shrestha, S.; Shahi, D. Comparative Metabolomic Profiling in the Roots and Leaves in Contrasting Genotypes Reveals Complex Mechanisms Involved in Post-Anthesis Drought Tolerance in Wheat. *PLoS ONE* **2019**, *14*, e0213502. [CrossRef]
6. Gu, H.; Yang, Y.; Xing, M.; Yue, C.; Wei, F.; Zhang, Y.; Zhao, W.; Huang, J. Physiological and Transcriptome Analyses of Opisthopappus Taihangensis in Response to Drought Stress. *Cell Biosci.* **2019**, *9*, 1–12. [CrossRef]
7. Li, P.; Lin, P.; Zhao, Z.; Li, Z.; Liu, Y.; Huang, C.; Huang, G.; Xu, L.; Deng, Z.; Zhang, Y.; et al. Gene Co-Expression Analysis Reveals Transcriptome Divergence between Wild and Cultivated Sugarcane under Drought Stress. *Int. J. Mol. Sci.* **2022**, *23*, 569. [CrossRef]
8. Liu, S.; Zenda, T.; Li, J.; Wang, Y.; Liu, X.; Duan, H. Comparative Transcriptomic Analysis of Contrasting Hybrid Cultivars Reveal Key Drought-Responsive Genes and Metabolic Pathways Regulating Drought Stress Tolerance in Maize at Various Stages. *PLoS ONE* **2020**, *15*, e0240468. [CrossRef]
9. Min, X.; Lin, X.; Ndayambaza, B.; Wang, Y.; Liu, W. Coordinated Mechanisms of Leaves and Roots in Response to Drought Stress Underlying Full-Length Transcriptome Profiling in *Vicia sativa*L. *BMC Plant Biol.* **2020**, *20*, 1–21. [CrossRef]
10. Jones, C.; De Vega, J.; Worthington, M.; Thomas, A.; Gasior, D.; Harper, J.; Doonan, J.; Fu, Y.; Bosch, M.; Corke, F.; et al. A Comparison of Differential Gene Expression in Response to the Onset of Water Stress Between Three Hybrid Brachiaria Genotypes. *Front. Plant Sci.* **2021**, *12*, 393. [CrossRef]
11. Fang, Y.; Xiong, L. General Mechanisms of Drought Response and Their Application in Drought Resistance Improvement in Plants. *Cell. Mol. Life Sci.* **2014**, *72*, 673–689. [CrossRef]
12. Schmutz, J.; McClean, P.E.; Mamidi, S.; Wu, G.A.; Cannon, S.B.; Grimwood, J.; Jenkins, J.; Shu, S.; Song, Q.; Chavarro, C.; et al. A Reference Genome for Common Bean and Genome-Wide Analysis of Dual Domestications. *Nat. Genet.* **2014**, *46*, 707–713. [CrossRef]
13. Vlasova, A.; Capella-Gutiérrez, S.; Rendón-Anaya, M.; Hernández-Oñate, M.; Minoche, A.E.; Erb, I.; Câmara, F.; Prieto-Barja, P.; Corvelo, A.; Sanseverino, W.; et al. Genome and Transcriptome Analysis of the Mesoamerican Common Bean and the Role of Gene Duplications in Establishing Tissue and Temporal Specialization of Genes. *Genome. Biol.* **2016**, *17*, 1–18. [CrossRef]
14. Müller, B.S.D.F.; Sakamoto, T.; Silveira, R.D.D.; Zambussi-Carvalho, P.F.; Pereira, M.; Pappas, G.J.; do Carmo Costa, M.M.; Guimarães, C.M.; Pereira, W.J.; Brondani, C.; et al. Differentially Expressed Genes during Flowering and Grain Filling in Common Bean (*Phaseolus vulgaris*) Grown under Drought Stress Conditions. *Plant Mol. Biol. Rep.* **2014**, *32*, 438–451. [CrossRef]
15. Wu, J.; Wang, L.; Li, L.; Wang, S. De Novo Assembly of the Common Bean Transcriptome Using Short Reads for the Discovery of Drought-Responsive Genes. *PLoS ONE* **2014**, *9*, e109262. [CrossRef]
16. Schafleitner, R.; Gutierrez Rosales, R.O.; Gaudin, A.; Alvarado Aliaga, C.A.; Martinez, G.N.; Tincopa Marca, L.R.; Bolivar, L.A.; Delgado, F.M.; Simon, R.; Bonierbale, M. Capturing Candidate Drought Tolerance Traits in Two Native Andean Potato Clones by Transcription Profiling of Field Grown Plants under Water Stress. *Plant Physiol. Biochem.* **2007**, *45*, 673–690. [CrossRef]
17. Lenka, S.K.; Katiyar, A.; Chinnusamy, V.; Bansal, K.C. Comparative Analysis of Drought-Responsive Transcriptome in Indica Rice Genotypes with Contrasting Drought Tolerance. *Plant Biotechnol. J.* **2011**, *9*, 315–327. [CrossRef]

18. Moenga, S.M.; Gai, Y.; Carrasquilla-Garcia, N.; Perilla-Henao, L.M.; Cook, D.R. Gene Co-Expression Analysis Reveals Transcriptome Divergence between Wild and Cultivated Chickpea under Drought Stress. *Plant J.* **2020**, *104*, 1195–1214. [CrossRef]
19. Zaman-Allah, M.; Jenkinson, D.M.; Vadez, V. Chickpea Genotypes Contrasting for Seed Yield under Terminal Drought Stress in the Field Differ for Traits Related to the Control of Water Use. *Funct. Plant Biol.* **2011**, *38*, 270–281. [CrossRef]
20. Waqas, M.; Azhar, M.T.; Rana, I.A.; Arif, A.; Atif, R.M. Drought Stress in Chickpea: Physiological, Breeding, and Omics Perspectives. In *Recent Approaches in Omics for Plant Resilience to Climate Change*; Springer: Berlin/Heidelberg, Germany, 2019; pp. 189–227. [CrossRef]
21. Rao Jagana, S.; Vadez, V.; Bhatnagar-Mathur, P.; Narasu, M.L.; Sharma, K.K. Better Root: Shoot Ratio Conferred Enhanced Harvest Index in Transgenic Groundnut Overexpressing the Rd29A: DREB1A Gene under Intermittent Drought Stress in an Outdoor Lysimetric Dry-down Trial. *J. SAT Agric. Res.* **2012**, *10*, 1–7.
22. Li, H.; Testerink, C.; Zhang, Y. How Roots and Shoots Communicate through Stressful Times. *Trends Plant Sci.* **2021**, *26*, 940–952. [CrossRef] [PubMed]
23. Vadez, V.; Kholova, J.; Zaman-Allah, M.; Belko, N.; Vadez, V.; Kholova, J.; Zaman-Allah, M.; Belko, N. Water: The Most Important 'Molecular' Component of Water Stress Tolerance Research. *Funct. Plant Biol.* **2013**, *40*, 1310–1322. [CrossRef] [PubMed]
24. Urrea, C.A.; Smith, J.R.; Porch, T.G. Release of Drought-Tolerant Pinto SB-DT2 and Small Red SB-DT3 Common Bean Germplasm from a Shuttle Breeding Program between Nebraska and Puerto Rico. *J. Plant Regist.* **2021**, *16*, 400–409. [CrossRef]
25. Kelly, J.D.; Hosfield, G.L.; Varner, G.V.; Uebersax, M.A.; Taylor, J. Registration of 'Matterhorn' Great Northern Bean. *Crop Sci.* **1999**, *39*, 589–590. [CrossRef]
26. Singh, S.P.; Terán, H.; Lema, M.; Dennis, M.F.; Hayes, R.; Robinson, C. Development of Large-Seeded High-Quality, High-Yielding Great Northern Dry Bean 'Hungerford' and 'Sawtooth'. *J. Plant Regist.* **2008**, *2*, 174–179. [CrossRef]
27. Hosfield, G.L.; Varner, G.V.; Uebersax, M.A.; Kelly, J.D. Registration of 'Merlot' Small Red Bean. *Crop Sci.* **2004**, *44*, 351. [CrossRef]
28. Osorno, J.M.; Grafton, K.F.; Rojas-Cifuentes, G.A.; Gelin, R.; Wal, A.J.V. Registration of 'Lariat' and 'Stampede' Pinto Beans. *J. Plant Regist.* **2010**, *4*, 5–11. [CrossRef]
29. Subramani, M.; Urrea, C.A.; Kalavacharla, V. Comparative Analysis of Untargeted Metabolomics in Tolerant and Sensitive Genotypes of Common Bean (*Phaseolus vulgaris* L.) Seeds Exposed to Terminal Drought Stress. *Metabolites* **2022**, *12*, 944. [CrossRef]
30. Dobin, A.; Davis, C.A.; Schlesinger, F.; Drenkow, J.; Zaleski, C.; Jha, S.; Batut, P.; Chaisson, M.; Gingeras, T.R. STAR: Ultrafast Universal RNA-Seq Aligner. *Bioinformatics* **2013**, *29*, 15–21. [CrossRef]
31. Anders, S.; Pyl, P.T.; Huber, W. HTSeq—A Python Framework to Work with High-Throughput Sequencing Data. *Bioinformatics* **2015**, *31*, 166–169. [CrossRef]
32. Love, M.I.; Huber, W.; Anders, S. Moderated Estimation of Fold Change and Dispersion for RNA-Seq Data with DESeq2. *Genome. Biol.* **2014**, *15*, 1–21. [CrossRef]
33. Yu, G.; Wang, L.G.; Han, Y.; He, Q.Y. ClusterProfiler: An R Package for Comparing Biological Themes among Gene Clusters. *Omics. J. Integr. Biol.* **2012**, *16*, 284–287. [CrossRef]
34. Tian, T.; Liu, Y.; Yan, H.; You, Q.; Yi, X.; Du, Z.; Xu, W.; Su, Z. AgriGO v2.0: A GO Analysis Toolkit for the Agricultural Community, 2017 Update. *Nucleic Acids Res.* **2017**, *45*, W122–W129. [CrossRef]
35. Kusi-Appiah Hayford, R.; Ligaba-Osena, A.; Subramani, M.; Brown, A.; Melmaiee, K.; Hossain, K.; Kalavacharla, V. Characterization and Expression Analysis of Common Bean Histone Deacetylase 6 during Development and Cold Stress Response. *Int. J. Genomics.* **2017**, *2017*, 2502691. [CrossRef]
36. Schmittgen, T.D.; Livak, K.J. Analyzing Real-Time PCR Data by the Comparative CT Method. *Nat. Protoc.* **2008**, *3*, 1101–1108. [CrossRef]
37. Tran, L.S.P.; Urao, T.; Qin, F.; Maruyama, K.; Kakimoto, T.; Shinozaki, K.; Yamaguchi-Shinozaki, K. Functional Analysis of AHK1/ATHK1 and Cytokinin Receptor Histidine Kinases in Response to Abscisic Acid, Drought, and Salt Stress in Arabidopsis. *Proc. Natl. Acad. Sci. USA* **2007**, *104*, 20623–20628. [CrossRef]
38. Kang, J.; Hwang, J.U.; Lee, M.; Kim, Y.Y.; Assmann, S.M.; Martinoia, E.; Lee, Y. PDR-Type ABC Transporter Mediates Cellular Uptake of the Phytohormone Abscisic Acid. *Proc. Natl. Acad. Sci. USA* **2010**, *107*, 2355–2360. [CrossRef]
39. Rao, I.M.; Beebe, S.E.; Polania, J.; Grajales, M.; Cajiao, C.; Ricaurte, J.; García, R.; Rivera, M. Evidence for Genotypic Differences among Elite Lines of Common Bean in the Ability to Remobilize Photosynthate to Increase Yield under Drought. *J. Agric. Sci.* **2017**, *155*, 857–875. [CrossRef]
40. Sofi, P.A.; Rehman, K.; Gull, M.; Kumari, J.; Djanaguiraman, M.; Prasad, P.V.V. Integrating Root Architecture and Physiological Approaches for Improving Drought Tolerance in Common Bean (*Phaseolus vulgaris* L.). *Plant Physiol. Rep.* **2021**, *26*, 4–22. [CrossRef]
41. Recchia, G.H.; Caldas, D.G.G.; Beraldo, A.L.A.; da Silva, M.J.; Tsai, S.M. Transcriptional Analysis of Drought-Induced Genes in the Roots of a Tolerant Genotype of the Common Bean (*Phaseolus vulgaris* L.). *Int. J. Mol. Sci.* **2013**, *14*, 7155–7179. [CrossRef]
42. Wu, J.; Chen, J.; Wang, L.; Wang, S. Genome-Wide Investigation of WRKY Transcription Factors Involved in Terminal Drought Stress Response in Common Bean. *Front. Plant Sci.* **2017**, *8*, 380. [CrossRef] [PubMed]
43. Gregorio Jorge, J.; Villalobos-López, M.A.; Chavarría-Alvarado, K.L.; Ríos-Meléndez, S.; López-Meyer, M.; Arroyo-Becerra, A. Genome-Wide Transcriptional Changes Triggered by Water Deficit on a Drought-Tolerant Common Bean Cultivar. *BMC Plant Biol.* **2020**, *20*, 1–20. [CrossRef] [PubMed]

44. López, C.M.; Pineda, M.; Alamillo, J.M. Differential Regulation of Drought Responses in Two Phaseolus Vulgaris Genotypes. *Plants* **2020**, *9*, 1815. [CrossRef] [PubMed]
45. Mahammed, F.; Babu, H.; Fakrudin, B.; Lakshmana, D.; Rakshith, M. In Silico Identification and Annotation of Drought Responsive Candidate Genes in Solanaceous Plants. Available online: https://papers.ssrn.com/sol3/papers.cfm?abstract_id=3777445 (accessed on 6 June 2022).
46. Vaid, N.; Pandey, P.K.; Tuteja, N. Genome-Wide Analysis of Lectin Receptor-like Kinase Family from Arabidopsis and Rice. *Plant Mol. Biol.* **2012**, *80*, 365–388. [CrossRef] [PubMed]
47. Cheng, M.C.; Liao, P.M.; Kuo, W.W.; Lin, T.P. The Arabidopsis ETHYLENE RESPONSE FACTOR1 Regulates Abiotic Stress-Responsive Gene Expression by Binding to Different Cis-Acting Elements in Response to Different Stress Signals. *Plant Physiol.* **2013**, *162*, 1566. [CrossRef]
48. Han, G.; Lu, C.; Guo, J.; Qiao, Z.; Sui, N.; Qiu, N.; Wang, B. C_2H_2 Zinc Finger Proteins: Master Regulators of Abiotic Stress Responses in Plants. *Front. Plant Sci.* **2020**, *11*, 115. [CrossRef]
49. Sakamoto, H.; Nakagawara, Y.; Oguri, S. The Expression of a Novel Gene Encoding an Ankyrin-Repeat Protein, DRA1, Is Regulated by Drought-Responsive Alternative Splicing. *Int. J. Bioeng. Life Sci.* **2014**, *7*, 1172–1175.
50. Wang, W.; Qin, Q.; Sun, F.; Wang, Y.; Xu, D.; Li, Z.; Fu, B. Genome-Wide Differences in DNA Methylation Changes in Two Contrasting Rice Genotypes in Response to Drought Conditions. *Front. Plant Sci.* **2016**, *7*, 1675. [CrossRef]
51. Penfield, S.; Clements, S.; Bailey, K.J.; Gilday, A.D.; Leegood, R.C.; Gray, J.E.; Graham, I.A. Expression and Manipulation of PHOSPHOENOLPYRUVATE CARBOXYKINASE 1 Identifies a Role for Malate Metabolism in Stomatal Closure. *Plant J.* **2012**, *69*, 679–688. [CrossRef]
52. Zeng, C.; Chen, Z.; Xia, J.; Zhang, K.; Chen, X.; Zhou, Y.; Bo, W.; Song, S.; Deng, D.; Guo, X.; et al. Chilling Acclimation Provides Immunity to Stress by Altering Regulatory Networks and Inducing Genes with Protective Functions in Cassava. *BMC Plant Biol.* **2014**, *14*, 1–16. [CrossRef]
53. Kim, D.Y.; Jin, J.Y.; Alejandro, S.; Martinoia, E.; Lee, Y. Overexpression of AtABCG36 Improves Drought and Salt Stress Resistance in Arabidopsis. *Physiol. Plant.* **2010**, *139*, 170–180. [CrossRef]
54. Gallardo, K.; Courty, P.E.; Le Signor, C.; Wipf, D.; Vernoud, V. Sulfate Transporters in the Plant's Response to Drought and Salinity: Regulation and Possible Functions. *Front. Plant Sci.* **2014**, *5*, 580. [CrossRef]
55. Ahmad, N.; Malagoli, M.; Wirtz, M.; Hell, R. Drought Stress in Maize Causes Differential Acclimation Responses of Glutathione and Sulfur Metabolism in Leaves and Roots. *BMC Plant Biol.* **2016**, *16*, 1–15. [CrossRef]
56. Bai, Z.Y.; Wang, T.; Wu, Y.H.; Wang, K.; Liang, Q.Y.; Pan, Y.Z.; Jiang, B.B.; Zhang, L.; Liu, G.L.; Jia, Y.; et al. Whole-Transcriptome Sequence Analysis of Differentially Expressed Genes in Phormium Tenax under Drought Stress. *Sci. Rep.* **2017**, *7*, 1–9. [CrossRef]
57. Du, M.; Ding, G.; Cai, Q. The Transcriptomic Responses of Pinus Massoniana to Drought Stress. *Forests* **2018**, *9*, 326. [CrossRef]
58. Sun, R.Z.; Liu, J.; Wang, Y.Y.; Deng, X. DNA Methylation-Mediated Modulation of Rapid Desiccation Tolerance Acquisition and Dehydration Stress Memory in the Resurrection Plant Boea Hygrometrica. *PLoS Genet.* **2021**, *17*, e1009549. [CrossRef]
59. Geilfus, C.-M. The PH of the Apoplast: Dynamic Factor with Functional Impact Under Stress. *Mol. Plant* **2017**, *10*, 1371–1386. [CrossRef]
60. Buljovcic, Z.; Engels, C. Nitrate Uptake Ability by Maize Roots during and after Drought Stress. *Plant Soil* **2001**, *229*, 125–135. [CrossRef]
61. Qi, J.; Song, C.P.; Wang, B.; Zhou, J.; Kangasjärvi, J.; Zhu, J.K.; Gong, Z. Reactive Oxygen Species Signaling and Stomatal Movement in Plant Responses to Drought Stress and Pathogen Attack. *J. Integr. Plant Biol.* **2018**, *60*, 805–826. [CrossRef]
62. Das, S.; Kar, R.K. Abscisic Acid Mediated Differential Growth Responses of Root and Shoot of *Vigna radiata* (L.) Wilczek Seedlings under Water Stress. *Plant Physiol. Biochem.* **2018**, *123*, 213–221. [CrossRef]
63. Brini, F.; Hanin, M.; Mezghani, I.; Berkowitz, G.A.; Masmoudi, K. Overexpression of Wheat Na^+/H^+ Antiporter TNHX1 and H^+-Pyrophosphatase TVP1 Improve Salt- and Drought-Stress Tolerance in Arabidopsis Thaliana Plants. *J. Exp. Bot.* **2007**, *58*, 301–308. [CrossRef] [PubMed]
64. Chaichi, M.; Sanjarian, F.; Razavi, K.; Gonzalez-Hernandez, J.L. Analysis of Transcriptional Responses in Root Tissue of Bread Wheat Landrace (*Triticum aestivum* L.) Reveals Drought Avoidance Mechanisms under Water Scarcity. *PLoS ONE* **2019**, *14*, e0212671. [CrossRef] [PubMed]
65. Lu, P.; Magwanga, R.O.; Guo, X.; Kirungu, J.N.; Lu, H.; Cai, X.; Zhou, Z.; Wei, Y.; Wang, X.; Zhang, Z.; et al. Genome-Wide Analysis of Multidrug and Toxic Compound Extrusion (MATE) Family in *Gossypium raimondii* and *Gossypium arboreum* and Its Expression Analysis Under Salt, Cadmium, and Drought Stress. *G3 Genes Genomes Genet.* **2018**, *8*, 2483. [CrossRef] [PubMed]
66. Ullah, A.; Manghwar, H.; Shaban, M.; Khan, A.H.; Akbar, A.; Ali, U.; Ali, E.; Fahad, S. Phytohormones Enhanced Drought Tolerance in Plants: A Coping Strategy. *Environ. Sci. Pollut. Res.* **2018**, *25*, 33103–33118. [CrossRef] [PubMed]
67. Govind, G.; Harshavardhan, V.T.; Hong, C.-Y. Phytohormone Signaling in Response to Drought. In *Salt and Drought Stress Tolerance in Plants*; Springer: Berlin/Heidelberg, Germany, 2020; pp. 315–335. [CrossRef]
68. Wang, J.; Zhang, S.; Fu, Y.; He, T.; Wang, X. Analysis of Dynamic Global Transcriptional Atlas Reveals Common Regulatory Networks of Hormones and Photosynthesis Across Nicotiana Varieties in Response to Long-Term Drought. *Front. Plant Sci.* **2020**, *11*, 672. [CrossRef]
69. Vogt, T. Phenylpropanoid Biosynthesis. *Mol. Plant* **2010**, *3*, 2–20. [CrossRef]

70. Yu, A.; Zhao, J.; Wang, Z.; Cheng, K.; Zhang, P.; Tian, G.; Liu, X.; Guo, E.; Du, Y.; Wang, Y. Transcriptome and Metabolite Analysis Reveal the Drought Tolerance of Foxtail Millet Significantly Correlated with Phenylpropanoids-Related Pathways during Germination Process under PEG Stress. *BMC Plant Biol.* **2020**, *20*, 1–17. [CrossRef]
71. Zhang, M.; Su, J.; Zhang, Y.; Xu, J.; Zhang, S. Conveying Endogenous and Exogenous Signals: MAPK Cascades in Plant Growth and Defense. *Curr. Opin. Plant Biol.* **2018**, *45*, 1–10. [CrossRef]
72. He, X.; Wang, C.; Wang, H.; Li, L.; Wang, C. The Function of MAPK Cascades in Response to Various Stresses in Horticultural Plants. *Front. Plant Sci.* **2020**, *11*, 952. [CrossRef]
73. Zhang, A.; Ji, Y.; Sun, M.; Lin, C.; Zhou, P.; Ren, J.; Luo, D.; Wang, X.; Ma, C.; Zhang, X.; et al. Research on the Drought Tolerance Mechanism of *Pennisetum glaucum* (L.) in the Root during the Seedling Stage. *BMC Genom.* **2021**, *22*, 1–14. [CrossRef]
74. Dudhate, A.; Shinde, H.; Tsugama, D.; Liu, S.; Takano, T. Transcriptomic Analysis Reveals the Differentially Expressed Genes and Pathways Involved in Drought Tolerance in Pearl Millet [*Pennisetum glaucum* (L.) R. Br]. *PLoS ONE* **2018**, *13*, e0195908. [CrossRef]
75. Tholl, D. Terpene Synthases and the Regulation, Diversity and Biological Roles of Terpene Metabolism. *Curr. Opin. Plant Biol.* **2006**, *9*, 297–304. [CrossRef]
76. Boncan, D.A.T.; Tsang, S.S.K.; Li, C.; Lee, I.H.T.; Lam, H.M.; Chan, T.F.; Hui, J.H.L. Terpenes and Terpenoids in Plants: Interactions with Environment and Insects. *Int. J. Mol. Sci.* **2020**, *21*, 7382. [CrossRef]
77. Yang, S.; Bai, M.; Hao, G.; Guo, H.; Fu, B. Transcriptomics Analysis of Field-Droughted Pear (*Pyrus* spp.) Reveals Potential Drought Stress Genes and Metabolic Pathways. *PeerJ* **2022**, *10*, e12921. [CrossRef]
78. Wan, L.; Li, Y.; Li, S.; Li, X. Transcriptomic Profiling Revealed Genes Involved in Response to Drought Stress in Alfalfa. *J. Plant Growth Regul.* **2022**, *41*, 92–112. [CrossRef]

plants

MDPI

Article

A Locus Controlling Leaf Rolling Degree in Wheat under Drought Stress Identified by Bulked Segregant Analysis

Xi Yang [1,2,†], Jingyi Wang [2,†], Xinguo Mao [2], Chaonan Li [2], Long Li [2], Yinghong Xue [2], Liheng He [1,*] and Ruilian Jing [2,*]

1 College of Agronomy, Shanxi Agricultural University, Jinzhong 030801, China
2 National Key Facility for Crop Gene Resources and Genetic Improvement, Institute of Crop Sciences, Chinese Academy of Agricultural Sciences, Beijing 100081, China
* Correspondence: sxndhlh@163.com (L.H.); jingruilian@caas.cn (R.J.)
† These authors contributed equally to this work.

Abstract: Drought stress frequently occurs, which seriously restricts the production of wheat (*Triticum aestivum* L.). Leaf rolling is a typical physiological phenomenon of plants during drought stress. To understand the genetic mechanism of wheat leaf rolling, we constructed an F_2 segregating population by crossing the slight-rolling wheat cultivar "Aikang 58" (AK58) with the serious-rolling wheat cultivar "Zhongmai 36" (ZM36). A combination of bulked segregant analysis (BSA) with Wheat 660K SNP Array was used to identify molecular markers linked to leaf rolling degree. A major locus for leaf rolling degree under drought stress was detected on chromosome 7A. We named this locus *LEAF ROLLING DEGREE 1* (*LERD1*), which was ultimately mapped to a region between 717.82 and 720.18 Mb. Twenty-one genes were predicted in this region, among which the basic helix-loop-helix (bHLH) transcription factor *TraesCS7A01G543300* was considered to be the most likely candidate gene for *LERD1*. The TraesCS7A01G543300 is highly homologous to the *Arabidopsis* ICE1 family proteins ICE/SCREAM, SCREAM2 and bHLH093, which control stomatal initiation and development. Two nucleotide variation sites were detected in the promoter region of *TraesCS7A01G543300* between the two wheat cultivars. Gene expression assays indicated that *TraesCS7A01G543300* was higher expressed in AK58 seedlings than that of ZM36. This research discovered a candidate gene related to wheat leaf rolling under drought stress, which may be helpful for understanding the leaf rolling mechanism and molecular breeding in wheat.

Keywords: *Triticum aestivum* L.; drought; leaf rolling; bulked segregant analysis; gene mapping

1. Introduction

Wheat (*Triticum aestivum* L.) is a staple food for more than 250 million people worldwide [1]. In addition to the benefit for nutrition and health, wheat provides about 21% of dietary calories and 20% of protein for humans, playing a prominent role in improving food security [2]. However, drought stress occurs frequently, which seriously restricts the production of wheat [3]. Therefore, improving wheat drought tolerance is an important approach to ensure food security [4,5]. By selecting and pyramiding favorable alleles related to drought tolerance traits in elite cultivars, crop performance in drought environments can be improved [6].

Leaf rolling is a typical physiological phenomenon of plants during drought stress, which is observed in various higher plants [7]. Leaf rolling is considered an adaptation to arid environments in wheat [8]. Moderate leaf rolling and erect leaf morphology are propitious to enhancing light capture, gas exchange for photosynthesis and carbon fixation [9]. Moreover, semi-rolling of leaves can reduce water loss by transpiration and interception of solar radiation by canopy; thus, it is of great significance to improve the adaptability of plants to environmental stress [10]. Consequently, the moderate rolling leaf trait is one of the aims of genetic improvement and molecular breeding in crops. Until now, more than 17 leaf rolling mutants, at least 70 genes/QTLs and 28 differentially expressed proteins related to leaf rolling traits have been reported in rice [11–16]. To date, at least five mutants with rolling leaves have been characterized in maize [17]. However, the genetic mechanism of leaf rolling under drought stress in wheat is rarely reported.

The stomatal is the main channel for water transpiration and gas exchange in plants, and changes in stomata density and morphology may affect plant drought tolerance [18]. An analysis of wilting mutant *multi-trait weakened* (*muw*) suggested that the increase in stomatal density can accelerate plant water loss and thus reduce drought tolerance [19]. In tomato, DELLA protein PROCERA (PRO) promotes the stomatal response to abscisic acid (ABA), and the pro mutant exhibited increased stomatal conductance and reduced drought tolerance [20]. Some proteins that are independent of the ABA signaling pathway can also play a function in drought tolerance by regulating stomatal closure, such as stress-responsive NAC 1 and SIMILAR TO RADICAL-INDUCED CELL DEATH1 [21,22].

In *Arabidopsis*, a genetic regulatory network has been identified that strictly controls stomatal development and pattern formation [23]. This includes three basic helix-loop-helix (bHLH) transcription factors that promote stomatal formation, SPEECHLESS (SPCH) is essential for the initiation of stomatal lineage, MUTE influences meristemoid to guard mother cell (GMC) conversion, FAMA determines the GMC to mature guard cell (GC) transition [24–26]. Two para-homologous protein ICE1/SCREAM (SCRM) and SCRM2 directly interact with SPCH, MUTE and FAMA, and then specify their sequential actions [27]. To prevent adjacent cells from becoming stomata, some extracellular and plasma membrane binding proteins are necessary to concert signals between developing stomatal and pavement cells [28], such as the EPIDERMAL PATTERNING FACTOR (EPF) and the Leu-rich repeat membrane protein TOO MANY MOUTHS (TMM) [29,30].

The fundamental mechanisms of stomatal formation in terrestrial plants are relatively conservative [23]. In rice, OsSPCH, OsMUTE, OsFAMA and OsICE1 are essential for the formation of mature stomata, which are orthologues of stomatal development regulators in Arabidopsis [31,32]. In moss, the most ancient extant stomata lineages, the formation of mature stomata requires PpSMF1 and PpSCRM1, which are also orthologous to the SPCH, MUTE, FAMA and ICE/SCRM in Arabidopsis [33,34]. The partnerships between ICE1/SCRM and SPCH, MUTE and FAMA are essential for the initiation and maturation of monocotyledonous stomata, but their protein function may be slightly different from that of Arabidopsis [31,35]. For example, in Brachypodium, the initiation of stomatal lineage requires the BdSCRM1, while the differentiation and function of stomatal complexes require BdSCRM2, which appears to be redundant in Arabidopsis [31,36]. A transcription factor BdMUTE has been proved to be necessary for subsidiary cell formation in the wheat relative Brachypodium [37].

The present study was based on the difference in leaf rolling degree under drought stress between two wheat cultivars, "Aikang 58" and "Zhongmai 36". We constructed an F2 population by crossing the two cultivars and used a combination of BSA with a Wheat 660K SNP array to identify molecular markers linked to leaf rolling degree. The objective of the present study was to map the locus that controls leaf rolling in wheat under drought stress.

2. Results

2.1. The Leaf Rolling Degree of AK58 Was Lower than That of ZM36 under Drought Stress

Two Chinese wheat cultivars, "Aikang 58" (AK58) and "Zhongmai 36" (ZM36), were used to compare the difference in leaf rolling degree. AK58 is a high-yield variety widely cultivated in irrigated areas of China, released by the Henan Institute of Science and Technology in 2005. ZM36 is a stable-yield variety suitable for rain-fed area cultivation, released by the Institute of Crop Sciences, Chinese Academy of Agricultural Sciences in 2018. The plants of AK58 and ZM36 grew well under well-watered conditions. On the 9th day of drought treatment, AK58 plants exhibited flat leaves, while ZM36 plant leaves showed a slight rolling phenotype. On the 12th day of drought treatment, the difference in leaf morphology became more obvious: the leaves of AK58 were wilted but remained flat, and the leaves of ZM36 were wilted and seriously rolled (Figure 1A). With the increase in drought stress treatment time, the leaf rolling index (LRI) of ZM36 leaves increased significantly, while AK58 leaves did not change significantly (Figure 1B). The leaf water contents of AK58 and ZM36 decreased; however, from the 6th day, the leaf water content of ZM36 was significantly lower than that of AK58 (Figure 1C). Therefore, wheat variety AK58 was identified as "slight-rolling" type, while ZM36 was "serious-rolling" type under drought stress at the seedling stage.

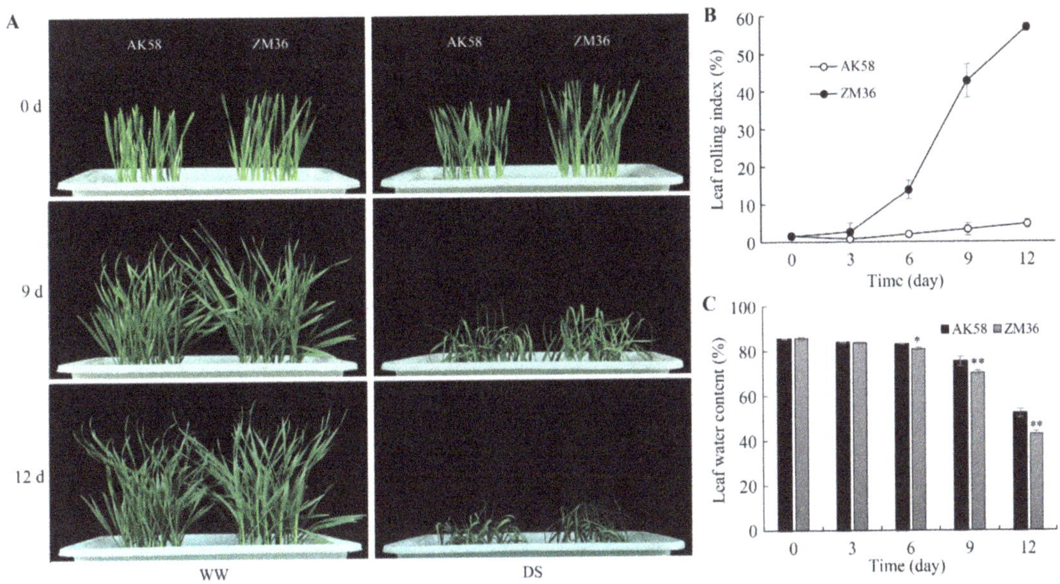

Figure 1. Under drought stress, the leaf rolling of ZM36 was more serious than that of AK58. (**A**) The seedling phenotype of the AK58 and ZM36 under WW (well-watered) and DS (drought stress) for 0, 9 and 12 d. (**B**) Leaf rolling index of AK58 and ZM36 under DS. (**C**) Leaf water contents of AK58 and ZM36 under DS. Data represent means ± SE. Error bars indicate SE. *, *t*-test with $p < 0.05$; **, *t*-test with $p < 0.01$.

2.2. AK58 Is More Tolerant to Drought than ZM36 at Seedling Stage

To investigate the relationship between seedling drought tolerance and leaf rolling degree in wheat, we evaluated the seedling drought tolerance of ZM36 and AK58. On the 11th day after re-watering, the survival rate of AK58 plants was 61.4%, and that of ZM36 was 19.6% (Figure 2A,B). The Malondialdehyde (MDA) content of ZM36 was significantly higher than that of AK58 on the 9th day of drought treatment (Figure 2D). We also performed PEG-600 treatment to simulate drought stress. Both AK58 and ZM36 plants exhibited turgid and flat leaves in the medium without PEG-6000 (Figure 2C left), but in mediums with 10% or 20% PEG-6000, ZM36 plants showed a rolling-leaf phenotype, while AK58 leaves remained flat (Figure 2C). These results demonstrated that AK58 has stronger drought tolerance than ZM36 at the seedling stage. In this experiment, the drought tolerance of wheat cultivars AK58 and ZM36 at the seedling stage was negatively correlated with the leaf rolling degree under drought stress.

Figure 2. Seedlings of wheat cultivar AK58 were more tolerant to dehydration than ZM36. (**A**) Performance of AK58 and ZM36 seedlings before drought (left), drought stress (DS) for 18 days (middle) and re-watered for 11 days (right). (**B**) Seedling survival rates of AK58 and ZM36 exposed to drought stress followed by re-watering. (**C**) Seedlings of AK58 and ZM36 grown in hydroponic culture under water (CK) or PEG-6000 treatment for 36 h. (**D**) MDA contents in leaves of AK58 and ZM36 seedlings grown under well-watered (WW) and DS. Values represent means ± SE. **, *t*-test with $p < 0.01$.

2.3. AK58 Leaves Exhibited Lower Stomatal Density than That of ZM36

To investigate the physiological basis of leaf rolling, we measured the rate of water loss (RWL) from excised-leaf of AK58 and ZM36. The RWL of ZM36 was significantly higher than that of AK58 from the 6th day (Figure 3A). Stomata on the leaf surface are the main channel for plants to discharge water [38]. Therefore, the difference in RWL between AK58 and ZM36 may be caused by the difference in stomatal morphology and/or density. There was no significant difference in the stomatal characteristics between AK58 and ZM36 leaves under drought stress (Figure S1). We then observed the leaf surface structures of AK58 and ZM36 plants by scanning electron microscopy (SEM). The stomatal densities on the adaxial (ad) or abaxial (ab) epidermis of ZM36 leaves were significantly higher than that of AK58 (Figure 3B,C); however, no other significant structural differences were observed on the AK58 and ZM36 leaf surface in this experiment.

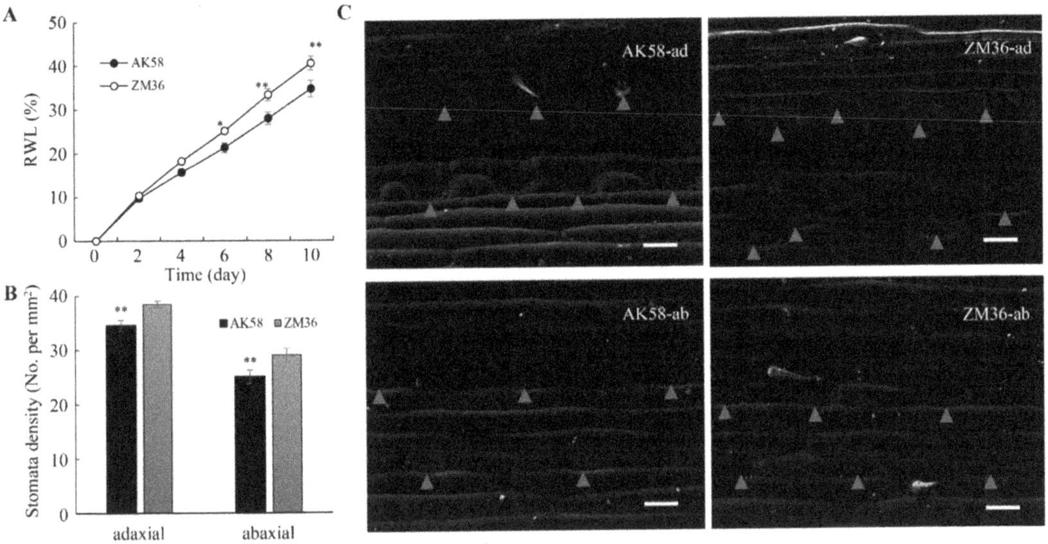

Figure 3. ZM36 has a higher leaf RWL and stomatal density than AK58. (**A**) RWL of AK58 and ZM36 plants. (**B**) Stomata density of AK58 and ZM36. Data represent means ± SE. *, *t*-test with $p < 0.05$; **, *t*-test with $p < 0.01$. (**C**) SEM analysis of the adaxial (ad) and abaxial (ab) epidermis of AK58 and ZM36. Triangles indicate stomata. Scale bars = 100 μm.

2.4. Genetic Analysis and Mapping of Leaf Rolling Degree Locus

To understand the genetics underlying the leaf rolling in wheat, we constructed a wheat F_2 segregating population by crossing a slight-rolling type cultivar AK58 with a serious-rolling cultivar ZM36. Under drought stress, seedlings of F_2 population individuals showed a continuous phenotypic distribution from slight-rolling to serious-rolling, which provided evidence that leaf rolling degree was a quantitative trait (Figure S2). A combination of BSA with Wheat 660K SNP Array was used to identify molecular markers linked to leaf rolling degree. Between the two parents and two bulks, a total of 721 SNP loci from the Wheat 660K SNP Array showed homozygous polymorphisms. The highest number and proportion of polymorphic SNPs were identified on chromosome 7A among all 21 chromosomes, indicating that 7A was likely the chromosome carrying the rolling degree locus, which was consequently named *LEAF ROLLING DEGREE 1* (*LERD1*) (Figure 4A). The chromosome interval 708–721 Mb was the predicted region for *LERD1* (Figure 4B).

To minimize the effects of errors caused by environmental effects and phenotyping errors, we chose 256 plants with serious-rolling phenotypes from the F_2 population of ~4000 individuals for further analysis. Seven polymorphic molecular markers were used to genotype individuals in the mapping population. According to the number of recombinants between *LERD1* and the molecular markers in the mapping population, *LERD1* was ultimately mapped to a region between 7A-29 (717.82 Mb) and 7A-12 (720.18 Mb) on chromosome 7A (Figure 4C).

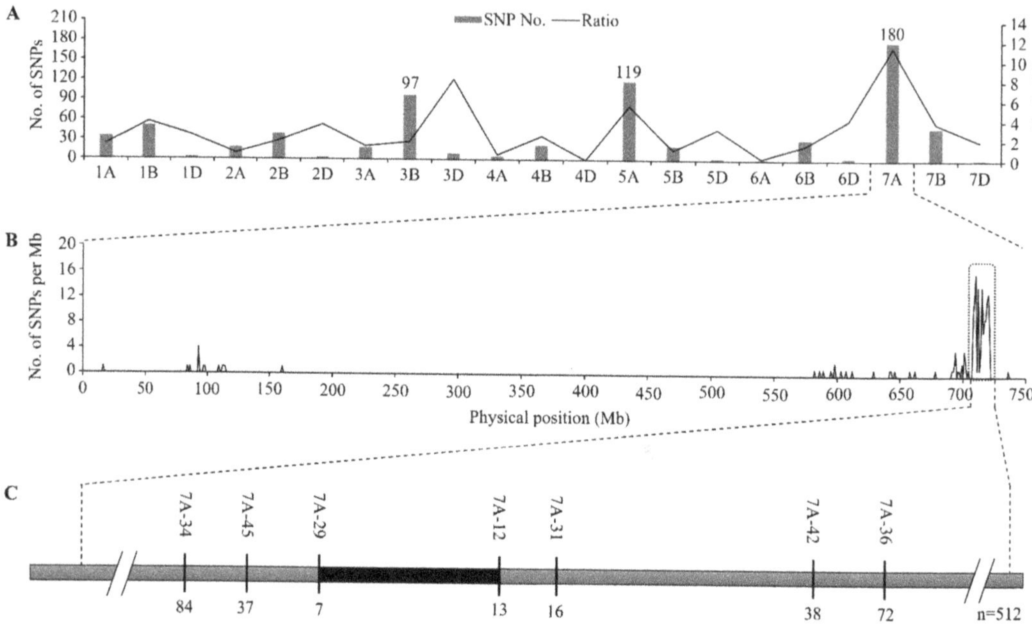

Figure 4. Mapping of locus controlling wheat leaf rolling degree. (**A**) The number and percentage of polymorphic SNP loci in each of the 21 wheat chromosomes. (**B**) The number of SNPs in each 1 Mb region of chromosome 7A identified using the Wheat 660K SNP array. (**C**) *LERD1* was mapped to a 2.36 Mb region between markers 7A-29 and 7A-12. The black rectangle represents the final target region harboring *LERD1*. The numbers below the bar indicate the number of recombinants between *LERD1* and the molecular marker.

2.5. Candidate Gene Prediction and Bioinformatics Analysis

A total of 21 genes were predicted in the target chromosome region, including a bHLH transcription factor TraesCS7A01G543300 (Table S1, Figure S3). TraesCS7A01G543300 encodes homologs of the ICE1/SCREAM, SCRM2 proteins and bHLH93 protein from Arabidopsis (Figure 5A), which were considered to be the major regulators of stomatal development in *Arabidopsis* [27]. Sequence polymorphism assays indicated that there was no nucleotide variation site in the coding region of *TraesCS7A01G543300*, but two SNPs were detected in the promoter region, one at −2278 bp and another at −974 bp between AK58 and ZM36 (Figure 5B). Differences in the promoter region may affect gene expression, so we measured the relative expression of genes at the seedling stage. The relative expression of *TraesCS7A01G543300* in AK58 seedling was 2.6 times that of ZM36 under the WW condition (Figure 5C). Therefore, *TraesCS7A01G543300* was considered the most likely candidate gene for *LERD1*.

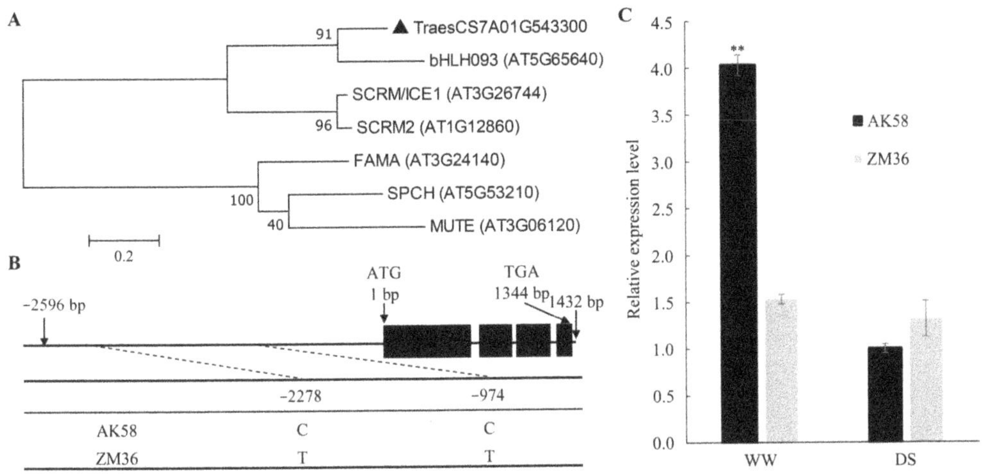

Figure 5. Candidate gene analysis. (**A**) Molecular phylogeny of TraesCS7A02G543300 and related bHLHs in *Arabidopsis*. (**B**) Schematic diagram of the *TraesCS7A02G543300* structure. The ATG start codon was designated as position 1 bp. Polymorphic sites were detected in the promoter region of *TraesCS7A02G543300*. (**C**) qPCR analysis of gene *TraesCS7A02G543300* expression. Data represent means ± SE (n = 3). **, $p < 0.01$.

3. Discussion

Traditional quantitative trait loci (QTL) mapping methods rely on phenotyping and genotyping of a large number of individuals from a mapping population, which is time-consuming and laborious [39]. BSA has been used to overcome this problem by only genotyping individuals with extreme phenotypes [40]. The advent and application of BSA and next-generation sequencing technologies have provided new opportunities for the rapid identification of QTLs [41]. Wang et al. developed a method to map QTLs by directly sequencing graded-pool samples from F_2 progeny using modified BSA [42]. Yu et al. used an F_2 population for Bulked Segregant RNA-seq (BSR-seq) and cloned a *knotted 1* homolog [43]. Wheat 660K SNP array is economical and reliable, which demonstrates great potential for marker-assisted selection [44]. In the present study, a combination of BSA and Wheat 660K SNP array was used to identify the locus affecting the leaf rolling of the wheat seedling. The *LERD1* has been mapped to a 2.36 Mb region, but further fine mapping requires more genetic populations, such as recombinant inbred lines (RIL) or near-isogenic lines (NIL) population.

Leaf rolling occurs when the uptake of soil water by the root system does not balance the need for evaporation [45]. Vascular plants maintain a balance between CO_2 absorption and water loss by stomatal movement, about 90% of leaf water loss in plants occurs through stomata [46,47]. Under drought stress, plants protect themselves from excessive water loss by diminishing stomatal aperture to reduce transpiration [48]. Previous studies have shown that wheat varieties that can maintain comparatively high leaf water potential and low transpiration rate under drought conditions tend to have stronger drought tolerance [49]. Leaf rolling in wheat can improve fog capturing and transport and enhance adaptation to drought stress in an arid climate [50,51].

In this study, the leaf water content of ZM36 was significantly lower than that of AK58 under drought stress, and the RWL of ZM36 was higher than that of AK58 (Figures 1C and 3A). We observed that there was no significant difference in stomatal aperture between AK58 and ZM36 under drought stress (Figure S1), while the stomatal density of ZM36 was higher than that of AK58 (Figure 3B,C). Therefore, we speculate the difference in RWL between ZM36 and AK58 is probably related to the difference in stomatal density.

The candidate genes may be involved in the regulation of stomatal density in wheat plants, thereby affecting the water loss rate of wheat plants under drought stress. An environmentally induced leaf-rolling phenotype is usually caused by abnormalities in the number, morphology, size or distribution of the bulliform cells [10]. Whether the bulliform cells of AK58 and ZM36 are different needs to be further analyzed.

The bHLH093 belongs to the ICE1 family bHLH-LZs, the expression of *bHLH093* may competitively inhibit the function of SCRM [27]. Overexpression of *bHLH093* with the 35S promoter resulted in a weak decreased number of mature stomata phenotypes [26]. In this study, *TraesCS7A02G543300* mapped within the target chromosome region encodes a bHLH family protein homologous to SCRM and bHLH093, which is more evolutionarily similar to *bHLH093* (Figure S3). The AK58 seedlings with higher *TraesCS7A02G543300* expression had a relatively lower stomatal density, while ZM36 with lower gene expression and relatively higher stomatal density. Our results suggested that *TraesCS7A02G543300* may have similar functions as *bHLH093*, but further detailed experimental evidence is needed.

In this study, we mapped a locus that controls leaf rolling of wheat seedlings under drought stress. We named it *LERD1*, which was finally mapped to a 2.36 Mb region between 717.82 and 720.18 Mb on chromosome 7A. A phenotypic analysis of two parents indicated that the RWL of ZM36 was higher than that of AK58, and the stomatal density of ZM36 was significantly higher than that of AK58. The bHLH transcription factor *TraesCS7A01G543300*, which is an ortholog of stomatal development regulators in *Arabidopsis*, was considered to be the most likely candidate gene for *LERD1*.

4. Materials and Methods

4.1. Plant Materials and Growth Condition

Two Chinese wheat cultivars, "Aikang 58" (AK58) and "Zhongmai 36" (ZM36) and an F_2 population derived from the cross of AK58 × ZM36 were used as the plant materials. AK58 is a high-yield cultivar widely cultivated in irrigated areas of China, released by the Henan Institute of Science and Technology in 2005 [52]. ZM36 is a stable-yield cultivar suitable for rain-fed area cultivation, released by the Institute of Crop Sciences, Chinese Academy of Agricultural Sciences in 2018. A total of about 4000 individuals of the F_2 population were used for genetic analysis and fine mapping.

For the pot experiment, wheat seeds were sown in containers (length 26 cm, width 19 cm, and height 9 cm), each containing 3 kg of mixed loam and organic fertilizer. The containers were placed horizontally in the field under rain-off shelter, which is located in the National Wheat Improvement Center in Beijing, China (116°28′ E, 39°48′ N).

4.2. Leaf Rolling Degree Assays

The leaf phenotype of wheat seedlings was observed under well-watered (WW) and drought stress (DS) treatments. WW refers to maintaining an adequate water supply, and DS means stop watering from the three-leaf stage. The day of the last watering was defined as the 0th day of drought treatment. On the 0th, 9th and 12th day of drought treatment, the soil moisture content was 21.8%, 7.2% and 5.8%, respectively.

For LRI (leaf rolling index), the largest leaf width (Lw) and the natural distance of the leaf margins (Ln) of the second fully expanded leaves were measured during 09:00–10:30. LRI was calculated using a formula LRI = (Lw − Ln)/Lw × 100% [53]. The experiment was set up in three biological replicates, and 10 leaves were measured in each replicate.

For leaf water content, fresh leaf weight (Fw) was measured immediately after sampling. Dry weight (Dw) was recorded after drying the leaves at 70 °C. Leaf water content was calculated as (Fw − Dw)/Fw × 100%.

According to the leaf morphology of plants treated with drought for 12 d, the leaf rolling degree of the F_2 population was classified into three grades: slight-rolling, leaf blade remained flat; moderate-rolling, leaf blade is partially contracted and curled; serious-rolling, leaf blade curled tightly (Figure S2).

4.3. Drought Tolerance Assays

For seedling survival rate, 14-day-old plants were stopped from water supply for 18 d, then re-watered, and the number of surviving plants was surveyed on the 11th day. On the 0th and 18th day of drought treatment, the soil moisture content was 21.8% and 3.9%, respectively. Statistical data were based on data obtained from three independent experiments.

For osmotic stress, 7-day-old seedlings were cultured in a hydroponic medium containing different concentrations of polyethylene glycol (PEG-6000) for 48 h. The plants were cultured in growth chambers with a 14 h light: 10 h dark period, 25 °C: 20 °C, and 70% relative humidity.

4.4. Rate of Water Loss (RWL) from Excised-Leaf

For detection of RWL, about 5 g of excised fresh leaves were incubated in the dark at 25 °C with 70% relative humidity and weighed every 2 h. RWL = $(W_0 - W_t)/W_0$. W_0: fresh weight of initial leaves. W_t: weight of leaves at t hours. The experiment was set up in three biological replicates.

4.5. Leaf Structure Assays

For light microscope analysis, the leaves of AK58 and ZM36 seedlings on the same leaf position were selected for stomatal aperture assay on the 5th day of drought treatment. The brush was dipped in gum Arabic solution and applied on the surface of leaves at 10:00. After drying, the film was removed with tweezers and placed on the slide, and one drop of distilled water was added to prepare the film and examined for stomatal opening using a light microscope (B5–223 IEP, Motic China Group, Xiamen, China).

For scanning electron microscopy (SEM) analysis, the leaves with the same position were collected from the three-leaf seedlings. Leaf samples were fixed on the sample table using conductive tape before being imaged by SEM (TM4000, Hitachi, Japan) at 5 kV.

4.6. Genetic Analysis of Leaf Rolling

The chromosome loci linked to leaf rolling was identified using the bulked segregant analysis (BSA) method, following the procedure of Wu et al. [54]. Two pools of extreme types of leaf rolling were constructed using the F_2 population. The slight-rolling pool was prepared by mixing equal amounts of leaves sampled from 20 individuals with slight-rolling. The serious-rolling pool was prepared by mixing equal amounts of leaves sampled from 20 serious-rolling individuals.

Genomic DNA was isolated from leaves using the CTAB method [55]. The two pools and two parents were genotyped with the Wheat 660K SNP Array (CapitalBio Technology Corporation, Beijing, China). The monomorphic, heterozygous and poor-quality SNP markers with ambiguous SNP signals were excluded, and homozygous polymorphism SNP markers linked to leaf rolling degree were used for further analysis. The physical locations of all SNP markers were searched in the IWGSC RefSeq v1.0, using BLAST.

4.7. Fine Mapping

To finely map the target locus, 18 SSR markers and 14 InDel (insertion/deletion) markers were developed in the predicted chromosomal region, of which four SSR markers and three InDel markers were polymorphic between two parents (Table S2). The mapping population was constructed from 256 individuals with serious-rolling phenotypes selected from the F_2 population of ~4000 individuals. The seven polymorphic markers and mapping population were used for genotyping analysis.

4.8. Bioinformatics Analysis and Expression Analysis of Candidate Gene

To predict functions of candidate genes, their amino acid sequences were used to blast against the NCBI database (https://www.ncbi.nlm.nih.gov/, accessed on 6 January 2021). The polymorphism assays of genomic DNA sequences were conducted using

Lasergene 7.1.0 (DNASTAR, Inc., Madison, WI, USA). A neighbor-joining phylogenetic tree was constructed using the MEGA software v5.2 (https://www.megasoftware.net/, https://www.ncbi.nlm.nih.gov/, accessed on 6 January 2021).

Total RNA was extracted from young leaves of 15-day-old seedlings using an RNAprep Pure Plant Kit (Tiangen, Beijing, China) and reverse transcription was performed using a FastQuant RT Kit with gDNase (Tiangen, Beijing, China). Real-time quantitative RT-PCR was performed on a Roche LightCycler® 96 Real-Time PCR System using the SYBR Green PCR Master Mix Reagent "Tli RNaseH Plus" (TaKaRa, Beijing, China) [56]. *TaTubulin* was used as the endogenous control. The primers are listed in Table S2.

Supplementary Materials: The following supporting information can be downloaded at: https://www.mdpi.com/article/10.3390/plants11162076/s1, Figure S1: Stomatal morphology on the abaxial surface of AK58 and ZM36 leaves under drought stress; Figure S2: Leaf rolling degree of the F_2 population; Figure S3. Molecular phylogeny of TraesCS7A02G543300 and related bHLHs; Table S1: Annotations of genes predicted in the candidate region on chromosome 7A; Table S2: Primers used in this study.

Author Contributions: J.W. and R.J. designed the experiments. X.Y. performed most of the experiments. Y.X., C.L. and L.L. participated in part of the phenotype evaluation. X.Y. and J.W. analyzed the results and wrote the manuscript. L.H. and X.M. provided advice on the experiments. The project was conceived, planned, and supervised by R.J. All authors have read and agreed to the published version of the manuscript.

Funding: This research was financially supported by the National Key R&D Program of China (2021YFD1200603), the National Natural Science Foundation of China (31571611) and the Research Program Sponsored by Ministerial and Provincial Co-Innovation Centre for Endemic Crops Production with High-quality and Efficiency in Loess Plateau (SBGJXTZX-29).

Institutional Review Board Statement: Not applicable.

Informed Consent Statement: Not applicable.

Data Availability Statement: Not applicable.

Conflicts of Interest: The authors declare no conflict of interest.

References

1. Shewry, P.R.; Hey, S.J. The contribution of wheat to human diet and health. *Food Energy Secur.* **2015**, *4*, 178–202. [CrossRef] [PubMed]
2. Juliana, P.; Poland, J.; Huerta-Espino, J.; Shrestha, S.; Crossa, J.; Crespo-Herrera, L.; Toledo, F.H.; Govindan, V.; Mondal, S.; Kumar, U.; et al. Improving grain yield, stress resilience and quality of bread wheat using large-scale genomics. *Nat. Genet.* **2019**, *51*, 1530–1539. [CrossRef] [PubMed]
3. Vaidya, A.S.; Helander, J.D.M.; Peterson, F.C.; Elzinga, D.; Dejonghe, W.; Kaundal, A.; Park, S.Y.; Xing, Z.; Mega, R.; Takeuchi, J.; et al. Dynamic control of plant water use using designed ABA receptor agonists. *Science* **2019**, *366*, 6464. [CrossRef] [PubMed]
4. Li, L.; Mao, X.G.; Wang, J.Y.; Chang, X.P.; Reynolds, M.; Jing, R.L. Genetic dissection of drought and heat-responsive agronomic traits in wheat. *Plant Cell Environ.* **2019**, *42*, 2540–2553. [CrossRef]
5. Mao, H.D.; Li, S.M.; Wang, Z.X.; Cheng, X.X.; Li, F.F.; Mei, F.M.; Chen, N.; Kang, Z.S. Regulatory changes in *TaSNAC8-6A* are associated with drought tolerance in wheat seedlings. *Plant Biotechnol. J.* **2020**, *18*, 1078–1092. [CrossRef]
6. Kumar, P.; Rouphael, Y.; Cardarelli, M.; Colla, G. Vegetable grafting as a tool to improve drought resistance and water use efficiency. *Front. Plant Sci.* **2017**, *8*, 1130. [CrossRef]
7. Myskow, B.; Goralska, M.; Lenarczyk, N.; Czyczylo-Mysza, I.; Stojalowski, S. Putative candidate genes responsible for leaf rolling in rye (*Secale cereale* L.). *BMC Genet.* **2018**, *19*, 57. [CrossRef]
8. Merrium, S.; Ali, Z.; Tahir, M.H.N.; Habib-Ur-Rahman, M.; Hakeem, S. Leaf rolling dynamics for atmospheric moisture harvesting in wheat plant as an adaptation to arid environments. *Environ. Sci. Pollut. Res. Int.* **2022**, *29*, 48995–49006. [CrossRef]
9. Zhang, G.H.; Hou, X.; Wang, L.; Xu, J.; Chen, J.; Fu, X.; Shen, N.W.; Nian, J.Q.; Jiang, Z.Z.; Hu, J.; et al. *PHOTO-SENSITIVE LEAF ROLLING 1* encodes a polygalacturonase that modifies cell wall structure and drought tolerance in rice. *New Phytol.* **2021**, *229*, 890–901. [CrossRef]
10. Li, W.Q.; Zhang, M.J.; Gan, P.F.; Qiao, L.; Yang, S.Q.; Miao, H.; Wang, G.F.; Zhang, M.M.; Liu, W.T.; Li, H.F.; et al. *CLD1/SRL1* modulates leaf rolling by affecting cell wall formation, epidermis integrity and water homeostasis in rice. *Plant J.* **2017**, *92*, 904–923. [CrossRef]

11. Liu, X.F.; Li, M.; Liu, K.; Tang, D.; Sun, M.F.; Li, Y.F.; Shen, Y.; Du, G.J.; Cheng, Z.K. Semi-Rolled Leaf2 modulates rice leaf rolling by regulating abaxial side cell differentiation. *J. Exp. Bot.* **2016**, *67*, 2139–2150. [CrossRef]
12. Zhang, Q.; Zheng, T.Q.; Hoang, L.; Wang, C.C.; Nafisah; Joseph, C.; Zhang, W.; Xu, J.; Li, Z. Joint mapping and allele mining of the rolled leaf trait in rice (*Oryza sativa* L.). *PLoS ONE* **2016**, *11*, e0158246. [CrossRef]
13. Zhu, Q.Q.; Yu, S.G.; Chen, G.S.; Ke, L.L.; Pan, D.R. Analysis of the differential gene and protein expression profile of the rolled leaf mutant of transgenic rice (*Oryza sativa* L.). *PLoS ONE* **2017**, *12*, e0181378. [CrossRef]
14. Chen, Z.H.; Teng, S.Z.; Liu, D.; Chang, Y.; Zhang, L.Y.; Cui, X.A.; Wu, J.X.; Ai, P.F.; Sun, X.H.; Lu, T.G.; et al. RLM1, encoding an R2R3 MYB transcription factor, regulates the development of secondary cell wall in rice. *Front. Plant Sci.* **2022**, *13*, 905111. [CrossRef]
15. Yu, N.; Liang, Y.P.; Wang, Q.P.; Peng, X.X.; He, Z.H.; Hou, X.W. Transcriptomic analysis of *OsRUS1* overexpression rice lines with rapid and dynamic leaf rolling morphology. *Sci. Rep.* **2022**, *12*, 6736. [CrossRef]
16. Liu, X.; Deng, X.J.; Li, C.Y.; Xiao, Y.K.; Zhao, K.; Guo, J.; Yang, X.R.; Zhang, H.S.; Chen, C.P.; Luo, Y.T.; et al. Mutation of protoporphyrinogen IX oxidase gene causes spotted and rolled leaf and its overexpression generates herbicide resistance in rice. *Int. J. Mol. Sci.* **2022**, *23*, 5781. [CrossRef]
17. Gao, L.L.; Yang, G.H.; Li, Y.F.; Fan, N.N.; Li, H.J.; Zhang, M.; Xu, R.B.; Zhang, M.Y.; Zhao, A.J.; Ni, Z.F.; et al. Fine mapping and candidate gene analysis of a QTL associated with leaf rolling index on chromosome 4 of maize (*Zea mays* L.). *Theor. Appl. Genet.* **2019**, *132*, 3047–3062. [CrossRef]
18. Rao, Y.C.; Yang, Y.L.; Xu, J.; Li, X.J.; Leng, Y.J.; Dai, L.P.; Huang, L.C.; Shao, G.S.; Ren, D.Y.; Hu, J.; et al. *EARLY SENESCENCE1* encodes a SCAR-LIKE PROTEIN2 that affects water loss in rice. *Plant Physiol.* **2015**, *169*, 1225–1239. [CrossRef]
19. Han, X.S.; Qin, Y.; Yu, F.; Ren, X.M.; Zhang, Z.X.; Qiu, F.Z. A megabase-scale deletion is associated with phenotypic variation of multiple traits in maize. *Genetics* **2019**, *211*, 305–316. [CrossRef]
20. Nir, I.; Shohat, H.; Panizel, I.; Olszewski, N.; Aharoni, A.; Weiss, D. The tomato DELLA protein PROCERA acts in guard cells to promote stomatal closure. *Plant Cell* **2017**, *29*, 3186–3197. [CrossRef]
21. You, J.; Zong, W.; Li, X.K.; Ning, J.; Hu, H.H.; Li, X.H.; Xiao, J.H.; Xiong, L.Z. The SNAC1-targeted gene *OsSRO1c* modulates stomatal closure and oxidative stress tolerance by regulating hydrogen peroxide in rice. *J. Exp. Bot.* **2013**, *64*, 569–583. [CrossRef]
22. Huang, X.Y.; Chao, D.Y.; Gao, J.P.; Zhu, M.Z.; Shi, M.; Lin, H.X. A previously unknown zinc finger protein, DST, regulates drought and salt tolerance in rice via stomatal aperture control. *Gene Dev.* **2009**, *23*, 1805–1817. [CrossRef]
23. Chater, C.C.C.; Caine, R.S.; Fleming, A.J.; Gray, J.E. Origins and evolution of stomatal development. *Plant Physiol.* **2017**, *174*, 624–638. [CrossRef]
24. MacAlister, C.A.; Ohashi-Ito, K.; Bergmann, D.C. Transcription factor control of asymmetric cell divisions that establish the stomatal lineage. *Nature* **2007**, *445*, 537–540. [CrossRef]
25. Pillitteri, L.J.; Bogenschutz, N.L.; Torii, K.U. The bHLH protein, MUTE, controls differentiation of stomata and the hydathode pore in Arabidopsis. *Plant Cell Physiol.* **2008**, *49*, 934–943. [CrossRef]
26. Ohashi-Ito, K.; Bergmann, D.C. *Arabidopsis* FAMA controls the final proliferation/differentiation switch during stomatal development. *Plant Cell* **2006**, *18*, 2493–2505. [CrossRef]
27. Kanaoka, M.M.; Pillitteri, L.J.; Fujii, H.; Yoshida, Y.; Bogenschutz, N.L.; Takabayashi, J.; Zhu, J.K.; Torii, K.U. *SCREAM/ICE1* and *SCREAM2* specify three cell-state transitional steps leading to *Arabidopsis* stomatal differentiation. *Plant Cell* **2008**, *20*, 1775–1785. [CrossRef]
28. Meng, X.Z.; Chen, X.; Mang, H.; Liu, C.L.; Yu, X.; Gao, X.Q.; Torii, K.U.; He, P.; Shan, L.B. Differential function of *Arabidopsis* SERK family receptor-like kinases in stomatal patterning. *Curr. Biol.* **2015**, *25*, 2361–2372. [CrossRef]
29. Qi, X.; Han, S.K.; Dang, J.H.; Garrick, J.M.; Ito, M.; Hofstetter, A.K.; Torii, K.U. Autocrine regulation of stomatal differentiation potential by EPF1 and ERECTA-LIKE1 ligand-receptor signaling. *eLife* **2017**, *6*, e24102. [CrossRef]
30. Wang, M.; Yang, K.Z.; Le, J. Organ-specific effects of brassinosteroids on stomatal production coordinate with the action of *TOO MANY MOUTHS*. *J. Integr. Plant Biol.* **2015**, *57*, 247–255. [CrossRef]
31. Liu, T.; Ohashi-Ito, K.; Bergmann, D.C. Orthologs of *Arabidopsis thaliana* stomatal bHLH genes and regulation of stomatal development in grasses. *Development* **2009**, *136*, 2265–2276. [CrossRef] [PubMed]
32. Wu, Z.L.; Chen, L.; Yu, Q.; Zhou, W.Q.; Gou, X.P.; Li, J.; Hou, S.W. Multiple transcriptional factors control stomata development in rice. *New Phytol.* **2019**, *223*, 220–232. [CrossRef] [PubMed]
33. Chater, C.C.; Caine, R.S.; Tomek, M.; Wallace, S.; Kamisugi, Y.; Cuming, A.C.; Lang, D.; MacAlister, C.A.; Casson, S.; Bergmann, D.C.; et al. Origin and function of stomata in the moss *Physcomitrella* patens. *Nat. Plants* **2016**, *2*, 16179. [CrossRef] [PubMed]
34. Caine, R.S.; Chater, C.C.; Kamisugi, Y.; Cuming, A.C.; Beerling, D.J.; Gray, J.E.; Fleming, A.J. An ancestral stomatal patterning module revealed in the non-vascular land plant *Physcomitrella* patens. *Development* **2016**, *143*, 3306–3314. [CrossRef]
35. Raissig, M.T.; Abrash, E.; Bettadapur, A.; Vogel, J.P.; Bergmann, D.C. Grasses use an alternatively wired bHLH transcription factor network to establish stomatal identity. *Proc. Natl. Acad. Sci. USA* **2016**, *113*, 8326–8331. [CrossRef]
36. Rudall, P.J.; Hilton, J.; Bateman, R.M. Several developmental and morphogenetic factors govern the evolution of stomatal patterning in land plants. *New Phytol.* **2013**, *200*, 598–614. [CrossRef]
37. Raissig, M.T.; Matos, J.L.; Anleu Gil, M.X.; Kornfeld, A.; Bettadapur, A.; Abrash, E.; Allison, H.R.; Badgley, G.; Vogel, J.P.; Berry, J.A.; et al. Mobile MUTE specifies subsidiary cells to build physiologically improved grass stomata. *Science* **2017**, *355*, 1215–1218. [CrossRef]

38. Fiorin, L.; Brodribb, T.J.; Anfodillo, T. Transport efficiency through uniformity: Organization of veins and stomata in angiosperm leaves. *New Phytol.* **2016**, *209*, 216–227. [CrossRef]
39. Deokar, A.; Sagi, M.; Daba, K.; Tar'an, B. QTL sequencing strategy to map genomic regions associated with resistance to ascochyta blight in chickpea. *Plant Biotechnol. J.* **2019**, *17*, 275–288. [CrossRef]
40. Huo, H.; Henry, I.M.; Coppoolse, E.R.; Verhoef-Post, M.; Schut, J.W.; de Rooij, H.; Vogelaar, A.; Joosen, R.V.; Woudenberg, L.; Comai, L.; et al. Rapid identification of lettuce seed germination mutants by bulked segregant analysis and whole genome sequencing. *Plant J.* **2016**, *88*, 345–360. [CrossRef]
41. Nguyen, K.L.; Grondin, A.; Courtois, B.; Gantet, P. Next-Generation Sequencing accelerates grop gene discovery. *Trends Plant Sci.* **2019**, *24*, 263–274. [CrossRef]
42. Wang, C.S.; Tang, S.C.; Zhan, Q.L.; Hou, Q.Q.; Zhao, Y.; Zhao, Q.; Feng, Q.; Zhou, C.C.; Lyu, D.F.; Cui, L.L.; et al. Dissecting a heterotic gene through GradedPool-Seq mapping informs a rice-improvement strategy. *Nat. Commun.* **2019**, *10*, 2982. [CrossRef]
43. Yu, C.C.; Yan, C.H.; Liu, Y.L.; Liu, Y.L.; Jia, Y.; Lavelle, D.; An, G.H.; Zhang, W.Y.; Zhang, L.; Han, R.K.; et al. Upregulation of a *KN1* homolog by transposon insertion promotes leafy head development in lettuce. *Proc. Natl. Acad. Sci. USA* **2020**, *117*, 33668–33678. [CrossRef]
44. Sun, C.W.; Dong, Z.D.; Zhao, L.; Ren, Y.; Zhang, N.; Chen, F. The Wheat 660K SNP array demonstrates great potential for marker-assisted selection in polyploid wheat. *Plant Biotechnol. J.* **2020**, *18*, 1354–1360. [CrossRef]
45. Baret, F.; Madec, S.; Irfan, K.; Lopez, J.; Comar, A.; Hemmerle, M.; Dutartre, D.; Praud, S.; Tixier, M.H. Leaf-rolling in maize crops: From leaf scoring to canopy-level measurements for phenotyping. *J. Exp. Bot.* **2018**, *69*, 2705–2716. [CrossRef]
46. Sussmilch, F.C.; Schultz, J.; Hedrich, R.; Roelfsema, M.R.G. Acquiring control: The evolution of stomatal signalling pathways. *Trends Plant Sci.* **2019**, *24*, 342–351. [CrossRef]
47. Buckley, T.N. The control of stomata by water balance. *New Phytol.* **2005**, *168*, 275–292. [CrossRef]
48. Ehonen, S.; Yarmolinsky, D.; Kollist, H.; Kangasjarvi, J. Reactive oxygen species, photosynthesis, and environment in the regulation of stomata. *Antioxid. Redox Signal.* **2019**, *30*, 1220–1237. [CrossRef]
49. Ahmed, K.; Shabbir, G.; Ahmed, M.; Shah, K.N. Phenotyping for drought resistance in bread wheat using physiological and biochemical traits. *Sci. Total Environ.* **2020**, *729*, 139082. [CrossRef]
50. Ali, Z.; Merrium, S.; Habib-Ur-Rahman, M.; Hakeem, S.; Saddique, M.A.B.; Sher, M.A. Wetting mechanism and morphological adaptation; leaf rolling enhancing atmospheric water acquisition in wheat crop-a review. *Environ. Sci. Pollut. Res. Int.* **2022**, *29*, 30967–30985. [CrossRef]
51. Merrium, S.; Ali, Z.; Habib-Ur-Rahman, M.; Hakeem, S.; Khalid, M.A. Leaf rolling and leaf angle improve fog capturing and transport in wheat; adaptation for drought stress in an arid climate. *Bot. Stud.* **2022**, *63*, 13. [CrossRef]
52. Jia, J.Z.; Xie, Y.L.; Cheng, J.F.; Kong, C.Z.; Wang, M.Y.; Gao, L.F.; Zhao, F.; Guo, J.Y.; Wang, K.; Li, G.W.; et al. Homology-mediated inter-chromosomal interactions in hexaploid wheat lead to specific subgenome territories following polyploidization and introgression. *Genome Biol.* **2021**, *22*, 26. [CrossRef]
53. Fang, L.K.; Zhao, F.M.; Cong, Y.F.; Sang, X.C.; Du, Q.; Wang, D.Z.; Li, Y.F.; Ling, Y.H.; Yang, Z.L.; He, G.H. Rolling-leaf14 is a 2OG-Fe (II) oxygenase family protein that modulates rice leaf rolling by affecting secondary cell wall formation in leaves. *Plant Biotechnol. J.* **2012**, *10*, 524–532. [CrossRef]
54. Wu, J.H.; Wang, Q.L.; Xu, L.S.; Chen, X.M.; Li, B.; Mu, J.M.; Zeng, Q.D.; Huang, L.L.; Han, D.J.; Kang, Z.S. Combining single nucleotide polymorphism genotyping array with bulked segregant analysis to map a gene controlling adult plant resistance to stripe rust in wheat line 03031-1-5 H62. *Phytopathology* **2018**, *108*, 103–113. [CrossRef]
55. Wang, J.Y.; Wang, R.T.; Mao, X.G.; Li, L.; Chang, X.P.; Zhang, X.Y.; Jing, R.L. *TaARF4* genes are linked to root growth and plant height in wheat. *Ann. Bot.* **2019**, *124*, 903–915. [CrossRef]
56. Wang, J.Y.; Wang, R.T.; Mao, X.G.; Zhang, J.L.; Liu, Y.N.; Xie, Q.; Yang, X.Y.; Chang, X.P.; Li, C.N.; Zhang, X.Y.; et al. RING finger ubiquitin E3 ligase gene *TaSDIR1-4A* contributes to determination of grain size in common wheat. *J. Exp. Bot.* **2020**, *71*, 5377–5388. [CrossRef]

MDPI AG
Grosspeteranlage 5
4052 Basel
Switzerland
Tel.: +41 61 683 77 34

Plants Editorial Office
E-mail: plants@mdpi.com
www.mdpi.com/journal/plants

www.ingramcontent.com/pod-product-compliance
Lightning Source LLC
Chambersburg PA
CBHW040801250926
51850CB00007BA/1052
* 9 7 8 3 7 2 5 8 6 8 1 2 4 *